LINEAR ALGEBRA
WITH
APPLICATIONS

SIXTH EDITION

Steven J. Leon
University of Massachusetts, Dartmouth

PRENTICE HALL
Upper Saddle River, NJ 07458

Library of Congress Cataloging-in-Publication Data

Leon, Steven J.
    Linear algebra with applications / Steven J. Leon.–6th ed.
        p. cm.
    Includes bibliographical references.
    ISBN 0-13-033781-1
    1. Algebras, Linear. I. Title.
    QA184 .L46 2002
    512′.5—dc21   2001040036

Acquisitions Editor: George Lobell
Editor-in-Chief: Sally Yagan
Associate Editor: Dawn Murrin
Vice-President/Director of Production and Manufacturing: David W. Riccardi
Executive Managing Editor: Kathleen Schiaparelli
Senior Managing Editor: Linda Mihatov Behrens
Production Editor: Bob Walters
Manufacturing Buyer: Alan Fischer
Manufacturing Manager: Trudy Pisciotti
Marketing Manager: Angela Battle
Marketing Assistant: Rachel Beckman
Assistant Editor of Media: Vince Jansen
Editorial Assistant/Supplements Editor: Melanie Van Benthuysen
Art Director: Maureen Eide
Assistant to the Art Director: John Christiana
Interior Designer: Jonathan Boylan
Cover Designer: Jonathan Boylan
Managing Editor, Audio/Video Assets: Grace Hazeldine
Creative Director: Carole Anson
Director of Creative Services: Paul Belfanti
Cover Photo: "Heaven and Earth", 1992/Daniel Goldstein Studio
Art Studio: Artworks
        Senior Manager: Patty Burns
        Production manager: Ronda Whitson
        Manager, Production Technologies: Matt Haas
        Project Coordinator: Jessica Einsig
        Illustrator: Steve McKinley

ⓒ 2002 by Prentice-Hall, Inc.
Upper Saddle River, New Jersey 07458

Printed in the United States of America

10   9   8   7   6   5   4   3   2

ISBN: 0-13-033781-1

Pearson Education Ltd., *London*
Pearson Education Australia PTY, Limited, *Sydney*
Pearson Education Singapore, Pte. Ltd.
Pearson Education North Asia Ltd., *Hong Kong*
Pearson Education Canada, Ltd., *Toronto*
Pearson Educaciûn de Mexico, S.A. de C.V.
Pearson Education – Japan, *Tokyo*
Pearson Education Malaysia, Pte. Ltd.

To Judith Russ Leon

CONTENTS

8 ITERATIVE METHODS *Web

9 JORDAN CANONICAL FORM *Web

---

*Web: The Supplemental Chapters 8 and 9 can be downloaded from the web page: (www.prenhall.com/Leon)
†Optional sections.

PREFACE

The author is pleased to see the text reach its sixth edition. While the continued support and enthusiasm of the many users has been most gratifying, this does not mean that a mild revision is in order. Linear algebra is more exciting now than at almost any time in the past. Its applications continue to spread to more and more fields. Largely due to the computer revolution of the last half century, linear algebra has risen to a role of prominence in the mathematical curriculum rivaling that of calculus. Modern software has also made it possible to dramatically improve the way the course is taught. The author teaches linear algebra every semester and continues to seek new ways to optimize student understanding. For this edition every chapter has been carefully scrutinized and enhanced. Additionally, many of the revisions in this edition are due to the helpful suggestions received from users and reviewers. Consequently, this new edition, while retaining the essence of previous editions, incorporates a wide array of substantive improvements.

WHAT'S NEW IN THE SIXTH EDITION?

1. Chapter Tests

New to this edition are chapter tests. At the end of each chapter there is a true-false exam testing the basic concepts covered in the chapter. Students are asked to prove or explain all of their answers.

2. Earlier Presentation of the Singular Value Decomposition

The singular value decomposition (SVD) has emerged as one of the most important tools in matrix applications. Unfortunately, the topic is often omitted from linear algebra textbooks. When covered, it usually appears near the end of the book and classes rarely have time to get that far. To remedy this, we have moved the singular value decomposition approximately 100 pages forward in the book. It is now covered in Section 5 of Chapter 6. In this section we also show the applications of the singular value decomposition to least squares problems, principal component analysis, information retrieval, numerical rank of a matrix, and digital imaging. The SVD section nicely ties together some of the major topics, such as fundamental subspaces, orthogonality, and eigenvalues. It provides an ideal climax to a linear algebra course.

3. New and Improved Applications

Eight applications were added to the previous edition. Some of these have been revised and improved in the current edition. A number of new applications have also been added. In Chapter 1 we show how matrices are used for search engines and information retrieval applications. This application is revisited in Chapters 5 and 6 after students have learned about orthogonality and singular values. Similarly, the statistical applications in Chapter 5 are revisited later in Chapter 6 after students have learned about the singular value decomposition.

4. New Computer Exercises Emphasizing Visualization

Chapter 6 has ten new MATLAB exercises to help students to visualize eigenvalues and singular values and to help them gain geometric insight into these subjects.

5. New and Improved Examples

Worked out examples make the textbook seem less abstract and more user friendly. Often students don't understand what a theorem says until they see a worked out example that illustrates the theorem. The impressive collection of examples was often cited as one of the strong points of the first edition of this book. This collection has continued to grow and improve with each new edition. More examples have been added throughout the sixth edition, and many of the previous examples have been revised and improved. Now, for example, the numbered of worked out examples in Chapter 1 has increased from 32 to 34. In a number of cases color shading is now used to emphasize how rows and columns are paired off in matrix computations.

6. New Theorem and Improved Nomenclature

Throughout this edition we have made a special effort to assign names to theorems so as to emphasize the importance of the results. Also, it is easier to refer back to a theorem if it has a name. We have added a new theorem to Chapter 6. This theorem does have a name, *The Principal Axes Theorem*.

7. Revised Organization of Chapter 5

In Chapter 5 the order of two of the sections has been reversed. Least squares problems are now covered before the section on general inner product spaces. To facilitate this change, some new material was added to Section 1 of the chapter. With this new ordering it is possible for classes that only treat Euclidean vector spaces to skip most of Section 4. These classes need only introduce the inner product notation in Section 4 and then move on to the next section or, if pressed for time, skip ahead to the next chapter.

8. New Subsection on Outer Products

A new subsection on outer product expansions has been added to Chapter 1. Outer product expansions are used in later chapters applications such as digital imaging.

9. Special Web Site and Supplemental Web Materials

Prentice Hall has developed a special Web site to accompany this book. This site includes a host of materials for both students and instructors. The Web pages are being extensively revised for the sixth edition and an exciting collection of new interactive course materials is currently being developed as we go to press. Some of the other features to be included on the Web pages are a collection of links with downloadable materials relating to each of the chapters in the book and a collection of application projects that are related to the topics covered in the book. You can also download two supplemental chapters for this book from the Prentice Hall site. The new chapters are:

- Chapter 8. Iterative Methods
- Chapter 9. Canonical Forms

The URL for the site is listed in the Supplementary Materials section of this Preface.

10. The ATLAST Companion Computer Manual Revision

ATLAST (Augmenting the Teaching of Linear Algebra through the use of Software Tools) is an NSF sponsored project to encourage and facilitate the use of software in the teaching of linear algebra. During a five year period, 1992–1997, the ATLAST Project conducted 18 faculty workshops using the MATLAB software package. Participants in these workshops designed computer exercises, projects, and lesson plans for software-based teaching of linear algebra. A selection of these materials has been published as a manual. *ATLAST Computer Exercises for Linear Algebra* (Prentice Hall, 1997). The ATLAST book is available as a *free* companion volume to this textbook when the two books are wrapped together for class orders. The ISBN for ordering the two-book bundle is given in the Supplementary Materials section of this Preface. The collection of software tools (M-files) developed to accompany the ATLAST book may be downloaded from the ATLAST Web site. You can link to the ATLAST site from the Prentice Hall Web page for this book. A second edition of the ATLAST book is in preparation and publication is expected in the fall of 2002. New developments related to the ATLAST Project and manual will be posted on the ATLAST Web site.

11. Mathematica Computer Exercises and Projects

A collection of ATLAST Mathematica Notebooks has been developed by Richard Neidinger of Davidson College. The collection contains Mathematical versions of the ATLAST projects and exercises. It can be downloaded for free from the ATLAST Web pages.

12. Maple Companion Manual

A new manual, *Visualizing Linear Algebra Using Maple*, by Sandra Keith, is now available as a companion volume to this book. The manual provides an ideal vehicle for those wishing to teach the course using Maple. The Keith manual is offered as a bundle with this book at a special discount price. The ISBN for ordering the two-book bundle is given in the Supplementary Materials section of this Preface.

13. MATLAB Companion Manual

A new manual, *Understanding Linear Algebra Using MATLAB*, by Irwin and Margaret Kleinfeld, is now available as a companion volume to this book. The book has MATLAB problems and projects suitable for a first course in linear algebra. The manual is offered as a bundle with this book at a special discount price. The ISBN for ordering the two-book bundle is given in the Supplementary Materials section of this Preface.

14. Student Guide to Linear Algebra with Applications

A new student study guide has been developed to accompany this edition. The guide is described in the Supplementary Materials section of this preface.

15. Other Changes

In preparing the sixth edition, the author has carefully reviewed every section of the book. In addition to the major changes that have been listed, many new exercises have been added and numerous minor improvements have been made throughout the text.

COMPUTER EXERCISES

This edition contains a section of computing exercises at the end of each chapter. These exercises are based on the software package MATLAB. The MATLAB Appendix in the book explains the basics of using the software. MATLAB has the advantage that it is a powerful tool for matrix computations and yet it is easy to learn. After reading the Appendix, students should be able to do the computing exercises without having to refer to any other software books or manuals. To help students get started we recommend one 50 minute classroom demonstration of the software. The assignments can be done either as ordinary homework assignments or as part of a formally scheduled computer laboratory course.

As mentioned previously, the ATLAST book is available as a companion volume to supplement the computer exercises in this book. Each of the eight chapters of the ATLAST book contains a section of short exercises and a section of longer projects.

While the course can be taught without any reference to the computer, we believe that computer exercises can greatly enhance student learning and provide a new dimension to linear algebra education. The Linear Algebra Curriculum Study Group has recommended that technology be used for a first course in linear algebra, and this view is generally accepted throughout the greater mathematics community.

OVERVIEW OF TEXT

This book is suitable for either a sophomore-level course or for a junior/senior-level course. The student should have some familiarity with the basics of differential and integral calculus. This prerequisite can be met by either one semester or two quarters of elementary calculus.

If the text is used for a sophomore-level course, the instructor should probably spend more time on the early chapters and omit many of the sections in the later chapters. For more advanced courses a quick review of many of the topics in the first two chapters and then a more complete coverage of the later chapters would be appropriate. The explanations in the text are given in sufficient detail so that beginning students should have little trouble reading and understanding the material. To further aid the student, a large number of examples have been worked out completely. Additionally, computer exercises at the end of each chapter give students the opportunity to perform numerical experiments and try to generalize the results. Applications are presented throughout the book. These applications can be used to motivate new material or to illustrate the relevance of material that has already been covered.

The text contains all the topics recommended by the National Science Foundation (NSF) sponsored Linear Algebra Curriculum Study Group (LACSG) and much more. Although there is more material than can be covered in a one-quarter or one-semester course, it is the author's feeling that it is easier for an instructor to leave out or skip material than it is to supplement a book with outside material. Even if many topics are omitted, the book should still provide students with a feeling for the overall scope of the subject matter. Furthermore, many students may use the book later as a reference and consequently may end up learning many of the omitted topics on their own.

In the next section of this preface a number of outlines are provided for one-semester courses at either the sophomore level or the junior/senior level and with either a matrix-oriented emphasis or a slightly more theoretical emphasis. To further aid the instructor in the choice of topics, three sections have been designated as optional and are marked with a dagger in the table of contents. These sections are not prerequisites for any of the following sections in the book. They may be skipped without any loss of continuity.

Ideally the entire book could be covered in a two-quarter or two-semester sequence. Although two semesters of linear algebra has been recommended by the LACSG, it is still not practical at many universities and colleges. At present there is no universal agreement on a core syllabus for a second course. Indeed, if all of the topics that instructors would like to see in a second course were included in a single volume, it would be a weighty (and expensive) book. An effort has been made in this text to cover all of the basic linear algebra topics that are necessary for modern applications. Furthermore, two additional chapters for a second course are available for downloading from the Internet. See the special Prentice Hall Web page discussed earlier.

SUGGESTED COURSE OUTLINES

I. Two-Semester Sequence

In a two semester sequence it is possible to cover all 39 sections of the book. Additional flexibility is possible by omitting any of the three optional sections in Chapters 2, 5, and 6. One could also include an extra lecture demonstrating how to use the MATLAB software.

II. One-Semester Sophomore-Level Course

A. A Basic Sophomore-Level Course

Chapter 1	Section 1–5	7 lectures
Chapter 2	Section 1–2	2 lectures
Chapter 3	Section 1–6	9 lectures
Chapter 4	Section 1–3	4 lectures
Chapter 5	Section 1–6	9 lectures
Chapter 6	Section 1–3	4 lectures
	Total	35 lectures

B. The LACSG Matrix Oriented Course

The core course recommended by the Linear Algebra Curriculum Study involves only the Euclidean vector spaces. Consequently, for this course you should omit Section 1 of Chapter 3 (on general vector spaces) and all references and exercises involving function spaces in Chapters 3 to 6. All of the topics in the LACSG core syllabus are included in the text. It is not necessary to introduce any supplementary materials. The LACSG recommended 28 lectures to cover the core material. This is possible if the class is taught in lecture format with an additional recitation section meeting once a week. If the course is taught with recitations, the author feels that the following schedule of 35 lectures is perhaps more reasonable.

Chapter 1	Sections 1–5	7 lectures
Chapter 2	Sections 1–2	2 lectures
Chapter 3	Sections 2–6	7 lectures
Chapter 4	Sections 1–3	2 lectures
Chapter 5	Sections 1–6	9 lectures
Chapter 6	Sections 1,3–5	8 lectures
	Total	35 lectures

III. One-Semester Junior/Senior-Level Courses

The coverage in an upper division course is dependent on the background of the students. Below are two possible courses with 35 lectures each.

A. Course 1

Chapter 1	Sections 1–5	6 lectures
Chapter 2	Sections 1–2	2 lectures
Chapter 3	Sections 1–6	7 lectures
Chapter 5	Sections 1–6	9 lectures
Chapter 6	Sections 1–7	10 lectures
	Section 8 if time allows	
Chapter 7	Section 4	1 lecture

B. Course 2

Review of Topics in Chapters 1–3		5 lectures
Chapter 4	Sections 1–3	2 lectures
Chapter 5	Sections 1–6	10 lectures
Chapter 6	Sections 1–7	11 lectures
	Section 8 if time allows	
Chapter 7	Sections 4–7	7 lectures
	If time allows, Sections 1–3	

SUPPLEMENTARY MATERIALS

- **ATLAST Computer Exercises for Linear Algebra.** The ATLAST book described earlier in this preface is available at no extra charge when ordered as a bundle with this textbook. The ISBN for ordering the two-book bundle is 0-13-096706-8. The ATLAST M-files may be downloaded for free from the ATLAST Web site. The *ATLAST Mathematica Notebooks* may also be downloaded for free from the ATLAST Web site.

- **Visualizing Linear Algebra using Maple,** by Sandra Keith, is available as a companion volume to this book. The manual is offered as a bundle with this book at a special discount price. The ISBN for ordering the two-book bundle is 0-13-074174-4.

- **Understanding Linear Algebra Using MATLAB,** by Irwin and Margaret Kleinfeld, is available as a companion volume to this book. This manual is also available as part of a bundle at a special discount price. This ISBN for ordering the two-book bundle is 0-13-060945-5.

- **Instructor's Solution Manual.** A solutions manual is available to all instructors teaching from this book. The manual contains complete solutions to all the nonroutine exercises in the book. The manual also contains answers to many of the elementary exercises that were not already listed in the answer key section of the book.

- **Student Guide to Linear Algebra with Applications.** The manual is available to students as a study to accompany this textbook. The manual summarizes important theorems, definitions, and concepts presented in the textbook. It provides solutions to some of the exercises and hints and suggestions on many other exercises.

- **Web Supplements.** The Prentice Hall Web site for this book has an impressive collection of supplementary materials. It is expected to be functional by October 15, 2001. The URL for the Web site is:

http://www.prenhall.com/leon

ACKNOWLEDGMENTS

The author would like to express his gratitude to the long list of reviewers that have contributed so much to all six editions of this book. Thanks also to the many users who have send in comments and suggestions. Thanks also to James Bunch, University of California, San Diego, for his many helpful suggestions.

A special thanks is due the reviewers of the sixth edition:

Steven Kaliszewski, Arizona State University
Anil Nerode, Cornell University
Lanita Presson, University of Alabama in Huntsville
Gilberto Schleiniger, University of Delaware
Carl Swenson, Seattle University

Many of the revisions and new exercises in this latest edition are a direct result of their comments and suggestions.

Thanks to Michael Downes for his help with the LaTeX cls and macro files and thanks to Judith Leon for all of her help and encouragement. Special thanks to Mathematics Editor, George Lobell, for his help in planning the new edition and to Production Manager Bob Walters for his work on coordinating production of the book. Thanks to the entire editorial, production, and sales staff at Prentice Hall for all their efforts.

Finally, the author would like to acknowledge the contributions of Gene Golub and Jim Wilkinson. Most of the first edition of the book was written in 1977–1978 while the author was a Visiting Scholar at Stanford University. During that period the author attended courses and lectures on numerical linear algebra given by Gene Golub and J. H. Wilkinson. Those lectures have greatly influenced this book.

Steven J. Leon
sleon@umassd.edu

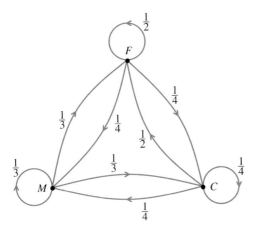

CHAPTER

1

MATRICES AND SYSTEMS OF EQUATIONS

Probably the most important problem in mathematics is that of solving a system of linear equations. Well over 75 percent of all mathematical problems encountered in scientific or industrial applications involve solving a linear system at some stage. By using the methods of modern mathematics, it is often possible to take a sophisticated problem and reduce it to a single system of linear equations. Linear systems arise in applications to such areas as business, economics, sociology, ecology, demography, genetics, electronics, engineering, and physics. Therefore, it seems appropriate to begin this book with a section on linear systems.

SYSTEMS OF LINEAR EQUATIONS

A *linear equation in n unknowns* is an equation of the form

$$a_1 x_1 + a_2 x_2 + \cdots + a_n x_n = b$$

where $a_1, a_2, \ldots, a_n$ and b are real numbers and $x_1, x_2, \ldots, x_n$ are variables. A *linear system* of m equations in n unknowns is then a system of the form

(1)
$$\begin{aligned}
a_{11}x_1 + a_{12}x_2 + \cdots + a_{1n}x_n &= b_1 \\
a_{21}x_1 + a_{22}x_2 + \cdots + a_{2n}x_n &= b_2 \\
&\vdots \\
a_{m1}x_1 + a_{m2}x_2 + \cdots + a_{mn}x_n &= b_m
\end{aligned}$$

where the a_{ij}'s and the b_i's are all real numbers. We will refer to systems of the form (1) as $m \times n$ linear systems. The following are examples of linear systems:

(a) $\begin{aligned} x_1 + 2x_2 &= 5 \\ 2x_1 + 3x_2 &= 8 \end{aligned}$

(b) $\begin{aligned} x_1 - x_2 + x_3 &= 2 \\ 2x_1 + x_2 - x_3 &= 4 \end{aligned}$

(c) $\begin{aligned} x_1 + x_2 &= 2 \\ x_1 - x_2 &= 1 \\ x_1 \quad\;\; &= 4 \end{aligned}$

System (a) is a 2×2 system, (b) is a 2×3 system, and (c) is a 3×2 system.

By a solution to an $m \times n$ system, we mean an ordered n-tuple of numbers $(x_1, x_2, \ldots , x_n)$ that satisfies all the equations of the system. For example, the ordered pair $(1, 2)$ is a solution to system (a), since

$$1 \cdot (1) + 2 \cdot (2) = 5$$
$$2 \cdot (1) + 3 \cdot (2) = 8$$

The ordered triple $(2, 0, 0)$ is a solution to system (b), since

$$1 \cdot (2) - 1 \cdot (0) + 1 \cdot (0) = 2$$
$$2 \cdot (2) + 1 \cdot (0) - 1 \cdot (0) = 4$$

Actually, system (b) has many solutions. If α is any real number, it is easily seen that the ordered triple $(2, \alpha, \alpha)$ is a solution. However, system (c) has no solution. It follows from the third equation that the first coordinate of any solution would have to be 4. Using $x_1 = 4$ in the first two equations, we see that the second coordinate must satisfy

$$4 + x_2 = 2$$
$$4 - x_2 = 1$$

Since there is no real number that satisfies both of these equations, the system has no solution. If a linear system has no solution, we say that the system is *inconsistent*. If the system has at least one solution, we say that it is *consistent*. Thus system (c) is inconsistent, while systems (a) and (b) are both consistent.

The set of all solutions to a linear system is called the *solution set* of the system. If a system is inconsistent, its solution set is empty. A consistent system will have a nonempty solution set. To solve a consistent system, we must find its solution set.

2 × 2 Systems

Let us examine geometrically a system of the form

$$a_{11}x_1 + a_{12}x_2 = b_1$$
$$a_{21}x_1 + a_{22}x_2 = b_2$$

Each equation can be represented graphically as a line in the plane. The ordered pair (x_1, x_2) will be a solution to the system if and only if it lies on both lines. For example, consider the three systems

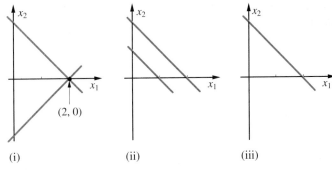

FIGURE I.I.I

(i) $x_1 + x_2 = 2$
     $x_1 - x_2 = 2$

(ii) $x_1 + x_2 = 2$
      $x_1 + x_2 = 1$

(iii) $x_1 + x_2 = 2$
        $-x_1 - x_2 = -2$

The two lines in system (i) intersect at the point $(2, 0)$. Thus $\{(2, 0)\}$ is the solution set to (i). In system (ii) the two lines are parallel. Therefore, system (ii) is inconsistent and hence its solution set is empty. The two equations in system (iii) both represent the same line. Any point on that line will be a solution to the system (see Figure 1.1.1).

In general, there are three possibilities: the lines intersect at a point, they are parallel, or both equations represent the same line. The solution set then contains either one, zero, or infinitely many points.

The situation is similar for $m \times n$ systems. An $m \times n$ system may or may not be consistent. If it is consistent, it must either have exactly one solution or infinitely many solutions. These are the only possibilities. We will see why this is so in Section 2 when we study the row echelon form. Of more immediate concern is the problem of finding all solutions to a given system. To tackle this problem, we introduce the notion of *equivalent systems*.

Equivalent Systems

Consider the two systems

(a) $3x_1 + 2x_2 - x_3 = -2$
           $x_2 = 3$
                $2x_3 = 4$

(b) $3x_1 + 2x_2 - x_3 = -2$
     $-3x_1 - x_2 + x_3 = 5$
      $3x_1 + 2x_2 + x_3 = 2$

System (a) is easy to solve because it is clear from the last two equations that $x_2 = 3$ and $x_3 = 2$. Using these values in the first equation, we get

$$3x_1 + 2 \cdot 3 - 2 = -2$$

$$x_1 = -2$$

Thus the solution to the system is $(-2, 3, 2)$. System (b) seems to be more difficult to solve. Actually, system (b) has the same solution as system (a). To see this, add the first two equations of the system:

$$3x_1 + 2x_2 - x_3 = -2$$
$$-3x_1 - x_2 + x_3 = 5$$
$$\overline{ x_2 = 3}$$

If (x_1, x_2, x_3) is any solution to (b), it must satisfy all the equations of the system. Thus it must satisfy any new equation formed by adding two of its equations. Therefore, x_2 must equal 3. Similarly, (x_1, x_2, x_3) must satisfy the new equation formed by subtracting the first equation from the third:

$$3x_1 + 2x_2 + x_3 = 2$$
$$3x_1 + 2x_2 - x_3 = -2$$
$$\overline{ 2x_3 = 4}$$

Therefore, any solution to system (b) must also be a solution to system (a). By a similar argument, it can be shown that any solution to (a) is also a solution to (b). This can be done by subtracting the first equation from the second:

$$x_2 = 3$$
$$3x_1 + 2x_2 - x_3 = -2$$
$$\overline{-3x_1 - x_2 + x_3 = 5}$$

Then add the first and third equations:

$$3x_1 + 2x_2 - x_3 = -2$$
$$2x_3 = 4$$
$$\overline{3x_1 + 2x_2 + x_3 = 2}$$

Thus (x_1, x_2, x_3) is a solution to system (b) if and only if it is a solution to system (a). Therefore, both systems have the same solution set, $\{(-2, 3, 2)\}$.

▶ **DEFINITION** Two systems of equations involving the same variables are said to be **equivalent** if they have the same solution set. ◀

Clearly, if we interchange the order in which two equations of a system are written, this will have no effect on the solution set. The reordered system will be equivalent to the original system. For example, the systems

$$x_1 + 2x_2 = 4 \qquad\qquad 4x_1 + x_2 = 6$$
$$3x_1 - x_2 = 2 \qquad \text{and} \qquad 3x_1 - x_2 = 2$$
$$4x_1 + x_2 = 6 \qquad\qquad x_1 + 2x_2 = 4$$

both involve the same three equations and, consequently, they must have the same solution set.

If one of the equations of a system is multiplied through by a nonzero real number, this will have no effect on the solution set, and the new system will be equivalent to the original system. For example, the systems

$$x_1 + x_2 + x_3 = 3$$
$$-2x_1 - x_2 + 4x_3 = 1$$

and

$$2x_1 + 2x_2 + 2x_3 = 6$$
$$-2x_1 - x_2 + 4x_3 = 1$$

are equivalent.

If a multiple of one equation is added to another equation, the new system will be equivalent to the original system. This follows since the n-tuple $(x_1, \ldots, x_n)$ will satisfy the two equations

$$a_{i1}x_1 + \cdots + a_{in}x_n = b_i$$
$$a_{j1}x_1 + \cdots + a_{jn}x_n = b_j$$

if and only if it satisfies the equations

$$a_{i1}x_1 + \cdots + a_{in}x_n = b_i$$
$$(a_{j1} + \alpha a_{i1})x_1 + \cdots + (a_{jn} + \alpha a_{in})x_n = b_j + \alpha b_i$$

To summarize, there are three operations that can be used on a system to obtain an equivalent system:

I.   The order in which any two equations are written may be interchanged.
II. Both sides of an equation may be multiplied by the same nonzero real number.
III. A multiple of one equation may be added to another.

Given a system of equations, we may use these operations to obtain an equivalent system that is easier to solve.

$n \times n$ Systems

Let us restrict ourselves to $n \times n$ systems for the remainder of this section. We will show that if an $n \times n$ system has exactly one solution, operations I and III can be used to obtain an equivalent "triangular system."

▶ **DEFINITION** A system is said to be in **triangular form** if in the kth equation the coefficients of the first $k - 1$ variables are all zero and the coefficient of x_k is nonzero ($k = 1, \ldots, n$). ◀

EXAMPLE I. The system

$$3x_1 + 2x_2 + x_3 = 1$$
$$x_2 - x_3 = 2$$
$$2x_3 = 4$$

is in triangular form, since in the second equation the coefficients are 0, 1, -1, respectively, and in the third equation the coefficients are 0, 0, 2, respectively. Because of the triangular form, this system is easy to solve. It follows from the third equation that $x_3 = 2$. Using this value in the second equation, we obtain

$$x_2 - 2 = 2 \quad \text{or} \quad x_2 = 4$$

Using $x_2 = 4$, $x_3 = 2$ in the first equation, we end up with

$$3x_1 + 2 \cdot 4 + 2 = 1$$
$$x_1 = -3$$

Thus the solution to the system is $(-3, 4, 2)$.   ◀

Any $n \times n$ triangular system can be solved in the same manner as the last example. First, the nth equation is solved for the value of x_n. This value is used in the $(n-1)$st equation to solve for x_{n-1}. The values x_n and x_{n-1} are used in the $(n-2)$nd equation to solve for x_{n-2}, and so on. We will refer to this method of solving a triangular system as *back substitution*.

EXAMPLE 2. Solve the system

$$2x_1 - x_2 + 3x_3 - 2x_4 = 1$$
$$x_2 - 2x_3 + 3x_4 = 2$$
$$4x_3 + 3x_4 = 3$$
$$4x_4 = 4$$

SOLUTION. Using back substitution, we obtain

$$4x_4 = 4 \qquad x_4 = 1$$
$$4x_3 + 3 \cdot 1 = 3 \qquad x_3 = 0$$
$$x_2 - 2 \cdot 0 + 3 \cdot 1 = 2 \qquad x_2 = -1$$
$$2x_1 - (-1) + 3 \cdot 0 - 2 \cdot 1 = 1 \qquad x_1 = 1$$

Thus the solution is $(1, -1, 0, 1)$.   ◀

If a system of equations is not triangular, we will use operations I and III to try to obtain an equivalent system that is in triangular form.

EXAMPLE 3. Solve the system

$$x_1 + 2x_2 + x_3 = 3$$
$$3x_1 - x_2 - 3x_3 = -1$$
$$2x_1 + 3x_2 + x_3 = 4$$

SOLUTION.   Subtracting 3 times the first row from the second yields

$$-7x_2 - 6x_3 = -10$$

Subtracting 2 times the first row from the third row yields

$$-x_2 - x_3 = -2$$

If the second and third equations of our system, respectively, are replaced by these new equations, we obtain the equivalent system

$$x_1 + 2x_2 + x_3 = 3$$
$$-7x_2 - 6x_3 = -10$$
$$-x_2 - x_3 = -2$$

If the third equation of this system is replaced by the sum of the third equation and $-\frac{1}{7}$ times the second equation, we end up with the following triangular system:

$$x_1 + 2x_2 + x_3 = 3$$
$$-7x_2 - 6x_3 = -10$$
$$-\frac{1}{7}x_3 = -\frac{4}{7}$$

Using back substitution, we get

$$x_3 = 4, \qquad x_2 = -2, \qquad x_1 = 3 \qquad \blacktriangleleft$$

Let us look back at the system of equations in the last example. We can associate with that system a 3×3 array of numbers whose entries are the coefficients of the x_i's.

$$\begin{bmatrix} 1 & 2 & 1 \\ 3 & -1 & -3 \\ 2 & 3 & 1 \end{bmatrix}$$

We will refer to this array as the *coefficient matrix* of the system. The term *matrix* means simply a rectangular array of numbers. A matrix having m rows and n columns is said to be $m \times n$.

If we attach to the coefficient matrix an additional column whose entries are the numbers on the right-hand side of the system, we obtain the new matrix

$$\left[\begin{array}{ccc|c} 1 & 2 & 1 & 3 \\ 3 & -1 & -3 & -1 \\ 2 & 3 & 1 & 4 \end{array} \right]$$

We will refer to this new matrix as the *augmented matrix*. In general, when an $m \times r$ matrix B is attached to an $m \times n$ matrix A in this way, the augmented matrix is denoted by $(A|B)$. Thus if

$$A = \begin{bmatrix} a_{11} & a_{12} & \cdots & a_{1n} \\ a_{21} & a_{22} & \cdots & a_{2n} \\ \vdots & & & \\ a_{m1} & a_{m2} & \cdots & a_{mn} \end{bmatrix}, \qquad B = \begin{bmatrix} b_{11} & b_{12} & \cdots & b_{1r} \\ b_{21} & b_{22} & \cdots & b_{2r} \\ \vdots & & & \\ b_{m1} & b_{m2} & \cdots & b_{mr} \end{bmatrix}$$

then

$$(A|B) = \begin{bmatrix} a_{11} & \cdots & a_{1n} & b_{11} & \cdots & b_{1r} \\ \vdots & & & \vdots & & \\ a_{m1} & \cdots & a_{mn} & b_{m1} & \cdots & b_{mr} \end{bmatrix}$$

With each system of equations we may associate an augmented matrix of the form

$$\begin{bmatrix} a_{11} & \cdots & a_{1n} & b_1 \\ \vdots & & & \vdots \\ a_{m1} & \cdots & a_{mn} & b_m \end{bmatrix}$$

The system can be solved by performing operations on the augmented matrix. The x_i's are place holders that can be omitted until the end of the computation. Corresponding to the three operations used to obtain equivalent systems, the following row operations may be applied to the augmented matrix.

Elementary Row Operations

   I. Interchange two rows.
   II. Multiply a row by a nonzero real number.
   III. Replace a row by its sum with a multiple of another row.

Returning to the example, we find that the first row is used to eliminate the elements in the first column of the remaining rows. The entries to be eliminated are given in color in the matrix below. We refer to the first row as the *pivotal row*. For emphasis, the entries in the pivotal row are all in bold type and the entire row is color shaded. The first nonzero entry in the pivotal row is called the *pivot*.

$$\left.\begin{array}{r} \text{(pivot } a_{11} = 1) \\ \text{entries to be eliminated} \\ a_{21} = 3 \text{ and } a_{31} = 2 \end{array}\right\} \rightarrow \begin{bmatrix} \mathbf{1} & \mathbf{2} & \mathbf{1} & \mathbf{3} \\ 3 & -1 & -3 & -1 \\ 2 & 3 & 1 & 4 \end{bmatrix} \begin{array}{l} \leftarrow \text{pivotal row} \\ \\ \end{array}$$

By using row operation **III**, 3 times the first row is subtracted from the second row and 2 times the first row is subtracted from the third. When this is done, we end up with the matrix

$$\begin{bmatrix} 1 & 2 & 1 & 3 \\ \mathbf{0} & \mathbf{-7} & \mathbf{-6} & \mathbf{-10} \\ 0 & -1 & -1 & -2 \end{bmatrix} \quad \leftarrow \text{pivotal row}$$

At this step we choose the second row as our new pivotal row and apply row operation **III** to eliminate the last element in the second column. This time the pivot is -7 and the quotient $\frac{-1}{-7} = \frac{1}{7}$ is the multiple of the pivotal row that is subtracted from the third row. We end up with the matrix

$$\begin{bmatrix} 1 & 2 & 1 & 3 \\ 0 & -7 & -6 & -10 \\ 0 & 0 & -\frac{1}{7} & -\frac{4}{7} \end{bmatrix}$$

This is the augmented matrix for the triangular system, which is equivalent to the original system. The solution to the system is easily obtained using back substitution.

EXAMPLE 4. Solve the system

$$\begin{aligned} - \ x_2 - \ x_3 + \ x_4 &= \ 0 \\ x_1 + \ x_2 + \ x_3 + \ x_4 &= \ 6 \\ 2x_1 + 4x_2 + \ x_3 - 2x_4 &= -1 \\ 3x_1 + \ x_2 - 2x_3 + 2x_4 &= \ 3 \end{aligned}$$

SOLUTION. The augmented matrix for this system is

$$\begin{bmatrix} 0 & -1 & -1 & 1 & 0 \\ 1 & 1 & 1 & 1 & 6 \\ 2 & 4 & 1 & -2 & -1 \\ 3 & 1 & -2 & 2 & 3 \end{bmatrix}$$

Since it is not possible to eliminate any entries using 0 as a pivot element, we will use row operation **I** to interchange the first two rows of the augmented matrix. The new first row will be the pivotal row and the pivot element will be 1.

$$(\text{pivot } a_{11} = 1) \quad \begin{bmatrix} \mathbf{1} & \mathbf{1} & \mathbf{1} & \mathbf{1} & \mathbf{6} \\ 0 & -1 & -1 & 1 & 0 \\ 2 & 4 & 1 & -2 & -1 \\ 3 & 1 & -2 & 2 & 3 \end{bmatrix} \quad \leftarrow \text{pivotal row}$$

Row operation **III** is then used twice to eliminate the two nonzero entries in the first column.

$$\left[\begin{array}{rrrr|r} 1 & 1 & 1 & 1 & 6 \\ \mathbf{0} & \mathbf{-1} & \mathbf{-1} & \mathbf{1} & \mathbf{0} \\ 0 & 2 & -1 & -4 & -13 \\ 0 & -2 & -5 & -1 & -15 \end{array}\right]$$

Next, the second row is used as the pivotal row to eliminate the entries in the second column below the pivot element -1.

$$\left[\begin{array}{rrrr|r} 1 & 1 & 1 & 1 & 6 \\ 0 & -1 & -1 & 1 & 0 \\ \mathbf{0} & \mathbf{0} & \mathbf{-3} & \mathbf{-2} & \mathbf{-13} \\ 0 & 0 & -3 & -3 & -15 \end{array}\right]$$

Finally, the third row is used as the pivotal row to eliminate the last element in the third column.

$$\left[\begin{array}{rrrr|r} 1 & 1 & 1 & 1 & 6 \\ 0 & -1 & -1 & 1 & 0 \\ 0 & 0 & -3 & -2 & -13 \\ 0 & 0 & 0 & -1 & -2 \end{array}\right]$$

This augmented matrix represents a triangular system. Solving by back substitution, we obtain the solution $(2, -1, 3, 2)$. ◄

In general, if an $n \times n$ linear system can be reduced to triangular form, then it will have a unique solution that can be obtained by performing back substitution on the triangular system. We can think of the reduction process as an algorithm involving $n - 1$ steps. At the first step, a pivot element is chosen from among the nonzero entries in the first column of the matrix. The row containing the pivot element is called the *pivotal row*. We interchange rows (if necessary) so that the pivotal row is the new first row. Multiples of the pivotal row are then subtracted from each of the remaining $n - 1$ rows so as to obtain 0's in the $(2, 1), \ldots, (n, 1)$ positions. At the second step, a pivot element is chosen from the nonzero entries in column 2, rows 2 through n of the matrix. The row containing the pivot is then interchanged with the second row of the matrix and is used as the new pivotal row. Multiples of the pivotal row are then subtracted from the remaining $n - 2$ rows so as to eliminate all entries below the pivot in the second column. The same procedure is repeated for columns 3 through $n - 1$. Note that at the second step row 1 and column 1 remain unchanged, at the third step the first two rows and first two columns remain unchanged, and

$$\text{Step 1} \quad \begin{pmatrix} x & x & x & x & | & x \\ x & x & x & x & | & x \\ x & x & x & x & | & x \\ x & x & x & x & | & x \end{pmatrix} \rightarrow \begin{pmatrix} x & x & x & x & | & x \\ 0 & x & x & x & | & x \\ 0 & x & x & x & | & x \\ 0 & x & x & x & | & x \end{pmatrix}$$

$$\text{Step 2} \quad \begin{pmatrix} x & x & x & x & | & x \\ 0 & x & x & x & | & x \\ 0 & x & x & x & | & x \\ 0 & x & x & x & | & x \end{pmatrix} \rightarrow \begin{pmatrix} x & x & x & x & | & x \\ 0 & x & x & x & | & x \\ 0 & 0 & x & x & | & x \\ 0 & 0 & x & x & | & x \end{pmatrix}$$

$$\text{Step 3} \quad \begin{pmatrix} x & x & x & x & | & x \\ 0 & x & x & x & | & x \\ 0 & 0 & x & x & | & x \\ 0 & 0 & x & x & | & x \end{pmatrix} \rightarrow \begin{pmatrix} x & x & x & x & | & x \\ 0 & x & x & x & | & x \\ 0 & 0 & x & x & | & x \\ 0 & 0 & 0 & x & | & x \end{pmatrix}$$

FIGURE 1.1.2

so on. At each step the overall dimensions of the system are effectively reduced by 1 (see Figure 1.1.2).

If the elimination process can be carried out as described, we will arrive at an equivalent triangular system after $n - 1$ steps. However, the procedure will break down if, at any step, all possible choices for a pivot element are equal to 0. When this happens, the alternative is to reduce the system to certain special echelon or staircase-shaped forms. These echelon forms will be studied in the next section. They will also be used for $m \times n$ systems, where $m \neq n$.

EXERCISES

1. Use back substitution to solve each of the following systems of equations.

(a) $x_1 - 3x_2 = 2$
    $2x_2 = 6$

(b) $x_1 + x_2 + x_3 = 8$
    $2x_2 + x_3 = 5$
    $3x_3 = 9$

(c) $x_1 + 2x_2 + 2x_3 + x_4 = 5$
    $3x_2 + x_3 - 2x_4 = 1$
    $-x_3 + 2x_4 = -1$
    $4x_4 = 4$

(d) $x_1 + x_2 + x_3 + x_4 + x_5 = 5$
    $2x_2 + x_3 - 2x_4 + x_5 = 1$
    $4x_3 + x_4 - 2x_5 = 1$
    $x_4 - 3x_5 = 0$
    $2x_5 = 2$

2. Write out the coefficient matrix for each of the systems in Exercise 1.

3. In each of the following systems, interpret each equation as a line in the plane. For each system, graph the lines and determine geometrically the number of solutions.

(a) $x_1 + x_2 = 4$
$x_1 - x_2 = 2$

(b) $x_1 + 2x_2 = 4$
$-2x_1 - 4x_2 = 4$

(c) $2x_1 - x_2 = 3$
$-4x_1 + 2x_2 = -6$

(d) $x_1 + x_2 = 1$
$x_1 - x_2 = 1$
$-x_1 + 3x_2 = 3$

4. Write an augmented matrix for each of the systems in Exercise 3.

5. Write out the system of equations that corresponds to each of the following augmented matrices.

(a) $\begin{bmatrix} 3 & 2 & | & 8 \\ 1 & 5 & | & 7 \end{bmatrix}$

(b) $\begin{bmatrix} 5 & -2 & 1 & | & 3 \\ 2 & 3 & -4 & | & 0 \end{bmatrix}$

(c) $\begin{bmatrix} 2 & 1 & 4 & | & -1 \\ 4 & -2 & 3 & | & 4 \\ 5 & 2 & 6 & | & -1 \end{bmatrix}$

(d) $\begin{bmatrix} 4 & -3 & 1 & 2 & | & 4 \\ 3 & 1 & -5 & 6 & | & 5 \\ 1 & 1 & 2 & 4 & | & 8 \\ 5 & 1 & 3 & -2 & | & 7 \end{bmatrix}$

6. Solve each of the following systems.

(a) $x_1 - 2x_2 = 5$
$3x_1 + x_2 = 1$

(b) $2x_1 + x_2 = 8$
$4x_1 - 3x_2 = 6$

(c) $4x_1 + 3x_2 = 4$
$\frac{2}{3}x_1 + 4x_2 = 3$

(d) $x_1 + 2x_2 - x_3 = 1$
$2x_1 - x_2 + x_3 = 3$
$-x_1 + 2x_2 + 3x_3 = 7$

(e) $2x_1 + x_2 + 3x_3 = 1$
$4x_1 + 3x_2 + 5x_3 = 1$
$6x_1 + 5x_2 + 5x_3 = -3$

(f) $3x_1 + 2x_2 + x_3 = 0$
$-2x_1 + x_2 - x_3 = 2$
$2x_1 - x_2 + 2x_3 = -1$

(g) $\frac{1}{3}x_1 + \frac{2}{3}x_2 + 2x_3 = -1$
$x_1 + 2x_2 + \frac{3}{2}x_3 = \frac{3}{2}$
$\frac{1}{2}x_1 + 2x_2 + \frac{12}{5}x_3 = \frac{1}{10}$

(h) $x_2 + x_3 + x_4 = 0$
$3x_1 + 3x_3 - 4x_4 = 7$
$x_1 + x_2 + x_3 + 2x_4 = 6$
$2x_1 + 3x_2 + x_3 + 3x_4 = 6$

7. The two systems

$$2x_1 + x_2 = 3$$
$$4x_1 + 3x_2 = 5$$

and

$$2x_1 + x_2 = -1$$
$$4x_1 + 3x_2 = 1$$

have the same coefficient matrix but different right-hand sides. Solve both systems simultaneously by eliminating the (2, 1) entry of the augmented matrix

$$\begin{bmatrix} 2 & 1 & 3 & -1 \\ 4 & 3 & 5 & 1 \end{bmatrix}$$

and then performing back substitutions for each of the columns corresponding to the right-hand sides.

8. Solve the two systems

$$x_1 + 2x_2 - 2x_3 = 1 \qquad\qquad x_1 + 2x_2 - 2x_3 = 9$$
$$2x_1 + 5x_2 + x_3 = 9 \qquad\qquad 2x_1 + 5x_2 + x_3 = 9$$
$$x_1 + 3x_2 + 4x_3 = 9 \qquad\qquad x_1 + 3x_2 + 4x_3 = -2$$

by doing elimination on a 3×5 augmented matrix and then performing two back substitutions.

9. Given a system of the form

$$-m_1 x_1 + x_2 = b_1$$
$$-m_2 x_1 + x_2 = b_2$$

where m_1, m_2, b_1, and b_2 are constants:

(a) Show that the system will have a unique solution if $m_1 \neq m_2$.
(b) If $m_1 = m_2$, show that the system will be consistent only if $b_1 = b_2$.
(c) Give a geometric interpretation to parts (a) and (b).

10. Consider a system of the form

$$a_{11} x_1 + a_{12} x_2 = 0$$
$$a_{21} x_1 + a_{22} x_2 = 0$$

where a_{11}, a_{12}, a_{21}, and a_{22} are constants. Explain why a system of this form must be consistent.

11. Give a geometrical interpretation of a linear equation in three unknowns. Give a geometrical description of the possible solution sets for a 3×3 linear system.

2 ROW ECHELON FORM

In Section 1 we learned a method for reducing an $n \times n$ linear system to triangular form. However, this method will fail if at any stage of the reduction process all the possible choices for a pivot element in a given column are 0.

EXAMPLE I. Consider the system represented by the augmented matrix

$$\left[\begin{array}{ccccc|c} \mathbf{1} & \mathbf{1} & \mathbf{1} & \mathbf{1} & \mathbf{1} & \mathbf{1} \\ -1 & -1 & 0 & 0 & 1 & -1 \\ -2 & -2 & 0 & 0 & 3 & 1 \\ 0 & 0 & 1 & 1 & 3 & -1 \\ 1 & 1 & 2 & 2 & 4 & 1 \end{array}\right] \leftarrow \text{pivotal row}$$

If row operation **III** is used to eliminate the nonzero entries in the last four rows of the first column, the resulting matrix will be

$$\left[\begin{array}{ccccc|c} 1 & 1 & 1 & 1 & 1 & 1 \\ \mathbf{0} & \mathbf{0} & \mathbf{1} & \mathbf{1} & \mathbf{2} & \mathbf{0} \\ 0 & 0 & 2 & 2 & 5 & 3 \\ 0 & 0 & 1 & 1 & 3 & -1 \\ 0 & 0 & 1 & 1 & 3 & 0 \end{array}\right] \leftarrow \text{pivotal row}$$

At this stage the reduction to triangular form breaks down. All four possible choices for the pivot element in the second column are 0. How do we proceed from here? Since our goal is to simplify the system as much as possible, it seems natural to move over to the third column and eliminate the last three entries.

$$\left[\begin{array}{ccccc|c} 1 & 1 & 1 & 1 & 1 & 1 \\ 0 & 0 & 1 & 1 & 2 & 0 \\ \mathbf{0} & \mathbf{0} & \mathbf{0} & \mathbf{0} & \mathbf{1} & \mathbf{3} \\ 0 & 0 & 0 & 0 & 1 & -1 \\ 0 & 0 & 0 & 0 & 1 & 0 \end{array}\right]$$

In the fourth column, all the choices for a pivot element are 0; so again we move on to the next column. If we use the third row as the pivotal row, the last two entries in the fifth column are eliminated.

$$\left[\begin{array}{ccccc|c} 1 & 1 & 1 & 1 & 1 & 1 \\ 0 & 0 & 1 & 1 & 2 & 0 \\ 0 & 0 & 0 & 0 & 1 & 3 \\ 0 & 0 & 0 & 0 & 0 & -4 \\ 0 & 0 & 0 & 0 & 0 & -3 \end{array}\right]$$

The equations represented by the last two rows are

$$0x_1 + 0x_2 + 0x_3 + 0x_4 + 0x_5 = -4$$

$$0x_1 + 0x_2 + 0x_3 + 0x_4 + 0x_5 = -3$$

Since there are no 5-tuples that could possibly satisfy these equations, the system is inconsistent. Note that the coefficient matrix that we end up with is not in triangular form; it is in staircase or echelon form.   ◀

Suppose now that we change the right-hand side of the system in the last example so as to obtain a consistent system. For example, if we start with

$$\left\{\begin{array}{ccccc|c} 1 & 1 & 1 & 1 & 1 & 1 \\ -1 & -1 & 0 & 0 & 1 & -1 \\ -2 & -2 & 0 & 0 & 3 & 1 \\ 0 & 0 & 1 & 1 & 3 & 3 \\ 1 & 1 & 2 & 2 & 4 & 4 \end{array}\right\}$$

then the reduction process will yield the augmented matrix

$$\left\{\begin{array}{ccccc|c} 1 & 1 & 1 & 1 & 1 & 1 \\ 0 & 0 & 1 & 1 & 2 & 0 \\ 0 & 0 & 0 & 0 & 1 & 3 \\ 0 & 0 & 0 & 0 & 0 & 0 \\ 0 & 0 & 0 & 0 & 0 & 0 \end{array}\right\}$$

The last two equations of the reduced system will be satisfied for any 5-tuple. Thus the solution set will be the set of all 5-tuples satisfying the first three equations.

$$\begin{aligned} x_1 + x_2 + x_3 + x_4 + x_5 &= 1 \\ x_3 + x_4 + 2x_5 &= 0 \\ x_5 &= 3 \end{aligned}$$

(1)

The variables corresponding to the first nonzero elements in each row of the augmented matrix will be referred to as *lead variables*. Thus x_1, x_3, and x_5 are the lead variables. The remaining variables corresponding to the columns skipped in the reduction process will be referred to as *free variables*. Thus x_2 and x_4 are the free variables. If we transfer the free variables over to the right-hand side in (1), we obtain the system

$$\begin{aligned} x_1 + x_3 + x_5 &= 1 - x_2 - x_4 \\ x_3 + 2x_5 &= -x_4 \\ x_5 &= 3 \end{aligned}$$

(2)

System (2) is triangular in the unknowns x_1, x_3, x_5. Thus, for each pair of values assigned to x_2 and x_4, there will be a unique solution. For example, if $x_2 = x_4 = 0$, then $x_5 = 3$, $x_3 = -6$, $x_1 = 4$, and hence $(4, 0, -6, 0, 3)$ is a solution to the system.

▶ **DEFINITION** A matrix is said to be in **row echelon form** if

   (i) The first nonzero entry in each row is 1.

  (ii) If row k does not consist entirely of zeros, the number of leading zero entries in row $k + 1$ is greater than the number of leading zero entries in row k.

 (iii) If there are rows whose entries are all zero, they are below the rows having nonzero entries. ◀

EXAMPLE 2. The following matrices are in row echelon form.

$$
\begin{bmatrix} 1 & 4 & 2 \\ 0 & 1 & 3 \\ 0 & 0 & 1 \end{bmatrix}, \quad
\begin{bmatrix} 1 & 2 & 3 \\ 0 & 0 & 1 \\ 0 & 0 & 0 \end{bmatrix}, \quad
\begin{bmatrix} 1 & 3 & 1 & 0 \\ 0 & 0 & 1 & 3 \\ 0 & 0 & 0 & 0 \end{bmatrix}
$$
◀

EXAMPLE 3. The following matrices are not in row echelon form.

$$
\begin{bmatrix} 2 & 4 & 6 \\ 0 & 3 & 5 \\ 0 & 0 & 4 \end{bmatrix}, \quad
\begin{bmatrix} 0 & 0 & 0 \\ 0 & 1 & 0 \end{bmatrix}, \quad
\begin{bmatrix} 0 & 1 \\ 1 & 0 \end{bmatrix}
$$

The first matrix does not satisfy condition (i). The second matrix fails to satisfy condition (iii), and the third matrix fails to satisfy the condition (ii). ◀

▶ **DEFINITION** The process of using row operations **I**, **II**, and **III** to transform a linear system into one whose augmented matrix is in row echelon form is called **Gaussian elimination**. ◀

Note that row operation **II** is necessary in order to scale the rows so that the lead coefficients are all 1. If the row echelon form of the augmented matrix contains a row of the form

$$
\begin{bmatrix} 0 & 0 & \cdots & 0 \,\big|\, 1 \end{bmatrix}
$$

the system is inconsistent. Otherwise, the system will be consistent. If the system is consistent and the nonzero rows of the row echelon of the matrix form a triangular system, the system will have a unique solution.

Overdetermined Systems

A linear system is said to be *overdetermined* if there are more equations than unknowns. Overdetermined systems are usually (but not always) inconsistent.

EXAMPLE 4.

(a)
$$\begin{aligned} x_1 + x_2 &= 1 \\ x_1 - x_2 &= 3 \\ -x_1 + 2x_2 &= -2 \end{aligned}$$

(b)
$$\begin{aligned} x_1 + 2x_2 + x_3 &= 1 \\ 2x_1 - x_2 + x_3 &= 2 \\ 4x_1 + 3x_2 + 3x_3 &= 4 \\ 2x_1 - x_2 + 3x_3 &= 5 \end{aligned}$$

(c)
$$\begin{aligned} x_1 + 2x_2 + x_3 &= 1 \\ 2x_1 - x_2 + x_3 &= 2 \\ 4x_1 + 3x_2 + 3x_3 &= 4 \\ 3x_1 + x_2 + 2x_3 &= 3 \end{aligned}$$

SOLUTION. By now the reader should be familiar enough with the elimination process that we can omit the intermediate steps in reducing each of these systems.

System (a):
$$\left[\begin{array}{cc|c} 1 & 1 & 1 \\ 1 & -1 & 3 \\ -1 & 2 & -2 \end{array} \right] \longrightarrow \left[\begin{array}{cc|c} 1 & 1 & 1 \\ 0 & 1 & -1 \\ 0 & 0 & 1 \end{array} \right]$$

It follows from the last row of the reduced matrix that the system is inconsistent. The three equations in system (a) represent lines in the plane. The first two lines intersect at the point $(2, -1)$. However, the third line does not pass through this point. Thus there are no points that lie on all three lines (see Figure 1.2.1).

FIGURE 1.2.1

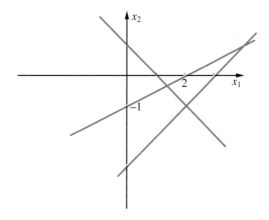

System (b):
$$\begin{Bmatrix} 1 & 2 & 1 & 1 \\ 2 & -1 & 1 & 2 \\ 4 & 3 & 3 & 4 \\ 2 & -1 & 3 & 5 \end{Bmatrix} \rightarrow \begin{Bmatrix} 1 & 2 & 1 & 1 \\ 0 & 1 & \frac{1}{5} & 0 \\ 0 & 0 & 1 & \frac{3}{2} \\ 0 & 0 & 0 & 0 \end{Bmatrix}$$

Using back substitution, we see that system (b) has exactly one solution $(0.1, -0.3, 1.5)$. The solution is unique because the nonzero rows of the reduced matrix form a triangular system.

System (c):
$$\begin{Bmatrix} 1 & 2 & 1 & 1 \\ 2 & -1 & 1 & 2 \\ 4 & 3 & 3 & 4 \\ 3 & 1 & 2 & 3 \end{Bmatrix} \rightarrow \begin{Bmatrix} 1 & 2 & 1 & 1 \\ 0 & 1 & \frac{1}{5} & 0 \\ 0 & 0 & 0 & 0 \\ 0 & 0 & 0 & 0 \end{Bmatrix}$$

Solving for x_2 and x_1 in terms of x_3, we obtain

$$x_2 = -0.2x_3$$
$$x_1 = 1 - 2x_2 - x_3 = 1 - 0.6x_3$$

It follows that the solution set is the set of all ordered triples of the form $(1 - 0.6\alpha, -0.2\alpha, \alpha)$, where α is a real number. This system is consistent and has infinitely many solutions because of the free variable x_3.   ◀

Underdetermined Systems

A system of m linear equations in n unknowns is said to be *underdetermined* if there are fewer equations than unknowns ($m < n$). Although it is possible for underdetermined systems to be inconsistent, they are usually consistent with infinitely many solutions. It is not possible for an underdetermined system to have only one solution. The reason for this is that any row echelon form of the coefficient matrix will involve $r \leq m$ nonzero rows. Thus there will be r lead variables and $n - r$ free variables, where $n - r \geq n - m > 0$. If the system is consistent, we can assign the free variables arbitrary values and solve for the lead variables. Therefore, a consistent underdetermined system will have infinitely many solutions.

EXAMPLE 5.

(a) $\begin{aligned} x_1 + 2x_2 + x_3 &= 1 \\ 2x_1 + 4x_2 + 2x_3 &= 3 \end{aligned}$

(b) $\begin{aligned} x_1 + x_2 + x_3 + x_4 + x_5 &= 2 \\ x_1 + x_2 + x_3 + 2x_4 + 2x_5 &= 3 \\ x_1 + x_2 + x_3 + 2x_4 + 3x_5 &= 2 \end{aligned}$

SOLUTION.

System (a):
$$\begin{bmatrix} 1 & 2 & 1 & | & 1 \\ 2 & 4 & 2 & | & 3 \end{bmatrix} \rightarrow \begin{bmatrix} 1 & 2 & 1 & | & 1 \\ 0 & 0 & 0 & | & 1 \end{bmatrix}$$

Clearly, system (a) is inconsistent. We can think of the two equations in system (a) as representing planes in 3-space. Usually, two planes intersect in a line; however, in this case the planes are parallel.

System (b):
$$\begin{bmatrix} 1 & 1 & 1 & 1 & 1 & | & 2 \\ 1 & 1 & 1 & 2 & 2 & | & 3 \\ 1 & 1 & 1 & 2 & 3 & | & 2 \end{bmatrix} \rightarrow \begin{bmatrix} 1 & 1 & 1 & 1 & 1 & | & 2 \\ 0 & 0 & 0 & 1 & 1 & | & 1 \\ 0 & 0 & 0 & 0 & 1 & | & -1 \end{bmatrix}$$

System (b) is consistent, and since there are two free variables, the system will have infinitely many solutions. Often with systems like this it is convenient to continue with the elimination process until all the terms above each leading 1 are eliminated. (These entries are printed in color.) Thus, for system (b), we will continue and eliminate the first two entries in the fifth column and then the first element in the fourth column.

$$\begin{bmatrix} 1 & 1 & 1 & 1 & 1 & | & 2 \\ 0 & 0 & 0 & 1 & 1 & | & 1 \\ 0 & 0 & 0 & 0 & 1 & | & -1 \end{bmatrix} \rightarrow \begin{bmatrix} 1 & 1 & 1 & 1 & 0 & | & 3 \\ 0 & 0 & 0 & 1 & 0 & | & 2 \\ 0 & 0 & 0 & 0 & 1 & | & -1 \end{bmatrix}$$
$$\rightarrow \begin{bmatrix} 1 & 1 & 1 & 0 & 0 & | & 1 \\ 0 & 0 & 0 & 1 & 0 & | & 2 \\ 0 & 0 & 0 & 0 & 1 & | & -1 \end{bmatrix}$$

If we put the free variables over on the right-hand side, it follows that

$$x_1 = 1 - x_2 - x_3$$
$$x_4 = 2$$
$$x_5 = -1$$

Thus, for any real numbers α and β, the 5-tuple

$$(1 - \alpha - \beta, \alpha, \beta, 2, -1)$$

is a solution to the system.   ◀

In the case where the row echelon form of a consistent system has free variables, it is convenient to continue the elimination process until all the entries above each lead 1 have been eliminated, as in system (b) of the previous example. The resulting reduced matrix is said to be in *reduced row echelon form*.

Reduced Row Echelon Form

▶ **DEFINITION** A matrix is said to be in **reduced row echelon form** if:

(i) The matrix is in row echelon form.

(ii) The first nonzero entry in each row is the only nonzero entry in its column.   ◀

The following matrices are in reduced row echelon form:

$$\begin{bmatrix} 1 & 0 \\ 0 & 1 \end{bmatrix}, \quad \begin{bmatrix} 1 & 0 & 0 & 3 \\ 0 & 1 & 0 & 2 \\ 0 & 0 & 1 & 1 \end{bmatrix}, \quad \begin{bmatrix} 0 & 1 & 2 & 0 \\ 0 & 0 & 0 & 1 \\ 0 & 0 & 0 & 0 \end{bmatrix}, \quad \begin{bmatrix} 1 & 2 & 0 & 1 \\ 0 & 0 & 1 & 3 \\ 0 & 0 & 0 & 0 \end{bmatrix}$$

The process of using elementary row operations to transform a matrix into reduced row echelon form is called *Gauss–Jordan reduction*.

EXAMPLE 6. Use Gauss–Jordan reduction to solve the system

$$\begin{aligned}
-x_1 + x_2 - x_3 + 3x_4 &= 0 \\
3x_1 + x_2 - x_3 - x_4 &= 0 \\
2x_1 - x_2 - 2x_3 - x_4 &= 0
\end{aligned}$$

SOLUTION.

$$\left[\begin{array}{cccc|c} -1 & 1 & -1 & 3 & 0 \\ 3 & 1 & -1 & -1 & 0 \\ 2 & -1 & -2 & -1 & 0 \end{array}\right] \rightarrow \left[\begin{array}{cccc|c} -1 & 1 & -1 & 3 & 0 \\ 0 & 4 & -4 & 8 & 0 \\ 0 & 1 & -4 & 5 & 0 \end{array}\right]$$

$$\rightarrow \left[\begin{array}{cccc|c} -1 & 1 & -1 & 3 & 0 \\ 0 & 4 & -4 & 8 & 0 \\ 0 & 0 & -3 & 3 & 0 \end{array}\right] \rightarrow \left[\begin{array}{cccc|c} 1 & -1 & 1 & -3 & 0 \\ 0 & 1 & -1 & 2 & 0 \\ 0 & 0 & 1 & -1 & 0 \end{array}\right] \begin{array}{l} \text{row} \\ \text{echelon} \\ \text{form} \end{array}$$

$$\rightarrow \left[\begin{array}{cccc|c} 1 & -1 & 0 & -2 & 0 \\ 0 & 1 & 0 & 1 & 0 \\ 0 & 0 & 1 & -1 & 0 \end{array}\right] \rightarrow \left[\begin{array}{cccc|c} 1 & 0 & 0 & -1 & 0 \\ 0 & 1 & 0 & 1 & 0 \\ 0 & 0 & 1 & -1 & 0 \end{array}\right] \begin{array}{l} \text{reduced} \\ \text{row echelon} \\ \text{form} \end{array}$$

If we set x_4 equal to any real number α, then $x_1 = \alpha$, $x_2 = -\alpha$, and $x_3 = \alpha$. Thus all ordered 4-tuples of the form $(\alpha, -\alpha, \alpha, \alpha)$ are solutions to the system. ◀

APPLICATION I: TRAFFIC FLOW

In the downtown section of a certain city, two sets of one-way streets intersect as shown in Figure 1.2.2. The average hourly volume of traffic entering and leaving this section during rush hour is given in the diagram. Determine the amount of traffic between each of the four intersections.

SOLUTION. At each intersection the number of automobiles entering must be the same as the number leaving. For example, at intersection A, the number of automobiles entering is $x_1 + 450$ and the number leaving is $x_2 + 610$. Thus

$$x_1 + 450 = x_2 + 610 \qquad \text{(intersection } A)$$

Similarly,

$$x_2 + 520 = x_3 + 480 \qquad \text{(intersection } B)$$

$$x_3 + 390 = x_4 + 600 \qquad \text{(intersection } C)$$

$$x_4 + 640 = x_1 + 310 \qquad \text{(intersection } D)$$

FIGURE 1.2.2

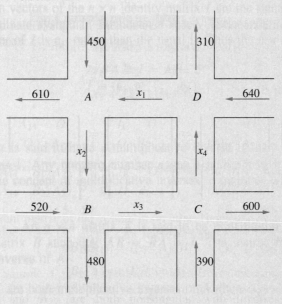

The augmented matrix for the system is

$$\left\{\begin{array}{cccc|c} 1 & -1 & 0 & 0 & 160 \\ 0 & 1 & -1 & 0 & -40 \\ 0 & 0 & 1 & -1 & 210 \\ -1 & 0 & 0 & 1 & -330 \end{array}\right\}$$

The reduced row echelon form for this matrix is

$$\left\{\begin{array}{cccc|c} 1 & 0 & 0 & -1 & 330 \\ 0 & 1 & 0 & -1 & 170 \\ 0 & 0 & 1 & -1 & 210 \\ 0 & 0 & 0 & 0 & 0 \end{array}\right\}$$

The system is consistent, and since there is a free variable, there are many possible solutions. The traffic flow diagram does not give enough information to determine x_1, x_2, x_3, x_4 uniquely. If the amount of traffic were known between any pair of intersections, the traffic on the remaining arteries could easily be calculated. For example, if the amount of traffic between intersections C and D averages 200 automobiles per hour, then $x_4 = 200$. Using this value, we can then solve for x_1, x_2, x_3.

$$x_1 = x_4 + 330 = 530$$
$$x_2 = x_4 + 170 = 370$$
$$x_3 = x_4 + 210 = 410 \qquad \blacktriangleleft$$

APPLICATION 2: ELECTRICAL NETWORKS

In an electrical network it is possible to determine the amount of current in each branch in terms of the resistances and the voltages. An example of a typical circuit is given in Figure 1.2.3.

FIGURE 1.2.3

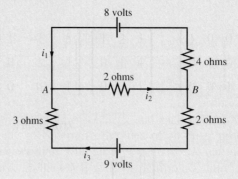

The symbols in the figure have the following meanings:

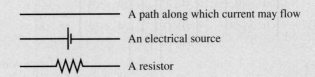

——————————— A path along which current may flow

———————|⊢——————— An electrical source

————-\/\/\/\————- A resistor

The electrical source is usually a battery (measured in volts) that drives a charge and produces a current. The current will flow out of the terminal of the battery represented by the longer vertical line. The resistances are measured in ohms. The letters represent nodes and the i's represent the currents between the nodes. The currents are measured in amperes. The arrows show the direction of the currents. If, however, one of the currents, say i_2, turns out to be negative, this would mean that the current along that branch is in the opposite direction of the arrow.

To determine the currents, the following rules are used:

Kirchhoff's Laws

1. At every node the sum of the incoming currents equals the sum of the outgoing currents.

2. Around every closed loop the algebraic sum of the voltage must equal the algebraic sum of the voltage drops.

The voltage drops E for each resistor are given by *Ohm's law*:

$$E = iR$$

where i represents the current in amperes and R the resistance in ohms.

Let us find the currents in the network pictured in Figure 1.2.3. From the first law, we have

$$i_1 - i_2 + i_3 = 0 \qquad \text{(node } A\text{)}$$
$$-i_1 + i_2 - i_3 = 0 \qquad \text{(node } B\text{)}$$

By the second law,

$$4i_1 + 2i_2 = 8 \qquad \text{(top loop)}$$
$$2i_2 + 5i_3 = 9 \qquad \text{(bottom loop)}$$

The network can be represented by the augmented matrix

$$\begin{bmatrix} 1 & -1 & 1 & 0 \\ -1 & 1 & -1 & 0 \\ 4 & 2 & 0 & 8 \\ 0 & 2 & 5 & 9 \end{bmatrix}$$

This matrix is easily reduced to row echelon form

$$\begin{bmatrix} 1 & -1 & 1 & 0 \\ 0 & 1 & -\frac{2}{3} & \frac{4}{3} \\ 0 & 0 & 1 & 1 \\ 0 & 0 & 0 & 0 \end{bmatrix}$$

Solving by back substitution, we see that $i_1 = 1$, $i_2 = 2$, and $i_3 = 1$.

Homogeneous Systems

A system of linear equations is said to be *homogeneous* if the constants on the right-hand side are all zero. Homogeneous systems are always consistent. It is a trivial matter to find a solution; just set all the variables equal to zero. Thus, if an $m \times n$ homogeneous system has a unique solution, it must be the trivial solution $(0, 0, \ldots, 0)$. The homogeneous system in Example 6 consisted of $m = 3$ equations in $n = 4$ unknowns. In the case that $n > m$, there will always be free variables and, consequently, additional nontrivial solutions. This result has essentially been proved in our discussion of underdetermined systems, but, because of its importance, we state it as a theorem.

▶ **THEOREM I.2.1** *An $m \times n$ homogeneous system of linear equations has a nontrivial solution if $n > m$.*

▶ *Proof.* A homogeneous system is always consistent. The row echelon form of the matrix can have at most m nonzero rows. Thus there are at most m lead variables. Since there are n variables altogether and $n > m$, there must be some free variables. The free variables can be assigned arbitrary values. For each assignment of values to the free variables, there is a solution to the system.   ◀

APPLICATION 3: CHEMICAL EQUATIONS

In the process of photosynthesis, plants use radiant energy from sunlight to convert carbon dioxide (CO_2) and water (H_2O) into glucose ($C_6H_{12}O_6$) and oxygen (O_2).

The chemical equation of the reaction is of the form

$$x_1CO_2 + x_2H_2O \rightarrow x_3O_2 + x_4C_6H_{12}O_6$$

In order to balance the equation we must choose x_1, x_2, x_3, and x_4 so that the number of carbon, hydrogen, and oxygen atoms are the same on each side of the equation. Since carbon dioxide contains one carbon atom and glucose contains six, then to balance the carbon atoms we require that

$$x_1 = 6x_4$$

Similarly, to balance the oxygen we need

$$2x_1 + x_2 = 2x_3 + 6x_4$$

and finally to balance the hydrogen we need

$$2x_2 = 12x_4$$

If we move all the unknowns to the left-hand sides of the three equations, we end up with the homogeneous linear system

$$
\begin{aligned}
x_1 \qquad\qquad\quad - \ 6x_4 &= 0 \\
2x_1 + \ x_2 - 2x_3 - \ 6x_4 &= 0 \\
2x_2 \qquad - 12x_4 &= 0
\end{aligned}
$$

By Theorem 1.2.1, the system has nontrivial solutions. To balance the equation, we must find solutions (x_1, x_2, x_3, x_4) whose entries are nonnegative integers. If we solve the system in the usual way, we see that x_4 is a free variable and

$$x_1 = x_2 = x_3 = 6x_4$$

In particular, if we take $x_4 = 1$, then $x_1 = x_2 = x_3 = 6$ and the equation takes the form

$$6CO_2 + 6H_2O \rightarrow 6O_2 + C_6H_{12}O_6$$

APPLICATION 4: ECONOMIC MODELS FOR EXCHANGE OF GOODS

Suppose that in a primitive society, the members of a tribe are engaged in three occupations: farming, manufacturing of tools and utensils, and weaving and sewing of clothing. Assume that initially the tribe has no monetary system and that all goods and services are bartered. Let us denote the three groups by F, M, and C, and suppose that the directed graph in Figure 1.2.4 indicates how the bartering system works in practice.

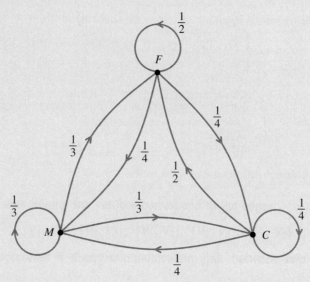

FIGURE 1.2.4

The figure indicates that the farmers keep half of their produce and give one-fourth of their produce to the manufacturers and one-fourth to the clothing producers. The manufacturers divide the goods evenly among the three groups, one-third going to each group. The group producing clothes gives half of the clothes to the farmers and divides the other half evenly between the manufacturers and themselves. The result is summarized in the following table.

	F	M	C
F	$\frac{1}{2}$	$\frac{1}{3}$	$\frac{1}{2}$
M	$\frac{1}{4}$	$\frac{1}{3}$	$\frac{1}{4}$
C	$\frac{1}{4}$	$\frac{1}{3}$	$\frac{1}{4}$

The first column of the table indicates the distribution of the goods produced by the farmers, the second column indicates the distribution of the manufactured goods, and the third column indicates the distribution of the clothing.

As the size of the tribe grows, the system of bartering becomes too cumbersome and, consequently, the tribe decides to institute a monetary system of exchange. For this simple economic system, we assume that there will be no accumulation of capital or debt and that the prices for each of the three types of goods will reflect the values of the existing bartering system. The question is how to assign values to the three types of goods that fairly represent the current bartering system.

The problem can be turned into a linear system of equations using an economic model that was originally developed by the Nobel prize winning economist Wassily

Leontief. For this model we will let x_1 be the monetary value of the goods produced by the farmers, x_2 be the value of the manufactured goods, and x_3 be the value of all the clothing produced. According to the first row of the table, the value of the goods received by the farmers amounts to half the value of the farm goods produced plus one-third the value of the manufactured products and half the value of the clothing goods. Thus the total value of goods received by the farmer is $\frac{1}{2}x_1 + \frac{1}{3}x_2 + \frac{1}{2}x_3$. If the system is fair, the total value of goods received by the farmers should equal x_1, the total value of the farm goods produced. Thus we have the linear equation

$$\frac{1}{2}x_1 + \frac{1}{3}x_2 + \frac{1}{2}x_3 = x_1$$

Using the second row of the table and equating the value of the goods produced and received by the manufacturers, we obtain a second equation:

$$\frac{1}{4}x_1 + \frac{1}{3}x_2 + \frac{1}{4}x_3 = x_2$$

Finally, using the third row of the table, we get

$$\frac{1}{4}x_1 + \frac{1}{3}x_2 + \frac{1}{4}x_3 = x_3$$

These equations can be rewritten as a homogeneous system:

$$-\tfrac{1}{2}x_1 + \tfrac{1}{3}x_2 + \tfrac{1}{2}x_3 = 0$$
$$\tfrac{1}{4}x_1 - \tfrac{2}{3}x_2 + \tfrac{1}{4}x_3 = 0$$
$$\tfrac{1}{4}x_1 + \tfrac{1}{3}x_2 - \tfrac{3}{4}x_3 = 0$$

The reduced row echelon form of the augmented matrix for this system is

$$\begin{bmatrix} 1 & 0 & -\frac{5}{3} & 0 \\ 0 & 1 & -1 & 0 \\ 0 & 0 & 0 & 0 \end{bmatrix}$$

There is one free variable x_3. Setting $x_3 = 3$, we obtain the solution $(5, 3, 3)$, and the general solution consists of all multiples of $(5, 3, 3)$. It follows that the variables x_1, x_2, x_3 should be assigned values in the ratio

$$x_1 : x_2 : x_3 = 5 : 3 : 3$$

This simple system is an example of the closed Leontief input-output model. Leontief's models are fundamental to our understanding of economic systems. Modern applications would involve thousands of industries and lead to very large linear systems. The Leontief models will be studied in greater detail later in Section 7 of Chapter 6.

EXERCISES

1. Which of the following matrices are in row echelon form? Which are in reduced row echelon form?

(a) $\begin{bmatrix} 1 & 2 & 3 & 4 \\ 0 & 0 & 1 & 2 \end{bmatrix}$

(b) $\begin{bmatrix} 1 & 0 & 0 \\ 0 & 0 & 0 \\ 0 & 0 & 1 \end{bmatrix}$

(c) $\begin{bmatrix} 1 & 3 & 0 \\ 0 & 0 & 1 \\ 0 & 0 & 0 \end{bmatrix}$

(d) $\begin{bmatrix} 0 & 1 \\ 0 & 0 \\ 0 & 0 \end{bmatrix}$

(e) $\begin{bmatrix} 1 & 1 & 1 \\ 0 & 1 & 2 \\ 0 & 0 & 3 \end{bmatrix}$

(f) $\begin{bmatrix} 1 & 4 & 6 \\ 0 & 0 & 1 \\ 0 & 1 & 3 \end{bmatrix}$

(g) $\begin{bmatrix} 1 & 0 & 0 & 1 & 2 \\ 0 & 1 & 0 & 2 & 4 \\ 0 & 0 & 1 & 3 & 6 \end{bmatrix}$

(h) $\begin{bmatrix} 0 & 1 & 3 & 4 \\ 0 & 0 & 1 & 3 \\ 0 & 0 & 0 & 0 \end{bmatrix}$

2. In each of the following, the augmented matrix is in row echelon form. For each case, indicate whether the corresponding linear system is consistent. If the system has a unique solution, find it.

(a) $\left[\begin{array}{cc|c} 1 & 2 & 4 \\ 0 & 1 & 3 \\ 0 & 0 & 1 \end{array}\right]$

(b) $\left[\begin{array}{cc|c} 1 & 3 & 1 \\ 0 & 1 & -1 \\ 0 & 0 & 0 \end{array}\right]$

(c) $\left[\begin{array}{ccc|c} 1 & -2 & 4 & 1 \\ 0 & 0 & 1 & 3 \\ 0 & 0 & 0 & 0 \end{array}\right]$

(d) $\left[\begin{array}{ccc|c} 1 & -2 & 2 & -2 \\ 0 & 1 & -1 & 3 \\ 0 & 0 & 1 & 2 \end{array}\right]$

(e) $\left[\begin{array}{ccc|c} 1 & 3 & 2 & -2 \\ 0 & 0 & 1 & 4 \\ 0 & 0 & 0 & 1 \end{array}\right]$

(f) $\left[\begin{array}{ccc|c} 1 & -1 & 3 & 8 \\ 0 & 1 & 2 & 7 \\ 0 & 0 & 1 & 2 \\ 0 & 0 & 0 & 0 \end{array}\right]$

3. In each of the following, the augmented matrix is in reduced row echelon form. In each case, find the solution set to the corresponding linear system.

(a) $\left[\begin{array}{ccc|c} 1 & 0 & 0 & -2 \\ 0 & 1 & 0 & 5 \\ 0 & 0 & 1 & 3 \end{array}\right]$

(b) $\left[\begin{array}{ccc|c} 1 & 4 & 0 & 2 \\ 0 & 0 & 1 & 3 \\ 0 & 0 & 0 & 1 \end{array}\right]$

(c) $\left[\begin{array}{ccc|c} 1 & -3 & 0 & 2 \\ 0 & 0 & 1 & -2 \\ 0 & 0 & 0 & 0 \end{array}\right]$

(d) $\left[\begin{array}{cccc|c} 1 & 2 & 0 & 1 & 5 \\ 0 & 0 & 1 & 3 & 4 \end{array}\right]$

(e) $\begin{bmatrix} 1 & 5 & -2 & 0 & 3 \\ 0 & 0 & 0 & 1 & 6 \\ 0 & 0 & 0 & 0 & 0 \\ 0 & 0 & 0 & 0 & 0 \end{bmatrix}$ (f) $\begin{bmatrix} 0 & 1 & 0 & 2 \\ 0 & 0 & 1 & -1 \\ 0 & 0 & 0 & 0 \end{bmatrix}$

4. For each of the systems in Exercise 3, make a list of the lead variables and a second list of the free variables.

5. For each of the following systems of equations, use Gaussian elimination to obtain an equivalent system whose coefficient matrix is in row echelon form. Indicate whether the system is consistent. If the system is consistent and involves no free variables, use back substitution to find the unique solution. If the system is consistent and there are free variables, transform it to reduced row echelon form and find all solutions.

(a) $x_1 - 2x_2 = 3$
    $2x_1 - x_2 = 9$

(b) $2x_1 - 3x_2 = 5$
    $-4x_1 + 6x_2 = 8$

(c) $x_1 + x_2 = 0$
    $2x_1 + 3x_2 = 0$
    $3x_1 - 2x_2 = 0$

(d) $3x_1 + 2x_2 - x_3 = 4$
    $x_1 - 2x_2 + 2x_3 = 1$
    $11x_1 + 2x_2 + x_3 = 14$

(e) $2x_1 + 3x_2 + x_3 = 1$
    $x_1 + x_2 + x_3 = 3$
    $3x_1 + 4x_2 + 2x_3 = 4$

(f) $x_1 - x_2 + 2x_3 = 4$
    $2x_1 + 3x_2 - x_3 = 1$
    $7x_1 + 3x_2 + 4x_3 = 7$

(g) $x_1 + x_2 + x_3 + x_4 = 0$
    $2x_1 + 3x_2 - x_3 - x_4 = 2$
    $3x_1 + 2x_2 + x_3 + x_4 = 5$
    $3x_1 + 6x_2 - x_3 - x_4 = 4$

(h) $x_1 - 2x_2 = 3$
    $2x_1 + x_2 = 1$
    $-5x_1 + 8x_2 = 4$

(i) $-x_1 + 2x_2 - x_3 = 2$
    $-2x_1 + 2x_2 + x_3 = 4$
    $3x_1 + 2x_2 + 2x_3 = 5$
    $-3x_1 + 8x_2 + 5x_3 = 17$

(j) $x_1 + 2x_2 - 3x_3 + x_4 = 1$
    $-x_1 - x_2 + 4x_3 - x_4 = 6$
    $-2x_1 - 4x_2 + 7x_3 - x_4 = 1$

(k) $x_1 + 3x_2 + x_3 + x_4 = 3$
    $2x_1 - 2x_2 + x_3 + 2x_4 = 8$
    $x_1 - 5x_2 + x_4 = 5$

(l) $x_1 - 3x_2 + x_3 = 1$
    $2x_1 + x_2 - x_3 = 2$
    $x_1 + 4x_2 - 2x_3 = 1$
    $5x_1 - 8x_2 + 2x_3 = 5$

6. Use Gauss–Jordan reduction to solve each of the following systems.

(a) $x_1 + x_2 = -1$
    $4x_1 - 3x_2 = 3$

(b) $x_1 + 3x_2 + x_3 + x_4 = 3$
    $2x_1 - 2x_2 + x_3 + 2x_4 = 8$
    $3x_1 + x_2 + 2x_3 - x_4 = -1$

(c) $x_1 + x_2 + x_3 = 0$
    $x_1 - x_2 - x_3 = 0$

(d) $x_1 + x_2 + x_3 + x_4 = 0$
    $2x_1 + x_2 - x_3 + 3x_4 = 0$
    $x_1 - 2x_2 + x_3 + x_4 = 0$

7. Give a geometric explanation of why a homogeneous linear system consisting of two equations in three unknowns must have infinitely many solutions. What are

the possible numbers of solutions for a nonhomogeneous 2×3 linear system? Give a geometric explanation of your answer.

8. Consider a linear system whose augmented matrix is of the form

$$\left(\begin{array}{rrr|r} 1 & 2 & 1 & 1 \\ -1 & 4 & 3 & 2 \\ 2 & -2 & a & 3 \end{array}\right)$$

For what values of a will the system have a unique solution?

9. Consider a linear system whose augmented matrix is of the form

$$\left(\begin{array}{rrr|r} 1 & 2 & 1 & 0 \\ 2 & 5 & 3 & 0 \\ -1 & 1 & \beta & 0 \end{array}\right)$$

(a) Is it possible for the system to be inconsistent? Explain.
(b) For what values of β will the system have infinitely many solutions?

10. Consider a linear system whose augmented matrix is of the form

$$\left(\begin{array}{rrr|r} 1 & 1 & 3 & 2 \\ 1 & 2 & 4 & 3 \\ 1 & 3 & a & b \end{array}\right)$$

(a) For what values of a and b will the system have infinitely many solutions?
(b) For what values of a and b will the system be inconsistent?

11. Given the linear systems

(a) $\begin{aligned} x_1 + 2x_2 &= 2 \\ 3x_1 + 7x_2 &= 8 \end{aligned}$
(b) $\begin{aligned} x_1 + 2x_2 &= 1 \\ 3x_1 + 7x_2 &= 7 \end{aligned}$

Solve both systems by incorporating the right-hand sides into a 2×2 matrix B and computing the reduced row echelon form of

$$(A|B) = \left(\begin{array}{rr|rr} 1 & 2 & 2 & 1 \\ 3 & 7 & 8 & 7 \end{array}\right)$$

12. Given the linear systems

(a) $\begin{aligned} x_1 + 2x_2 + x_3 &= 2 \\ -x_1 - x_2 + 2x_3 &= 3 \\ 2x_1 + 3x_2 &= 0 \end{aligned}$
(b) $\begin{aligned} x_1 + 2x_2 + x_3 &= -1 \\ -x_1 - x_2 + 2x_3 &= 2 \\ 2x_1 + 3x_2 &= -2 \end{aligned}$

Solve both systems by computing the row echelon form of an augmented matrix $(A|B)$ and performing back substitution twice.

13. Determine the values of x_1, x_2, x_3, x_4 for the following traffic flow diagram.

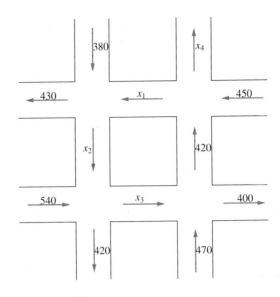

14. Consider the following traffic flow diagram, where a_1, a_2, a_3, a_4, b_1, b_2, b_3, b_4 are fixed positive integers. Set up a linear system in the unknowns x_1, x_2, x_3, x_4 and show that the system will be consistent if and only if

$$a_1 + a_2 + a_3 + a_4 = b_1 + b_2 + b_3 + b_4$$

What can you conclude about the number of automobiles entering and leaving the traffic network?

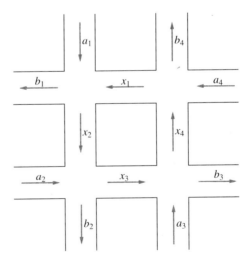

15. Let (c_1, c_2) be a solution to the 2×2 system

$$a_{11}x_1 + a_{12}x_2 = 0$$
$$a_{21}x_1 + a_{22}x_2 = 0$$

Show that for any real number α the ordered pair $(\alpha c_1, \alpha c_2)$ is also a solution.

16. In Application 3 the solution $(6, 6, 6, 1)$ was obtained by setting the free variable $x_4 = 1$.

(a) Determine the solution corresponding to $x_4 = 0$. What information, if any, does this solution give about the chemical reaction? Is the term "trivial solution" appropriate in this case?

(b) Choose some other values of x_4, such as 2, 4, or 5, and determine the corresponding solutions. How are these nontrivial solutions related?

17. Liquid benzene burns in the atmosphere. If a cold object is placed directly over the benzene, water will condense on the object and a deposit of soot (carbon) will also form on the object. The chemical equation for this reaction is of the form

$$x_1 C_6 H_6 + x_2 O_2 \rightarrow x_3 C + x_4 H_2 O$$

Determine values of x_1, x_2, x_3, and x_4 to balance the equation.

18. Nitric acid is prepared commercially by a series of three chemical reactions. In the first reaction, nitrogen (N_2) is combined with hydrogen (H_2) to form ammonia (NH_3). Next the ammonia is combined with oxygen (O_2) to form nitrogen dioxide (NO_2) and water. Finally, the NO_2 reacts with some of the water to form nitric acid (HNO_3) and nitrogen oxide (NO). The amounts of each of the components of these reactions are measured in moles (a standard unit of measurement for chemical reactions). How many moles of nitrogen, hydrogen, and oxygen are necessary in order to produce eight moles of nitric acid?

19. In Application 4, determine the relative values of x_1, x_2, and x_3 if the distribution of goods is as described in the following table.

	F	M	C
F	$\frac{1}{3}$	$\frac{1}{3}$	$\frac{1}{3}$
M	$\frac{1}{3}$	$\frac{1}{2}$	$\frac{1}{6}$
C	$\frac{1}{3}$	$\frac{1}{6}$	$\frac{1}{2}$

20. Determine the amount of each current for the following networks.

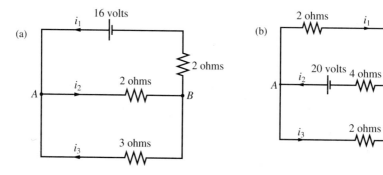

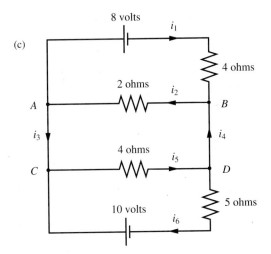

(c)

3 MATRIX ALGEBRA

In this section we define arithmetic operations with matrices and look at some of their algebraic properties. Matrices are one of the most powerful tools in mathematics. To use matrices effectively, we must be adept at matrix arithmetic.

The entries of a matrix are called *scalars*. They are usually either real or complex numbers. For the most part we will be working with matrices whose entries are real numbers. Throughout the first five chapters of the book the reader may assume that the term *scalar* refers to a real number. However, in Chapter 6 there will be occasions when we will use the set of complex numbers as our scalar field.

Matrix Notation

If we wish to refer to matrices without specifically writing out all their entries, we will use capital letters A, B, C, and so on. In general, a_{ij} will denote the entry of the matrix A that is in the ith row and the jth column. Thus, if A is an $m \times n$ matrix, then

$$A = \begin{bmatrix} a_{11} & a_{12} & \cdots & a_{1n} \\ a_{21} & a_{22} & \cdots & a_{2n} \\ \vdots & & & \\ a_{m1} & a_{m2} & \cdots & a_{mn} \end{bmatrix}$$

We will sometimes shorten this to $A = (a_{ij})$. Similarly, a matrix B may be referred to as (b_{ij}), a matrix C as (c_{ij}), and so on.

Vectors

Matrices that have only one row or one column are of special interest since they are used to represent solutions to linear systems. A solution to a system of m linear equations in n unknowns is an n-tuple of real numbers. We will refer to an n-tuple of real numbers as a *vector*. If an n-tuple is represented in terms of a $1 \times n$ matrix, then we will refer to it as a *row vector*. Alternatively, if the n-tuple is represented by an $n \times 1$ matrix, then we will refer to it as a *column vector*. For example, the solution to the linear system

$$x_1 + x_2 = 3$$
$$x_1 - x_2 = 1$$

can be represented by the row vector $(2, 1)$ or the column vector $\begin{pmatrix} 2 \\ 1 \end{pmatrix}$.

In working with matrix equations, it is generally more convenient to represent the solutions in terms of column vectors ($n \times 1$ matrices). The set of all $n \times 1$ matrices of real numbers is called *Euclidean n-space* and is usually denoted by R^n. Since we will be working almost exclusively with column vectors in the future, we will generally omit the word "column" and refer to the elements of R^n as simply *vectors*, rather than as column vectors.

The standard notation for a column vector is a boldface lowercase letter.

$$\mathbf{x} = \begin{bmatrix} x_1 \\ x_2 \\ \vdots \\ x_n \end{bmatrix}$$

Given an $m \times n$ matrix A, it is often necessary to refer to a particular row or column. The ith row vector of A will be denoted by $\mathbf{a}(i, :)$ and the jth column vector will be denoted by $\mathbf{a}(:, j)$. Since we will be working primarily with column vectors, it is convenient to have a shorthand notation for the jth column vector. We will use $\mathbf{a}_j$ in place of $\mathbf{a}(:, j)$. References to row vectors are far less frequent, so we will not use any shorthand notation to represent them.

If A is an $m \times n$ matrix, then the row vectors of A are given by

$$\mathbf{a}(i, :) = (a_{i1}, a_{i2}, \dots, a_{in}) \qquad i = 1, \dots, m$$

and the column vectors are given by

$$\mathbf{a}_j = \mathbf{a}(:, j) = \begin{bmatrix} a_{1j} \\ a_{2j} \\ \vdots \\ a_{mj} \end{bmatrix} \qquad j = 1, \dots, n$$

The matrix A can be represented either in terms of its column vectors or in terms of its row vectors.

$$A = (\mathbf{a}_1, \mathbf{a}_2, \ldots, \mathbf{a}_n) \qquad \text{or} \qquad A = \begin{bmatrix} \mathbf{a}(1,:) \\ \mathbf{a}(2,:) \\ \vdots \\ \mathbf{a}(m,:) \end{bmatrix}$$

Similarly, if B is an $n \times r$ matrix, then $B = (\mathbf{b}_1, \mathbf{b}_2, \ldots, \mathbf{b}_r)$.

Equality

In order for two matrices to be equal, they must have the same dimensions and their corresponding entries must agree.

▶ **DEFINITION** Two $m \times n$ matrices A and B are said to be **equal** if $a_{ij} = b_{ij}$ for each i and j. ◀

Scalar Multiplication

If A is a matrix and α is a scalar, then αA is the matrix formed by multiplying each of the entries of A by α.

▶ **DEFINITION** If A is an $m \times n$ matrix and α is a scalar, then αA is the $m \times n$ matrix whose (i, j) entry is αa_{ij}. ◀

For example, if

$$A = \begin{bmatrix} 4 & 8 & 2 \\ 6 & 8 & 10 \end{bmatrix}$$

then

$$\frac{1}{2}A = \begin{bmatrix} 2 & 4 & 1 \\ 3 & 4 & 5 \end{bmatrix} \qquad \text{and} \qquad 3A = \begin{bmatrix} 12 & 24 & 6 \\ 18 & 24 & 30 \end{bmatrix}$$

Matrix Addition

Two matrices with the same dimensions can be added by adding their corresponding entries.

▶ **DEFINITION** If $A = (a_{ij})$ and $B = (b_{ij})$ are both $m \times n$ matrices, then the **sum** $A + B$ is the $m \times n$ matrix whose (i, j) entry is $a_{ij} + b_{ij}$ for each ordered pair (i, j). ◀

For example,

$$\begin{bmatrix} 3 & 2 & 1 \\ 4 & 5 & 6 \end{bmatrix} + \begin{bmatrix} 2 & 2 & 2 \\ 1 & 2 & 3 \end{bmatrix} = \begin{bmatrix} 5 & 4 & 3 \\ 5 & 7 & 9 \end{bmatrix}$$

$$\begin{bmatrix} 2 \\ 1 \\ 8 \end{bmatrix} + \begin{bmatrix} -8 \\ 3 \\ 2 \end{bmatrix} = \begin{bmatrix} -6 \\ 4 \\ 10 \end{bmatrix}$$

If we define $A - B$ to be $A + (-1)B$, then it turns out that $A - B$ is formed by subtracting the corresponding entry of B from each entry of A. Thus

$$\begin{bmatrix} 2 & 4 \\ 3 & 1 \end{bmatrix} - \begin{bmatrix} 4 & 5 \\ 2 & 3 \end{bmatrix} = \begin{bmatrix} 2 & 4 \\ 3 & 1 \end{bmatrix} + (-1)\begin{bmatrix} 4 & 5 \\ 2 & 3 \end{bmatrix}$$

$$= \begin{bmatrix} 2 & 4 \\ 3 & 1 \end{bmatrix} + \begin{bmatrix} -4 & -5 \\ -2 & -3 \end{bmatrix}$$

$$= \begin{bmatrix} 2-4 & 4-5 \\ 3-2 & 1-3 \end{bmatrix}$$

$$= \begin{bmatrix} -2 & -1 \\ 1 & -2 \end{bmatrix}$$

If O represents a matrix, with the same dimensions as A, whose entries are all 0, then

$$A + O = O + A = A$$

That is, the zero matrix acts as an additive identity on the set of all $m \times n$ matrices. Furthermore, each $m \times n$ matrix A has an additive inverse. Indeed,

$$A + (-1)A = O = (-1)A + A$$

It is customary to denote the additive inverse by $-A$. Thus

$$-A = (-1)A$$

Matrix Multiplication and Linear Systems

We have yet to define the most important operation, the multiplication of two matrices. Much of the motivation behind the definition comes from the applications to linear systems of equations. If we have a system of one linear equation in one unknown, it can be written in the form

(1) $$ax = b$$

We generally think of a, x, and b as being scalars; however, they could also be treated as 1×1 matrices. Our goal now is to generalize equation (1) so that we can represent an $m \times n$ linear system by a single matrix equation

$$A\mathbf{x} = \mathbf{b}$$

where A is an $m \times n$ matrix, $\mathbf{x}$ is in R^n, and $\mathbf{b}$ is in R^m. We consider first the case of one equation in several unknowns.

CASE I One Equation in Several Unknowns

Let us begin by examining the case of one equation in several variables. Consider for example the equation

$$3x_1 + 2x_2 + 5x_3 = 4$$

If we set

$$A = \begin{bmatrix} 3 & 2 & 5 \end{bmatrix} \qquad \text{and} \qquad \mathbf{x} = \begin{bmatrix} x_1 \\ x_2 \\ x_3 \end{bmatrix}$$

and define the product $A\mathbf{x}$ by

$$A\mathbf{x} = \begin{bmatrix} 3 & 2 & 5 \end{bmatrix} \begin{bmatrix} x_1 \\ x_2 \\ x_3 \end{bmatrix} = 3x_1 + 2x_2 + 5x_3$$

then the equation $3x_1 + 2x_2 + 5x_3 = 4$ can be written as the matrix equation

$$A\mathbf{x} = 4$$

In general, if

$$A = \begin{bmatrix} a_1 & a_2 & \ldots & a_n \end{bmatrix}$$

and $\mathbf{x}$ is a column vector with n entries, then the product $A\mathbf{x}$ is defined by

$$A\mathbf{x} = a_1 x_1 + a_2 x_2 + \cdots + a_n x_n$$

For example, if

$$A = \begin{bmatrix} 2 & 1 & -3 & 4 \end{bmatrix} \qquad \text{and} \qquad \mathbf{x} = \begin{bmatrix} 3 \\ 2 \\ 1 \\ -2 \end{bmatrix}$$

then

$$A\mathbf{x} = 2 \cdot 3 + 1 \cdot 2 + (-3) \cdot 1 + 4 \cdot (-2) = -3$$

Note that the result of multiplying a row vector on the left times a column vector on the right is a scalar. Consequently, this type of multiplication is often referred to as a *scalar product*.

CASE 2 *M* Equations in *N* Unknowns

Consider now an $m \times n$ linear system

(2)
$$
\begin{aligned}
a_{11}x_1 + a_{12}x_2 + \cdots + a_{1n}x_n &= b_1 \\
a_{21}x_1 + a_{22}x_2 + \cdots + a_{2n}x_n &= b_2 \\
&\vdots \\
a_{m1}x_1 + a_{m2}x_2 + \cdots + a_{mn}x_n &= b_m
\end{aligned}
$$

It is desirable to write the system in a form similar to (1), that is, as a matrix equation

(3)
$$
A\mathbf{x} = \mathbf{b}
$$

where $A = (a_{ij})$ is known, $\mathbf{x}$ is an $n \times 1$ matrix of unknowns, and $\mathbf{b}$ is an $m \times 1$ matrix representing the right-hand side of the system. Thus, if we set

$$
A = \begin{bmatrix} a_{11} & a_{12} & \cdots & a_{1n} \\ a_{21} & a_{22} & \cdots & a_{2n} \\ \vdots & & & \\ a_{m1} & a_{m2} & \cdots & a_{mn} \end{bmatrix}, \quad
\mathbf{x} = \begin{bmatrix} x_1 \\ x_2 \\ \vdots \\ x_n \end{bmatrix}, \quad
\mathbf{b} = \begin{bmatrix} b_1 \\ b_2 \\ \vdots \\ b_m \end{bmatrix}
$$

and define the product $A\mathbf{x}$ by

(4)
$$
A\mathbf{x} = \begin{bmatrix} a_{11}x_1 + a_{12}x_2 + \cdots + a_{1n}x_n \\ a_{21}x_1 + a_{22}x_2 + \cdots + a_{2n}x_n \\ \vdots \\ a_{m1}x_1 + a_{m2}x_2 + \cdots + a_{mn}x_n \end{bmatrix}
$$

then the linear system of equations (2) is equivalent to the matrix equation (3).

Given an $m \times n$ matrix A and a vector $\mathbf{x}$ in R^n, it is possible to compute a product $A\mathbf{x}$ by (4). The product $A\mathbf{x}$ will be an $m \times 1$ matrix, that is, a vector in R^m. The rule for determining the ith entry of $A\mathbf{x}$ is

$$
a_{i1}x_1 + a_{i2}x_2 + \cdots + a_{in}x_n
$$

which is equal to $\mathbf{a}(i, :)\mathbf{x}$, the scalar product of the ith row vector of A times the column vector $\mathbf{x}$. Thus

$$
A\mathbf{x} = \begin{bmatrix} \mathbf{a}(1, :)\mathbf{x} \\ \mathbf{a}(2, :)\mathbf{x} \\ \vdots \\ \mathbf{a}(n, :)\mathbf{x} \end{bmatrix}
$$

EXAMPLE I.

$$A = \begin{bmatrix} 4 & 2 & 1 \\ 5 & 3 & 7 \end{bmatrix}, \quad \mathbf{x} = \begin{bmatrix} x_1 \\ x_2 \\ x_3 \end{bmatrix}$$

$$A\mathbf{x} = \begin{bmatrix} 4x_1 + 2x_2 + x_3 \\ 5x_1 + 3x_2 + 7x_3 \end{bmatrix}$$ ◀

EXAMPLE 2.

$$A = \begin{bmatrix} -3 & 1 \\ 2 & 5 \\ 4 & 2 \end{bmatrix}, \quad \mathbf{x} = \begin{bmatrix} 2 \\ 4 \end{bmatrix}$$

$$A\mathbf{x} = \begin{bmatrix} -3 \cdot 2 + 1 \cdot 4 \\ 2 \cdot 2 + 5 \cdot 4 \\ 4 \cdot 2 + 2 \cdot 4 \end{bmatrix} = \begin{bmatrix} -2 \\ 24 \\ 16 \end{bmatrix}$$ ◀

EXAMPLE 3. Write the following system of equations as a matrix equation $A\mathbf{x} = \mathbf{b}$.

$$\begin{aligned} 3x_1 + 2x_2 + x_3 &= 5 \\ x_1 - 2x_2 + 5x_3 &= -2 \\ 2x_1 + x_2 - 3x_3 &= 1 \end{aligned}$$

SOLUTION.

$$\begin{bmatrix} 3 & 2 & 1 \\ 1 & -2 & 5 \\ 2 & 1 & -3 \end{bmatrix} \begin{bmatrix} x_1 \\ x_2 \\ x_3 \end{bmatrix} = \begin{bmatrix} 5 \\ -2 \\ 1 \end{bmatrix}$$ ◀

APPLICATION I: WEIGHT REDUCTION

Bob weighs 178 pounds. He wishes to lose weight through a program of dieting and exercise. After consulting Table 1, he sets up the exercise schedule in Table 2. How many calories will he burn up exercising each day if he follows this program?

TABLE I

CALORIES BURNED PER HOUR

Exercise Activity	Weight in lb			
	152	161	170	178
Walking 2 mph	213	225	237	249
Running 5.5 mph	651	688	726	764
Bicycling 5.5 mph	304	321	338	356
Tennis (moderate)	420	441	468	492

TABLE 2

HOURS PER DAY FOR EACH ACTIVITY

	Exercise Schedule			
	Walking	Running	Bicycling	Tennis
Monday	1.0	0.0	1.0	0.0
Tuesday	0.0	0.0	0.0	2.0
Wednesday	0.4	0.5	0.0	0.0
Thursday	0.0	0.0	0.5	2.0
Friday	0.4	0.5	0.0	0.0

SOLUTION. The information pertaining to Bob is located in column four of Table 1. This information can be represented by a column vector $\mathbf{x}$. The information in Table 2 can be represented by a 5×4 matrix A. To answer the question, we simply calculate $A\mathbf{x}$.

$$
\begin{bmatrix}
1.0 & 0.0 & 1.0 & 0.0 \\
0.0 & 0.0 & 0.0 & 2.0 \\
0.4 & 0.5 & 0.0 & 0.0 \\
0.0 & 0.0 & 0.5 & 2.0 \\
0.4 & 0.5 & 0.0 & 0.0
\end{bmatrix}
\begin{bmatrix}
249 \\
764 \\
356 \\
492
\end{bmatrix}
=
\begin{bmatrix}
605.0 \\
984.0 \\
481.6 \\
1162.0 \\
481.6
\end{bmatrix}
\begin{matrix}
Monday \\
Tuesday \\
Wednesday \\
Thursday \\
Friday
\end{matrix}
\qquad \blacktriangleleft
$$

An alternative way to represent the linear system (2) as a matrix equation is to express the product $A\mathbf{x}$ as a sum of column vectors:

$$
A\mathbf{x} =
\begin{bmatrix}
a_{11}x_1 + a_{12}x_2 + \cdots + a_{1n}x_n \\
a_{21}x_1 + a_{22}x_2 + \cdots + a_{2n}x_n \\
\vdots \\
a_{m1}x_1 + a_{m2}x_2 + \cdots + a_{mn}x_n
\end{bmatrix}
$$

$$= x_1 \begin{bmatrix} a_{11} \\ a_{21} \\ \vdots \\ a_{m1} \end{bmatrix} + x_2 \begin{bmatrix} a_{12} \\ a_{22} \\ \vdots \\ a_{m2} \end{bmatrix} + \cdots + x_n \begin{bmatrix} a_{1n} \\ a_{2n} \\ \vdots \\ a_{mn} \end{bmatrix}$$

Thus we have

(5) $$A\mathbf{x} = x_1\mathbf{a}_1 + x_2\mathbf{a}_2 + \cdots + x_n\mathbf{a}_n$$

Using this formula we can represent the system of equations (2) as a matrix equation of the form

(6) $$x_1\mathbf{a}_1 + x_2\mathbf{a}_2 + \cdots + x_n\mathbf{a}_n = \mathbf{b}$$

EXAMPLE 4.   The linear system

$$2x_1 + 3x_2 - 2x_3 = 5$$
$$5x_1 - 4x_2 + 2x_3 = 6$$

can be written as a matrix equation

$$x_1 \begin{bmatrix} 2 \\ 5 \end{bmatrix} + x_2 \begin{bmatrix} 3 \\ -4 \end{bmatrix} + x_3 \begin{bmatrix} -2 \\ 2 \end{bmatrix} = \begin{bmatrix} 5 \\ 6 \end{bmatrix}$$ ◀

▶ **DEFINITION** If $\mathbf{a}_1, \mathbf{a}_2, \ldots, \mathbf{a}_n$ are vectors in R^m and $c_1, c_2, \ldots, c_n$ are scalars, then a sum of the form

$$c_1\mathbf{a}_1 + c_2\mathbf{a}_2 + \cdots + c_n\mathbf{a}_n$$

is said to be a **linear combination** of the vectors $\mathbf{a}_1, \mathbf{a}_2, \ldots, \mathbf{a}_n$.   ◀

It follows from equation (5) that the product $A\mathbf{x}$ is a linear combination of the columns vectors of A. Some books even use this linear combination representation as the definition of matrix vector multiplication.

If A is an $m \times n$ matrix and $\mathbf{x}$ is a vector in R^n, then

$$A\mathbf{x} = x_1\mathbf{a}_1 + x_2\mathbf{a}_2 + \cdots + x_n\mathbf{a}_n$$

EXAMPLE 5.   If we choose $x_1 = 2$, $x_2 = 3$, $x_3 = 4$ in Example 4, then

$$\begin{bmatrix} 5 \\ 6 \end{bmatrix} = 2 \begin{bmatrix} 2 \\ 5 \end{bmatrix} + 3 \begin{bmatrix} 3 \\ -4 \end{bmatrix} + 4 \begin{bmatrix} -2 \\ 2 \end{bmatrix}$$

Thus the vector $\begin{pmatrix} 5 \\ 6 \end{pmatrix}$ is a linear combination of the three column vectors of the coefficient matrix. It follows that the linear system in Example 4 is consistent and

$$\mathbf{x} = \begin{pmatrix} 2 \\ 3 \\ 4 \end{pmatrix}$$

is a solution to the system.   ◀

The matrix equation (6) provides a nice way of characterizing whether a linear system of equations is consistent. Indeed, the following theorem is a direct consequence of (6).

▶ **THEOREM 1.3.1 (Consistency Theorem for Linear Systems)** *A linear system* $A\mathbf{x} = \mathbf{b}$ *is consistent if and only if* $\mathbf{b}$ *can be written as a linear combination of the column vectors of* A.   ◀

EXAMPLE 6. The linear system

$$x_1 + 2x_2 = 1$$
$$2x_1 + 4x_2 = 1$$

is inconsistent since the vector $\begin{pmatrix} 1 \\ 1 \end{pmatrix}$ cannot be written as a linear combination of the column vectors $\begin{pmatrix} 1 \\ 2 \end{pmatrix}$ and $\begin{pmatrix} 2 \\ 4 \end{pmatrix}$. Note that any linear combination of these vectors would be of the form

$$x_1 \begin{pmatrix} 1 \\ 2 \end{pmatrix} + x_2 \begin{pmatrix} 2 \\ 4 \end{pmatrix} = \begin{pmatrix} x_1 + 2x_2 \\ 2x_1 + 4x_2 \end{pmatrix}$$

and hence the second entry of the vector must be double the first entry.   ◀

Matrix Multiplication

More generally, it is possible to multiply a matrix A times a matrix B if the number of columns of A equals the number of rows of B. The first column of the product is determined by the first column of B; that is, the first column of AB is $A\mathbf{b}_1$, the second column of AB is $A\mathbf{b}_2$, and so on. Thus the product AB is the matrix whose columns are $A\mathbf{b}_1, A\mathbf{b}_2, \ldots, A\mathbf{b}_n$.

$$AB = (A\mathbf{b}_1, A\mathbf{b}_2, \ldots, A\mathbf{b}_n)$$

The (i, j) entry of AB is the ith entry of the column vector $A\mathbf{b}_j$. It is determined by multiplying the ith row vector of A times the jth column vector of B.

▶ **DEFINITION** If $A = (a_{ij})$ is an $m \times n$ matrix and $B = (b_{ij})$ is an $n \times r$ matrix, then the product $AB = C = (c_{ij})$ is the $m \times r$ matrix whose entries are defined by

$$c_{ij} = \mathbf{a}(i, :)\mathbf{b}_j = \sum_{k=1}^{n} a_{ik} b_{kj}$$

EXAMPLE 7. If

$$A = \begin{bmatrix} 3 & -2 \\ 2 & 4 \\ 1 & -3 \end{bmatrix} \quad \text{and} \quad B = \begin{bmatrix} -2 & 1 & 3 \\ 4 & 1 & 6 \end{bmatrix}$$

then

$$AB = \begin{bmatrix} 3 & -2 \\ 2 & 4 \\ 1 & -3 \end{bmatrix} \begin{bmatrix} -2 & 1 & 3 \\ 4 & 1 & 6 \end{bmatrix}$$

$$= \begin{bmatrix} 3 \cdot (-2) - 2 \cdot 4 & 3 \cdot 1 - 2 \cdot 1 & 3 \cdot 3 - 2 \cdot 6 \\ 2 \cdot (-2) + 4 \cdot 4 & 2 \cdot 1 + 4 \cdot 1 & 2 \cdot 3 + 4 \cdot 6 \\ 1 \cdot (-2) - 3 \cdot 4 & 1 \cdot 1 - 3 \cdot 1 & 1 \cdot 3 - 3 \cdot 6 \end{bmatrix}$$

$$= \begin{bmatrix} -14 & 1 & -3 \\ 12 & 6 & 30 \\ -14 & -2 & -15 \end{bmatrix}$$

The shading indicates how the $(2, 3)$ entry of the product is computed using the second row of A and the third column of B. It is also possible to multiply BA; however, the resulting matrix is not equal to AB. In fact, AB and BA do not even have the same dimensions.

$$BA = \begin{bmatrix} -2 \cdot 3 + 1 \cdot 2 + 3 \cdot 1 & -2 \cdot (-2) + 1 \cdot 4 + 3 \cdot (-3) \\ 4 \cdot 3 + 1 \cdot 2 + 6 \cdot 1 & 4 \cdot (-2) + 1 \cdot 4 + 6 \cdot (-3) \end{bmatrix}$$

$$= \begin{bmatrix} -1 & -1 \\ 20 & -22 \end{bmatrix}$$

EXAMPLE 8. If

$$A = \begin{bmatrix} 3 & 4 \\ 1 & 2 \end{bmatrix} \quad \text{and} \quad B = \begin{bmatrix} 1 & 2 \\ 4 & 5 \\ 3 & 6 \end{bmatrix}$$

then it is impossible to multiply A times B, since the number of columns of A does not equal the number of rows of B. However, it is possible to multiply B times A.

$$BA = \begin{bmatrix} 1 & 2 \\ 4 & 5 \\ 3 & 6 \end{bmatrix} \begin{bmatrix} 3 & 4 \\ 1 & 2 \end{bmatrix} = \begin{bmatrix} 5 & 8 \\ 17 & 26 \\ 15 & 24 \end{bmatrix}$$ ◀

If A and B are both $n \times n$ matrices, then AB and BA will also be $n \times n$ matrices, but in general they will not be equal. *Multiplication of matrices is not commutative.*

EXAMPLE 9. If

$$A = \begin{bmatrix} 1 & 1 \\ 0 & 0 \end{bmatrix} \quad \text{and} \quad B = \begin{bmatrix} 1 & 1 \\ 2 & 2 \end{bmatrix}$$

then

$$AB = \begin{bmatrix} 1 & 1 \\ 0 & 0 \end{bmatrix} \begin{bmatrix} 1 & 1 \\ 2 & 2 \end{bmatrix} = \begin{bmatrix} 3 & 3 \\ 0 & 0 \end{bmatrix}$$

and

$$BA = \begin{bmatrix} 1 & 1 \\ 2 & 2 \end{bmatrix} \begin{bmatrix} 1 & 1 \\ 0 & 0 \end{bmatrix} = \begin{bmatrix} 1 & 1 \\ 2 & 2 \end{bmatrix}$$

and hence $AB \neq BA$. ◀

APPLICATION 2: PRODUCTION COSTS

A company manufactures three products. Its production expenses are divided into three categories. In each category, an estimate is given for the cost of producing a single item of each product. An estimate is also made of the amount of each product to be produced per quarter. These estimates are given in Tables 3 and 4. The company would like to present at their stockholders' meeting a single table showing the total costs for each quarter in each of the three categories: raw materials, labor, and overhead.

TABLE 3

PRODUCTION COSTS PER ITEM (dollars)

	Product		
Expenses	A	B	C
Raw materials	0.10	0.30	0.15
Labor	0.30	0.40	0.25
Overhead and miscellaneous	0.10	0.20	0.15

TABLE 4			
AMOUNT PRODUCED PER QUARTER			

	Season			
Product	Summer	Fall	Winter	Spring
A	4000	4500	4500	4000
B	2000	2600	2400	2200
C	5800	6200	6000	6000

SOLUTION. Let us consider the problem in terms of matrices. Each of the two tables can be represented by a matrix.

$$M = \begin{bmatrix} 0.10 & 0.30 & 0.15 \\ 0.30 & 0.40 & 0.25 \\ 0.10 & 0.20 & 0.15 \end{bmatrix}$$

and

$$P = \begin{bmatrix} 4000 & 4500 & 4500 & 4000 \\ 2000 & 2600 & 2400 & 2200 \\ 5800 & 6200 & 6000 & 6000 \end{bmatrix}$$

If we form the product MP, the first column of MP will represent the costs for the summer quarter.

Raw materials: $(0.10)(4000) + (0.30)(2000) + (0.15)(5800) = 1870$
Labor: $(0.30)(4000) + (0.40)(2000) + (0.25)(5800) = 3450$
Overhead and
miscellaneous: $(0.10)(4000) + (0.20)(2000) + (0.15)(5800) = 1670$

The costs for the fall quarter are given in the second column of MP.

Raw materials: $(0.10)(4500) + (0.30)(2600) + (0.15)(6200) = 2160$
Labor: $(0.30)(4500) + (0.40)(2600) + (0.25)(6200) = 3940$
Overhead and
miscellaneous: $(0.10)(4500) + (0.20)(2600) + (0.15)(6200) = 1900$

Columns 3 and 4 of MP represent the costs for the winter and spring quarters.

$$MP = \begin{bmatrix} 1870 & 2160 & 2070 & 1960 \\ 3450 & 3940 & 3810 & 3580 \\ 1670 & 1900 & 1830 & 1740 \end{bmatrix}$$

TABLE 5

	Summer	Fall	Winter	Spring	Year
			Season		
Raw materials	1,870	2,160	2,070	1,960	8,060
Labor	3,450	3,940	3,810	3,580	14,780
Overhead and miscellaneous	1,670	1,900	1,830	1,740	7,140
Total production costs	6,990	8,000	7,710	7,280	29,980

The entries in row 1 of *MP* represent the total cost of raw materials for each of the four quarters. The entries in rows 2 and 3 represent total cost for labor and overhead, respectively, for each of the four quarters. The yearly expenses in each category may be obtained by adding the entries in each row. The numbers in each of the columns may be added to obtain the total production costs for each quarter. Table 5 summarizes the total production costs. ◀

Notational Rules

Just as in ordinary algebra, if an expression involves both multiplication and addition and there are no parentheses to indicate the order of the operations, multiplications are carried out before additions. This is true for both scalar and matrix multiplications. For example, if

$$A = \begin{pmatrix} 3 & 4 \\ 1 & 2 \end{pmatrix}, \qquad B = \begin{pmatrix} 1 & 3 \\ 2 & 1 \end{pmatrix}, \qquad C = \begin{pmatrix} -2 & 1 \\ 3 & 2 \end{pmatrix}$$

then

$$A + BC = \begin{pmatrix} 3 & 4 \\ 1 & 2 \end{pmatrix} + \begin{pmatrix} 7 & 7 \\ -1 & 4 \end{pmatrix} = \begin{pmatrix} 10 & 11 \\ 0 & 6 \end{pmatrix}$$

and

$$3A + B = \begin{pmatrix} 9 & 12 \\ 3 & 6 \end{pmatrix} + \begin{pmatrix} 1 & 3 \\ 2 & 1 \end{pmatrix} = \begin{pmatrix} 10 & 15 \\ 5 & 7 \end{pmatrix}$$

Algebraic Rules

The following theorem provides some useful rules for doing matrix arithmetic.

▶ **THEOREM 1.3.2** *Each of the following statements is valid for any scalars α and β and for any matrices A, B, and C for which the indicated operations are defined.*

1. $A + B = B + A$
2. $(A + B) + C = A + (B + C)$

3. $(AB)C = A(BC)$

4. $A(B + C) = AB + AC$

5. $(A + B)C = AC + BC$

6. $(\alpha\beta)A = \alpha(\beta A)$

7. $\alpha(AB) = (\alpha A)B = A(\alpha B)$

8. $(\alpha + \beta)A = \alpha A + \beta A$

9. $\alpha(A + B) = \alpha A + \alpha B$ ◀

We will prove two of the rules and leave the rest for the reader to verify.

▶**Proof of (4).** Assume that $A = (a_{ij})$ is an $m \times n$ matrix and $B = (b_{ij})$ and $C = (c_{ij})$ are both $n \times r$ matrices. Let $D = A(B + C)$ and $E = AB + AC$. It follows that

$$d_{ij} = \sum_{k=1}^{n} a_{ik}(b_{kj} + c_{kj})$$

and

$$e_{ij} = \sum_{k=1}^{n} a_{ik}b_{kj} + \sum_{k=1}^{n} a_{ik}c_{kj}$$

But

$$\sum_{k=1}^{n} a_{ik}(b_{kj} + c_{kj}) = \sum_{k=1}^{n} a_{ik}b_{kj} + \sum_{k=1}^{n} a_{ik}c_{kj}$$

so that $d_{ij} = e_{ij}$ and hence $A(B + C) = AB + AC$. ◀

▶**Proof of (3).** Let A be an $m \times n$ matrix, B an $n \times r$ matrix, and C an $r \times s$ matrix. Let $D = AB$ and $E = BC$. We must show that $DC = AE$. By the definition of matrix multiplication,

$$d_{il} = \sum_{k=1}^{n} a_{ik}b_{kl} \qquad \text{and} \qquad e_{kj} = \sum_{l=1}^{r} b_{kl}c_{lj}$$

The ijth term of DC is

$$\sum_{l=1}^{r} d_{il}c_{lj} = \sum_{l=1}^{r} \left(\sum_{k=1}^{n} a_{ik}b_{kl} \right) c_{lj}$$

and the (i, j) entry of AE is

$$\sum_{k=1}^{n} a_{ik}e_{kj} = \sum_{k=1}^{n} a_{ik} \left(\sum_{l=1}^{r} b_{kl}c_{lj} \right)$$

Since

$$\sum_{l=1}^{r} \left(\sum_{k=1}^{n} a_{ik}b_{kl} \right) c_{lj} = \sum_{l=1}^{r} \left(\sum_{k=1}^{n} a_{ik}b_{kl}c_{lj} \right) = \sum_{k=1}^{n} a_{ik} \left(\sum_{l=1}^{r} b_{kl}c_{lj} \right)$$

it follows that

$$(AB)C = DC = AE = A(BC)$$ ◀

The arithmetic rules given in Theorem 1.3.2 seem quite natural since they are similar to the rules that we use with real numbers. However, there are some important differences between the rules for matrix arithmetic and those for real number arithmetic. In particular, multiplication of real numbers is commutative; however, we saw in Example 6 that matrix multiplication is not commutative. This difference warrants special emphasis.

WARNING: In general, $AB \neq BA$. Matrix multiplication is *not* commutative.

Some of the other differences between matrix arithmetic and real number arithmetic are illustrated in Exercises 19 through 22.

EXAMPLE 10. If

$$A = \begin{bmatrix} 1 & 2 \\ 3 & 4 \end{bmatrix}, \qquad B = \begin{bmatrix} 2 & 1 \\ -3 & 2 \end{bmatrix}, \qquad \text{and} \qquad C = \begin{bmatrix} 1 & 0 \\ 2 & 1 \end{bmatrix}$$

verify that $A(BC) = (AB)C$ and $A(B + C) = AB + AC$.

SOLUTION.

$$A(BC) = \begin{bmatrix} 1 & 2 \\ 3 & 4 \end{bmatrix} \begin{bmatrix} 4 & 1 \\ 1 & 2 \end{bmatrix} = \begin{bmatrix} 6 & 5 \\ 16 & 11 \end{bmatrix}$$

$$(AB)C = \begin{bmatrix} -4 & 5 \\ -6 & 11 \end{bmatrix} \begin{bmatrix} 1 & 0 \\ 2 & 1 \end{bmatrix} = \begin{bmatrix} 6 & 5 \\ 16 & 11 \end{bmatrix}$$

Thus

$$A(BC) = \begin{bmatrix} 6 & 5 \\ 16 & 11 \end{bmatrix} = (AB)C$$

$$A(B + C) = \begin{bmatrix} 1 & 2 \\ 3 & 4 \end{bmatrix} \begin{bmatrix} 3 & 1 \\ -1 & 3 \end{bmatrix} = \begin{bmatrix} 1 & 7 \\ 5 & 15 \end{bmatrix}$$

$$AB + AC = \begin{bmatrix} -4 & 5 \\ -6 & 11 \end{bmatrix} + \begin{bmatrix} 5 & 2 \\ 11 & 4 \end{bmatrix} = \begin{bmatrix} 1 & 7 \\ 5 & 15 \end{bmatrix}$$

Therefore,

$$A(B + C) = AB + AC \qquad \blacktriangleleft$$

Notation   Since $(AB)C = A(BC)$, we may simply omit the parentheses and write ABC. The same is true for a product of four or more matrices. In the case where an $n \times n$ matrix is multiplied by itself a number of times, it is convenient to use exponential notation. Thus, if k is a positive integer, then

$$A^k = \underbrace{AA \cdots A}_{k \text{ times}}$$

EXAMPLE 11. If

$$A = \begin{bmatrix} 1 & 1 \\ 1 & 1 \end{bmatrix}$$

then

$$A^2 = \begin{bmatrix} 1 & 1 \\ 1 & 1 \end{bmatrix} \begin{bmatrix} 1 & 1 \\ 1 & 1 \end{bmatrix} = \begin{bmatrix} 2 & 2 \\ 2 & 2 \end{bmatrix}$$

$$A^3 = AAA = AA^2 = \begin{bmatrix} 1 & 1 \\ 1 & 1 \end{bmatrix} \begin{bmatrix} 2 & 2 \\ 2 & 2 \end{bmatrix} = \begin{bmatrix} 4 & 4 \\ 4 & 4 \end{bmatrix}$$

and in general

$$A^n = \begin{bmatrix} 2^{n-1} & 2^{n-1} \\ 2^{n-1} & 2^{n-1} \end{bmatrix}$$ ◀

APPLICATION 3: A SIMPLE MODEL FOR MARITAL STATUS COMPUTATIONS

In a certain town, 30 percent of the married women get divorced each year and 20 percent of the single women get married each year. There are 8000 married women and 2000 single women. Assuming that the total population of women remains constant, how many married women and how many single women will there be after 1 year? After 2 years?

SOLUTION. Form a matrix A as follows. The entries in the first row of A will be the percent of married and single women, respectively, that are married after 1 year. The entries in the second row will be the percent of women who are single after 1 year. Thus

$$A = \begin{bmatrix} 0.70 & 0.20 \\ 0.30 & 0.80 \end{bmatrix}$$

If we let $\mathbf{x} = \begin{bmatrix} 8000 \\ 2000 \end{bmatrix}$, the number of married and single women after 1 year can be computed by multiplying A times $\mathbf{x}$.

$$A\mathbf{x} = \begin{bmatrix} 0.70 & 0.20 \\ 0.30 & 0.80 \end{bmatrix} \begin{bmatrix} 8000 \\ 2000 \end{bmatrix} = \begin{bmatrix} 6000 \\ 4000 \end{bmatrix}$$

After 1 year there will be 6000 married women and 4000 single women. To find the number of married and single women after 2 years, compute

$$A^2\mathbf{x} = A(A\mathbf{x}) = \begin{bmatrix} 0.70 & 0.20 \\ 0.30 & 0.80 \end{bmatrix} \begin{bmatrix} 6000 \\ 4000 \end{bmatrix} = \begin{bmatrix} 5000 \\ 5000 \end{bmatrix}$$

After 2 years, half of the women will be married and half will be single. In general, the number of married and single women after n years can be determined by computing $A^n\mathbf{x}$. ◀

APPLICATION 4: ECOLOGY: DEMOGRAPHICS OF THE LOGGERHEAD SEA TURTLE

Management and preservation of many wildlife species depend on our ability to model population dynamics. A standard modeling technique is to divide the life cycle of a species into a number of stages. The models assume that the population sizes for each stage depend only on the female population and that the probability of survival of an individual female from one year to the next depends only on the stage of the life cycle and not on the actual age of an individual. For example, let us consider a four-stage model for analyzing the population dynamics of the loggerhead turtle. At each stage we estimate the probability of survival over a 1 year period. We also estimate the ability to reproduce in terms of the expected number of eggs laid in a given year. The results are summarized in Table 6. The approximate ages for each stage are listed in parentheses next to the stage description.

If d_i represents the duration of the ith stage, and s_i is the annual survivorship rate for that stage, then it can be shown that the proportion remaining in stage i the

FIGURE 1.3.1

TABLE 6

FOUR-STAGE MODEL FOR LOGGERHEAD TURTLE DEMOGRAPHICS

Stage number	Description (age in years)	Annual survivorship	Eggs laid per year
1	Eggs, hatchlings (<1)	0.67	0
2	Juveniles and subadults (1–21)	0.74	0
3	Novice breeders (22)	0.81	127
4	Mature breeders (23–54)	0.81	79

following year will be

$$(7) \qquad p_i = \left(\frac{1 - s_i^{d_i - 1}}{1 - s_i^{d_i}} \right) s_i$$

and the proportion of the population that will survive and move into stage $i + 1$ the following year will be

$$(8) \qquad q_i = \frac{s_i^{d_i}(1 - s_i)}{1 - s_i^{d_i}}$$

If we let e_i denote the average number of eggs laid by a member of stage i $(i = 2, 3, 4)$ in 1 year and form the matrix

$$(9) \qquad L = \begin{bmatrix} p_1 & e_2 & e_3 & e_4 \\ q_1 & p_2 & 0 & 0 \\ 0 & q_2 & p_3 & 0 \\ 0 & 0 & q_3 & p_4 \end{bmatrix}$$

then L can be used to predict the turtle populations for each stage in future years. A matrix of the form (9) is called a *Leslie matrix*, and the corresponding population model is sometimes referred to as a *Leslie population model*. Using the figures from Table 6, the Leslie matrix for our model is

$$L = \begin{bmatrix} 0 & 0 & 127 & 79 \\ 0.67 & 0.7394 & 0 & 0 \\ 0 & 0.0006 & 0 & 0 \\ 0 & 0 & 0.81 & 0.8077 \end{bmatrix}$$

Suppose that the initial populations at each stage were 200,000, 300,000, 500, and 1500, respectively. If we represent these initial populations by a vector $\mathbf{x}_0$, the

TABLE 7
LOGGERHEAD TURTLE POPULATION PROJECTIONS

Stage number	Initial population	10 years	25 years	50 years
1	200,000	114,264	74,039	35,966
2	300,000	329,212	213,669	103,795
3	500	214	139	68
4	1,500	1,061	687	334

populations at each stage after 1 year are determined by computing

$$\mathbf{x}_1 = L\mathbf{x}_0 = \begin{bmatrix} 0 & 0 & 127 & 79 \\ 0.67 & 0.7394 & 0 & 0 \\ 0 & 0.0006 & 0 & 0 \\ 0 & 0 & 0.81 & 0.8077 \end{bmatrix} \begin{bmatrix} 200,000 \\ 300,000 \\ 500 \\ 1,500 \end{bmatrix} = \begin{bmatrix} 182,000 \\ 355,820 \\ 180 \\ 1,617 \end{bmatrix}$$

(The computations have been rounded to the nearest integer.) To determine the population vector after 2 years, we multiply again by the matrix L.

$$\mathbf{x}_2 = L\mathbf{x}_1 = L^2\mathbf{x}_0$$

In general, the population after k years is determined by computing $\mathbf{x}_k = L^k\mathbf{x}_0$. To see longer-range trends, we compute $\mathbf{x}_{10}$, $\mathbf{x}_{25}$, $\mathbf{x}_{50}$. The results are summarized in Table 7. The model predicts that the total number of breeding-age turtles will decrease by 80 percent over a 50-year period.

A seven-stage model describing the population dynamics is presented in reference [1] below. We will use the seven-stage model in the computer exercises at the end of this chapter. Reference [2] is the original paper by Leslie.

REFERENCES

1. Crouse, Deborah T., Larry B. Crowder, and Hal Caswell, "A Stage-Based Population Model for Loggerhead Sea Turtles and Implications for Conservation," *Ecology*, 68(5), 1987.
2. Leslie, P. H., "On the Use of Matrices in Certain Population Mathematics," *Biometrika*, 33, 1945.

The Identity Matrix

Just as the number 1 acts as an identity for the multiplication of real numbers, there is a special matrix I that acts as an identity for matrix multiplication, that is,

(10) $$IA = AI = A$$

for any $n \times n$ matrix A. It is easy to verify that, if we define I to be an $n \times n$ matrix with 1's on the main diagonal and 0's elsewhere, then I satisfies equation (10) for any $n \times n$ matrix A. More formally, we have the following definition.

▶ **DEFINITION** The $n \times n$ **identity matrix** is the matrix $I = (\delta_{ij})$, where

$$\delta_{ij} = \begin{cases} 1 & \text{if } i = j \\ 0 & \text{if } i \neq j \end{cases}$$ ◀

As an example, let us verify equation (10) in the case $n = 3$.

$$\begin{pmatrix} 1 & 0 & 0 \\ 0 & 1 & 0 \\ 0 & 0 & 1 \end{pmatrix} \begin{pmatrix} 3 & 4 & 1 \\ 2 & 6 & 3 \\ 0 & 1 & 8 \end{pmatrix} = \begin{pmatrix} 3 & 4 & 1 \\ 2 & 6 & 3 \\ 0 & 1 & 8 \end{pmatrix}$$

and

$$\begin{pmatrix} 3 & 4 & 1 \\ 2 & 6 & 3 \\ 0 & 1 & 8 \end{pmatrix} \begin{pmatrix} 1 & 0 & 0 \\ 0 & 1 & 0 \\ 0 & 0 & 1 \end{pmatrix} = \begin{pmatrix} 3 & 4 & 1 \\ 2 & 6 & 3 \\ 0 & 1 & 8 \end{pmatrix}$$

In general, if B is any $m \times n$ matrix and C is any $n \times r$ matrix, then

$$BI = B \quad \text{and} \quad IC = C$$

The column vectors of the $n \times n$ identity matrix I are the standard vectors used to define a coordinate system in Euclidean n-space. The standard notation for the jth column vector of I is $\mathbf{e}_j$, rather than the usual $\mathbf{i}_j$. Thus the $n \times n$ identity matrix can be written

$$I = (\mathbf{e}_1, \mathbf{e}_2, \ldots, \mathbf{e}_n)$$

Matrix Inversion

A real number a is said to have a multiplicative inverse if there exists a number b such that $ab = 1$. Any nonzero number a has a multiplicative inverse $b = \frac{1}{a}$. We generalize the concept of multiplicative inverses to matrices with the following definition.

▶ **DEFINITION** An $n \times n$ matrix A is said to be **nonsingular** or **invertible** if there exists a matrix B such that $AB = BA = I$. The matrix B is said to be a **multiplicative inverse** of A. ◀

If B and C are both multiplicative inverses of A, then

$$B = BI = B(AC) = (BA)C = IC = C$$

Thus a matrix can have at most one multiplicative inverse. We will refer to the multiplicative inverse of a nonsingular matrix A as simply the *inverse* of A and denote it by A^{-1}.

EXAMPLE 12. The matrices

$$\begin{pmatrix} 2 & 4 \\ 3 & 1 \end{pmatrix} \quad \text{and} \quad \begin{pmatrix} -\frac{1}{10} & \frac{2}{5} \\ \frac{3}{10} & -\frac{1}{5} \end{pmatrix}$$

are inverses of each other, since

$$\begin{pmatrix} 2 & 4 \\ 3 & 1 \end{pmatrix} \begin{pmatrix} -\frac{1}{10} & \frac{2}{5} \\ \frac{3}{10} & -\frac{1}{5} \end{pmatrix} = \begin{pmatrix} 1 & 0 \\ 0 & 1 \end{pmatrix}$$

and

$$\begin{pmatrix} -\frac{1}{10} & \frac{2}{5} \\ \frac{3}{10} & -\frac{1}{5} \end{pmatrix} \begin{pmatrix} 2 & 4 \\ 3 & 1 \end{pmatrix} = \begin{pmatrix} 1 & 0 \\ 0 & 1 \end{pmatrix} \qquad \blacktriangleleft$$

EXAMPLE 13. The 3×3 matrices

$$\begin{pmatrix} 1 & 2 & 3 \\ 0 & 1 & 4 \\ 0 & 0 & 1 \end{pmatrix} \quad \text{and} \quad \begin{pmatrix} 1 & -2 & 5 \\ 0 & 1 & -4 \\ 0 & 0 & 1 \end{pmatrix}$$

are inverses, since

$$\begin{pmatrix} 1 & 2 & 3 \\ 0 & 1 & 4 \\ 0 & 0 & 1 \end{pmatrix} \begin{pmatrix} 1 & -2 & 5 \\ 0 & 1 & -4 \\ 0 & 0 & 1 \end{pmatrix} = \begin{pmatrix} 1 & 0 & 0 \\ 0 & 1 & 0 \\ 0 & 0 & 1 \end{pmatrix}$$

and

$$\begin{pmatrix} 1 & -2 & 5 \\ 0 & 1 & -4 \\ 0 & 0 & 1 \end{pmatrix} \begin{pmatrix} 1 & 2 & 3 \\ 0 & 1 & 4 \\ 0 & 0 & 1 \end{pmatrix} = \begin{pmatrix} 1 & 0 & 0 \\ 0 & 1 & 0 \\ 0 & 0 & 1 \end{pmatrix} \qquad \blacktriangleleft$$

EXAMPLE 14. The matrix

$$A = \begin{pmatrix} 1 & 0 \\ 0 & 0 \end{pmatrix}$$

has no inverse. Indeed, if B is any 2×2 matrix, then

$$BA = \begin{bmatrix} b_{11} & b_{12} \\ b_{21} & b_{22} \end{bmatrix} \begin{bmatrix} 1 & 0 \\ 0 & 0 \end{bmatrix} = \begin{bmatrix} b_{11} & 0 \\ b_{21} & 0 \end{bmatrix}$$

Thus BA cannot equal I.   ◀

▶ **DEFINITION** An $n \times n$ matrix is said to be **singular** if it does not have a multiplicative inverse.   ◀

In the next section we will learn how to determine whether a matrix has a multiplicative inverse. We will also learn a method for computing the inverse of a nonsingular matrix.

The Transpose of a Matrix

Given an $m \times n$ matrix A, it is often useful to form a new $n \times m$ matrix whose columns are the rows of A.

▶ **DEFINITION** The **transpose** of an $m \times n$ matrix A is the $n \times m$ matrix B defined by

(11) $$b_{ji} = a_{ij}$$

for $j = 1, \dots, n$ and $i = 1, \dots, m$. The transpose of A is denoted by A^T.   ◀

It follows from (11) that the jth row of A^T has the same entries, respectively, as the jth column of A, and the ith column of A^T has the same entries, respectively, as the ith row of A.

EXAMPLE 15.

A. If $A = \begin{bmatrix} 1 & 2 & 3 \\ 4 & 5 & 6 \end{bmatrix}$, then $A^T = \begin{bmatrix} 1 & 4 \\ 2 & 5 \\ 3 & 6 \end{bmatrix}$.

B. If $B = \begin{bmatrix} -3 & 2 & 1 \\ 4 & 3 & 2 \\ 1 & 2 & 5 \end{bmatrix}$, then $B^T = \begin{bmatrix} -3 & 4 & 1 \\ 2 & 3 & 2 \\ 1 & 2 & 5 \end{bmatrix}$.

C. If $C = \begin{bmatrix} 1 & 2 \\ 2 & 3 \end{bmatrix}$, then $C^T = \begin{bmatrix} 1 & 2 \\ 2 & 3 \end{bmatrix}$.   ◀

There are four basic algebraic rules involving transposes.

Algebraic Rules for Transposes

1. $(A^T)^T = A$
2. $(\alpha A)^T = \alpha A^T$
3. $(A + B)^T = A^T + B^T$
4. $(AB)^T = B^T A^T$

The first three rules are straightforward. We leave it to the reader to verify that they are valid. To prove the fourth rule, we need only show that the (i, j) entries of $(AB)^T$ and $B^T A^T$ are equal. If A is an $m \times n$ matrix, then, in order for the multiplications to be possible, B must have n rows. The (i, j) entry of $(AB)^T$ is the (j, i) entry of AB. It is computed by multiplying the jth row vector of A times the i column vector of B.

$$(12) \quad \mathbf{a}(j, :)\mathbf{b}_i = (a_{j1}, a_{j2}, \ldots, a_{jn}) \begin{bmatrix} b_{1i} \\ b_{2i} \\ \vdots \\ b_{ni} \end{bmatrix} = a_{j1}b_{1i} + a_{j2}b_{2i} + \cdots + a_{jn}b_{ni}$$

The (i, j) entry of $B^T A^T$ is computed by the ith row of B^T times the jth column of A^T. Since the ith row of B^T is the transpose of the ith column of B and the j column of A^T is the transpose as the jth row of A, it follows that the (i, j) entry of $B^T A^T$ is given by

$$(13) \quad \mathbf{b}_i^T \mathbf{a}(j, :)^T = (b_{1i}, b_{2i}, \ldots, b_{ni}) \begin{bmatrix} a_{j1} \\ a_{j2} \\ \vdots \\ a_{jn} \end{bmatrix} = b_{1i}a_{j1} + b_{2i}a_{j2} + \cdots + b_{ni}a_{jn}$$

It follows from (12) and (13) that the (i, j) entries of $(AB)^T$ and $B^T A^T$ are equal. The following example illustrates the idea behind the last proof.

EXAMPLE 16. Let

$$A = \begin{bmatrix} 1 & 2 & 1 \\ 3 & 3 & 5 \\ 2 & 4 & 1 \end{bmatrix}, \qquad B = \begin{bmatrix} 1 & 0 & 2 \\ 2 & 1 & 1 \\ 5 & 4 & 1 \end{bmatrix}$$

Note that the $(3, 2)$ entry of AB is computed using the third row of A and the second column of B.

$$AB = \begin{bmatrix} 1 & 2 & 1 \\ 3 & 3 & 5 \\ 2 & 4 & 1 \end{bmatrix} \begin{bmatrix} 1 & 0 & 2 \\ 2 & 1 & 1 \\ 5 & 4 & 1 \end{bmatrix} = \begin{bmatrix} 10 & 6 & 5 \\ 34 & 23 & 14 \\ 15 & 8 & 9 \end{bmatrix}$$

When the product is transposed, the $(3, 2)$ entry of AB becomes the $(2, 3)$ entry of $(AB)^T$.

$$(AB)^T = \begin{bmatrix} 10 & 34 & 15 \\ 6 & 23 & 8 \\ 5 & 14 & 9 \end{bmatrix}$$

On the other hand, the $(2, 3)$ entry of $B^T A^T$ is computed using the second row of B^T and the third column of A^T.

$$B^T A^T = \begin{bmatrix} 1 & 2 & 5 \\ 0 & 1 & 4 \\ 2 & 1 & 1 \end{bmatrix} \begin{bmatrix} 1 & 3 & 2 \\ 2 & 3 & 4 \\ 1 & 5 & 1 \end{bmatrix} = \begin{bmatrix} 10 & 34 & 15 \\ 6 & 23 & 8 \\ 5 & 14 & 9 \end{bmatrix}$$

In both cases the arithmetic for computing the $(3, 2)$ entry is the same.   ◀

The matrix C in Example 15 is its own transpose. This often happens with matrices that arise in applications.

▶ **DEFINITION** An $n \times n$ matrix A is said to be **symmetric** if $A^T = A$.   ◀

The following are some examples of symmetric matrices:

$$\begin{bmatrix} 1 & 0 \\ 0 & -4 \end{bmatrix} \qquad \begin{bmatrix} 2 & 3 & 4 \\ 3 & 1 & 5 \\ 4 & 5 & 3 \end{bmatrix} \qquad \begin{bmatrix} 0 & 1 & 2 \\ 1 & 1 & -2 \\ 2 & -2 & -3 \end{bmatrix}$$

One type of application that leads to symmetric matrices is problems involving networks. These problems are often solved using the techniques of an area of mathematics called *graph theory*.

APPLICATION 5: NETWORKS AND GRAPHS

Graph theory is one of the important areas of applied mathematics. It is used to model problems in virtually all the applied sciences. Graph theory is particularly useful in applications involving communication networks.

A *graph* is defined to be a set of points called *vertices* together with a set of unordered pairs of vertices, which are referred to as *edges*. Figure 1.3.2 gives a geometrical representation of a graph. We can think of the vertices V_1, V_2, V_3, V_4, V_5 as corresponding to the nodes in a communications network.

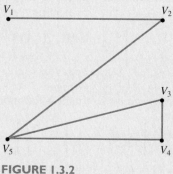

FIGURE 1.3.2

The line segments joining the vertices correspond to the edges:

$$\{V_1, V_2\}, \{V_2, V_5\}, \{V_3, V_4\}, \{V_3, V_5\}, \{V_4, V_5\}$$

Each edge represents a direct communication link between two nodes of the network.

An actual communications network could involve a large number of vertices and edges. Indeed, if there are millions of vertices, a graphical picture of the network would be quite confusing. An alternative is to use a matrix representation for the network. If the graph contains a total of n vertices, we can define an $n \times n$ matrix A by

$$a_{ij} = \begin{cases} 1 & \text{if } \{V_i, V_j\} \text{ is an edge of the graph} \\ 0 & \text{if there is no edge joining } V_i \text{ and } V_j \end{cases}$$

The matrix A is called the *adjacency matrix* of the graph. The adjacency matrix for the graph in Figure 1.3.2 is given by

$$A = \begin{bmatrix} 0 & 1 & 0 & 0 & 0 \\ 1 & 0 & 0 & 0 & 1 \\ 0 & 0 & 0 & 1 & 1 \\ 0 & 0 & 1 & 0 & 1 \\ 0 & 1 & 1 & 1 & 0 \end{bmatrix}$$

Note that the matrix A is symmetric. Indeed, any adjacency matrix must be symmetric, for if $\{V_i, V_j\}$ is an edge of the graph, then $a_{ij} = a_{ji} = 1$ and $a_{ij} = a_{ji} = 0$ if there is no edge joining V_i and V_j. In either case, $a_{ij} = a_{ji}$.

We can think of a *walk* in a graph as a sequence of edges linking one vertex to another. For example, in Figure 1.3.2 the edges $\{V_1, V_2\}$, $\{V_2, V_5\}$ represent a walk from vertex V_1 to vertex V_5. The length of the walk is said to be 2 since it consists of two edges. A simple way to describe the walk is to indicate the movement between vertices using arrows. Thus $V_1 \rightarrow V_2 \rightarrow V_5$ denotes a walk of length 2 from V_1 to V_5. Similarly, $V_4 \rightarrow V_5 \rightarrow V_2 \rightarrow V_1$ represents a walk of length 3 from V_4 to V_1. It is possible to traverse the same edges more than once in a walk. For example, $V_5 \rightarrow V_3 \rightarrow V_5 \rightarrow V_3$ is a walk of length 3 from V_5 to V_3. In general, by taking powers of the adjacency matrix we can determine the number of walks of any specified length between two vertices.

▶ **THEOREM 1.3.3** *If A is an n×n adjacency matrix of a graph and $a_{ij}^{(k)}$ represents the ijth entry of A^k, then $a_{ij}^{(k)}$ is equal to the number of walks of length k from V_i to V_j.*

▶ *Proof.* The proof is by mathematical induction. In the case $k = 1$, it follows from the definition of the adjacency matrix that a_{ij} represents the number of walks of length 1 from V_i to V_j. Assume for some m that each entry of A^m is equal to the number of walks of length m between the corresponding vertices. Thus $a_{il}^{(m)}$ is the number of walks of length m from V_i to V_l. If there is an edge $\{V_l, V_j\}$, then $a_{il}^{(m)} a_{lj} = a_{il}^{(m)}$ is the number of walks of length $m + 1$ from V_i to V_j of the form

$$V_i \to \cdots \to V_l \to V_j$$

It follows that the total number of walks of length $m + 1$ from V_i to V_j is given by

$$a_{i1}^{(m)} a_{1j} + a_{i2}^{(m)} a_{2j} + \cdots + a_{in}^{(m)} a_{nj}$$

But this is just the (i, j) entry of A^{m+1}.   ◀

EXAMPLE 17. To determine the number of walks of length 3 between any two vertices of the graph in Figure 1.3.2, we need only compute

$$A^3 = \begin{bmatrix} 0 & 2 & 1 & 1 & 0 \\ 2 & 0 & 1 & 1 & 4 \\ 1 & 1 & 2 & 3 & 4 \\ 1 & 1 & 3 & 2 & 4 \\ 0 & 4 & 4 & 4 & 2 \end{bmatrix}$$

Thus the number of walks of length 3 from V_3 to V_5 is $a_{35}^{(3)} = 4$. Note that the matrix A^3 is symmetric. This reflects the fact that there are the same number of walks of length 3 from V_i to V_j as there are from V_j to V_i.   ◀

APPLICATION 6: INFORMATION RETRIEVAL

The growth of digital libraries on the Internet has led to dramatic improvements in the storage and retrieval of information. Modern retrieval methods are based on matrix theory and linear algebra.

In a typical situation, a database consists of a collection of documents and we wish to search the collection and find the documents that best match some particular search conditions. Depending on the type of database, we could search for such items as research articles in journals, Web pages on the Internet, books in a library, or movies in a film collection.

To see how the searches are done, let us assume that our database consists of n documents and that there are m key words that can be used for searches. Not all words are allowable since it would not be practical to search for common words such as articles or prepositions. If the key words are ordered alphabetically, then we can represent the database by an $m \times n$ matrix A. Each document is represented by a column of the matrix. The first entry in the jth column of A would be a number representing the relative frequency of the first key word in the jth document. The entry a_{2j} represents the relative frequency of the second key word, and so on. The list of key words to be used in the search is represented by a vector $\mathbf{x}$ in R^m. The ith entry of $\mathbf{x}$ is taken to be 1 if the ith word in the list of key words is on our search list; otherwise, we set $x_i = 0$. To carry out the search, we simply multiply A^T times $\mathbf{x}$.

For example, suppose that our database, consists of these book titles:

B1. *Applied Linear Algebra*
B2. *Elementary Linear Algebra*
B3. *Elementary Linear Algebra with Applications*
B4. *Linear Algebra and Its Applications*
B5. *Linear Algebra with Applications*
B6. *Matrix Algebra with Applications*
B7. *Matrix Theory*

The collection of key words is given by the following alphabetical list:

algebra, application, elementary, linear, matrix, theory

None of the titles repeats any key word more than once, so for simplicity we will just use 0's and 1's for the entries of the database matrix, rather than relative frequencies. Thus the (i, j) entry of the matrix will be 1 if the ith word appears in the title of the jth book and 0 if it does not. We will assume that our search engine is sophisticated enough to equate various forms of a word. So, for example, in our list of titles the words *applied* and *applications* are both counted as forms of the word *application*. The database matrix for our list of books is the array defined by Table 8.

TABLE 8

ARRAY REPRESENTATION FOR
DATABASE OF LINEAR ALGEBRA
BOOKS

Key Words	Books						
	B1	B2	B3	B4	B5	B6	B7
algebra	1	1	1	1	1	1	0
application	1	0	1	1	1	1	0
elementary	0	1	1	0	0	0	0
linear	1	1	1	1	1	0	0
matrix	0	0	0	0	0	1	1
theory	0	0	0	0	0	0	1

If the words we are searching for are *applied, linear,* and *algebra,* then the database matrix and search vector are given by

$$
A = \begin{bmatrix} 1 & 1 & 1 & 1 & 1 & 1 & 0 \\ 1 & 0 & 1 & 1 & 1 & 1 & 0 \\ 0 & 1 & 1 & 0 & 0 & 0 & 0 \\ 1 & 1 & 1 & 1 & 1 & 0 & 0 \\ 0 & 0 & 0 & 0 & 0 & 1 & 1 \\ 0 & 0 & 0 & 0 & 0 & 0 & 1 \end{bmatrix} \qquad \mathbf{x} = \begin{bmatrix} 1 \\ 1 \\ 0 \\ 1 \\ 0 \\ 0 \end{bmatrix}
$$

If we set $\mathbf{y} = A^T \mathbf{x}$, then

$$
\mathbf{y} = \begin{bmatrix} 1 & 1 & 0 & 1 & 0 & 0 \\ 1 & 0 & 1 & 1 & 0 & 0 \\ 1 & 1 & 1 & 1 & 0 & 0 \\ 1 & 1 & 0 & 1 & 0 & 0 \\ 1 & 1 & 0 & 1 & 0 & 0 \\ 1 & 1 & 0 & 0 & 1 & 0 \\ 0 & 0 & 0 & 0 & 1 & 1 \end{bmatrix} \begin{bmatrix} 1 \\ 1 \\ 0 \\ 1 \\ 0 \\ 0 \end{bmatrix} = \begin{bmatrix} 3 \\ 2 \\ 3 \\ 3 \\ 3 \\ 2 \\ 0 \end{bmatrix}
$$

The value of y_1 is the number of search word matches in the title of the first book, the value of y_2 is the number of matches in the second book title, and so on. Since $y_1 = y_3 = y_4 = y_5 = 3$, the titles of books B1, B3, B4, and B5 must contain all three search words. If the search is set up to find titles matching all search words, then the search engine will report the titles of the first, third, fourth, and fifth books. If the search is set up to match one or more search words, then the search engine will report first the four titles that match all the words and then the two titles that match two of the words.

Modern searches could easily involve millions of documents with hundreds of thousands of possible key words. Indeed, as of January 2001, there were more than 1.3 billion Web pages on the Internet, and it is not uncommon for search engines to acquire or update as many as 10 million Web pages in a single day. The database matrix in this case would be very large, but searches can be simplified dramatically since the matrices and search vectors are *sparse,* that is, most of the entries in any column are 0's. If we did a search to find all Web pages on the Internet containing the key words *linear* and *algebra,* we could easily turn up thousands of pages, some of which may not even be about linear algebra. If we were to increase the number of search words and require that all search words be matched, then we would run a risk of excluding some crucial linear algebra pages. For Web pages, the entries of the database matrix should represent the relative frequencies of the key words in the documents. Rather than match all words of the expanded search list, our Web search should give priority to those pages that match most of the key words with high relative frequencies. To accomplish this, we need to find the columns of the database matrix A that are "closest" to the search vector $\mathbf{x}$. One way to measure how close two vectors are is to define *the angle between the vectors*. We will do this in Section 1 of Chapter 5.

We will also revisit the information retrieval application again after we have learned about the *singular value decomposition* (Chapter 6, Section 5). This decomposition can be used to find a simpler approximation to the database matrix which will speed up the searches dramatically. Often it has the added advantage of filtering out *noise*; that is, using the approximate version of the database matrix may automatically have the effect of eliminating pages that use key words in unwanted contexts. For example, a dental student and a mathematics student could both use *calculus* as one of their search words. Since the list of mathematics search words does not contain any other dental terms, a mathematics search using an approximation database matrix is likely to eliminate all pages relating to dentistry. Similarly, the mathematics pages would be filtered out in the dental student's search.

REFERENCES

1. Berry, Michael W., and Murray Browne, *Understanding Search Engines – Mathematical Modeling and Text Retrieval*, SIAM, Philadelphia, 1999.

EXERCISES

1. If

$$A = \begin{bmatrix} 3 & 1 & 4 \\ -2 & 0 & 1 \\ 1 & 2 & 2 \end{bmatrix} \quad \text{and} \quad B = \begin{bmatrix} 1 & 0 & 2 \\ -3 & 1 & 1 \\ 2 & -4 & 1 \end{bmatrix}$$

compute:

(a) $2A$ (b) $A + B$ (c) $2A - 3B$ (d) $(2A)^T - (3B)^T$
(e) AB (f) BA (g) $A^T B^T$ (h) $(BA)^T$

2. For each of the following pairs of matrices, determine whether it is possible to multiply the first matrix times the second. If it is possible, perform the multiplication.

(a) $\begin{bmatrix} 3 & 5 & 1 \\ -2 & 0 & 2 \end{bmatrix} \begin{bmatrix} 2 & 1 \\ 1 & 3 \\ 4 & 1 \end{bmatrix}$ (b) $\begin{bmatrix} 4 & -2 \\ 6 & -4 \\ 8 & -6 \end{bmatrix} \begin{bmatrix} 1 & 2 & 3 \end{bmatrix}$

(c) $\begin{bmatrix} 1 & 4 & 3 \\ 0 & 1 & 4 \\ 0 & 0 & 2 \end{bmatrix} \begin{bmatrix} 3 & 2 \\ 1 & 1 \\ 4 & 5 \end{bmatrix}$ (d) $\begin{bmatrix} 4 & 6 \\ 2 & 1 \end{bmatrix} \begin{bmatrix} 3 & 1 & 5 \\ 4 & 1 & 6 \end{bmatrix}$

$2 \times 2 \quad 2 \times 3$

(e) $\begin{bmatrix} 4 & 6 & 1 \\ 2 & 1 & 1 \end{bmatrix} \begin{bmatrix} 3 & 1 & 5 \\ 4 & 1 & 6 \end{bmatrix}$ (f) $\begin{bmatrix} 2 \\ -1 \\ 3 \end{bmatrix} \begin{bmatrix} 3 & 2 & 4 & 5 \end{bmatrix}$

2×3 2×3

3. For which of the pairs in Exercise 2 is it possible to multiply the second matrix times the first, and what would the dimension of the product be?

4. Write each of the following systems of equations as a matrix equation.

(a) $\begin{aligned} 3x_1 + 2x_2 &= 1 \\ 2x_1 - 3x_2 &= 5 \end{aligned}$
(b) $\begin{aligned} x_1 + x_2 &= 5 \\ 2x_1 + x_2 - x_3 &= 6 \\ 3x_1 - 2x_2 + 2x_3 &= 7 \end{aligned}$
(c) $\begin{aligned} 2x_1 + x_2 + x_3 &= 4 \\ x_1 - x_2 + 2x_3 &= 2 \\ 3x_1 - 2x_2 - x_3 &= 0 \end{aligned}$

5. If

$$A = \begin{bmatrix} 3 & 4 \\ 1 & 1 \\ 2 & 7 \end{bmatrix}$$

verify that:

(a) $5A = 3A + 2A$ (b) $6A = 3(2A)$ (c) $(A^T)^T = A$

6. If

$$A = \begin{bmatrix} 4 & 1 & 6 \\ 2 & 3 & 5 \end{bmatrix} \quad \text{and} \quad B = \begin{bmatrix} 1 & 3 & 0 \\ -2 & 2 & -4 \end{bmatrix}$$

verify that:

(a) $A + B = B + A$ (b) $3(A + B) = 3A + 3B$

(c) $(A + B)^T = A^T + B^T$

7. If

$$A = \begin{bmatrix} 2 & 1 \\ 6 & 3 \\ -2 & 4 \end{bmatrix} \quad \text{and} \quad B = \begin{bmatrix} 2 & 4 \\ 1 & 6 \end{bmatrix}$$

verify that:

(a) $3(AB) = (3A)B = A(3B)$ (b) $(AB)^T = B^T A^T$

8. If

$$A = \begin{bmatrix} 2 & 4 \\ 1 & 3 \end{bmatrix}, \quad B = \begin{bmatrix} -2 & 1 \\ 0 & 4 \end{bmatrix}, \quad C = \begin{bmatrix} 3 & 1 \\ 2 & 1 \end{bmatrix}$$

verify that:

(a) $(A + B) + C = A + (B + C)$ (b) $(AB)C = A(BC)$

(c) $A(B + C) = AB + AC$ (d) $(A + B)C = AC + BC$

9. Prove the associative law of multiplication for 2×2 matrices; that is, let

$$A = \begin{bmatrix} a_{11} & a_{12} \\ a_{21} & a_{22} \end{bmatrix}, \qquad B = \begin{bmatrix} b_{11} & b_{12} \\ b_{21} & b_{22} \end{bmatrix}, \qquad C = \begin{bmatrix} c_{11} & c_{12} \\ c_{21} & c_{22} \end{bmatrix}$$

and show that

$$(AB)C = A(BC)$$

10. Let

$$A = \begin{bmatrix} \frac{1}{2} & -\frac{1}{2} \\ -\frac{1}{2} & \frac{1}{2} \end{bmatrix}$$

Compute A^2 and A^3. What will A^n turn out to be?

11. Let

$$A = \begin{bmatrix} \frac{1}{2} & -\frac{1}{2} & -\frac{1}{2} & -\frac{1}{2} \\ -\frac{1}{2} & \frac{1}{2} & -\frac{1}{2} & -\frac{1}{2} \\ -\frac{1}{2} & -\frac{1}{2} & \frac{1}{2} & -\frac{1}{2} \\ -\frac{1}{2} & -\frac{1}{2} & -\frac{1}{2} & \frac{1}{2} \end{bmatrix}$$

Compute A^2 and A^3. What will A^{2n} and A^{2n+1} turn out to be?

12. Let

$$A = \begin{bmatrix} 0 & 1 & 0 & 0 \\ 0 & 0 & 1 & 0 \\ 0 & 0 & 0 & 1 \\ 0 & 0 & 0 & 0 \end{bmatrix}$$

Show that $A^n = O$ for $n \geq 4$.

13. Given

$$A = \begin{bmatrix} 1 & 2 \\ 1 & -2 \end{bmatrix}, \qquad \mathbf{b} = \begin{bmatrix} 4 \\ 0 \end{bmatrix}, \qquad \mathbf{c} = \begin{bmatrix} -3 \\ -2 \end{bmatrix}$$

(a) Write $\mathbf{b}$ as a linear combination of the column vectors $\mathbf{a}_1$ and $\mathbf{a}_2$.

(b) Use the result from part (a) to determine a solution to the linear system $A\mathbf{x} = \mathbf{b}$. Does the system have any other solutions? Explain.

(c) Write $\mathbf{c}$ as a linear combination of the column vectors $\mathbf{a}_1$ and $\mathbf{a}_2$.

14. For each of the following choices of A and $\mathbf{b}$, determine whether the system $A\mathbf{x} = \mathbf{b}$ is consistent by examining how $\mathbf{b}$ relates to the column vectors of A. Explain your answers in each case.

(a) $A = \begin{bmatrix} 2 & 1 \\ -2 & -1 \end{bmatrix}$, $\mathbf{b} = \begin{bmatrix} 3 \\ 1 \end{bmatrix}$ (b) $A = \begin{bmatrix} 1 & 4 \\ 2 & 3 \end{bmatrix}$, $\mathbf{b} = \begin{bmatrix} 5 \\ 5 \end{bmatrix}$

(c) $A = \begin{bmatrix} 3 & 2 & 1 \\ 3 & 2 & 1 \\ 3 & 2 & 1 \end{bmatrix}$, $\mathbf{b} = \begin{bmatrix} 1 \\ 0 \\ -1 \end{bmatrix}$

15. Let

$$A = \begin{bmatrix} a_{11} & a_{12} \\ a_{21} & a_{22} \end{bmatrix}$$

Show that if $d = a_{11}a_{22} - a_{21}a_{12} \neq 0$ then

$$A^{-1} = \frac{1}{d} \begin{bmatrix} a_{22} & -a_{12} \\ -a_{21} & a_{11} \end{bmatrix}$$

16. Let A be a nonsingular matrix. Show that A^{-1} is also nonsingular and $(A^{-1})^{-1} = A$.

17. Prove that if A is nonsingular then A^T is nonsingular and

$$(A^T)^{-1} = (A^{-1})^T$$

[*Hint:* $(AB)^T = B^T A^T$.]

18. Let A be a nonsingular $n \times n$ matrix. Use mathematical induction to prove that A^m is nonsingular and

$$(A^m)^{-1} = (A^{-1})^m$$

for $m = 1, 2, 3, \ldots$.

19. Explain why each of the following arithmetic rules will not work in general when the real numbers a and b are replaced by $n \times n$ matrices A and B.

(a) $(a + b)^2 = a^2 + 2ab + b^2$
(b) $(a + b)(a - b) = a^2 - b^2$

20. Find 2×2 matrices A and B that both are not the zero matrix for which $AB = O$.

21. Find nonzero matrices A, B, C such that

$$AC = BC \quad \text{and} \quad A \neq B$$

22. The matrix

$$A = \begin{bmatrix} 1 & -1 \\ 1 & -1 \end{bmatrix}$$

has the property that $A^2 = O$. Is it possible for a nonzero symmetric 2×2 matrix to have this property? Prove your answer.

23. Is the product of two symmetric matrices necessarily symmetric? Prove your answer.

24. Let A be an $m \times n$ matrix.

   (a) Explain why the matrix multiplications $A^T A$ and $A A^T$ are possible.
   (b) Show that $A^T A$ and $A A^T$ are both symmetric.

25. Let A and B be symmetric $n \times n$ matrices. Prove that $AB = BA$ if and only if AB is also symmetric.

26. A matrix A is said to be *skew-symmetric* if $A^T = -A$. Show that if a matrix is skew-symmetric then its diagonal entries must all be 0.

27. Let A be an $n \times n$ matrix and let

$$B = A + A^T \qquad \text{and} \qquad C = A - A^T$$

   (a) Show that B is symmetric and C is skew-symmetric.
   (b) Show that every $n \times n$ matrix can be represented as a sum of a symmetric matrix and a skew-symmetric matrix.

28. Suppose that in Application 1 Bob loses 8 pounds. If he continues the same exercise program, how many calories will he burn up each day?

29. In Application 3, how many married women and how many single women will there be after 3 years?

30. In Application 6, suppose that we are searching the database of seven linear algebra books for the search words *elementary, matrix, algebra*. Form a search vector **x** and then compute a vector **y** that represents the results of the search. Explain the significance of the entries of the vector **y**.

31. Given the matrix

$$A = \begin{bmatrix} 0 & 1 & 0 & 1 & 1 \\ 1 & 0 & 1 & 1 & 0 \\ 0 & 1 & 0 & 0 & 1 \\ 1 & 1 & 0 & 0 & 1 \\ 1 & 0 & 1 & 1 & 0 \end{bmatrix}$$

   (a) Draw a graph that has A as its adjacency matrix. Be sure to label the vertices of the graph.
   (b) By inspecting the graph, determine the number of walks of length 2 from V_2 to V_3 and from V_2 to V_5.
   (c) Compute the second row of A^3 and use it to determine the number of walks of length 3 from V_2 to V_3 and from V_2 to V_5.

32. Given the graph

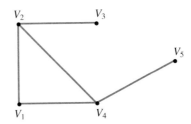

(a) Determine the adjacency matrix A of the graph.

(b) Compute A^2. What do the entries in the first row of A^2 tell you about walks of length 2 that start from V_1?

(c) Compute A^3. How many walks of length 3 are there from V_2 to V_4? How many walks of length less than or equal to 3 are there from V_2 to V_4?

33. Let A be a 2×2 matrix with $a_{11} \neq 0$ and let $\alpha = a_{21}/a_{11}$. Show that A can be factored into a product of the form

$$\begin{bmatrix} 1 & 0 \\ \alpha & 1 \end{bmatrix} \begin{bmatrix} a_{11} & a_{12} \\ 0 & b \end{bmatrix}$$

What is the value of b?

4   ELEMENTARY MATRICES

In this section we view the process of solving a linear system in terms of matrix multiplications, rather than row operations. Given a linear system $A\mathbf{x} = \mathbf{b}$, we can multiply both sides by a sequence of special matrices to obtain an equivalent system in row echelon form. The special matrices we will use are called *elementary matrices*. We will use them to see how to compute the inverse of a nonsingular matrix and also to obtain an important matrix factorization. We begin by considering the effects of multiplying both sides of a linear system by a nonsingular matrix.

Equivalent Systems

Given an $m \times n$ linear system $A\mathbf{x} = \mathbf{b}$, we can obtain an equivalent system by multiplying both sides of the equation by a nonsingular $m \times m$ matrix M.

(1) $A\mathbf{x} = \mathbf{b}$

(2) $M A\mathbf{x} = M\mathbf{b}$

Clearly, any solution to (1) will also be a solution to (2). On the other hand, if $\hat{\mathbf{x}}$ is a solution to (2), then

$$M^{-1}(MA\hat{\mathbf{x}}) = M^{-1}(M\mathbf{b})$$
$$A\hat{\mathbf{x}} = \mathbf{b}$$

so the two systems are equivalent.

To obtain an equivalent system that is easier to solve, we can apply a sequence of nonsingular matrices $E_1, \ldots, E_k$ to both sides of the equation $A\mathbf{x} = \mathbf{b}$ to obtain a simpler system:

$$U\mathbf{x} = \mathbf{c}$$

where $U = E_k \cdots E_1 A$ and $\mathbf{c} = E_k \cdots E_2 E_1 \mathbf{b}$. The new system will be equivalent to the original provided that $M = E_k \cdots E_1$ is nonsingular. However, M is the product of nonsingular matrices. The following theorem shows that any product of nonsingular matrices is nonsingular.

▶ **THEOREM 1.4.1** *If A and B are nonsingular $n \times n$ matrices, then AB is also nonsingular and $(AB)^{-1} = B^{-1}A^{-1}$.*

▶*Proof.*

$$(B^{-1}A^{-1})AB = B^{-1}(A^{-1}A)B = B^{-1}B = I$$
$$(AB)(B^{-1}A^{-1}) = A(BB^{-1})A^{-1} = AA^{-1} = I$$ ◀

It follows by induction that if $E_1, \ldots, E_k$ are all nonsingular then the product $E_1 E_2 \cdots E_k$ is nonsingular and

$$(E_1 E_2 \cdots E_k)^{-1} = E_k^{-1} \cdots E_2^{-1} E_1^{-1}$$

We will show next that any of the three elementary row operations can be accomplished by multiplying A on the left by a nonsingular matrix.

Elementary Matrices

If we start with the identity matrix I and then perform exactly one elementary row operation, the resulting matrix is called an *elementary* matrix.

There are three types of elementary matrices corresponding to the three types of elementary row operations.

Type I An elementary matrix of type I is a matrix obtained by interchanging two rows of I.

EXAMPLE 1. Let

$$E_1 = \begin{bmatrix} 0 & 1 & 0 \\ 1 & 0 & 0 \\ 0 & 0 & 1 \end{bmatrix}$$

E_1 is an elementary matrix of type I, since it was obtained by interchanging the first two rows of I. Let A be a 3×3 matrix.

$$E_1 A = \begin{bmatrix} 0 & 1 & 0 \\ 1 & 0 & 0 \\ 0 & 0 & 1 \end{bmatrix} \begin{bmatrix} a_{11} & a_{12} & a_{13} \\ a_{21} & a_{22} & a_{23} \\ a_{31} & a_{32} & a_{33} \end{bmatrix} = \begin{bmatrix} a_{21} & a_{22} & a_{23} \\ a_{11} & a_{12} & a_{13} \\ a_{31} & a_{32} & a_{33} \end{bmatrix}$$

$$A E_1 = \begin{bmatrix} a_{11} & a_{12} & a_{13} \\ a_{21} & a_{22} & a_{23} \\ a_{31} & a_{32} & a_{33} \end{bmatrix} \begin{bmatrix} 0 & 1 & 0 \\ 1 & 0 & 0 \\ 0 & 0 & 1 \end{bmatrix} = \begin{bmatrix} a_{12} & a_{11} & a_{13} \\ a_{22} & a_{21} & a_{23} \\ a_{32} & a_{31} & a_{33} \end{bmatrix}$$

Multiplying A on the left by E_1 interchanges the first and second rows of A. Right multiplication of A by E_1 is equivalent to the elementary column operation of interchanging the first and second columns. ◀

Type II An elementary matrix of type II is a matrix obtained by multiplying a row of I by a nonzero constant.

EXAMPLE 2.

$$E_2 = \begin{bmatrix} 1 & 0 & 0 \\ 0 & 1 & 0 \\ 0 & 0 & 3 \end{bmatrix}$$

is an elementary matrix of type II.

$$\begin{bmatrix} 1 & 0 & 0 \\ 0 & 1 & 0 \\ 0 & 0 & 3 \end{bmatrix} \begin{bmatrix} a_{11} & a_{12} & a_{13} \\ a_{21} & a_{22} & a_{23} \\ a_{31} & a_{32} & a_{33} \end{bmatrix} = \begin{bmatrix} a_{11} & a_{12} & a_{13} \\ a_{21} & a_{22} & a_{23} \\ 3a_{31} & 3a_{32} & 3a_{33} \end{bmatrix}$$

$$\begin{bmatrix} a_{11} & a_{12} & a_{13} \\ a_{21} & a_{22} & a_{23} \\ a_{31} & a_{32} & a_{33} \end{bmatrix} \begin{bmatrix} 1 & 0 & 0 \\ 0 & 1 & 0 \\ 0 & 0 & 3 \end{bmatrix} = \begin{bmatrix} a_{11} & a_{12} & 3a_{13} \\ a_{21} & a_{22} & 3a_{23} \\ a_{31} & a_{32} & 3a_{33} \end{bmatrix}$$

Multiplication on the left by E_2 performs the elementary row operation of multiplying the third row by 3, while multiplication on the right by E_2 performs the elementary column operation of multiplying the third column by 3. ◀

Type III An elementary matrix of type III is a matrix obtained from I by adding a multiple of one row to another row.

EXAMPLE 3.

$$E_3 = \begin{bmatrix} 1 & 0 & 3 \\ 0 & 1 & 0 \\ 0 & 0 & 1 \end{bmatrix}$$

is an elementary matrix of type III. If A is a 3×3 matrix, then

$$E_3 A = \begin{bmatrix} a_{11} + 3a_{31} & a_{12} + 3a_{32} & a_{13} + 3a_{33} \\ a_{21} & a_{22} & a_{23} \\ a_{31} & a_{32} & a_{33} \end{bmatrix}$$

$$A E_3 = \begin{bmatrix} a_{11} & a_{12} & 3a_{11} + a_{13} \\ a_{21} & a_{22} & 3a_{21} + a_{23} \\ a_{31} & a_{32} & 3a_{31} + a_{33} \end{bmatrix}$$

Multiplication on the left by E_3 adds 3 times the third row to the first row. Multiplication on the right adds 3 times the first column to the third column. ◀

In general, suppose that E is an $n \times n$ elementary matrix. We can think of E as being obtained from I by either a row operation or a column operation. If A is an $n \times r$ matrix, *premultiplying A by E has the effect of performing that same row operation on A.* If B is an $m \times n$ matrix, *postmultiplying B by E is equivalent to performing that same column operation on B.*

▶ **THEOREM 1.4.2** *If E is an elementary matrix, then E is nonsingular and E^{-1} is an elementary matrix of the same type.*

▶ *Proof.* If E is the elementary matrix of type I formed from I by interchanging the ith and jth rows, then E can be transformed back into I by interchanging these same rows again. Thus $EE = I$ and hence E is its own inverse. If E is the elementary matrix of type II formed by multiplying the ith row of I by a nonzero scalar α, then E can be transformed into the identity by multiplying either its ith row or its ith column by $1/\alpha$. Thus

$$E^{-1} = \begin{bmatrix} 1 & & & & & & \\ & \ddots & & & & O & \\ & & 1 & & & & \\ & & & 1/\alpha & & & \\ & & & & 1 & & \\ & O & & & & \ddots & \\ & & & & & & 1 \end{bmatrix} \quad i\text{th row}$$

Finally, suppose that E is the elementary matrix of type III formed from I by adding m times the ith row to the jth row.

$$E = \begin{pmatrix} 1 & & & & & & & \\ \vdots & \ddots & & & & & O & \\ 0 & \cdots & 1 & & & & & \\ \vdots & & & \ddots & & & & \\ 0 & \cdots & m & \cdots & 1 & & & \\ \vdots & & & & & \ddots & & \\ 0 & \cdots & 0 & \cdots & 0 & \cdots & 1 \end{pmatrix} \quad \begin{matrix} \\ \\ i\text{th row} \\ \\ j\text{th row} \\ \\ \end{matrix}$$

E can be transformed back into I by either subtracting m times the ith row from the jth row or by subtracting m times the jth column from the ith column. Thus

$$E^{-1} = \begin{pmatrix} 1 & & & & & & & \\ \vdots & \ddots & & & & & O & \\ 0 & \cdots & 1 & & & & & \\ \vdots & & & \ddots & & & & \\ 0 & \cdots & -m & \cdots & 1 & & & \\ \vdots & & & & & \ddots & & \\ 0 & \cdots & 0 & \cdots & 0 & \cdots & 1 \end{pmatrix} \qquad \blacktriangleleft$$

▶ **DEFINITION** A matrix B is **row equivalent** to A if there exists a finite sequence $E_1, E_2, \ldots, E_k$ of elementary matrices such that

$$B = E_k E_{k-1} \cdots E_1 A \qquad \blacktriangleleft$$

In other words, B is row equivalent to A if B can be obtained from A by a finite number of row operations. In particular, two augmented matrices $(A \,|\, \mathbf{b})$ and $(B \,|\, \mathbf{c})$ are row equivalent if and only if $A\mathbf{x} = \mathbf{b}$ and $B\mathbf{x} = \mathbf{c}$ are equivalent systems. The following properties of row equivalent matrices are easily established.

   I. If A is row equivalent to B, then B is row equivalent to A.

   II. If A is row equivalent to B, and B is row equivalent to C, then A is row equivalent to C.

Property (I) can be proved using Theorem 1.4.2. The details of the proofs of (I) and (II) are left as an exercise for the reader.

▶ **THEOREM 1.4.3 (Equivalent Conditions for Nonsingularity)** *Let* A *be an* $n \times n$ *matrix. The following are equivalent:*

   (a) *A is nonsingular.*

(b) $A\mathbf{x} = \mathbf{0}$ *has only the trivial solution* $\mathbf{0}$.

(c) A *is row equivalent to* I.

▶ *Proof.* We prove first that statement (a) implies statement (b). If A is nonsingular and $\hat{\mathbf{x}}$ is a solution to $A\mathbf{x} = \mathbf{0}$, then

$$\hat{\mathbf{x}} = I\hat{\mathbf{x}} = (A^{-1}A)\hat{\mathbf{x}} = A^{-1}(A\hat{\mathbf{x}}) = A^{-1}\mathbf{0} = \mathbf{0}$$

Thus $A\mathbf{x} = \mathbf{0}$ has only the trivial solution. Next we show that statement (b) implies statement (c). If we use elementary row operations, the system can be transformed into the form $U\mathbf{x} = \mathbf{0}$, where U is in row echelon form. If one of the diagonal elements of U were 0, the last row of U would consist entirely of 0's. But then $A\mathbf{x} = \mathbf{0}$ would be equivalent to a system with more unknowns than equations and hence by Theorem 1.2.1 would have a nontrivial solution. Thus U must be a triangular matrix with diagonal elements all equal to 1. It follows then that I is the reduced row echelon form of A and hence A is row equivalent to I.

Finally, we will show that statement (c) implies statement (a). If A is row equivalent to I, there exist elementary matrices $E_1, E_2, \ldots, E_k$ such that

$$A = E_k E_{k-1} \cdots E_1 I = E_k E_{k-1} \cdots E_1$$

But since E_i is invertible, $i = 1, \ldots, k$, the product $E_k E_{k-1} \cdots E_1$ is also invertible. Hence A is nonsingular and

$$A^{-1} = (E_k E_{k-1} \cdots E_1)^{-1} = E_1^{-1} E_2^{-1} \cdots E_k^{-1} \qquad ◀$$

▶ **COROLLARY 1.4.4** *The system of n linear equations in n unknowns* $A\mathbf{x} = \mathbf{b}$ *has a unique solution if and only if A is nonsingular.*

▶ *Proof.* If A is nonsingular, then $A^{-1}\mathbf{b}$ is the only solution to $A\mathbf{x} = \mathbf{b}$. Conversely, suppose that $A\mathbf{x} = \mathbf{b}$ has a unique solution $\hat{\mathbf{x}}$. If A is singular, $A\mathbf{x} = \mathbf{0}$ has a solution $\mathbf{z} \neq \mathbf{0}$. Let $\mathbf{y} = \hat{\mathbf{x}} + \mathbf{z}$. Clearly, $\mathbf{y} \neq \hat{\mathbf{x}}$ and

$$A\mathbf{y} = A(\hat{\mathbf{x}} + \mathbf{z}) = A\hat{\mathbf{x}} + A\mathbf{z} = \mathbf{b} + \mathbf{0} = \mathbf{b}$$

Thus $\mathbf{y}$ is also a solution to $A\mathbf{x} = \mathbf{b}$, which is a contradiction. Therefore, if $A\mathbf{x} = \mathbf{b}$ has a unique solution, A must be nonsingular. ◀

If A is nonsingular, A is row equivalent to I, so there exist elementary matrices $E_1, \ldots, E_k$ such that

$$E_k E_{k-1} \cdots E_1 A = I$$

Multiplying both sides of this equation on the right by A^{-1}, we obtain

$$E_k E_{k-1} \cdots E_1 I = A^{-1}.$$

Thus the same series of elementary row operations that transforms a nonsingular matrix A into I will transform I into A^{-1}. This gives us a method for computing A^{-1}. If we augment A by I and perform the elementary row operations that transform A into I on the augmented matrix, then I will be transformed into A^{-1}. That is, the reduced row echelon form of the augmented matrix $(A|I)$ will be $(I|A^{-1})$.

EXAMPLE 4. Compute A^{-1} if

$$A = \begin{pmatrix} 1 & 4 & 3 \\ -1 & -2 & 0 \\ 2 & 2 & 3 \end{pmatrix}$$

SOLUTION.

$$\begin{pmatrix} 1 & 4 & 3 & | & 1 & 0 & 0 \\ -1 & -2 & 0 & | & 0 & 1 & 0 \\ 2 & 2 & 3 & | & 0 & 0 & 1 \end{pmatrix} \rightarrow \begin{pmatrix} 1 & 4 & 3 & | & 1 & 0 & 0 \\ 0 & 2 & 3 & | & 1 & 1 & 0 \\ 0 & -6 & -3 & | & -2 & 0 & 1 \end{pmatrix}$$

$$\rightarrow \begin{pmatrix} 1 & 4 & 3 & | & 1 & 0 & 0 \\ 0 & 2 & 3 & | & 1 & 1 & 0 \\ 0 & 0 & 6 & | & 1 & 3 & 1 \end{pmatrix} \rightarrow \begin{pmatrix} 1 & 4 & 0 & | & \frac{1}{2} & -\frac{3}{2} & -\frac{1}{2} \\ 0 & 2 & 0 & | & \frac{1}{2} & -\frac{1}{2} & -\frac{1}{2} \\ 0 & 0 & 6 & | & 1 & 3 & 1 \end{pmatrix}$$

$$\rightarrow \begin{pmatrix} 1 & 0 & 0 & | & -\frac{1}{2} & -\frac{1}{2} & \frac{1}{2} \\ 0 & 2 & 0 & | & \frac{1}{2} & -\frac{1}{2} & -\frac{1}{2} \\ 0 & 0 & 6 & | & 1 & 3 & 1 \end{pmatrix} \rightarrow \begin{pmatrix} 1 & 0 & 0 & | & -\frac{1}{2} & -\frac{1}{2} & \frac{1}{2} \\ 0 & 1 & 0 & | & \frac{1}{4} & -\frac{1}{4} & -\frac{1}{4} \\ 0 & 0 & 1 & | & \frac{1}{6} & \frac{1}{2} & \frac{1}{6} \end{pmatrix}$$

Thus

$$A^{-1} = \begin{pmatrix} -\frac{1}{2} & -\frac{1}{2} & \frac{1}{2} \\ \frac{1}{4} & -\frac{1}{4} & -\frac{1}{4} \\ \frac{1}{6} & \frac{1}{2} & \frac{1}{6} \end{pmatrix} \qquad \blacktriangleleft$$

EXAMPLE 5. Solve the system

$$\begin{aligned} x_1 + 4x_2 + 3x_3 &= 12 \\ -x_1 - 2x_2 &= -12 \\ 2x_1 + 2x_2 + 3x_3 &= 8 \end{aligned}$$

The coefficient matrix of this system is the matrix A of the last example. The solution to the system then is

$$\mathbf{x} = A^{-1}\mathbf{b} = \begin{pmatrix} -\frac{1}{2} & -\frac{1}{2} & \frac{1}{2} \\ \frac{1}{4} & -\frac{1}{4} & -\frac{1}{4} \\ \frac{1}{6} & \frac{1}{2} & \frac{1}{6} \end{pmatrix} \begin{pmatrix} 12 \\ -12 \\ 8 \end{pmatrix} = \begin{pmatrix} 4 \\ 4 \\ -\frac{8}{3} \end{pmatrix} \qquad \blacktriangleleft$$

Diagonal and Triangular Matrices

An $n \times n$ matrix A is said to be *upper triangular* if $a_{ij} = 0$ for $i > j$ and *lower triangular* if $a_{ij} = 0$ for $i < j$. Also, A is said to be *triangular* if it is either upper triangular or lower triangular. For example, the 3×3 matrices

$$\begin{bmatrix} 3 & 2 & 1 \\ 0 & 2 & 1 \\ 0 & 0 & 5 \end{bmatrix} \qquad \text{and} \qquad \begin{bmatrix} 1 & 0 & 0 \\ 6 & 2 & 0 \\ 1 & 4 & 3 \end{bmatrix}$$

are both triangular. The first is upper triangular and the second is lower triangular.

A triangular matrix may have 0's on the diagonal. However, for a linear system $A\mathbf{x} = \mathbf{b}$ to be in triangular form, the coefficient matrix A must be triangular with nonzero diagonal entries.

An $n \times n$ matrix A is *diagonal* if $a_{ij} = 0$ whenever $i \neq j$. The matrices

$$\begin{bmatrix} 1 & 0 \\ 0 & 2 \end{bmatrix} \qquad \begin{bmatrix} 1 & 0 & 0 \\ 0 & 3 & 0 \\ 0 & 0 & 1 \end{bmatrix} \qquad \begin{bmatrix} 0 & 0 & 0 \\ 0 & 2 & 0 \\ 0 & 0 & 0 \end{bmatrix}$$

are all diagonal. A diagonal matrix is both upper triangular and lower triangular.

Triangular Factorization

If an $n \times n$ matrix A can be reduced to upper triangular form using only row operation **III**, then it is possible to represent the reduction process in terms of a matrix factorization. We illustrate how this is done in the following example.

EXAMPLE 6.   Let

$$A = \begin{bmatrix} 2 & 4 & 2 \\ 1 & 5 & 2 \\ 4 & -1 & 9 \end{bmatrix}$$

and let us carry out the reduction process using only row operation **III**. At the first step we subtract $\frac{1}{2}$ times the first row from the second and then we subtract twice the first row from the third.

$$\begin{bmatrix} 2 & 4 & 2 \\ 1 & 5 & 2 \\ 4 & -1 & 9 \end{bmatrix} \rightarrow \begin{bmatrix} 2 & 4 & 2 \\ 0 & 3 & 1 \\ 0 & -9 & 5 \end{bmatrix}$$

To keep track of the multiples of the first row that were subtracted, we set $l_{21} = \frac{1}{2}$ and $l_{31} = 2$. We complete the elimination process by eliminating the -9 in the (3,2) position.

$$\begin{bmatrix} 2 & 4 & 2 \\ 0 & 3 & 1 \\ 0 & -9 & 5 \end{bmatrix} \rightarrow \begin{bmatrix} 2 & 4 & 2 \\ 0 & 3 & 1 \\ 0 & 0 & 8 \end{bmatrix}$$

Let $l_{32} = -3$, the multiple of the second row subtracted from row three. If we call the resulting matrix U and set

$$L = \begin{bmatrix} 1 & 0 & 0 \\ l_{21} & 1 & 0 \\ l_{31} & l_{32} & 1 \end{bmatrix} = \begin{bmatrix} 1 & 0 & 0 \\ \frac{1}{2} & 1 & 0 \\ 2 & -3 & 1 \end{bmatrix}$$

then it is easily verified that

$$LU = \begin{bmatrix} 1 & 0 & 0 \\ \frac{1}{2} & 1 & 0 \\ 2 & -3 & 1 \end{bmatrix} \begin{bmatrix} 2 & 4 & 2 \\ 0 & 3 & 1 \\ 0 & 0 & 8 \end{bmatrix} = \begin{bmatrix} 2 & 4 & 2 \\ 1 & 5 & 2 \\ 4 & -1 & 9 \end{bmatrix} = A \quad \blacktriangleleft$$

The matrix L in the previous example is lower triangular with 1's on the diagonal. We say that L is *unit lower triangular*. The factorization of the matrix A into a product of a unit lower triangular matrix L times an upper triangular matrix U is often referred to as an *LU factorization*.

To see why the factorization in Example 6 works, let us view the reduction process in terms of elementary matrices. The three row operations that were applied to the matrix A can be represented in terms of multiplications by elementary matrices

(3) $$E_3 E_2 E_1 A = U$$

where

$$E_1 = \begin{bmatrix} 1 & 0 & 0 \\ -\frac{1}{2} & 1 & 0 \\ 0 & 0 & 1 \end{bmatrix}, \quad E_2 = \begin{bmatrix} 1 & 0 & 0 \\ 0 & 1 & 0 \\ -2 & 0 & 1 \end{bmatrix}, \quad E_3 = \begin{bmatrix} 1 & 0 & 0 \\ 0 & 1 & 0 \\ 0 & 3 & 1 \end{bmatrix}$$

correspond to the row operations in the reduction process. Since each of the elementary matrices is nonsingular, we can multiply equation (3) by their inverses.

$$A = E_1^{-1} E_2^{-1} E_3^{-1} U$$

[We multiply in reverse order since $(E_3 E_2 E_1)^{-1} = E_1^{-1} E_2^{-1} E_3^{-1}$.] However, when the inverses are multiplied in this order, the multipliers l_{21}, l_{31}, l_{32} fill in below the diagonal in the product.

$$E_1^{-1} E_2^{-1} E_3^{-1} = \begin{bmatrix} 1 & 0 & 0 \\ \frac{1}{2} & 1 & 0 \\ 0 & 0 & 1 \end{bmatrix} \begin{bmatrix} 1 & 0 & 0 \\ 0 & 1 & 0 \\ 2 & 0 & 1 \end{bmatrix} \begin{bmatrix} 1 & 0 & 0 \\ 0 & 1 & 0 \\ 0 & -3 & 1 \end{bmatrix} = L$$

In general, if an $n \times n$ matrix A can be reduced to upper triangular form using only row operation **III**, then A has an LU factorization. The matrix L is unit lower triangular, and if $i > j$, then l_{ij} is the multiple of the jth row subtracted from the ith row during the reduction process.

The LU factorization is a very useful way of viewing the elimination process. We will find it particularly useful in Chapter 7 when we study computer methods for solving linear systems. Many of the major topics in linear algebra can be viewed in terms of matrix factorizations. We will study other interesting and important factorizations in Chapters 5 through 7.

EXERCISES

1. Which of the following are elementary matrices? Classify each elementary matrix by type.

(a) $\begin{bmatrix} 0 & 1 \\ 1 & 0 \end{bmatrix}$
(b) $\begin{bmatrix} 2 & 0 \\ 0 & 3 \end{bmatrix}$
(c) $\begin{bmatrix} 1 & 0 & 0 \\ 0 & 1 & 0 \\ 5 & 0 & 1 \end{bmatrix}$
(d) $\begin{bmatrix} 1 & 0 & 0 \\ 0 & 5 & 0 \\ 0 & 0 & 1 \end{bmatrix}$

2. Find the inverse of each matrix in Exercise 1. For each elementary matrix, verify that its inverse is an elementary matrix of the same type.

3. For each of the following pairs of matrices, find an elementary matrix E such that $EA = B$.

(a) $A = \begin{bmatrix} 2 & -1 \\ 5 & 3 \end{bmatrix}$ $B = \begin{bmatrix} -4 & 2 \\ 5 & 3 \end{bmatrix}$

(b) $A = \begin{bmatrix} 2 & 1 & 3 \\ -2 & 4 & 5 \\ 3 & 1 & 4 \end{bmatrix}$ $B = \begin{bmatrix} 2 & 1 & 3 \\ 3 & 1 & 4 \\ -2 & 4 & 5 \end{bmatrix}$

(c) $A = \begin{bmatrix} 4 & -2 & 3 \\ 1 & 0 & 2 \\ -2 & 3 & 1 \end{bmatrix}$ $B = \begin{bmatrix} 4 & -2 & 3 \\ 1 & 0 & 2 \\ 0 & 3 & 5 \end{bmatrix}$

4. For each of the following pairs of matrices, find an elementary matrix E such that $AE = B$.

(a) $A = \begin{bmatrix} 4 & 1 & 3 \\ 2 & 1 & 4 \\ 1 & 3 & 2 \end{bmatrix}$, $B = \begin{bmatrix} 3 & 1 & 4 \\ 4 & 1 & 2 \\ 2 & 3 & 1 \end{bmatrix}$

(b) $A = \begin{bmatrix} 2 & 4 \\ 1 & 6 \end{bmatrix}$, $B = \begin{bmatrix} 2 & -2 \\ 1 & 3 \end{bmatrix}$

(c) $A = \begin{bmatrix} 4 & -2 & 3 \\ -2 & 4 & 2 \\ 6 & 1 & -2 \end{bmatrix}$, $B = \begin{bmatrix} 2 & -2 & 3 \\ -1 & 4 & 2 \\ 3 & 1 & -2 \end{bmatrix}$

5. Given

$$A = \begin{bmatrix} 1 & 2 & 4 \\ 2 & 1 & 3 \\ 1 & 0 & 2 \end{bmatrix}, \quad B = \begin{bmatrix} 1 & 2 & 4 \\ 2 & 1 & 3 \\ 2 & 2 & 6 \end{bmatrix}, \quad C = \begin{bmatrix} 1 & 2 & 4 \\ 0 & -1 & -3 \\ 2 & 2 & 6 \end{bmatrix}$$

(a) Find an elementary matrix E such that $EA = B$.

(b) Find an elementary matrix F such that $FB = C$.

(c) Is C row equivalent to A? Explain.

6. Given

$$A = \begin{bmatrix} 2 & 1 & 1 \\ 6 & 4 & 5 \\ 4 & 1 & 3 \end{bmatrix}$$

(a) Find elementary matrices E_1, E_2, E_3 such that

$$E_3 E_2 E_1 A = U$$

where U is an upper triangular matrix.

(b) Determine the inverses of E_1, E_2, E_3 and set $L = E_1^{-1} E_2^{-1} E_3^{-1}$. What type of matrix is L? Verify that $A = LU$.

7. Compute the LU factorization of each of the following matrices.

(a) $\begin{bmatrix} 3 & 1 \\ 9 & 5 \end{bmatrix}$ (b) $\begin{bmatrix} 2 & 4 \\ -2 & 1 \end{bmatrix}$

(c) $\begin{bmatrix} 1 & 1 & 1 \\ 3 & 5 & 6 \\ -2 & 2 & 7 \end{bmatrix}$ (d) $\begin{bmatrix} -2 & 1 & 2 \\ 4 & 1 & -2 \\ -6 & -3 & 4 \end{bmatrix}$

8. Let

$$A = \begin{bmatrix} 1 & 0 & 1 \\ 3 & 3 & 4 \\ 2 & 2 & 3 \end{bmatrix}$$

(a) Verify that

$$A^{-1} = \begin{bmatrix} 1 & 2 & -3 \\ -1 & 1 & -1 \\ 0 & -2 & 3 \end{bmatrix}$$

(b) Use A^{-1} to solve $Ax = b$ for the following choices of **b**.

(i) $\mathbf{b} = (1, 1, 1)^T$ (ii) $\mathbf{b} = (1, 2, 3)^T$ (iii) $\mathbf{b} = (-2, 1, 0)^T$

9. Find the inverse of each of the following matrices.

(a) $\begin{bmatrix} -1 & 1 \\ 1 & 0 \end{bmatrix}$ (b) $\begin{bmatrix} 2 & 5 \\ 1 & 3 \end{bmatrix}$ (c) $\begin{bmatrix} 2 & 6 \\ 3 & 8 \end{bmatrix}$

(d) $\begin{bmatrix} 3 & 0 \\ 9 & 3 \end{bmatrix}$ (e) $\begin{bmatrix} 1 & 1 & 1 \\ 0 & 1 & 1 \\ 0 & 0 & 1 \end{bmatrix}$ (f) $\begin{bmatrix} 2 & 0 & 5 \\ 0 & 3 & 0 \\ 1 & 0 & 3 \end{bmatrix}$

(g) $\begin{bmatrix} -1 & -3 & -3 \\ 2 & 6 & 1 \\ 3 & 8 & 3 \end{bmatrix}$ (h) $\begin{bmatrix} 1 & 0 & 1 \\ -1 & 1 & 1 \\ -1 & -2 & -3 \end{bmatrix}$

10. Given

$$A = \begin{bmatrix} 3 & 1 \\ 5 & 2 \end{bmatrix} \quad \text{and} \quad B = \begin{bmatrix} 1 & 2 \\ 3 & 4 \end{bmatrix}$$

compute A^{-1} and use it to:

(a) Find a 2×2 matrix X such that $AX = B$.
(b) Find a 2×2 matrix Y such that $YA = B$.

11. Given

$$A = \begin{bmatrix} 5 & 3 \\ 3 & 2 \end{bmatrix} \quad B = \begin{bmatrix} 6 & 2 \\ 2 & 4 \end{bmatrix}, \quad C = \begin{bmatrix} 4 & -2 \\ -6 & 3 \end{bmatrix}$$

Solve each of the following matrix equations.

(a) $AX + B = C$ (b) $XA + B = C$ (c) $AX + B = X$ (d) $XA + C = X$

12. Is the transpose of an elementary matrix an elementary matrix of the same type? Is the product of two elementary matrices an elementary matrix?

13. Let U and R be $n \times n$ upper triangular matrices and set $T = UR$. Show that T is also upper triangular and that $t_{jj} = u_{jj}r_{jj}$ for $j = 1, \ldots, n$.

14. Let A be a 3×3 matrix and suppose that

$$2\mathbf{a}_1 + \mathbf{a}_2 - 4\mathbf{a}_3 = \mathbf{0}$$

How many solutions will the system $A\mathbf{x} = \mathbf{0}$ have? Explain. Is A nonsingular? Explain.

15. Let A and B be $n \times n$ matrices and let $C = AB$. Prove that if B is singular then C must be singular. [*Hint:* Use Theorem 1.4.3.]

16. Let U be an $n \times n$ upper triangular matrix with nonzero diagonal entries.

(a) Explain why U must be nonsingular.
(b) Explain why U^{-1} must be upper triangular.

17. Let A be a nonsingular $n \times n$ matrix and let B be an $n \times r$ matrix. Show that the reduced row echelon form of $(A|B)$ is $(I|C)$, where $C = A^{-1}B$.

18. In general, matrix multiplication is not commutative (i.e., $AB \neq BA$). However, in certain special cases the commutative property does hold. Show that:

(a) If D_1 and D_2 are $n \times n$ diagonal matrices, then $D_1 D_2 = D_2 D_1$.
(b) If A is an $n \times n$ matrix and

$$B = a_0 I + a_1 A + a_2 A^2 + \cdots + a_k A^k$$

where $a_0, a_1, \ldots, a_k$ are scalars, then $AB = BA$.

19. Show that if A is a symmetric nonsingular matrix then A^{-1} is also symmetric.

20. Prove that if A is row equivalent to B then B is row equivalent to A.

21. (a) Prove that, if A is row equivalent to B and B is row equivalent to C, then A is row equivalent to C.

(b) Prove that any two nonsingular $n \times n$ matrices are row equivalent.

22. Prove that B is row equivalent to A if and only if there exists a nonsingular matrix M such that $B = MA$.

23. Given a vector $\mathbf{x} \in R^{n+1}$, the $(n+1) \times (n+1)$ matrix V defined by

$$v_{ij} = \begin{cases} 1 & \text{if } j = 1 \\ x_i^{j-1} & \text{for } j = 2, \ldots, n+1 \end{cases}$$

is called the Vandermonde matrix.

(a) Show that if

$$V\mathbf{c} = \mathbf{y}$$

and

$$p(x) = c_1 + c_2 x + \cdots + c_{n+1} x^n$$

then

$$p(x_i) = y_i, \qquad i = 1, 2, \ldots, n+1$$

(b) Suppose that $x_1, x_2, \ldots, x_{n+1}$ are all distinct. Show that if $\mathbf{c}$ is a solution to $V\mathbf{x} = \mathbf{0}$ then the coefficients $c_1, c_2, \ldots, c_n$ must all be zero, and hence V must be nonsingular.

5 PARTITIONED MATRICES

Often it is useful to think of a matrix as being composed of a number of submatrices. A matrix C can be partitioned into smaller matrices by drawing horizontal lines between the rows and vertical lines between the columns. The smaller matrices are often referred to as *blocks*. For example, let

$$C = \begin{pmatrix} 1 & -2 & 4 & 1 & 3 \\ 2 & 1 & 1 & 1 & 1 \\ 3 & 3 & 2 & -1 & 2 \\ 4 & 6 & 2 & 2 & 4 \end{pmatrix}$$

If lines are drawn between the second and third rows and between the third and fourth columns, then C will be divided into four submatrices, C_{11}, C_{12}, C_{21}, and C_{22}.

$$\begin{pmatrix} C_{11} & C_{12} \\ C_{21} & C_{22} \end{pmatrix} = \left(\begin{array}{ccc|cc} 1 & -2 & 4 & 1 & 3 \\ 2 & 1 & 1 & 1 & 1 \\ \hline 3 & 3 & 2 & -1 & 2 \\ 4 & 6 & 2 & 2 & 4 \end{array} \right)$$

One useful way of partitioning a matrix is into columns. For example, if

$$B = \begin{pmatrix} -1 & 2 & 1 \\ 2 & 3 & 1 \\ 1 & 4 & 1 \end{pmatrix}$$

we can partition B into three column submatrices:

$$B = (\mathbf{b}_1, \mathbf{b}_2, \mathbf{b}_3) = \left(\begin{array}{c|c|c} -1 & 2 & 1 \\ 2 & 3 & 1 \\ 1 & 4 & 1 \end{array} \right)$$

Suppose that we are given a matrix A with three columns, then the product AB can be viewed as a block multiplication. Each of the blocks of B is multiplied by A and the result is a matrix with three blocks, $A\mathbf{b}_1$, $A\mathbf{b}_2$, $A\mathbf{b}_3$, that is,

$$AB = A(\mathbf{b}_1, \mathbf{b}_2, \mathbf{b}_3) = (A\mathbf{b}_1, A\mathbf{b}_2, A\mathbf{b}_3)$$

For example, if

$$A = \begin{pmatrix} 1 & 3 & 1 \\ 2 & 1 & -2 \end{pmatrix}$$

$$A\mathbf{b}_1 = \begin{pmatrix} 6 \\ -2 \end{pmatrix}, \quad A\mathbf{b}_2 = \begin{pmatrix} 15 \\ -1 \end{pmatrix}, \quad A\mathbf{b}_3 = \begin{pmatrix} 5 \\ 1 \end{pmatrix}$$

and hence

$$A(\mathbf{b}_1, \mathbf{b}_2, \mathbf{b}_3) = \left(\begin{array}{c|c|c} 6 & 15 & 5 \\ -2 & -1 & 1 \end{array} \right)$$

In general, if A is an $m \times n$ matrix and B is an $n \times r$ that has been partitioned into columns $(\mathbf{b}_1, \dots, \mathbf{b}_r)$, then the block multiplication of A times B is given by

$$AB = (A\mathbf{b}_1, A\mathbf{b}_2, \dots, A\mathbf{b}_r)$$

In particular,

$$(\mathbf{a}_1, \dots, \mathbf{a}_n) = A = AI = (A\mathbf{e}_1, \dots, A\mathbf{e}_n)$$

Let A be an $m \times n$ matrix. If we partition A into rows, then

$$A = \begin{pmatrix} \mathbf{a}(1,:) \\ \mathbf{a}(2,:) \\ \vdots \\ \mathbf{a}(m,:) \end{pmatrix}$$

If B is an $n \times r$ matrix, the ith row of the product AB is determined by multiplying the ith row of A times B. Thus the ith row of AB is $\mathbf{a}(i,:)B$. In general, the product AB can be partitioned into rows as follows:

$$AB = \begin{bmatrix} \mathbf{a}(1,:)B \\ \mathbf{a}(2,:)B \\ \vdots \\ \mathbf{a}(m,:)B \end{bmatrix}$$

To illustrate this result, let us look at an example. If

$$A = \begin{bmatrix} 2 & 5 \\ 3 & 4 \\ 1 & 7 \end{bmatrix} \quad \text{and} \quad B = \begin{bmatrix} 3 & 2 & -3 \\ -1 & 1 & 1 \end{bmatrix}$$

then

$$\mathbf{a}(1,:)B = \begin{bmatrix} 1 & 9 & -1 \end{bmatrix}$$

$$\mathbf{a}(2,:)B = \begin{bmatrix} 5 & 10 & -5 \end{bmatrix}$$

$$\mathbf{a}(3,:)B = \begin{bmatrix} -4 & 9 & 4 \end{bmatrix}$$

These are the row vectors of the product AB.

$$AB = \begin{bmatrix} \mathbf{a}(1,:)B \\ \mathbf{a}(2,:)B \\ \mathbf{a}(3,:)B \end{bmatrix} = \left[\begin{array}{ccc} 1 & 9 & -1 \\ \hline 5 & 10 & -5 \\ \hline -4 & 9 & 4 \end{array} \right]$$

Next we consider how to compute the product AB in terms of more general partitions of A and B.

Block Multiplication

Let A be an $m \times n$ matrix and B an $n \times r$ matrix. It is often useful to partition A and B and express the product in terms of the submatrices of A and B. Consider the following four cases.

CASE I

$B = \begin{bmatrix} B_1 & B_2 \end{bmatrix}$, where B_1 is an $n \times t$ matrix and B_2 is an $n \times (r-t)$ matrix.

$$\begin{aligned} AB &= A(\mathbf{b}_1, \ldots, \mathbf{b}_t, \mathbf{b}_{t+1}, \ldots, \mathbf{b}_r) \\ &= (A\mathbf{b}_1, \ldots, A\mathbf{b}_t, A\mathbf{b}_{t+1}, \ldots, A\mathbf{b}_r) \\ &= (A(\mathbf{b}_1, \ldots, \mathbf{b}_t), A(\mathbf{b}_{t+1}, \ldots, \mathbf{b}_r)) \\ &= \begin{bmatrix} AB_1 & AB_2 \end{bmatrix} \end{aligned}$$

Thus,

$$A \begin{bmatrix} B_1 & B_2 \end{bmatrix} = \begin{bmatrix} AB_1 & AB_2 \end{bmatrix}$$

CASE 2

$A = \begin{pmatrix} A_1 \\ A_2 \end{pmatrix}$, where A_1 is a $k \times n$ matrix and A_2 is an $(m - k) \times n$ matrix.

$$\begin{bmatrix} A_1 \\ A_2 \end{bmatrix} B = \begin{bmatrix} \mathbf{a}(1, :) \\ \vdots \\ \mathbf{a}(k, :) \\ \hline \mathbf{a}(k+1, :) \\ \vdots \\ \mathbf{a}(m, :) \end{bmatrix} B = \begin{bmatrix} \mathbf{a}(1, :)B \\ \vdots \\ \mathbf{a}(k, :)B \\ \hline \mathbf{a}(k+1, :)B \\ \vdots \\ \mathbf{a}(m, :)B \end{bmatrix}$$

$$= \begin{bmatrix} \begin{bmatrix} \mathbf{a}(1, :) \\ \vdots \\ \mathbf{a}(k, :) \end{bmatrix} B \\ \begin{bmatrix} \mathbf{a}(k+1, :) \\ \vdots \\ \mathbf{a}(m, :) \end{bmatrix} B \end{bmatrix} = \begin{bmatrix} A_1 B \\ A_2 B \end{bmatrix}$$

Thus,

$$\begin{bmatrix} A_1 \\ A_2 \end{bmatrix} B = \begin{bmatrix} A_1 B \\ A_2 B \end{bmatrix}$$

CASE 3

$A = \begin{bmatrix} A_1 & A_2 \end{bmatrix}$ and $B = \begin{bmatrix} B_1 \\ B_2 \end{bmatrix}$, where A_1 is an $m \times s$ matrix, A_2 is an $m \times (n-s)$ matrix, B_1 is an $s \times r$ matrix, and B_2 is an $(n - s) \times r$ matrix. If $C = AB$, then

$$c_{ij} = \sum_{l=1}^{n} a_{il}b_{lj} = \sum_{l=1}^{s} a_{il}b_{lj} + \sum_{l=s+1}^{n} a_{il}b_{lj}$$

Thus c_{ij} is the sum of the (i, j) entry of $A_1 B_1$ and the (i, j) entry of $A_2 B_2$. Therefore,

$$AB = C = A_1 B_1 + A_2 B_2$$

and it follows that

$$\begin{bmatrix} A_1 & A_2 \end{bmatrix} \begin{bmatrix} B_1 \\ B_2 \end{bmatrix} = A_1 B_1 + A_2 B_2$$

CASE 4

Let A and B both be partitioned as follows:

$$A = \left[\begin{array}{c|c} A_{11} & A_{12} \\ \hline A_{21} & A_{22} \end{array} \right] \begin{array}{c} k \\ m - k \end{array}, \qquad B = \left[\begin{array}{c|c} B_{11} & B_{12} \\ \hline B_{21} & B_{22} \end{array} \right] \begin{array}{c} s \\ n - s \end{array}$$

$$ \begin{array}{cc} s & n - s \end{array} \qquad\qquad \begin{array}{cc} t & r - t \end{array}$$

Let

$$A_1 = \begin{bmatrix} A_{11} \\ A_{21} \end{bmatrix} \qquad A_2 = \begin{bmatrix} A_{12} \\ A_{22} \end{bmatrix}$$

$$B_1 = \begin{bmatrix} B_{11} & B_{12} \end{bmatrix} \qquad B_2 = \begin{bmatrix} B_{21} & B_{22} \end{bmatrix}$$

It follows from case 3 that

$$AB = \begin{bmatrix} A_1 & A_2 \end{bmatrix} \begin{bmatrix} B_1 \\ B_2 \end{bmatrix} = A_1 B_1 + A_2 B_2$$

It follows from cases 1 and 2 that

$$A_1 B_1 = \begin{bmatrix} A_{11} \\ A_{21} \end{bmatrix} B_1 = \begin{bmatrix} A_{11} B_1 \\ A_{21} B_1 \end{bmatrix} = \begin{bmatrix} A_{11} B_{11} & A_{11} B_{12} \\ A_{21} B_{11} & A_{21} B_{12} \end{bmatrix}$$

$$A_2 B_2 = \begin{bmatrix} A_{12} \\ A_{22} \end{bmatrix} B_2 = \begin{bmatrix} A_{12} B_2 \\ A_{22} B_2 \end{bmatrix} = \begin{bmatrix} A_{12} B_{21} & A_{12} B_{22} \\ A_{22} B_{21} & A_{22} B_{22} \end{bmatrix}$$

Therefore,

$$\begin{pmatrix} A_{11} & A_{12} \\ A_{21} & A_{22} \end{pmatrix} \begin{pmatrix} B_{11} & B_{12} \\ B_{21} & B_{22} \end{pmatrix} = \begin{pmatrix} A_{11} B_{11} + A_{12} B_{21} & A_{11} B_{12} + A_{12} B_{22} \\ A_{21} B_{11} + A_{22} B_{21} & A_{21} B_{12} + A_{22} B_{22} \end{pmatrix}$$

In general, if the blocks have the proper dimensions, the block multiplication can be carried out in the same manner as ordinary matrix multiplication. If

$$A = \begin{bmatrix} A_{11} & \cdots & A_{1t} \\ \vdots & & \\ A_{s1} & \cdots & A_{st} \end{bmatrix} \qquad \text{and} \qquad B = \begin{bmatrix} B_{11} & \cdots & B_{1r} \\ \vdots & & \\ B_{t1} & \cdots & B_{tr} \end{bmatrix}$$

then

$$AB = \begin{pmatrix} C_{11} & \cdots & C_{1r} \\ \vdots & & \\ C_{s1} & \cdots & C_{sr} \end{pmatrix}$$

where

$$C_{ij} = \sum_{k=1}^{t} A_{ik} B_{kj}$$

The multiplication can be carried out in this manner only if the number of columns of A_{ik} equals the number of rows of B_{kj} for each k.

EXAMPLE I. Let

$$A = \begin{pmatrix} 1 & 1 & 1 & 1 \\ 2 & 2 & 1 & 1 \\ 3 & 3 & 2 & 2 \end{pmatrix}$$

and

$$B = \begin{pmatrix} B_{11} & B_{12} \\ B_{21} & B_{22} \end{pmatrix} = \left(\begin{array}{cc|cc} 1 & 1 & 1 & 1 \\ 1 & 2 & 1 & 1 \\ \hline 3 & 1 & 1 & 1 \\ 3 & 2 & 1 & 2 \end{array} \right)$$

Partition A into four blocks and perform the block multiplication.

SOLUTION. Since each B_{kj} has two rows, the A_{ik}'s must each have two columns. Thus there are two possibilities:

(i) $$\begin{pmatrix} A_{11} & A_{12} \\ A_{21} & A_{22} \end{pmatrix} = \left(\begin{array}{cc|cc} 1 & 1 & 1 & 1 \\ \hline 2 & 2 & 1 & 1 \\ 3 & 3 & 2 & 2 \end{array} \right)$$

in which case

$$\left(\begin{array}{cc|cc} 1 & 1 & 1 & 1 \\ \hline 2 & 2 & 1 & 1 \\ 3 & 3 & 2 & 2 \end{array} \right) \left(\begin{array}{cc|cc} 1 & 1 & 1 & 1 \\ 1 & 2 & 1 & 1 \\ \hline 3 & 1 & 1 & 1 \\ 3 & 2 & 1 & 2 \end{array} \right) = \left(\begin{array}{cc|cc} 8 & 6 & 4 & 5 \\ \hline 10 & 9 & 6 & 7 \\ 18 & 15 & 10 & 12 \end{array} \right)$$

or

(ii) $$\begin{pmatrix} A_{11} & A_{12} \\ A_{21} & A_{22} \end{pmatrix} = \left(\begin{array}{cc|cc} 1 & 1 & 1 & 1 \\ 2 & 2 & 1 & 1 \\ \hline 3 & 3 & 2 & 2 \end{array} \right)$$

in which case

$$\begin{pmatrix} 1 & 1 & 1 & 1 \\ 2 & 2 & 1 & 1 \\ 3 & 3 & 2 & 2 \end{pmatrix} \begin{pmatrix} 1 & 1 & 1 & 1 \\ 1 & 2 & 1 & 1 \\ 3 & 1 & 1 & 1 \\ 3 & 2 & 1 & 2 \end{pmatrix} = \begin{pmatrix} 8 & 6 & 4 & 5 \\ 10 & 9 & 6 & 7 \\ 18 & 15 & 10 & 12 \end{pmatrix}$$

◀

EXAMPLE 2. Let A be an $n \times n$ matrix of the form

$$\begin{pmatrix} A_{11} & O \\ O & A_{22} \end{pmatrix}$$

where A_{11} is a $k \times k$ matrix ($k < n$). Show that A is nonsingular if and only if A_{11} and A_{22} are nonsingular.

SOLUTION. If A_{11} and A_{22} are nonsingular, then

$$\begin{pmatrix} A_{11}^{-1} & O \\ O & A_{22}^{-1} \end{pmatrix} \begin{pmatrix} A_{11} & O \\ O & A_{22} \end{pmatrix} = \begin{pmatrix} I_k & O \\ O & I_{n-k} \end{pmatrix} = I$$

and

$$\begin{pmatrix} A_{11} & O \\ O & A_{22} \end{pmatrix} \begin{pmatrix} A_{11}^{-1} & O \\ O & A_{22}^{-1} \end{pmatrix} = \begin{pmatrix} I_k & O \\ O & I_{n-k} \end{pmatrix} = I$$

so A is nonsingular and

$$A^{-1} = \begin{pmatrix} A_{11}^{-1} & O \\ O & A_{22}^{-1} \end{pmatrix}$$

Conversely, if A is nonsingular, then let $B = A^{-1}$ and partition B in the same manner as A. Since

$$BA = I = AB$$

it follows that

$$\begin{pmatrix} B_{11} & B_{12} \\ B_{21} & B_{22} \end{pmatrix} \begin{pmatrix} A_{11} & O \\ O & A_{22} \end{pmatrix} = \begin{pmatrix} I_k & O \\ O & I_{n-k} \end{pmatrix} = \begin{pmatrix} A_{11} & O \\ O & A_{22} \end{pmatrix} \begin{pmatrix} B_{11} & B_{12} \\ B_{21} & B_{22} \end{pmatrix}$$

$$\begin{pmatrix} B_{11}A_{11} & B_{12}A_{22} \\ B_{21}A_{11} & B_{22}A_{22} \end{pmatrix} = \begin{pmatrix} I_k & O \\ O & I_{n-k} \end{pmatrix} = \begin{pmatrix} A_{11}B_{11} & A_{11}B_{12} \\ A_{22}B_{21} & A_{22}B_{22} \end{pmatrix}$$

Thus

$$B_{11}A_{11} = I_k = A_{11}B_{11}$$

$$B_{22}A_{22} = I_{n-k} = A_{22}B_{22}$$

and hence A_{11} and A_{22} are both nonsingular with inverses B_{11} and B_{22}, respectively.   ◀

Outer Product Expansions

Given two vectors $\mathbf{x}$ and $\mathbf{y}$ in R^n, we can multiply them together if we transpose one of the vectors first. The matrix product $\mathbf{x}^T\mathbf{y}$ is the product of a row vector (a $1 \times n$ matrix) times a column vector (an $n \times 1$ matrix). The result will be a 1×1 matrix or simply a scalar.

$$\mathbf{x}^T\mathbf{y} = (x_1, x_2, \ldots, x_n)\begin{pmatrix} y_1 \\ y_2 \\ \vdots \\ y_n \end{pmatrix} = x_1 y_1 + x_2 y_2 + \cdots + x_n y_n$$

This type of product is referred to as a *scalar product* or an *inner product*. The scalar product is one of the most commonly performed operations. For example, when we multiply two matrices, each entry of the product is computed as a scalar product (a row vector times a column vector).

It is also useful to multiply a column vector times a row vector. The matrix product $\mathbf{x}\mathbf{y}^T$ is the product of an $n \times 1$ matrix times an $1 \times n$ matrix. The result is a full $n \times n$ matrix.

$$\mathbf{x}\mathbf{y}^T = \begin{pmatrix} x_1 \\ x_2 \\ \vdots \\ x_n \end{pmatrix}(y_1, y_2, \ldots, y_n) = \begin{pmatrix} x_1 y_1 & x_1 y_2 & \cdots & x_1 y_n \\ x_2 y_1 & x_2 y_2 & \cdots & x_2 y_n \\ \vdots & & & \\ x_n y_1 & x_n y_2 & \cdots & x_n y_n \end{pmatrix}$$

The product $\mathbf{x}\mathbf{y}^T$ is referred to as the *outer product* of $\mathbf{x}$ and $\mathbf{y}$. The outer product matrix has special structure in that each of its rows is a multiple of $\mathbf{y}^T$ and each of its column vectors is a multiple of $\mathbf{x}$. For example, if

$$\mathbf{x} = \begin{pmatrix} 4 \\ 1 \\ 3 \end{pmatrix} \quad \text{and} \quad \mathbf{y} = \begin{pmatrix} 3 \\ 5 \\ 2 \end{pmatrix}$$

then

$$\mathbf{x}\mathbf{y}^T = \begin{pmatrix} 4 \\ 1 \\ 3 \end{pmatrix}(3, 5, 2) = \begin{pmatrix} 12 & 20 & 8 \\ 3 & 5 & 2 \\ 9 & 15 & 6 \end{pmatrix}$$

Note that each row is a multiple of $(3, 5, 2)$ and each column is a multiple of $\mathbf{x}$.

We are now ready to generalize the idea of an outer product from vectors to matrices. Suppose that we start with an $m \times n$ matrix X and a $k \times n$ matrix Y. We can then form a matrix product XY^T. If we partition X into columns and Y^T into rows and perform the block multiplication, we see that XY^T can be represented as a sum of outer products of vectors.

$$XY^T = (\mathbf{x}_1, \mathbf{x}_2, \ldots, \mathbf{x}_n) \begin{pmatrix} \mathbf{y}_1^T \\ \mathbf{y}_2^T \\ \vdots \\ \mathbf{y}_n^T \end{pmatrix} = \mathbf{x}_1\mathbf{y}_1^T + \mathbf{x}_2\mathbf{y}_2^T + \cdots + \mathbf{x}_n\mathbf{y}_n^T$$

This representation is referred to as an *outer product expansion*. These types of expansions play an important part in many applications. In Section 5 of Chapter 6 we will see how outer product expansions are used in digital imaging and in information retrieval applications.

EXAMPLE 3. Given

$$X = \begin{pmatrix} 3 & 1 \\ 2 & 4 \\ 1 & 2 \end{pmatrix} \qquad \text{and} \qquad Y = \begin{pmatrix} 1 & 2 \\ 2 & 4 \\ 3 & 1 \end{pmatrix}$$

compute the outer product expansion of XY^T.

SOLUTION.

$$\begin{aligned} XY^T &= \begin{pmatrix} 3 & 1 \\ 2 & 4 \\ 1 & 2 \end{pmatrix} \begin{pmatrix} 1 & 2 & 3 \\ 2 & 4 & 1 \end{pmatrix} \\ &= \begin{pmatrix} 3 \\ 2 \\ 1 \end{pmatrix} (1, 2, 3) + \begin{pmatrix} 1 \\ 4 \\ 2 \end{pmatrix} (2, 4, 1) \\ &= \begin{pmatrix} 3 & 6 & 9 \\ 2 & 4 & 6 \\ 1 & 2 & 3 \end{pmatrix} + \begin{pmatrix} 2 & 4 & 1 \\ 8 & 16 & 4 \\ 4 & 8 & 2 \end{pmatrix} \end{aligned}$$

◀

EXERCISES

1. Let A be a nonsingular $n \times n$ matrix. Perform the following multiplications.

(a) $A^{-1}(A \quad I)$

(b) $\begin{pmatrix} A \\ I \end{pmatrix} A^{-1}$

(c) $(A \quad I)^T (A \quad I)$

(d) $(A \quad I)(A \quad I)^T$

(e) $\begin{pmatrix} A^{-1} \\ I \end{pmatrix} (A \quad I)$

2. Let $B = A^T A$. Show that $b_{ij} = \mathbf{a}_i^T \mathbf{a}_j$.

3. Let

$$A = \begin{pmatrix} 1 & 1 \\ 2 & -1 \end{pmatrix} \qquad \text{and} \qquad B = \begin{pmatrix} 2 & 1 \\ 1 & 3 \end{pmatrix}$$

(a) Calculate $A\mathbf{b}_1$ and $A\mathbf{b}_2$.

(b) Calculate $\mathbf{a}(1,:)B$ and $\mathbf{a}(2,:)B$.

(c) Multiply AB and verify that its column vectors are the vectors in part (a) and its row vectors are the vectors in part (b).

4. Let

$$I = \begin{bmatrix} 1 & 0 \\ 0 & 1 \end{bmatrix}, \qquad E = \begin{bmatrix} 0 & 1 \\ 1 & 0 \end{bmatrix}, \qquad O = \begin{bmatrix} 0 & 0 \\ 0 & 0 \end{bmatrix}$$

$$C = \begin{bmatrix} 1 & 0 \\ -1 & 1 \end{bmatrix}, \qquad D = \begin{bmatrix} 2 & 0 \\ 0 & 2 \end{bmatrix}$$

and

$$B = \begin{bmatrix} B_{11} & B_{12} \\ B_{21} & B_{22} \end{bmatrix} = \begin{bmatrix} 1 & 1 & 1 & 1 \\ 1 & 2 & 1 & 1 \\ 3 & 1 & 1 & 1 \\ 3 & 2 & 1 & 2 \end{bmatrix}$$

Perform each of the following block multiplications.

(a) $\begin{bmatrix} O & I \\ I & O \end{bmatrix} \begin{bmatrix} B_{11} & B_{12} \\ B_{21} & B_{22} \end{bmatrix}$

(b) $\begin{bmatrix} C & O \\ O & C \end{bmatrix} \begin{bmatrix} B_{11} & B_{12} \\ B_{21} & B_{22} \end{bmatrix}$

(c) $\begin{bmatrix} D & O \\ O & I \end{bmatrix} \begin{bmatrix} B_{11} & B_{12} \\ B_{21} & B_{22} \end{bmatrix}$

(d) $\begin{bmatrix} E & O \\ O & E \end{bmatrix} \begin{bmatrix} B_{11} & B_{12} \\ B_{21} & B_{22} \end{bmatrix}$

5. Perform each of the following block multiplications.

(a) $\begin{bmatrix} 1 & 1 & 1 & -1 \\ 2 & 1 & 2 & -1 \end{bmatrix} \begin{bmatrix} 4 & -2 & 1 \\ 2 & 3 & 1 \\ 1 & 1 & 2 \\ 1 & 2 & 3 \end{bmatrix}$

(b) $\begin{bmatrix} 4 & -2 \\ 2 & 3 \\ 1 & 1 \\ 1 & 2 \end{bmatrix} \begin{bmatrix} 1 & 1 & 1 & -1 \\ 2 & 1 & 2 & -1 \end{bmatrix}$

(c) $\begin{bmatrix} \frac{3}{5} & -\frac{4}{5} & 0 & 0 \\ \frac{4}{5} & \frac{3}{5} & 0 & 0 \\ 0 & 0 & 1 & 0 \end{bmatrix} \begin{bmatrix} \frac{3}{5} & \frac{4}{5} & 0 \\ -\frac{4}{5} & \frac{3}{5} & 0 \\ 0 & 0 & 1 \\ 0 & 0 & 0 \end{bmatrix}$

(d) $\begin{bmatrix} 0 & 0 & 1 & 0 & 0 \\ 0 & 1 & 0 & 0 & 0 \\ 1 & 0 & 0 & 0 & 0 \\ 0 & 0 & 0 & 0 & 1 \\ 0 & 0 & 0 & 1 & 0 \end{bmatrix} \begin{bmatrix} 1 & -1 \\ 2 & -2 \\ 3 & -3 \\ 4 & -4 \\ 5 & -5 \end{bmatrix}$

6. Given

$$X = \begin{bmatrix} 2 & 1 & 5 \\ 4 & 2 & 3 \end{bmatrix} \qquad Y = \begin{bmatrix} 1 & 2 & 4 \\ 2 & 3 & 1 \end{bmatrix}$$

(a) Compute the outer product expansion of XY^T.

(b) Compute the outer product expansion of YX^T. How is the outer product expansion of YX^T related to the outer product expansion of XY^T?

7. Let

$$A = \begin{bmatrix} A_{11} & A_{12} \\ A_{21} & A_{22} \end{bmatrix} \quad \text{and} \quad A^T = \begin{bmatrix} A_{11}^T & A_{21}^T \\ A_{12}^T & A_{22}^T \end{bmatrix}$$

Is it possible to perform the block multiplications of AA^T and A^TA? Explain.

8. Let A be an $m \times n$ matrix, X an $n \times r$ matrix, and B an $m \times r$ matrix. Show that

$$AX = B$$

if and only if

$$A\mathbf{x}_j = \mathbf{b}_j, \quad j = 1, \dots, r$$

9. Let A be an $n \times n$ matrix and let D be a $n \times n$ diagonal matrix.

(a) Show that $D = (d_{11}\mathbf{e}_1, d_{22}\mathbf{e}_2, \dots, d_{nn}\mathbf{e}_n)$.

(b) Show that $AD = (d_{11}\mathbf{a}_1, d_{22}\mathbf{a}_2, \dots, d_{nn}\mathbf{a}_n)$.

10. Let U be an $m \times m$ matrix, V be an $n \times n$ matrix, and let

$$\Sigma = \begin{bmatrix} \Sigma_1 \\ O \end{bmatrix}$$

where Σ_1 is an $n \times n$ diagonal matrix with diagonal entries $\sigma_1, \sigma_2, \dots, \sigma_n$ and O is the $(m-n) \times n$ zero matrix.

(a) If $U = (U_1, U_2)$, where U_1 has n columns, show that

$$U\Sigma = U_1\Sigma_1$$

(b) Show that if $A = U\Sigma V^T$ then A can be expressed as an outer product expansion of the form

$$A = \sigma_1\mathbf{u}_1\mathbf{v}_1^T + \sigma_2\mathbf{u}_2\mathbf{v}_2^T + \cdots + \sigma_n\mathbf{u}_n\mathbf{v}_n^T$$

11. Let

$$A = \begin{bmatrix} A_{11} & A_{12} \\ O & A_{22} \end{bmatrix}$$

where all four blocks are $n \times n$ matrices.

(a) If A_{11} and A_{22} are nonsingular, show that A must also be nonsingular and that A^{-1} must be of the form

$$\left[\begin{array}{c|c} A_{11}^{-1} & C \\ \hline O & A_{22}^{-1} \end{array} \right]$$

(b) Determine C.

12. Let

$$A = \begin{bmatrix} O & I \\ B & O \end{bmatrix}$$

where all four submatrices are $k \times k$. Determine A^2 and A^4.

13. Let I denote the $n \times n$ identity matrix. Find a block form for the inverse of each of the following $2n \times 2n$ matrices.

(a) $\begin{bmatrix} O & I \\ I & O \end{bmatrix}$
(b) $\begin{bmatrix} I & O \\ B & I \end{bmatrix}$

14. Let A and B be $m \times n$ and $n \times r$ matrices, respectively, and define $(m+n) \times (m+n)$ matrices S and M by

$$S = \begin{bmatrix} I & A \\ O & I \end{bmatrix}, \qquad M = \begin{bmatrix} AB & O \\ B & O \end{bmatrix}$$

Determine the block form of S^{-1} and use it to compute the block form of the product $S^{-1}MS$.

15. Let

$$A = \begin{bmatrix} A_{11} & A_{12} \\ A_{21} & A_{22} \end{bmatrix}$$

where A_{11} is a $k \times k$ nonsingular matrix. Show that A can be factored into a product

$$\begin{bmatrix} I & O \\ B & I \end{bmatrix} \begin{bmatrix} A_{11} & A_{12} \\ O & C \end{bmatrix}$$

where

$$B = A_{21} A_{11}^{-1} \qquad \text{and} \qquad C = A_{22} - A_{21} A_{11}^{-1} A_{12}$$

16. Given A, B, L, M, S, T are $n \times n$ matrices with A, B, M nonsingular and L, S, T singular. Is it possible to find matrices X and Y such that

$$\begin{bmatrix} O & I & O & O & O & O \\ O & O & I & O & O & O \\ O & O & O & I & O & O \\ O & O & O & O & I & O \\ O & O & O & O & O & X \\ Y & O & O & O & O & O \end{bmatrix} \begin{bmatrix} M \\ A \\ T \\ L \\ A \\ B \end{bmatrix} = \begin{bmatrix} A \\ T \\ L \\ A \\ S \\ T \end{bmatrix}$$

If so, show how; if not, explain why.

17. Let A be an $n \times n$ matrix and $\mathbf{x} \in R^n$.

(a) A scalar c can also be considered as a 1×1 matrix $C = (c)$ and a vector $\mathbf{b} \in R^n$ can be considered as an $n \times 1$ matrix B. Show that, although the matrix multiplication CB is not defined, the matrix product BC is equal to $c\mathbf{b}$, the scalar multiplication of c times $\mathbf{b}$.

(b) Partition A into columns and $\mathbf{x}$ into rows and perform the block multiplication of A times $\mathbf{x}$.

(c) Show that

$$A\mathbf{x} = x_1 \mathbf{a}_1 + x_2 \mathbf{a}_2 + \cdots + x_n \mathbf{a}_n$$

18. Show that, if A is an $n \times n$ matrix with the property $A\mathbf{x} = \mathbf{0}$ for all $\mathbf{x} \in R^n$, then $A = O$. [*Hint:* Let $\mathbf{x} = \mathbf{e}_j$ for $j = 1, \dots, n$.]

19. Let B and C be $n \times n$ matrices with the property $B\mathbf{x} = C\mathbf{x}$ for all $\mathbf{x} \in R^n$. Show that $B = C$.

20. Consider a system of the form

$$\begin{pmatrix} A & \mathbf{a} \\ \mathbf{c}^T & \beta \end{pmatrix} \begin{pmatrix} \mathbf{x} \\ x_{n+1} \end{pmatrix} = \begin{pmatrix} \mathbf{b} \\ b_{n+1} \end{pmatrix}$$

where A is a nonsingular $n \times n$ matrix and $\mathbf{a}$, $\mathbf{b}$, and $\mathbf{c}$ are vectors in R^n.

(a) Multiply both sides of the system by

$$\begin{pmatrix} A^{-1} & \mathbf{0} \\ -\mathbf{c}^T A^{-1} & 1 \end{pmatrix}$$

to obtain an equivalent triangular system.

(b) Set $\mathbf{y} = A^{-1}\mathbf{a}$ and $\mathbf{z} = A^{-1}\mathbf{b}$. Show that if $\beta - \mathbf{c}^T\mathbf{y} \neq 0$ then the solution of the system can be determined by letting

$$x_{n+1} = \frac{b_{n+1} - \mathbf{c}^T\mathbf{z}}{\beta - \mathbf{c}^T\mathbf{y}}$$

and then setting

$$\mathbf{x} = \mathbf{z} - x_{n+1}\mathbf{y}$$

MATLAB EXERCISES

The following exercises are to be solved computationally using the software package MATLAB, which is described in the Appendix of this book. The exercises also contain questions that should be answered relating to the underlying mathematical principles illustrated in the computations. Save a record of your session in a file. After editing and printing out the file, the answers to the questions can then be filled in directly on the printout.

MATLAB has a help facility that explains all its operations and commands. For example, to obtain information on the MATLAB command **rand** you need only type: **help rand**. The commands used in the MATLAB exercises for this chapter are **inv**, **round**, **rand**, **flops**, **rref**, **format**, **sum**, **eye**, **triu**, **ones**, **zeros**, **magic**. The operations introduced are $+, -, *, ', \backslash$. The $+$ and $-$ represent the usual addition and subtraction operations for both scalars and matrices. The $*$ corresponds to multiplication of either scalars or matrices, and the $'$ operation corresponds to the transpose operation. If A is a nonsingular $n \times n$ matrix and B is any $n \times r$ matrix, the operation $A\backslash B$ is equivalent to computing $A^{-1}B$.

1. Use MATLAB to generate random 4×4 matrices A and B. For each of the following compute $A1$, $A2$, $A3$, $A4$ as indicated and determine which of the matrices are equal. You can use MATLAB to test whether two matrices are equal by computing their difference.

(a) $A1 = A * B$, $A2 = B * A$, $A3 = (A' * B')'$, $A4 = (B' * A')'$
(b) $A1 = A' * B'$, $A2 = (A * B)'$, $A3 = B' * A'$, $A4 = (B * A)'$
(c) $A1 = \textbf{inv}(A * B)$, $A2 = \textbf{inv}(A) * \textbf{inv}(B)$, $A3 = \textbf{inv}(B * A)$, $A4 = \textbf{inv}(B) * \textbf{inv}(A)$
(d) $A1 = \textbf{inv}((A * B)')$, $A2 = \textbf{inv}(A' * B')$, $A3 = \textbf{inv}(A') * \textbf{inv}(B')$, $A4 = (\textbf{inv}(A) * \textbf{inv}(B))'$

2. Generate an 8×8 matrix and a vector in R^8, both having integer entries, by setting

$$A = \textbf{round}(10 * \textbf{rand}(8)) \quad \text{and} \quad b = \textbf{round}(10 * \textbf{rand}(8, 1))$$

(a) We can estimate the amount of arithmetic operations involved in solving the system $Ax = b$ by using the MATLAB function **flops**. The value of the variable **flops** is MATLAB's estimate of the total number of floating-point arithmetic operations that have been carried out so far in the current MATLAB session. The value of **flops** can be reset to 0 by typing **flops**(0). Reset **flops** to 0 and then find the solution **x** to the system $Ax = b$ using the "\" operation. Now type **flops** to determine approximately how many arithmetic operations were carried out in solving the system.

(b) Next let us solve the system using Gauss–Jordan reduction. First reset **flops** to 0. Compute the reduced row echelon form of the augmented matrix $(A \quad b)$. This can be done with the MATLAB command

$$U = \textbf{rref}([A \quad b])$$

In exact arithmetic the last column of the reduced row echelon form of the augmented matrix should be the solution to the system. Why? Explain. Set **y** equal to the last column of U and type **flops** to estimate the number of floating operations that were used in computing **y**. Which method of solving the system was more efficient? Using the "\" operation or the Gauss–Jordan reduction? (*Note:* The "\" operation is essentially reduction to triangular form followed by back substitution.)

(c) The solutions **x** and **y** obtained from the two methods look to be the same, but, if you examine more digits of the vectors using MATLAB's **format long**, you will see that they are not identical. To how many digits do the two vectors agree? An easier way to compare the two vectors is to use **format short** and to look at the difference $\mathbf{x} - \mathbf{y}$.

(d) Which of the two computed solutions **x** and **y** is more accurate? To answer this, compare each of the products $A\mathbf{x}$ and $A\mathbf{y}$ to the right-hand side **b**. The simplest way to do this is to look at the differences $\mathbf{r} = \mathbf{b} - A\mathbf{x}$ and $\mathbf{s} = \mathbf{b} - A\mathbf{y}$. The vectors **r** and **s** are called the residual vectors for the computed solutions **x** and **y**, respectively. Which of the computed solutions has the smallest residual vector?

3. Set $A = \textbf{round}(10 * \textbf{rand}(6))$. By construction, the matrix A will have integer entries. Let us change the sixth column of A so as to make the matrix singular. Set

$$B = A', \qquad A(:, 6) = -\textbf{sum}(B(1 : 5, :))'$$

(a) Set $\mathbf{x} = \mathbf{ones}(6, 1)$ and use MATLAB to compute $A * \mathbf{x}$. Why do we know that A must be singular? Explain. Check that A is singular by computing its reduced row echelon form.

(b) Set

$$B = \mathbf{x} * [1 : 6]$$

The product AB should equal the zero matrix. Why? Explain. Verify that this is so by computing AB using the MATLAB operation $*$.

(c) Set

$$C = \mathbf{round}(10 * \mathbf{rand}(6)) \qquad \text{and} \qquad D = B + C$$

Although $C \neq D$, the products AC and AD should be equal. Why? Explain. Compute $A * C$ and $A * D$ and verify that they are indeed equal.

4. Construct a matrix as follows. Set

$$B = \mathbf{eye}(10) - \mathbf{triu}(\mathbf{ones}(10), 1)$$

Why do we know that B must be nonsingular? Set

$$C = \mathbf{inv}(B) \qquad \text{and} \qquad \mathbf{x} = C(:, 10)$$

Now change B slightly by setting $B(10, 1) = -1/256$. Use MATLAB to compute the product $B\mathbf{x}$. From the result of this computation, what can you conclude about the new matrix B? Is it still nonsingular? Explain. Use MATLAB to compute its reduced row echelon form.

5. Generate a matrix A by setting

$$A = \mathbf{round}(10 * \mathbf{rand}(6))$$

and generate a vector $\mathbf{b}$ by setting

$$\mathbf{b} = \mathbf{round}(20 * \mathbf{rand}(6, 1)) - 10$$

(a) Since A was generated randomly, we would expect it to be nonsingular. The system $A\mathbf{x} = \mathbf{b}$ should have a unique solution. Find the solution using the "\" operation. Use MATLAB to compute the reduced row echelon form U of $[A \ \ \mathbf{b}]$. How does the last column of U compare with the solution $\mathbf{x}$? In exact arithmetic they should be the same. Why? Explain. To compare the two, compute the difference $U(:, 7) - \mathbf{x}$ or examine both using **format long.**

(b) Let us now change A so as to make it singular. Set

$$A(:, 3) = A(:, 1 : 2) * [4 \ \ 3]'$$

Use MATLAB to compute $\mathbf{rref}([A \ \ \mathbf{b}])$. How many solutions will the system $A\mathbf{x} = \mathbf{b}$ have? Explain.

(c) Set

$$\mathbf{y} = \mathbf{round}(20 * \mathbf{rand}(6, 1)) - 10 \quad \text{and} \quad \mathbf{c} = A * \mathbf{y}$$

Why do we know that the system $A\mathbf{x} = \mathbf{c}$ must be consistent? Explain. Compute the reduced row echelon form U of $[A \ \ \mathbf{c}]$. How many solutions does the system $A\mathbf{x} = \mathbf{c}$ have? Explain.

(d) The free variable determined by the echelon form should be x_3. By examining the system corresponding to the matrix U, you should be able to determine the solution corresponding to $x_3 = 0$. Enter this solution in MATLAB as a column vector **w**. To check that $A\mathbf{w} = \mathbf{c}$, compute the residual vector $\mathbf{c} - A\mathbf{w}$.

(e) Set $U(:, 7) = \texttt{zeros}(6, 1)$. The matrix U should now correspond to the reduced row echelon form of $(A \mid \mathbf{0})$. Use U to determine the solution to the homogeneous system when the free variable $x_3 = 1$ (do this by hand) and enter your result as a vector **z**. Check your answer by computing $A * \mathbf{z}$.

(f) Set $\mathbf{v} = \mathbf{w} + 3 * \mathbf{z}$. The vector **v** should be a solution to the system $A\mathbf{x} = \mathbf{c}$. Why? Explain. Verify that **v** is a solution by using MATLAB to compute the residual vector $\mathbf{c} - A\mathbf{v}$. What is the value of the free variable x_3 for this solution? How could we determine all possible solutions to the system in terms of the vectors **w** and **z**? Explain.

6. Consider the graph

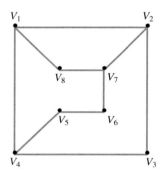

(a) Determine the adjacency matrix A for the graph and enter it in MATLAB.

(b) Compute A^2 and determine the number of walks of length 2 from (i) V_1 to V_7, (ii) V_4 to V_8, (iii) V_5 to V_6, (iv) V_8 to V_3.

(c) Compute A^4, A^6, A^8 and answer the questions in part (b) for walks of lengths 4, 6, and 8. Make a conjecture as to when there will be no walks of even length from vertex V_i to vertex V_j.

(d) Compute A^3, A^5, A^7 and answer the questions from part (b) for walks of lengths 3, 5, and 7. Does your conjecture from part (c) hold for walks of odd length? Explain. Make a conjecture as to whether there are any walks of length k from V_i to V_j based on whether $i + j + k$ is odd or even.

(e) If we add the edges $\{V_3, V_6\}$, $\{V_5, V_8\}$ to the graph, the adjacency matrix B for the new graph can be generated by setting $B = A$ and then setting

$$B(3, 6) = 1, \qquad B(6, 3) = 1, \qquad B(5, 8) = 1, \qquad B(8, 5) = 1$$

Compute B^k, for $k = 2, 3, 4, 5$. Is your conjecture from part (d) still valid for the new graph?

(f) Add the edge $\{V_6, V_8\}$ to the figure and construct the adjacency matrix C for the resulting graph. Compute powers of C to determine whether your conjecture from part (d) will still hold for this new graph.

7. In Application 3 of Section 3, the numbers of married and single women after 1 and 2 years were determined by computing the products AX and A^2X for the given matrices A and X. Use **format long** and enter these matrices in MATLAB. Compute A^k and A^kX for $k = 5, 10, 15, 20$. What is happening to A^k as k gets large? What is the long-run distribution of married and single women in the town?

8. The following table describes a seven-stage model for the life cycle of the loggerhead turtle.

SEVEN-STAGE MODEL FOR LOGGERHEAD TURTLE DEMOGRAPHICS

Stage Number	Description (age in years)	Annual Survivorship	Eggs Laid per Year
1	Eggs, hatchlings (<1)	0.6747	0
2	Small juveniles (1–7)	0.7857	0
3	Large juveniles (8–15)	0.6758	0
4	Subadults (16–21)	0.7425	0
5	Novice breeders (22)	0.8091	127
6	First-year remigrants (23)	0.8091	4
7	Mature breeders (24–54)	0.8091	80

The corresponding Leslie matrix is given by

$$L = \begin{pmatrix} 0 & 0 & 0 & 0 & 127 & 4 & 80 \\ 0.6747 & 0.7370 & 0 & 0 & 0 & 0 & 0 \\ 0 & 0.0486 & 0.6610 & 0 & 0 & 0 & 0 \\ 0 & 0 & 0.0147 & 0.6907 & 0 & 0 & 0 \\ 0 & 0 & 0 & 0.0518 & 0 & 0 & 0 \\ 0 & 0 & 0 & 0 & 0.8091 & 0 & 0 \\ 0 & 0 & 0 & 0 & 0 & 0.8091 & 0.8089 \end{pmatrix}$$

Suppose that the number of turtles in each stage of the initial turtle population is described by the vector

$$x_0 = (200{,}000 \quad 130{,}000 \quad 100{,}000 \quad 70{,}000 \quad 500 \quad 400 \quad 1100)^T$$

(a) Enter L in MATLAB and then set

$$x0 = (200000, 130000, 100000, 70000, 500, 400, 1100)'$$

Use the command

$$x50 = \text{round}(L^{\wedge}50*x0)$$

to compute x_{50}. Compute also the values of x_{100}, x_{150}, x_{200}, x_{250}, and x_{300}.

(b) Loggerhead turtles lay their eggs on land. Suppose that conservationists take special measures to protect these eggs and as a result the survival rate for eggs and hatchlings increases to 77 percent. To incorporate this change into our model, we need only change the (2,1) entry of L to 0.77. Make this modification to the matrix L and repeat part (a). Has the survival potential of the loggerhead turtle improved significantly?

(c) Suppose that instead of improving the survival rate for eggs and hatchlings we could devise a means of protecting the small juveniles so that their survival rate increases to 88 percent. Use equations (7) and (8) from Application 4 of Section 3 to determine the proportion of small juveniles that survive and remain in the same stage and the proportion that survive and grow to the next stage. Modify your original matrix L accordingly and repeat part (a) using the new matrix. Has the survival potential of the loggerhead turtle improved significantly?

9. Set $A = \mathbf{magic}(8)$ and then compute its reduced row echelon form. The lead 1's should correspond to the first three variables x_1, x_2, x_3 and the remaining five variables are all free.

(a) Set $\mathbf{c} = [1 : 8]'$ and determine whether the system $A\mathbf{x} = \mathbf{c}$ is consistent by computing the reduced row echelon form of $[A \quad \mathbf{c}]$. Does the system turn out to be consistent? Explain.

(b) Set

$$\mathbf{b} = [\,8 \quad -8 \quad -8 \quad 8 \quad 8 \quad -8 \quad -8 \quad 8\,]';$$

and consider the system $A\mathbf{x} = \mathbf{b}$. This system should be consistent. Verify that it is by computing $U = \mathbf{rref}([A \ \mathbf{b}])$. We should be able to find a solution for any choice of the five free variables. Indeed, set $\mathbf{x2} = \mathbf{round}(10 * \mathbf{rand}(5, 1))$. If $\mathbf{x2}$ represents the last five coordinates of a solution to the system, then we should be able to determine $\mathbf{x1} = (x_1, x_2, x_3)^T$ in terms of $\mathbf{x2}$. To do this, set $U = \mathbf{rref}([A \quad \mathbf{b}])$. The nonzero rows of U correspond to a linear system with block form

(1) $$\begin{pmatrix} I & V \end{pmatrix} \begin{bmatrix} \mathbf{x1} \\ \mathbf{x2} \end{bmatrix} = \mathbf{c}$$

To solve equation (1), set

$$V = U(1 : 3, \ 4 : 8), \quad \mathbf{c} = U(1 : 3, \ 9)$$

and use MATLAB to compute $\mathbf{x1}$ in terms of $\mathbf{x2}$, $\mathbf{c}$, and V. Set $\mathbf{x} = [\mathbf{x1}; \ \mathbf{x2}]$ and verify that $\mathbf{x}$ is a solution to the system.

10. Set

$$B = [-1, -1; \ 1, 1] \quad \text{and} \quad A = [\mathbf{zeros}(2), \mathbf{eye}(2); \ \mathbf{eye}(2), B]$$

and verify that $B^2 = O$.

(a) Use MATLAB to compute A^2, A^4, A^6, and A^8. Make a conjecture as to what the block form of A^{2k} will be in terms of the submatrices I, O, and B. Use mathematical induction to prove that your conjecture is true for any positive integer k.

(b) Use MATLAB to compute A^3, A^5, A^7, and A^9. Make a conjecture as to what the block form of A^{2k-1} will be in terms of the submatrices I, O, and B. Prove your conjecture.

11. (a) The MATLAB commands

$$A = \mathbf{round}(10 * \mathbf{rand}(6)), \quad B = A' * A$$

will result in a symmetric matrix with integer entries. Why? Explain. Compute B in this way and verify these claims. Next, partition B into four 3×3 submatrices. To determine the submatrices in MATLAB, set

$$B11 = B(1:3, \; 1:3), \quad B12 = B(1:3, \; 4:6)$$

and define $B21$ and $B22$ in a similar manner using rows 4 through 6 of B.

(b) Set $C = \mathbf{inv}(B11)$. It should be the case that $C^T = C$ and $B21^T = B12$. Why? Explain. Use the MATLAB operation ' to compute the transposes and verify these claims. Next, set

$$E = B21 * C \quad \text{and} \quad F = B22 - B21 * C * B21'$$

and use the MATLAB functions **eye** and **zeros** to construct

$$L = \begin{bmatrix} I & O \\ E & I \end{bmatrix}, \quad D = \begin{bmatrix} B11 & O \\ O & F \end{bmatrix}$$

Compute $H = L * D * L'$ and compare it to B by computing $H - B$. Prove that if all computations had been done in exact arithmetic LDL^T would equal B exactly.

CHAPTER TEST

This chapter test consists of 10 true–false questions. In each case answer *true* if the statement is always true and *false* otherwise. In the case of a true statement, explain or prove your answer. In the case of a false statement, give an example to show that the statement is not always true. For example, consider the following statements about $n \times n$ matrices A and B.

(i) $A + B = B + A$

(ii) $AB = BA$

Statement **(i)** is always *true*. Explanation: The (i, j) entry of $A + B$ is $a_{ij} + b_{ij}$ and the (i, j) entry of $B + A$ is $b_{ij} + a_{ij}$. Since $a_{ij} + b_{ij} = b_{ij} + a_{ij}$ for each i and j, it follows that $A + B = B + A$.

The answer for statement **(ii)** is *false*. Although the statement may be true in some cases, it is not always true. To show this, we need only exhibit one instance where equality fails to hold. Thus, for example, if

$$A = \begin{bmatrix} 1 & 2 \\ 3 & 1 \end{bmatrix} \quad \text{and} \quad B = \begin{bmatrix} 2 & 3 \\ 1 & 1 \end{bmatrix}$$

then

$$AB = \begin{bmatrix} 4 & 5 \\ 7 & 10 \end{bmatrix} \quad \text{and} \quad BA = \begin{bmatrix} 11 & 7 \\ 4 & 3 \end{bmatrix}$$

This proves that statement **(ii)** is false.

1. If the row echelon form of A involves free variables, then the system $A\mathbf{x} = \mathbf{b}$ will have infinitely many solutions.

2. A homogeneous linear system is always consistent.

3. An $n \times n$ matrix A is nonsingular if and only if the reduced row echelon form of A is I (the identity matrix).

4. If A and B are nonsingular $n \times n$ matrices, then $A + B$ is also nonsingular and $(A + B)^{-1} = A^{-1} + B^{-1}$.

5. If A and B are nonsingular $n \times n$ matrices, then AB is also nonsingular and $(AB)^{-1} = A^{-1}B^{-1}$.

6. If A and B are $n \times n$ matrices, then $(A - B)^2 = A^2 - 2AB + B^2$.

7. If $AB = AC$ and $A \neq O$ (the zero matrix), then $B = C$.

8. The product of two elementary matrices is an elementary matrix.

9. If $\mathbf{x}$ and $\mathbf{y}$ are nonzero vectors in R^n and $A = \mathbf{x}\mathbf{y}^T$, then the row echelon form of A will have exactly one nonzero row.

10. Let A be an 4×3 matrix with $\mathbf{a}_2 = \mathbf{a}_3$. If $\mathbf{b} = \mathbf{a}_1 + \mathbf{a}_2 + \mathbf{a}_3$, then the system $A\mathbf{x} = \mathbf{b}$ will have infinitely many solutions.

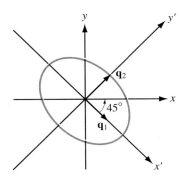

CHAPTER

2

DETERMINANTS

With each square matrix it is possible to associate a real number called the determinant of the matrix. The value of this number will tell us whether the matrix is singular.

In Section 1 the definition of the determinant of a matrix is given. In Section 2 we study properties of determinants and derive an elimination method for evaluating determinants. The elimination method is generally the simplest method to use for evaluating the determinant of an $n \times n$ matrix when $n > 3$. In Section 3 we see how determinants can be applied to solving $n \times n$ linear systems and how they can be used to calculate the inverse of a matrix. An application involving cryptography is also presented in Section 3. Further applications of determinants are presented in Chapters 3 and 6.

1 THE DETERMINANT OF A MATRIX

With each $n \times n$ matrix A it is possible to associate a scalar, $\det(A)$, whose value will tell us whether the matrix is nonsingular. Before proceeding to the general definition, let us consider the following cases.

CASE 1 1 × 1 Matrices

If $A = (a)$ is a 1×1 matrix, then A will have a multiplicative inverse if and only if $a \neq 0$. Thus, if we define

$$\det(A) = a$$

then A will be nonsingular if and only if $\det(A) \neq 0$.

CASE 2 2 × 2 Matrices

Let

$$A = \begin{bmatrix} a_{11} & a_{12} \\ a_{21} & a_{22} \end{bmatrix}$$

By Theorem 1.4.3, A will be nonsingular if and only if it is row equivalent to I. Then, if $a_{11} \neq 0$, we can test whether A is row equivalent to I by performing the following operations:

1. Multiply the second row of A by a_{11}

$$\begin{bmatrix} a_{11} & a_{12} \\ a_{11}a_{21} & a_{11}a_{22} \end{bmatrix}$$

2. Subtract a_{21} times the first row from the new second row

$$\begin{bmatrix} a_{11} & a_{12} \\ 0 & a_{11}a_{22} - a_{21}a_{12} \end{bmatrix}$$

Since $a_{11} \neq 0$, the resulting matrix will be row equivalent to I if and only if

(1) $$a_{11}a_{22} - a_{21}a_{12} \neq 0$$

If $a_{11} = 0$, we can switch the two rows of A. The resulting matrix

$$\begin{bmatrix} a_{21} & a_{22} \\ 0 & a_{12} \end{bmatrix}$$

will be row equivalent to I if and only if $a_{21}a_{12} \neq 0$. This requirement is equivalent to condition (1) when $a_{11} = 0$. Thus, if A is any 2 × 2 matrix and we define

$$\det(A) = a_{11}a_{22} - a_{12}a_{21}$$

then A is nonsingular if and only if $\det(A) \neq 0$.

Notation We can refer to the determinant of a specific matrix by enclosing the array between vertical lines. For example, if

$$A = \begin{bmatrix} 3 & 4 \\ 2 & 1 \end{bmatrix}$$

then

$$\begin{vmatrix} 3 & 4 \\ 2 & 1 \end{vmatrix}$$

represents the determinant of A.

CASE 3 3×3 Matrices

We can test whether a 3×3 matrix is nonsingular by performing row operations to see if the matrix is row equivalent to the identity matrix I. To carry out the elimination in the first column of an arbitrary 3×3 matrix A, let us first assume that $a_{11} \neq 0$. The elimination can then be performed by subtracting a_{21}/a_{11} times the first row from the second and a_{31}/a_{11} times the first row from the third.

$$
\begin{pmatrix} a_{11} & a_{12} & a_{13} \\ a_{21} & a_{22} & a_{23} \\ a_{31} & a_{32} & a_{33} \end{pmatrix} \rightarrow
\begin{pmatrix} a_{11} & a_{12} & a_{13} \\ 0 & \dfrac{a_{11}a_{22} - a_{21}a_{12}}{a_{11}} & \dfrac{a_{11}a_{23} - a_{21}a_{13}}{a_{11}} \\ 0 & \dfrac{a_{11}a_{32} - a_{31}a_{12}}{a_{11}} & \dfrac{a_{11}a_{33} - a_{31}a_{13}}{a_{11}} \end{pmatrix}
$$

The matrix on the right will be row equivalent to I if and only if

$$
a_{11} \begin{vmatrix} \dfrac{a_{11}a_{22} - a_{21}a_{12}}{a_{11}} & \dfrac{a_{11}a_{23} - a_{21}a_{13}}{a_{11}} \\ \dfrac{a_{11}a_{32} - a_{31}a_{12}}{a_{11}} & \dfrac{a_{11}a_{33} - a_{31}a_{13}}{a_{11}} \end{vmatrix} \neq 0
$$

Although the algebra is somewhat messy, this condition can be simplified to

(2) $\qquad a_{11}a_{22}a_{33} - a_{11}a_{32}a_{23} - a_{12}a_{21}a_{33} + a_{12}a_{31}a_{23}$

$$
+ a_{13}a_{21}a_{32} - a_{13}a_{31}a_{22} \neq 0
$$

Thus, if we define

(3) $\qquad \det(A) = a_{11}a_{22}a_{33} - a_{11}a_{32}a_{23} - a_{12}a_{21}a_{33}$

$$
+ a_{12}a_{31}a_{23} + a_{13}a_{21}a_{32} - a_{13}a_{31}a_{22}
$$

then for the case $a_{11} \neq 0$ the matrix will be nonsingular if and only if $\det(A) \neq 0$.
What if $a_{11} = 0$? Consider the following possibilities:

(i) $a_{11} = 0,\ a_{21} \neq 0$
(ii) $a_{11} = a_{21} = 0,\ a_{31} \neq 0$
(iii) $a_{11} = a_{21} = a_{31} = 0$

In case (i), it is not difficult to show that A is row equivalent to I if and only if

$$
-a_{12}a_{21}a_{33} + a_{12}a_{31}a_{23} + a_{13}a_{21}a_{32} - a_{13}a_{31}a_{22} \neq 0
$$

But this condition is the same as condition (2) with $a_{11} = 0$. The details of case (i) are left as an exercise for the reader (see Exercise 7).
In case (ii) it follows that

$$
A = \begin{pmatrix} 0 & a_{12} & a_{13} \\ 0 & a_{22} & a_{23} \\ a_{31} & a_{32} & a_{33} \end{pmatrix}
$$

is row equivalent to I if and only if

$$a_{31}(a_{12}a_{23} - a_{22}a_{13}) \neq 0$$

Again this is a special case of condition (2) with $a_{11} = a_{21} = 0$.

Clearly, in case (iii) the matrix A cannot be row equivalent to I and hence must be singular. In this case, if we set a_{11}, a_{21}, and a_{31} equal to 0 in formula (3), the result will be $\det(A) = 0$.

In general, then, formula (2) gives a necessary and sufficient condition for a 3×3 matrix A to be nonsingular (regardless of the value of a_{11}).

We would now like to define the determinant of an $n \times n$ matrix. To see how to do this, note that the determinant of a 2×2 matrix

$$A = \begin{bmatrix} a_{11} & a_{12} \\ a_{21} & a_{22} \end{bmatrix}$$

can be defined in terms of the two 1×1 matrices

$$M_{11} = (a_{22}) \quad \text{and} \quad M_{12} = (a_{21})$$

The matrix M_{11} is formed from A by deleting its first row and first column, and M_{12} is formed from A by deleting its first row and second column.

The determinant of A can be expressed in the form

(4) $$\det(A) = a_{11}a_{22} - a_{12}a_{21} = a_{11}\det(M_{11}) - a_{12}\det(M_{12})$$

For a 3×3 matrix A, we can rewrite equation (3) in the form

$$\det(A) = a_{11}(a_{22}a_{33} - a_{32}a_{23}) - a_{12}(a_{21}a_{33} - a_{31}a_{23}) + a_{13}(a_{21}a_{32} - a_{31}a_{22})$$

For $j = 1, 2, 3$, let M_{1j} denote the 2×2 matrix formed from A by deleting its first row and jth column. The determinant of A can then be represented in the form

(5) $$\det(A) = a_{11}\det(M_{11}) - a_{12}\det(M_{12}) + a_{13}\det(M_{13})$$

where

$$M_{11} = \begin{bmatrix} a_{22} & a_{23} \\ a_{32} & a_{33} \end{bmatrix}, \quad M_{12} = \begin{bmatrix} a_{21} & a_{23} \\ a_{31} & a_{33} \end{bmatrix}, \quad M_{13} = \begin{bmatrix} a_{21} & a_{22} \\ a_{31} & a_{32} \end{bmatrix}$$

To see how to generalize (4) and (5) to the case $n > 3$, we introduce the following definition.

▶ **DEFINITION** Let $A = (a_{ij})$ be an $n \times n$ matrix and let M_{ij} denote the $(n-1) \times (n-1)$ matrix obtained from A by deleting the row and column containing a_{ij}. The determinant of M_{ij} is called the **minor** of a_{ij}. We define the **cofactor** A_{ij} of a_{ij} by

$$A_{ij} = (-1)^{i+j}\det(M_{ij})$$ ◀

In view of this definition, for a 2×2 matrix A, we may rewrite equation (4) in the form

(6) $$\det(A) = a_{11}A_{11} + a_{12}A_{12} \qquad (n = 2)$$

Equation (6) is called the *cofactor expansion* of $\det(A)$ along the first row of A. Note that we could also write

(7) $$\det(A) = a_{21}(-a_{12}) + a_{22}a_{11} = a_{21}A_{21} + a_{22}A_{22}$$

Equation (7) expresses $\det(A)$ in terms of the entries of the second row of A and their cofactors. Actually, there is no reason why we must expand along a row of the matrix; the determinant could just as well be represented by the cofactor expansion along one of the columns.

$$\det(A) = a_{11}a_{22} + a_{21}(-a_{12})$$
$$= a_{11}A_{11} + a_{21}A_{21} \qquad \text{(first column)}$$
$$\det(A) = a_{12}(-a_{21}) + a_{22}a_{11}$$
$$= a_{12}A_{12} + a_{22}A_{22} \qquad \text{(second column)}$$

For a 3×3 matrix A, we have

(8) $$\det(A) = a_{11}A_{11} + a_{12}A_{12} + a_{13}A_{13}$$

Thus the determinant of a 3×3 matrix can be defined in terms of the elements in the first row of the matrix and their corresponding cofactors.

EXAMPLE I.   If

$$A = \begin{bmatrix} 2 & 5 & 4 \\ 3 & 1 & 2 \\ 5 & 4 & 6 \end{bmatrix}$$

then

$$\det(A) = a_{11}A_{11} + a_{12}A_{12} + a_{13}A_{13}$$
$$= (-1)^2 a_{11}\det(M_{11}) + (-1)^3 a_{12}\det(M_{12}) + (-1)^4 a_{13}\det(M_{13})$$
$$= 2\begin{vmatrix} 1 & 2 \\ 4 & 6 \end{vmatrix} - 5\begin{vmatrix} 3 & 2 \\ 5 & 6 \end{vmatrix} + 4\begin{vmatrix} 3 & 1 \\ 5 & 4 \end{vmatrix}$$
$$= 2(6 - 8) - 5(18 - 10) + 4(12 - 5)$$
$$= -16 \qquad \blacktriangleleft$$

As in the case of 2×2 matrices, the determinant of a 3×3 matrix can be represented as a cofactor expansion using any row or column. For example, equation (3) can be rewritten in the form

$$\det(A) = a_{12}a_{31}a_{23} - a_{13}a_{31}a_{22} - a_{11}a_{32}a_{23} + a_{13}a_{21}a_{32} + a_{11}a_{22}a_{33} - a_{12}a_{21}a_{33}$$
$$= a_{31}(a_{12}a_{23} - a_{31}a_{22}) - a_{32}(a_{11}a_{23} - a_{13}a_{21}) + a_{33}(a_{11}a_{22} - a_{12}a_{21})$$
$$= a_{31}A_{31} + a_{32}A_{32} + a_{33}A_{33}$$

This is the cofactor expansion along the third row of A.

EXAMPLE 2. Let A be the matrix in Example 1. The cofactor expansion of $\det(A)$ along the second column is given by

$$\det(A) = -5 \begin{vmatrix} 3 & 2 \\ 5 & 6 \end{vmatrix} + 1 \begin{vmatrix} 2 & 4 \\ 5 & 6 \end{vmatrix} - 4 \begin{vmatrix} 2 & 4 \\ 3 & 2 \end{vmatrix}$$
$$= -5(18 - 10) + 1(12 - 20) - 4(4 - 12) = -16 \qquad \blacktriangleleft$$

The determinant of a 4×4 matrix can be defined in terms of a cofactor expansion along any row or column. To compute the value of the 4×4 determinant, we would have to evaluate four 3×3 determinants.

▶ **DEFINITION** The **determinant** of an $n \times n$ matrix A, denoted $\det(A)$, is a scalar associated with the matrix A that is defined inductively as follows:

$$\det(A) = \begin{cases} a_{11} & \text{if } n = 1 \\ a_{11}A_{11} + a_{12}A_{12} + \cdots + a_{1n}A_{1n} & \text{if } n > 1 \end{cases}$$

where

$$A_{1j} = (-1)^{1+j} \det(M_{1j}) \qquad j = 1, \dots, n$$

are the cofactors associated with the entries in the first row of A.   ◀

As we have seen, it is not necessary to limit ourselves to using the first row for the cofactor expansion. We state the following theorem without proof.

▶ **THEOREM 2.1.1** *If A is an $n \times n$ matrix with $n \geq 2$, then $\det(A)$ can be expressed as a cofactor expansion using any row or column of A.*

$$\det(A) = a_{i1}A_{i1} + a_{i2}A_{i2} + \cdots + a_{in}A_{in}$$
$$= a_{1j}A_{1j} + a_{2j}A_{2j} + \cdots + a_{nj}A_{nj}$$

for $i = 1, \dots, n$ and $j = 1, \dots, n$.   ◀

The cofactor expansion of a 4×4 determinant will involve four 3×3 determinants. We can often save work by expanding along the row or column that contains the most zeros. For example, to evaluate

$$\begin{vmatrix} 0 & 2 & 3 & 0 \\ 0 & 4 & 5 & 0 \\ 0 & 1 & 0 & 3 \\ 2 & 0 & 1 & 3 \end{vmatrix}$$

we would expand down the first column. The first three terms will drop out, leaving

$$-2 \begin{vmatrix} 2 & 3 & 0 \\ 4 & 5 & 0 \\ 1 & 0 & 3 \end{vmatrix} = -2 \cdot 3 \cdot \begin{vmatrix} 2 & 3 \\ 4 & 5 \end{vmatrix} = 12$$

The cofactor expansion can be used to establish some important results about determinants. These results are given in the following theorems.

▶ **THEOREM 2.1.2** *If A is an $n \times n$ matrix, then $\det(A^T) = \det(A)$.*

▶*Proof.* The proof is by induction on n. Clearly, the result holds if $n = 1$, since a 1×1 matrix is necessarily symmetric. Assume that the result holds for all $k \times k$ matrices and that A is a $(k + 1) \times (k + 1)$ matrix. Expanding $\det(A)$ along the first row of A, we get

$$\det(A) = a_{11} \det(M_{11}) - a_{12} \det(M_{12}) + - \cdots \pm a_{1,k+1} \det(M_{1,k+1})$$

Since the M_{ij}'s are all $k \times k$ matrices, it follows from the induction hypothesis that

$$(9) \qquad \det(A) = a_{11} \det(M_{11}^T) - a_{12} \det(M_{12}^T) + - \cdots \pm a_{1,k+1} \det(M_{1,k+1}^T)$$

The right-hand side of (9) is just the expansion by minors of $\det(A^T)$ using the first column of A^T. Therefore,

$$\det(A^T) = \det(A)$$ ◀

▶ **THEOREM 2.1.3** *If A is an $n \times n$ triangular matrix, the determinant of A equals the product of the diagonal elements of A.*

▶*Proof.* In view of Theorem 2.1.2, it suffices to prove the theorem for lower triangular matrices. The result follows easily using the cofactor expansion and induction on n. The details of this are left for the reader (see Exercise 8). ◀

▶ **THEOREM 2.1.4** *Let A be an $n \times n$ matrix.*

(i) *If A has a row or column consisting entirely of zeros, then $\det(A) = 0$.*
(ii) *If A has two identical rows or two identical columns, then $\det(A) = 0$.* ◀

Both of these results can be easily proved using the cofactor expansion. The proofs are left for the reader (see Exercises 9 and 10).

In the next section we look at the effect of row operations on the value of the determinant. This will allow us to make use of Theorem 2.1.3 to derive a more efficient method for computing the value of a determinant.

EXERCISES

1. Given

$$A = \begin{bmatrix} 3 & 2 & 4 \\ 1 & -2 & 3 \\ 2 & 3 & 2 \end{bmatrix}$$

(a) Find the values of det(M_{21}), det(M_{22}), and det(M_{23}).

(b) Find the values of A_{21}, A_{22}, and A_{23}.

(c) Use your answers from part (b) to compute det(A).

2. Use determinants to determine whether the following 2×2 matrices are nonsingular.

(a) $\begin{bmatrix} 3 & 5 \\ 2 & 4 \end{bmatrix}$ (b) $\begin{bmatrix} 3 & 6 \\ 2 & 4 \end{bmatrix}$ (c) $\begin{bmatrix} 3 & -6 \\ 2 & 4 \end{bmatrix}$

3. Evaluate the following determinants.

(a) $\begin{vmatrix} 3 & 5 \\ -2 & -3 \end{vmatrix}$ (b) $\begin{vmatrix} 5 & -2 \\ -8 & 4 \end{vmatrix}$ (c) $\begin{vmatrix} 3 & 1 & 2 \\ 2 & 4 & 5 \\ 2 & 4 & 5 \end{vmatrix}$

(d) $\begin{vmatrix} 4 & 3 & 0 \\ 3 & 1 & 2 \\ 5 & -1 & -4 \end{vmatrix}$ (e) $\begin{vmatrix} 1 & 3 & 2 \\ 4 & 1 & -2 \\ 2 & 1 & 3 \end{vmatrix}$ (f) $\begin{vmatrix} 2 & -1 & 2 \\ 1 & 3 & 2 \\ 5 & 1 & 6 \end{vmatrix}$

(g) $\begin{vmatrix} 2 & 0 & 0 & 1 \\ 0 & 1 & 0 & 0 \\ 1 & 6 & 2 & 0 \\ 1 & 1 & -2 & 3 \end{vmatrix}$ (h) $\begin{vmatrix} 2 & 1 & 2 & 1 \\ 3 & 0 & 1 & 1 \\ -1 & 2 & -2 & 1 \\ -3 & 2 & 3 & 1 \end{vmatrix}$

4. Evaluate the following determinants by inspection.

(a) $\begin{vmatrix} 3 & 5 \\ 2 & 4 \end{vmatrix}$ (b) $\begin{vmatrix} 2 & 0 & 0 \\ 4 & 1 & 0 \\ 7 & 3 & -2 \end{vmatrix}$ (c) $\begin{vmatrix} 3 & 0 & 0 \\ 2 & 1 & 1 \\ 1 & 2 & 2 \end{vmatrix}$ (d) $\begin{vmatrix} 4 & 0 & 2 & 1 \\ 5 & 0 & 4 & 2 \\ 2 & 0 & 3 & 4 \\ 1 & 0 & 2 & 3 \end{vmatrix}$

5. Evaluate the following determinant. Write your answer as a polynomial in x.

$$\begin{vmatrix} a-x & b & c \\ 1 & -x & 0 \\ 0 & 1 & -x \end{vmatrix}$$

6. Find all values of λ for which the following determinant will equal 0.

$$\begin{vmatrix} 2-\lambda & 4 \\ 3 & 3-\lambda \end{vmatrix}$$

7. Let A be a 3×3 matrix with $a_{11} = 0$ and $a_{21} \neq 0$. Show that A is row equivalent to I if and only if

$$-a_{12}a_{21}a_{33} + a_{12}a_{31}a_{23} + a_{13}a_{21}a_{32} - a_{13}a_{31}a_{22} \neq 0$$

8. Write out the details of the proof of Theorem 2.1.3.

9. Prove that if a row or a column of an $n \times n$ matrix A consists entirely of zeros then $\det(A) = 0$.

10. Use mathematical induction to prove that if A is an $(n + 1) \times (n + 1)$ matrix with two identical rows then $\det(A) = 0$.

11. Let A and B be 2×2 matrices.

   (a) Does $\det(A + B) = \det(A) + \det(B)$?
   (b) Does $\det(AB) = \det(A)\det(B)$?
   (c) Does $\det(AB) = \det(BA)$?

   Justify your answers.

12. Let A and B be 2×2 matrices and let

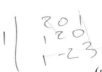

$$C = \begin{bmatrix} a_{11} & a_{12} \\ b_{21} & b_{22} \end{bmatrix}, \quad D = \begin{bmatrix} b_{11} & b_{12} \\ a_{21} & a_{22} \end{bmatrix}, \quad E = \begin{bmatrix} 0 & \alpha \\ \beta & 0 \end{bmatrix}$$

   (a) Show that $\det(A + B) = \det(A) + \det(B) + \det(C) + \det(D)$.
   (b) Show that if $B = EA$ then $\det(A + B) = \det(A) + \det(B)$.

13. Let A be a symmetric tridiagonal matrix (i.e., A is symmetric and $a_{ij} = 0$ whenever $|i - j| > 1$). Let B be the matrix formed from A by deleting the first two rows and columns. Show that

$$\det(A) = a_{11}\det(M_{11}) - a_{12}^2\det(B)$$

2 PROPERTIES OF DETERMINANTS

In this section we consider the effects of row operations on the determinant of a matrix. Once these effects have been established, we will prove that a matrix A is singular if and only if its determinant is zero, and we will develop a method for evaluating determinants using row operations. Also, we will establish an important theorem about the determinant of the product of two matrices. We begin with the following lemma.

▶ **LEMMA 2.2.1** *Let A be an $n \times n$ matrix. If A_{jk} denotes the cofactor of a_{jk} for $k = 1, \ldots, n$, then*

(1) $$a_{i1}A_{j1} + a_{i2}A_{j2} + \cdots + a_{in}A_{jn} = \begin{cases} \det(A) & \text{if } i = j \\ 0 & \text{if } i \neq j \end{cases}$$

▶ *Proof.* If $i = j$, (1) is just the cofactor expansion of $\det(A)$ along the ith row of A. To prove (1) in the case $i \neq j$, let A^* be the matrix obtained by replacing the jth row of A by the ith row of A.

$$A^* = \begin{bmatrix} a_{11} & a_{12} & \cdots & a_{1n} \\ & \vdots & & \\ a_{i1} & a_{i2} & \cdots & a_{in} \\ & \vdots & & \\ a_{i1} & a_{i2} & \cdots & a_{in} \\ & \vdots & & \\ a_{n1} & a_{n2} & \cdots & a_{nn} \end{bmatrix} \quad j\text{th row}$$

Since two rows of A^* are the same, its determinant must be zero. It follows from the cofactor expansion of $\det(A^*)$ along the jth row that

$$0 = \det(A^*) = a_{i1}A^*_{j1} + a_{i2}A^*_{j2} + \cdots + a_{in}A^*_{jn}$$
$$= a_{i1}A_{j1} + a_{i2}A_{j2} + \cdots + a_{in}A_{jn} \qquad \blacktriangleleft$$

Let us now consider the effects of each of the three row operations on the value of the determinant.

Row Operation I

Two rows of A are interchanged.
If A is a 2×2 matrix and

$$E = \begin{bmatrix} 0 & 1 \\ 1 & 0 \end{bmatrix}$$

then

$$\det(EA) = \begin{vmatrix} a_{21} & a_{22} \\ a_{11} & a_{12} \end{vmatrix} = a_{21}a_{12} - a_{22}a_{11} = -\det(A)$$

For $n > 2$, let E_{ij} be the elementary matrix that switches rows i and j of A. It is a simple induction proof to show that $\det(E_{ij}A) = -\det(A)$. We illustrate the idea behind the proof for the case $n = 3$. Suppose that the first and third rows of a 3×3 matrix A have been interchanged. Expanding $\det(E_{13}A)$ along the second row and making use of the result for 2×2 matrices, we see that

$$\det(E_{13}A) = \begin{vmatrix} a_{31} & a_{32} & a_{33} \\ a_{21} & a_{22} & a_{23} \\ a_{11} & a_{12} & a_{13} \end{vmatrix}$$

$$= -a_{21}\begin{vmatrix} a_{32} & a_{33} \\ a_{12} & a_{13} \end{vmatrix} + a_{22}\begin{vmatrix} a_{31} & a_{33} \\ a_{11} & a_{13} \end{vmatrix} - a_{23}\begin{vmatrix} a_{31} & a_{31} \\ a_{11} & a_{12} \end{vmatrix}$$

$$= a_{21}\begin{vmatrix} a_{12} & a_{13} \\ a_{32} & a_{33} \end{vmatrix} - a_{22}\begin{vmatrix} a_{11} & a_{13} \\ a_{31} & a_{33} \end{vmatrix} + a_{23}\begin{vmatrix} a_{11} & a_{12} \\ a_{31} & a_{32} \end{vmatrix}$$

$$= -\det(A)$$

In general, if A is an $n \times n$ matrix and E_{ij} is the $n \times n$ elementary matrix formed by interchanging the ith and jth rows of I, then

$$\det(E_{ij}A) = -\det(A)$$

In particular,

$$\det(E_{ij}) = \det(E_{ij}I) = -\det(I) = -1$$

Thus, for any elementary matrix E of type I,

$$\det(EA) = -\det(A) = \det(E)\det(A)$$

Row Operation II

A row of A is multiplied by a nonzero constant.

Let E denote the elementary matrix of type II formed from I by multiplying the ith row by the nonzero constant α. If $\det(EA)$ is expanded by cofactors along the ith row, then

$$\det(EA) = \alpha a_{i1} A_{i1} + \alpha a_{i2} A_{i2} + \cdots + \alpha a_{in} A_{in}$$
$$= \alpha(a_{i1} A_{i1} + a_{i2} A_{i2} + \cdots + a_{in} A_{in})$$
$$= \alpha \det(A)$$

In particular,

$$\det(E) = \det(EI) = \alpha \det(I) = \alpha$$

and hence

$$\det(EA) = \alpha \det(A) = \det(E)\det(A)$$

Row Operation III

A multiple of one row is added to another row.

Let E be the elementary matrix of type III formed from I by adding c times the ith row to the jth row. Since E is triangular and its diagonal elements are all 1, it follows that $\det(E) = 1$. We will show that

$$\det(EA) = \det(A) = \det(E)\det(A)$$

If $\det(EA)$ is expanded by cofactors along the jth row, it follows from Lemma 2.2.1 that

$$\det(EA) = (a_{j1} + ca_{i1})A_{j1} + (a_{j2} + ca_{i2})A_{j2} + \cdots + (a_{jn} + ca_{in})A_{jn}$$
$$= (a_{j1} A_{j1} + \cdots + a_{jn} A_{jn}) + c(a_{i1} A_{j1} + \cdots + a_{in} A_{jn})$$
$$= \det(A)$$

Thus,

$$\det(EA) = \det(A) = \det(E)\det(A)$$

Summary

In summation, if E is an elementary matrix, then

$$\det(EA) = \det(E)\det(A)$$

where

(2)
$$\det(E) = \begin{cases} -1 & \text{if } E \text{ is of type I} \\ \alpha \neq 0 & \text{if } E \text{ is of type II} \\ 1 & \text{if } E \text{ is of type III} \end{cases}$$

Similar results hold for column operations. Indeed, if E is an elementary matrix, then

$$\det(AE) = \det\left((AE)^T\right) = \det\left(E^T A^T\right)$$
$$= \det\left(E^T\right)\det\left(A^T\right) = \det(E)\det(A)$$

Thus, the effects that row or column operations have on the value of the determinant can be summarized as follows:

  I. Interchanging two rows (or columns) of a matrix changes the sign of the determinant.
  II. Multiplying a single row or column of a matrix by a scalar has the effect of multiplying the value of the determinant by that scalar.
  III. Adding a multiple of one row (or column) to another does not change the value of the determinant.

NOTE As a consequence of III, if one row (or column) of a matrix is a multiple of another, the determinant of the matrix must equal zero.

Main Results

We can now make use of the effects of row operations on determinants to prove two major theorems and to establish a simpler method of computing determinants. It follows from (2) that all elementary matrices have nonzero determinants. This observation can be used to prove the following theorem.

▶ **THEOREM 2.2.2** *An $n \times n$ matrix A is singular if and only if*

$$\det(A) = 0$$

▶ *Proof.* The matrix A can be reduced to row echelon form with a finite number of row operations. Thus

$$U = E_k E_{k-1} \cdots E_1 A$$

where U is in row echelon form and the E_i's are all elementary matrices.

$$\det(U) = \det(E_k E_{k-1} \cdots E_1 A)$$
$$= \det(E_k)\det(E_{k-1}) \cdots \det(E_1)\det(A)$$

Since the determinants of the E_i's are all nonzero, it follows that $\det(A) = 0$ if and only if $\det(U) = 0$. If A is singular, then U has a row consisting entirely of zeros and hence $\det(U) = 0$. If A is nonsingular, U is triangular with 1's along the diagonal and hence $\det(U) = 1$. ◀

From the proof of Theorem 2.2.2 we can obtain a method for computing $\det(A)$. Reduce A to row echelon form.

$$U = E_k E_{k-1} \cdots E_1 A$$

If the last row of U consists entirely of zeros, A is singular and $\det(A) = 0$. Otherwise, A is nonsingular and

$$\det(A) = \left[\det(E_k)\det(E_{k-1}) \cdots \det(E_1)\right]^{-1}$$

Actually, if A is nonsingular, it is simpler to reduce A to triangular form. This can be done using only row operations I and III. Thus

$$T = E_m E_{m-1} \cdots E_1 A$$

and hence

$$\det(A) = \pm\det(T) = \pm t_{11}t_{22} \cdots t_{nn}$$

where the t_{ii}'s are the diagonal entries of T. The sign will be positive if row operation I has been used an even number of times and negative otherwise.

EXAMPLE 1. Evaluate

$$\begin{vmatrix} 2 & 1 & 3 \\ 4 & 2 & 1 \\ 6 & -3 & 4 \end{vmatrix}$$

SOLUTION.

$$\begin{vmatrix} 2 & 1 & 3 \\ 4 & 2 & 1 \\ 6 & -3 & 4 \end{vmatrix} = \begin{vmatrix} 2 & 1 & 3 \\ 0 & 0 & -5 \\ 0 & -6 & -5 \end{vmatrix} = (-1)\begin{vmatrix} 2 & 1 & 3 \\ 0 & -6 & -5 \\ 0 & 0 & -5 \end{vmatrix}$$

$$= (-1)(2)(-6)(-5)$$

$$= -60 \qquad ◀$$

We now have two methods for evaluating the determinant of an $n \times n$ matrix A. If $n > 3$ and A has nonzero entries, elimination is the most efficient method in the sense that it involves fewer arithmetic operations. In Table 1 the number of arithmetic operations involved in each method is given for $n = 2, 3, 4, 5, 10$. It is not difficult to derive general formulas for the number of operations in each of the methods (see Exercises 17 and 18).

We have seen that, for any elementary matrix E,

$$\det(EA) = \det(E)\det(A) = \det(AE)$$

This is a special case of the following theorem.

	Cofactors		Elimination	
n	Additions	Multiplications	Additions	Multiplications and Divisions
2	1	2	1	3
3	5	9	5	10
4	23	40	14	23
5	119	205	30	45
10	3,628,799	6,235,300	285	339

TABLE 1

▶ **THEOREM 2.2.3** *If A and B are n × n matrices, then*
$$\det(AB) = \det(A)\det(B)$$

▶*Proof.* If B is singular, it follows from Theorem 1.4.3 that AB is also singular (see Exercise 14 of Chapter 1, Section 4), and therefore
$$\det(AB) = 0 = \det(A)\det(B)$$
If B is nonsingular, B can be written as a product of elementary matrices. We have already seen that the result holds for elementary matrices. Thus

$$\begin{aligned}
\det(AB) &= \det(AE_k E_{k-1} \cdots E_1) \\
&= \det(A)\det(E_k)\det(E_{k-1}) \cdots \det(E_1) \\
&= \det(A)\det(E_k E_{k-1} \cdots E_1) \\
&= \det(A)\det(B) \qquad\qquad\qquad\blacktriangleleft
\end{aligned}$$

If A is singular, the computed value of $\det(A)$ using exact arithmetic must be 0. However, this result is unlikely if the computations are done by computer. Since computers use a finite number system, roundoff errors are usually unavoidable. Consequently, it is more likely that the computed value of $\det(A)$ will only be near 0. Because of roundoff errors, it is virtually impossible to determine computationally whether a matrix is exactly singular. In computer applications it is often more meaningful to ask whether a matrix is "close" to being singular. In general, the value of $\det(A)$ is not a good indicator of nearness to singularity. In Section 5 of Chapter 6 we will discuss how to determine whether a matrix is close to being singular.

EXERCISES

1. Evaluate each of the following determinants by inspection.

(a) $\begin{vmatrix} 0 & 0 & 3 \\ 0 & 4 & 1 \\ 2 & 3 & 1 \end{vmatrix}$

(b) $\begin{vmatrix} 1 & 1 & 1 & 3 \\ 0 & 3 & 1 & 1 \\ 0 & 0 & 2 & 2 \\ -1 & -1 & -1 & 2 \end{vmatrix}$

(c) $\begin{vmatrix} 0 & 0 & 0 & 1 \\ 1 & 0 & 0 & 0 \\ 0 & 1 & 0 & 0 \\ 0 & 0 & 1 & 0 \end{vmatrix}$

2. Let

$$A = \begin{pmatrix} 0 & 1 & 2 & 3 \\ 1 & 1 & 1 & 1 \\ -2 & -2 & 3 & 3 \\ 1 & 2 & -2 & -3 \end{pmatrix}$$

(a) Use the elimination method to evaluate $\det(A)$.

(b) Use the value of $\det(A)$ to evaluate

$$\begin{vmatrix} 0 & 1 & 2 & 3 \\ -2 & -2 & 3 & 3 \\ 1 & 2 & -2 & -3 \\ 1 & 1 & 1 & 1 \end{vmatrix} + \begin{vmatrix} 0 & 1 & 2 & 3 \\ 1 & 1 & 1 & 1 \\ -1 & -1 & 4 & 4 \\ 2 & 3 & -1 & -2 \end{vmatrix}$$

3. For each of the following, compute the determinant and state whether the matrix is singular or nonsingular.

(a) $\begin{bmatrix} 3 & 1 \\ 6 & 2 \end{bmatrix}$ (b) $\begin{bmatrix} 3 & 1 \\ 4 & 2 \end{bmatrix}$ (c) $\begin{bmatrix} 3 & 3 & 1 \\ 0 & 1 & 2 \\ 0 & 2 & 3 \end{bmatrix}$

(d) $\begin{bmatrix} 2 & 1 & 1 \\ 4 & 3 & 5 \\ 2 & 1 & 2 \end{bmatrix}$ (e) $\begin{bmatrix} 2 & -1 & 3 \\ -1 & 2 & -2 \\ 1 & 4 & 0 \end{bmatrix}$ (f) $\begin{bmatrix} 1 & 1 & 1 & 1 \\ 2 & -1 & 3 & 2 \\ 0 & 1 & 2 & 1 \\ 0 & 0 & 7 & 3 \end{bmatrix}$

4. Find all possible choices of c that would make the following matrix singular.

$$\begin{bmatrix} 1 & 1 & 1 \\ 1 & 9 & c \\ 1 & c & 3 \end{bmatrix}$$

5. Let A be an $n \times n$ matrix and α a scalar. Show that

$$\det(\alpha A) = \alpha^n \det(A)$$

6. Let A be a nonsingular matrix. Show that

$$\det(A^{-1}) = \frac{1}{\det(A)}$$

7. Let A and B be 3×3 matrices with $\det(A) = 4$ and $\det(B) = 5$. Find the value of:

(a) $\det(AB)$ (b) $\det(3A)$ (c) $\det(2AB)$ (d) $\det(A^{-1}B)$

8. Let E_1, E_2, E_3 be 3×3 elementary matrices of types I, II, and III, respectively, and let A be a 3×3 matrix with $\det(A) = 6$. Assume, additionally, that E_2 was formed from I by multiplying its second row by 3. Find the values of each of the following.

   (a) $\det(E_1 A)$ (b) $\det(E_2 A)$ (c) $\det(E_3 A)$

   (d) $\det(A E_1)$ (e) $\det(E_1^2)$ (f) $\det(E_1 E_2 E_3)$

9. Let A and B be row equivalent matrices and suppose that B can be obtained from A using only row operations I and III. How do the values of $\det(A)$ and $\det(B)$ compare? How will the values compare if B can be obtained from A using only row operation III? Explain your answers.

10. Consider the 3×3 Vandermonde matrix

$$V = \begin{bmatrix} 1 & x_1 & x_1^2 \\ 1 & x_2 & x_2^2 \\ 1 & x_3 & x_3^2 \end{bmatrix}$$

   (a) Show that $\det(V) = (x_2 - x_1)(x_3 - x_1)(x_3 - x_2)$. [*Hint:* Make use of row operation III.]

   (b) What conditions must the scalars x_1, x_2, x_3 satisfy in order for V to be nonsingular?

11. Suppose that a 3×3 matrix A factors into a product

$$\begin{bmatrix} 1 & 0 & 0 \\ l_{21} & 1 & 0 \\ l_{31} & l_{32} & 1 \end{bmatrix} \begin{bmatrix} u_{11} & u_{12} & u_{13} \\ 0 & u_{22} & u_{23} \\ 0 & 0 & u_{33} \end{bmatrix}$$

   Determine the value of $\det(A)$.

12. Let A and B be $n \times n$ matrices. Prove that the product AB is nonsingular if and only if A and B are both nonsingular.

13. Let A and B be $n \times n$ matrices. Prove that if $AB = I$ then $BA = I$. What is the significance of this result in terms of the definition of a nonsingular matrix?

14. A matrix A is said to be *skew symmetric* if $A^T = -A$. For example,

$$A = \begin{bmatrix} 0 & 1 \\ -1 & 0 \end{bmatrix}$$

   is skew symmetric since

$$A^T = \begin{bmatrix} 0 & -1 \\ 1 & 0 \end{bmatrix} = -A$$

Show that, if A is an $n \times n$ skew symmetric matrix and n is odd, then A must be singular.

15. Let A be a nonsingular $n \times n$ matrix with a nonzero cofactor A_{nn} and set

$$c = \frac{\det(A)}{A_{nn}}$$

Show that if we subtract c from A_{nn} then the resulting matrix will be singular.

16. Let $\mathbf{x}$ and $\mathbf{y}$ be elements of R^3, and let $\mathbf{z}$ be the vector in R^3 whose coordinates are defined by

$$z_1 = \begin{vmatrix} x_2 & x_3 \\ y_2 & y_3 \end{vmatrix}, \qquad z_2 = - \begin{vmatrix} x_1 & x_3 \\ y_1 & y_3 \end{vmatrix}, \qquad z_3 = \begin{vmatrix} x_1 & x_2 \\ y_1 & y_2 \end{vmatrix}$$

Let

$$X = (\mathbf{x}, \mathbf{x}, \mathbf{y})^T \qquad \text{and} \qquad Y = (\mathbf{x}, \mathbf{y}, \mathbf{y})^T$$

Show that

$$\mathbf{x}^T \mathbf{z} = \det(X) = 0 \qquad \text{and} \qquad \mathbf{y}^T \mathbf{z} = \det(Y) = 0$$

17. Show that evaluating the determinant of an $n \times n$ matrix by cofactors involves $(n! - 1)$ additions and $\displaystyle\sum_{k=1}^{n-1} n!/k!$ multiplications.

18. Show that the elimination method of computing the value of the determinant of an $n \times n$ matrix involves $[n(n-1)(2n-1)]/6$ additions and $[(n-1)(n^2+n+3)]/3$ multiplications and divisions. [*Hint:* At the ith step of the reduction process it takes $n - i$ divisions to calculate the multiples of the ith row that are to be subtracted from the remaining rows below the pivot. We must then calculate new values for the $(n-i)^2$ entries in rows $i+1$ through n and columns $i+1$ through n.]

3 CRAMER'S RULE

In this section we learn a method for computing the inverse of a nonsingular matrix A using determinants. We also learn a method for solving $A\mathbf{x} = \mathbf{b}$ using determinants. Both methods depend on Lemma 2.2.1.

The Adjoint of a Matrix

Let A be an $n \times n$ matrix. We define a new matrix called the *adjoint* of A by

$$\text{adj } A = \begin{bmatrix} A_{11} & A_{21} & \cdots & A_{n1} \\ A_{12} & A_{22} & \cdots & A_{n2} \\ \vdots & & & \\ A_{1n} & A_{2n} & \cdots & A_{nn} \end{bmatrix}$$

Thus, to form the adjoint, we must replace each term by its cofactor and then transpose the resulting matrix. By Lemma 2.2.1

$$a_{i1}A_{j1} + a_{i2}A_{j2} + \cdots + a_{in}A_{jn} = \begin{cases} \det(A) & \text{if } i = j \\ 0 & \text{if } i \neq j \end{cases}$$

and hence it follows that

$$A(\operatorname{adj} A) = \det(A)I$$

If A is nonsingular, $\det(A)$ is a nonzero scalar, and we may write

$$A\left(\frac{1}{\det(A)}\operatorname{adj} A\right) = I$$

Thus

$$A^{-1} = \frac{1}{\det(A)}\operatorname{adj} A$$

EXAMPLE 1.   For a 2×2 matrix

$$\operatorname{adj} A = \begin{bmatrix} a_{22} & -a_{12} \\ -a_{21} & a_{11} \end{bmatrix}$$

If A is nonsingular, then

$$A^{-1} = \frac{1}{a_{11}a_{22} - a_{12}a_{21}} \begin{bmatrix} a_{22} & -a_{12} \\ -a_{21} & a_{11} \end{bmatrix}$$

◀

EXAMPLE 2.   Let

$$A = \begin{bmatrix} 2 & 1 & 2 \\ 3 & 2 & 2 \\ 1 & 2 & 3 \end{bmatrix}$$

Compute $\operatorname{adj} A$ and A^{-1}.

SOLUTION.

$$\operatorname{adj} A = \begin{bmatrix} \begin{vmatrix} 2 & 2 \\ 2 & 3 \end{vmatrix} & -\begin{vmatrix} 3 & 2 \\ 1 & 3 \end{vmatrix} & \begin{vmatrix} 3 & 2 \\ 1 & 2 \end{vmatrix} \\ -\begin{vmatrix} 1 & 2 \\ 2 & 3 \end{vmatrix} & \begin{vmatrix} 2 & 2 \\ 1 & 3 \end{vmatrix} & -\begin{vmatrix} 2 & 1 \\ 1 & 2 \end{vmatrix} \\ \begin{vmatrix} 1 & 2 \\ 2 & 2 \end{vmatrix} & -\begin{vmatrix} 2 & 2 \\ 3 & 2 \end{vmatrix} & \begin{vmatrix} 2 & 1 \\ 3 & 2 \end{vmatrix} \end{bmatrix}^T = \begin{bmatrix} 2 & 1 & -2 \\ -7 & 4 & 2 \\ 4 & -3 & 1 \end{bmatrix}$$

$$A^{-1} = \frac{1}{\det(A)} \text{ adj } A = \frac{1}{5} \begin{bmatrix} 2 & 1 & -2 \\ -7 & 4 & 2 \\ 4 & -3 & 1 \end{bmatrix}$$

◀

Using the formula

$$A^{-1} = \frac{1}{\det(A)} \text{ adj } A$$

we can derive a rule for representing the solution to the system $A\mathbf{x} = \mathbf{b}$ in terms of determinants.

▶ **THEOREM 2.3.1 (Cramer's Rule)** *Let A be an $n \times n$ nonsingular matrix, and let $\mathbf{b} \in R^n$. Let A_i be the matrix obtained by replacing the ith column of A by $\mathbf{b}$. If $\mathbf{x}$ is the unique solution to $A\mathbf{x} = \mathbf{b}$, then*

$$x_i = \frac{\det(A_i)}{\det(A)} \qquad for \quad i = 1, 2, \dots, n$$

▶ *Proof.* Since

$$\mathbf{x} = A^{-1}\mathbf{b} = \frac{1}{\det(A)}(\text{adj } A)\mathbf{b}$$

it follows that

$$x_i = \frac{b_1 A_{1i} + b_2 A_{2i} + \cdots + b_n A_{ni}}{\det(A)}$$

$$= \frac{\det(A_i)}{\det(A)}$$

◀

EXAMPLE 3. Use Cramer's rule to solve

$$x_1 + 2x_2 + x_3 = 5$$
$$2x_1 + 2x_2 + x_3 = 6$$
$$x_1 + 2x_2 + 3x_3 = 9$$

SOLUTION.

$$\det(A) = \begin{vmatrix} 1 & 2 & 1 \\ 2 & 2 & 1 \\ 1 & 2 & 3 \end{vmatrix} = -4 \qquad \det(A_1) = \begin{vmatrix} 5 & 2 & 1 \\ 6 & 2 & 1 \\ 9 & 2 & 3 \end{vmatrix} = -4$$

$$\det(A_2) = \begin{vmatrix} 1 & 5 & 1 \\ 2 & 6 & 1 \\ 1 & 9 & 3 \end{vmatrix} = -4 \qquad \det(A_3) = \begin{vmatrix} 1 & 2 & 5 \\ 2 & 2 & 6 \\ 1 & 2 & 9 \end{vmatrix} = -8$$

Therefore,

$$x_1 = \frac{-4}{-4} = 1, \qquad x_2 = \frac{-4}{-4} = 1, \qquad x_3 = \frac{-8}{-4} = 2$$

◀

Cramer's rule gives us a convenient method for writing the solution to an $n \times n$ system of linear equations in terms of determinants. To compute the solution, however, we must evaluate $n + 1$ determinants of order n. Evaluating even two of these determinants generally involves more computation than solving the system using Gaussian elimination.

APPLICATION 1: CODED MESSAGES

A common way of sending a coded message is to assign an integer value to each letter of the alphabet and to send the message as a string of integers. For example, the message

SEND MONEY

might be coded as

5, 8, 10, 21, 7, 2, 10, 8, 3

Here the S is represented by a 5, the E by an 8, and so on. Unfortunately, this type of code is generally quite easy to break. In a longer message we might be able to guess which letter is represented by a number based on the relative frequency of occurrence for that number. Thus, for example, if 8 is the most frequently occurring number in the coded message, then it is likely that it represents the letter E, the letter that occurs most frequently in the English language.

We can disguise the message further using matrix multiplications. If A is a matrix whose entries are all integers and whose determinant is ± 1, then, since $A^{-1} = \pm \operatorname{adj} A$, the entries of A^{-1} will be integers. We can use such a matrix to transform the message. The transformed message will be more difficult to decipher. To illustrate the technique, let

$$A = \begin{bmatrix} 1 & 2 & 1 \\ 2 & 5 & 3 \\ 2 & 3 & 2 \end{bmatrix}$$

The coded message is put into the columns of a matrix B having three rows.

$$B = \begin{bmatrix} 5 & 21 & 10 \\ 8 & 7 & 8 \\ 10 & 2 & 3 \end{bmatrix}$$

The product

$$AB = \begin{bmatrix} 1 & 2 & 1 \\ 2 & 5 & 3 \\ 2 & 3 & 2 \end{bmatrix} \begin{bmatrix} 5 & 21 & 10 \\ 8 & 7 & 8 \\ 10 & 2 & 3 \end{bmatrix} = \begin{bmatrix} 31 & 37 & 29 \\ 80 & 83 & 69 \\ 54 & 67 & 50 \end{bmatrix}$$

gives the coded message to be sent:

$$31, 80, 54, 37, 83, 67, 29, 69, 50$$

The person receiving the message can decode it by multiplying by A^{-1}

$$\begin{bmatrix} 1 & -1 & 1 \\ 2 & 0 & -1 \\ -4 & 1 & 1 \end{bmatrix} \begin{bmatrix} 31 & 37 & 29 \\ 80 & 83 & 69 \\ 54 & 67 & 50 \end{bmatrix} = \begin{bmatrix} 5 & 21 & 10 \\ 8 & 7 & 8 \\ 10 & 2 & 3 \end{bmatrix}$$

To construct a coding matrix A, we can begin with the identity I and successively apply row operation III, being careful to add integer multiples of one row to another. Row operation I can also be used. The resulting matrix A will have integer entries, and since

$$\det(A) = \pm \det(I) = \pm 1$$

A^{-1} will also have integer entries.

REFERENCES

1. Hansen, Robert, *Two-Year College Mathematics Journal*, 13(1), 1982.

EXERCISES

1. For each of the following, compute (i) $\det(A)$, (ii) adj A, and (iii) A^{-1}.

(a) $A = \begin{bmatrix} 1 & 2 \\ 3 & -1 \end{bmatrix}$

(b) $A = \begin{bmatrix} 3 & 1 \\ 2 & 4 \end{bmatrix}$

(c) $A = \begin{bmatrix} 1 & 3 & 1 \\ 2 & 1 & 1 \\ -2 & 2 & -1 \end{bmatrix}$

(d) $A = \begin{bmatrix} 1 & 1 & 1 \\ 0 & 1 & 1 \\ 0 & 0 & 1 \end{bmatrix}$

2. Use Cramer's rule to solve each of the following systems.

(a) $\begin{aligned} x_1 + 2x_2 &= 3 \\ 3x_1 - x_2 &= 1 \end{aligned}$

(b) $\begin{aligned} 2x_1 + 3x_2 &= 2 \\ 3x_1 + 2x_2 &= 5 \end{aligned}$

(c) $\begin{aligned} 2x_1 + x_2 - 3x_3 &= 0 \\ 4x_1 + 5x_2 + x_3 &= 8 \\ -2x_1 - x_2 + 4x_3 &= 2 \end{aligned}$

(d) $\begin{aligned} x_1 + 3x_2 + x_3 &= 1 \\ 2x_1 + x_2 + x_3 &= 5 \\ -2x_1 + 2x_2 - x_3 &= -8 \end{aligned}$

(e) $\begin{aligned} x_1 + x_2 \qquad\qquad &= 0 \\ x_2 + x_3 - 2x_4 &= 1 \\ x_1 \qquad + 2x_3 + x_4 &= 0 \\ x_1 + x_2 \qquad + x_4 &= 0 \end{aligned}$

3. Given

$$A = \begin{bmatrix} 1 & 2 & 1 \\ 0 & 4 & 3 \\ 1 & 2 & 2 \end{bmatrix}$$

Determine the $(2, 3)$ entry of A^{-1} by computing a quotient of two determinants.

4. Let A be the matrix in Exercise 3. Compute the third column of A^{-1} by using Cramer's rule to solve $A\mathbf{x} = \mathbf{e}_3$.

5. Given

$$A = \begin{bmatrix} 1 & 2 & 3 \\ 2 & 3 & 4 \\ 3 & 4 & 5 \end{bmatrix}$$

(a) Compute the determinant of A. Is A nonsingular?

(b) Compute adj A and the product A adj A.

6. If A is singular, what can you say about the product A adj A?

7. Let B_j denote the matrix obtained by replacing the jth column of the identity matrix with a vector $\mathbf{b} = (b_1, \dots, b_n)^T$. Use Cramer's rule to show that

$$b_j = \det(B_j) \qquad \text{for} \quad j = 1, \dots, n$$

8. Let A be a nonsingular $n \times n$ matrix with $n > 1$. Show that

$$\det(\text{adj } A) = (\det(A))^{n-1}$$

9. Let A be a 4×4 matrix. If

$$\text{adj } A = \begin{bmatrix} 2 & 0 & 0 & 0 \\ 0 & 2 & 1 & 0 \\ 0 & 4 & 3 & 2 \\ 0 & -2 & -1 & 2 \end{bmatrix}$$

(a) Calculate the value of $\det(\text{adj } A)$. What should the value of $\det(A)$ be? [*Hint:* Use the result from Exercise 8.]

(b) Find A.

10. Show that if A is nonsingular then adj A is nonsingular and

$$(\text{adj } A)^{-1} = \det(A^{-1})A = \text{adj } A^{-1}$$

11. Show that if A is singular then adj A is also singular.

12. Show that if $\det(A) = 1$ then

$$\text{adj}(\text{adj } A) = A$$

13. Suppose that Q is a matrix with the property $Q^{-1} = Q^T$. Show that

$$q_{ij} = \frac{Q_{ij}}{\det(Q)}$$

14. In coding a message, a blank space was represented by 0, an A by 1, a B by 2, a C by 3, and so on. The message was transformed using the matrix

$$A = \begin{bmatrix} -1 & -1 & 2 & 0 \\ 1 & 1 & -1 & 0 \\ 0 & 0 & -1 & 1 \\ 1 & 0 & 0 & -1 \end{bmatrix}$$

and sent as

$$-19, 19, 25, -21, 0, 18, -18, 15, 3, 10, -8, 3, -2, 20, -7, 12$$

What was the message?

MATLAB EXERCISES

The first four exercises involve integer matrices and illustrate some of the properties of determinants that were covered in this chapter. The last two exercises illustrate some of the differences that may arise when we work with determinants in floating-point arithmetic.

In theory, the value of the determinant should tell us whether the matrix is nonsingular. However, if the matrix is singular and its determinant is computed using finite precision arithmetic, then, because of roundoff errors, the computed value of the determinant may not equal zero. A computed value near zero does not necessarily mean that the matrix is singular or even close to being singular. Furthermore, a matrix may be singular or nearly singular and have a determinant that is not even close to zero (see Exercise 6).

1. Generate random 5×5 matrices with integer entries by setting

$$A = \textbf{round}(10 * \textbf{rand}(5)) \quad \text{and} \quad B = \textbf{round}(20 * \textbf{rand}(5)) - 10$$

Use MATLAB to compute each of the following pairs of numbers. In each case check whether the first is equal to the second.

(a) $\det(A)$ $\det(A^T)$ (b) $\det(A + B)$ $\det(A) + \det(B)$
(c) $\det(AB)$ $\det(A)\det(B)$ (d) $\det(A^T B^T)$ $\det(A^T)\det(B^T)$
(e) $\det(A^{-1})$ $1/\det(A)$ (f) $\det(AB^{-1})$ $\det(A)/\det(B)$

2. Are $n \times n$ magic squares nonsingular? Use MATLAB to compute $\textbf{det}(\textbf{magic}(n))$ in the cases $n = 3, 4, \dots, 10$. What seems to be happening? Check the cases $n = 24$ and 25 to see if the pattern still holds.

3. Set $A = \textbf{round}(10 * \textbf{rand}(6))$. In each of the following, use MATLAB to compute a second matrix as indicated. State how the second matrix is related to A and compute the determinants of both matrices. How are the determinants related?

(a) $B = A$; $B(2, :) = A(1, :)$; $B(1, :) = A(2, :)$
(b) $C = A$; $C(3, :) = 4 * A(3, :)$
(c) $D = A$; $D(5, :) = A(5, :) + 2 * A(4, :)$

4. We can generate a random 6×6 matrix A whose entries consist entirely of zeros and ones by setting

$$A = \textbf{round}(\textbf{rand}(6))$$

(a) What percentage of these random 0–1 matrices are singular? You can estimate the percentage using MATLAB by setting

$$y = \textbf{zeros}(1, 100);$$

and then generating 100 test matrices and setting $y(j) = 1$ if the jth matrix is singular and 0 otherwise. The easy way to do this in MATLAB is to use a *for loop*. Generate the loop as follows:

$$\textbf{for} \quad j = 1 : 100$$
$$A = \textbf{round}(\textbf{rand}(6));$$
$$y(j) = (\textbf{det}(A) == 0);$$
$$\textbf{end}$$

(*Note*: A semicolon at the end of a line suppresses printout. It is recommended that you include one at the end of each line of calculation that occurs inside a for loop.) To determine how many singular matrices were generated, use the MATLAB command **sum(y)**. What percentage of the matrices generated were singular?

(b) For any positive integer n, we can generate a random 6×6 matrix A whose entries are integers from 0 to n by setting

$$A = \textbf{round}(n * \textbf{rand}(6))$$

What percentage of random integer matrices generated in this manner will be singular if $n = 3$? If $n = 6$? If $n = 10$? We can estimate the answers to these questions using MATLAB. In each case, generate 100 test matrices and determine how many of the matrices are singular.

5. If a matrix is sensitive to roundoff errors, the computed value of its determinant may differ drastically from the exact value. For an example of this, set

$$U = \textbf{round}(100 * \textbf{rand}(10)); \quad U = \textbf{triu}(U, 1) + 0.1 * \textbf{eye}(10)$$

In theory,

$$\det(U) = \det(U^T) = 10^{-10}$$

and

$$\det(UU^T) = \det(U)\det(U^T) = 10^{-20}$$

Compute $\det(U)$, $\det(U')$, and $\det(U * U')$ using MATLAB. Do the computed values match the theoretical values?

6. Use MATLAB to construct a matrix A by setting

$$A = \textbf{vander}(1:6); \quad A = A - \textbf{diag}(\textbf{sum}(A'))$$

(a) By construction, the entries in each row of A should all add up to zero. To check this, set $\mathbf{x} = \textbf{ones}(6, 1)$ and use MATLAB to compute the product $A\mathbf{x}$. The matrix A should be singular. Why? Explain. Use the MATLAB functions **det** and **inv** to compute the values of $\det(A)$ and A^{-1}. Which MATLAB function is a more reliable indicator of singularity?

(b) Use MATLAB to compute $\det(A^T)$. Are the computed values of $\det(A)$ and $\det(A^T)$ equal? Another way to check if a matrix is singular is to compute its reduced row echelon form. Use MATLAB to compute the reduced row echelon forms of A and A^T.

(c) To see what is going wrong, it helps to know how MATLAB computes determinants. The MATLAB routine for determinants first computes a form of the LU factorization of the matrix. The determinant of the L is ± 1 depending on the whether an even or odd number of row interchanges were used in the computation. The computed value of the determinant of A is the product of the diagonal entries of U multiplied by $\det(L) = \pm 1$. In the special case that the original matrix has integer entries, the exact determinant should take on an integer value. So in this case MATLAB will round its decimal answer to the nearest integer. To see what is happening for our original matrix, use the following commands to compute and display the factor U.

format short e
$[L, U] = \textbf{lu}(A); \ U$

In exact arithmetic U should be singular. Is the computed matrix U singular? If not, what goes wrong? Use the following commands to see the rest of the computation of $d = \det(A)$.

format short
$d = \textbf{prod}(\textbf{diag}(U))$
$d = \textbf{round}(d)$

CHAPTER TEST

In each of the following answer *true* if the statement is always true and *false* otherwise. In the case of a true statement, explain or prove your answer. In the case of a false statement, give an example to show that the statement is not always true. In each of the following, assume that all the matrices are $n \times n$.

1. $\det(AB) = \det(BA)$

2. $\det(A + B) = \det(A) + \det(B)$

3. $\det(cA) = c \det(A)$

4. $\det((AB)^T) = \det(A) \det(B)$

5. $\det(A) = \det(B)$ implies $A = B$.

6. $\det(A^k) = \det(A)^k$

7. A triangular matrix is nonsingular if and only if its diagonal entries are all nonzero.

8. If $\mathbf{x}$ is a nonzero vector in R^n and $A\mathbf{x} = \mathbf{0}$, then $\det(A) = 0$.

9. If A and B are row equivalent matrices, then their determinants are equal.

10. If $A \neq O$, but $A^k = O$ (where O denotes the zero matrix) for some positive integer k, then A must be singular.

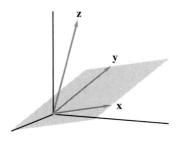

CHAPTER
3

VECTOR SPACES

The operations of addition and scalar multiplication are used in many diverse contexts in mathematics. Regardless of the context, however, these operations usually obey the same set of arithmetic rules. Thus a general theory of mathematical systems involving addition and scalar multiplication will have application to many areas in mathematics. Mathematical systems of this form are called vector spaces or linear spaces. In this chapter the definition of a vector space is given and some of the general theory of vector spaces is developed.

1 DEFINITION AND EXAMPLES

In this section we present the formal definition of a vector space. Before doing this, however, it is instructive to look at a number of examples. We begin with the Euclidean vector spaces R^n.

Euclidean Vector Spaces

Perhaps the most elementary vector spaces are the Euclidean vector spaces R^n, $n = 1, 2, \ldots$. For simplicity, let us consider first R^2. Nonzero vectors in R^2 can be represented geometrically by directed line segments. This geometric representation will help us visualize how the operations of scalar multiplication and addition work in R^2. Given a nonzero vector $\mathbf{x} = \begin{bmatrix} x_1 \\ x_2 \end{bmatrix}$, we can associate it with the line segment in the plane from $(0, 0)$ to (x_1, x_2) (see Figure 3.1.1). If we equate line segments that have the same length and direction (Figure 3.1.2), $\mathbf{x}$ can be represented by any

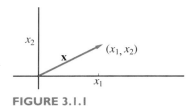

FIGURE 3.1.1

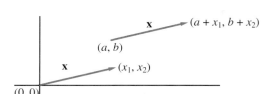

FIGURE 3.1.2

line segment from (a, b) to $(a + x_1, b + x_2)$. For example, the vector $\mathbf{x} = \begin{bmatrix} 2 \\ 1 \end{bmatrix}$ in R^2 could just as well be represented by the directed line segment from $(2, 2)$ to $(4, 3)$ or from $(-1, -1)$ to $(1, 0)$, as shown in Figure 3.1.3.

We can think of the Euclidean length of a vector $\mathbf{x} = \begin{bmatrix} x_1 \\ x_2 \end{bmatrix}$ as the length of any directed line segment representing $\mathbf{x}$. The length of the line segment from $(0, 0)$ to (x_1, x_2) is $\sqrt{x_1^2 + x_2^2}$ (see Figure 3.1.4). For each vector $\mathbf{x} = \begin{bmatrix} x_1 \\ x_2 \end{bmatrix}$ and each scalar α, the product $\alpha\mathbf{x}$ is defined by

$$\alpha \begin{bmatrix} x_1 \\ x_2 \end{bmatrix} = \begin{bmatrix} \alpha x_1 \\ \alpha x_2 \end{bmatrix}$$

For example, as shown in Figure 3.1.5, if $\mathbf{x} = \begin{bmatrix} 2 \\ 1 \end{bmatrix}$, then

$$3\mathbf{x} = \begin{bmatrix} 6 \\ 3 \end{bmatrix}, \qquad -\mathbf{x} = \begin{bmatrix} -2 \\ -1 \end{bmatrix}, \qquad -2\mathbf{x} = \begin{bmatrix} -4 \\ -2 \end{bmatrix}$$

FIGURE 3.1.3

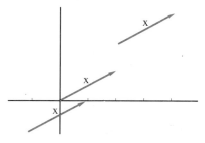

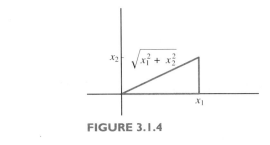

FIGURE 3.1.4

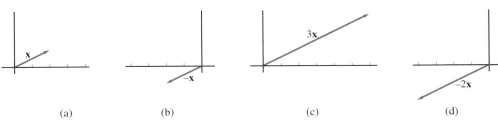

(a)   (b)   (c)   (d)

FIGURE 3.1.5

The vector $3\mathbf{x}$ is in the same direction as $\mathbf{x}$, but its length is three times that of $\mathbf{x}$. The vector $-\mathbf{x}$ has the same length as $\mathbf{x}$, but it points in the opposite direction. The vector $-2\mathbf{x}$ is twice as long as $\mathbf{x}$ and it points in the same direction as $-\mathbf{x}$. The sum of two vectors

$$\mathbf{u} = \begin{bmatrix} u_1 \\ u_2 \end{bmatrix} \qquad \text{and} \qquad \mathbf{v} = \begin{bmatrix} v_1 \\ v_2 \end{bmatrix}$$

is defined by

$$\mathbf{u} + \mathbf{v} = \begin{bmatrix} u_1 + v_1 \\ u_2 + v_2 \end{bmatrix}$$

Note that, if $\mathbf{v}$ is placed at the terminal point of $\mathbf{u}$, then $\mathbf{u} + \mathbf{v}$ is represented by the directed line segment from the initial point of $\mathbf{u}$ to the terminal point of $\mathbf{v}$ (Figure 3.1.6). If both $\mathbf{u}$ and $\mathbf{v}$ are placed at the origin and a parallelogram is formed as in Figure 3.1.7, the diagonals of the parallelogram will represent the sum $\mathbf{u} + \mathbf{v}$ and the difference $\mathbf{v} - \mathbf{u}$. In a similar manner, vectors in R^3 can be represented by directed line segments in 3-space (see Figure 3.1.8).

In general, scalar multiplication and addition in R^n are defined by

$$\alpha\mathbf{x} = \begin{bmatrix} \alpha x_1 \\ \alpha x_2 \\ \vdots \\ \alpha x_n \end{bmatrix} \qquad \text{and} \qquad \mathbf{x} + \mathbf{y} = \begin{bmatrix} x_1 + y_1 \\ x_2 + y_2 \\ \vdots \\ x_n + y_n \end{bmatrix}$$

for any $\mathbf{x}, \mathbf{y} \in R^n$ and any scalar α.

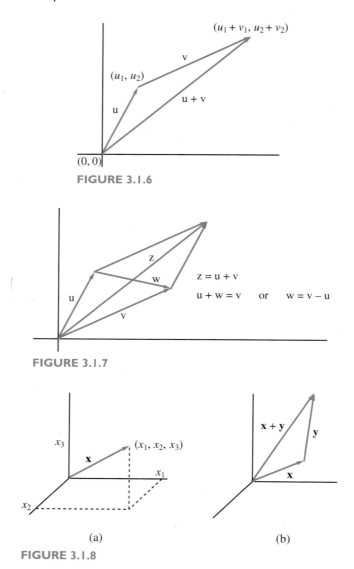

FIGURE 3.1.6

FIGURE 3.1.7

$$z = u + v$$

$$u + w = v \quad \text{or} \quad w = v - u$$

(a) $\quad$ (b)

FIGURE 3.1.8

The Vector Space $R^{m \times n}$

We can also view R^n as the set of all $n \times 1$ matrices with real entries. The addition and scalar multiplication of vectors in R^n is just the usual addition and scalar multiplication of matrices. More generally, let $R^{m \times n}$ denote the set of all $m \times n$ matrices with real entries. If $A = (a_{ij})$ and $B = (b_{ij})$, the sum $A + B$ is defined to be the $m \times n$ matrix $C = (c_{ij})$, where $c_{ij} = a_{ij} + b_{ij}$. Given a scalar α, we can define αA to be the $m \times n$ matrix whose ijth entry is αa_{ij}. Thus, by defining operations on the set $R^{m \times n}$, we have created a mathematical system. The operations of addition and scalar multiplication of $R^{m \times n}$ obey certain arithmetic rules. These rules form the axioms that are used to define the concept of a vector space.

Vector Space Axioms

▶ **DEFINITION** Let V be a set on which the operations of addition and scalar multiplication are defined. By this we mean that, with each pair of elements $\mathbf{x}$ and $\mathbf{y}$ in V, we can associate a unique element $\mathbf{x} + \mathbf{y}$ that is also in V, and with each element $\mathbf{x}$ in V and each scalar α, we can associate a unique element $\alpha\mathbf{x}$ in V. The set V together with the operations of addition and scalar multiplication is said to form a **vector space** if the following axioms are satisfied.

A1. $\mathbf{x} + \mathbf{y} = \mathbf{y} + \mathbf{x}$ for any $\mathbf{x}$ and $\mathbf{y}$ in V.
A2. $(\mathbf{x} + \mathbf{y}) + \mathbf{z} = \mathbf{x} + (\mathbf{y} + \mathbf{z})$ for any $\mathbf{x}, \mathbf{y}, \mathbf{z}$ in V.
A3. There exists an element $\mathbf{0}$ in V such that $\mathbf{x} + \mathbf{0} = \mathbf{x}$ for each $\mathbf{x} \in V$.
A4. For each $\mathbf{x} \in V$, there exists an element $-\mathbf{x}$ in V such that $\mathbf{x} + (-\mathbf{x}) = \mathbf{0}$.
A5. $\alpha(\mathbf{x} + \mathbf{y}) = \alpha\mathbf{x} + \alpha\mathbf{y}$ for each scalar α and any $\mathbf{x}$ and $\mathbf{y}$ in V.
A6. $(\alpha + \beta)\mathbf{x} = \alpha\mathbf{x} + \beta\mathbf{x}$ for any scalars α and β and any $\mathbf{x} \in V$.
A7. $(\alpha\beta)\mathbf{x} = \alpha(\beta\mathbf{x})$ for any scalars α and β and any $\mathbf{x} \in V$.
A8. $1 \cdot \mathbf{x} = \mathbf{x}$ for all $\mathbf{x} \in V$. ◀

The elements of V are called **vectors** and are usually denoted by letters from the end of the alphabet: $\mathbf{u}, \mathbf{v}, \mathbf{w}, \mathbf{x}, \mathbf{y}$, and $\mathbf{z}$. The term *scalar* will generally refer to a real number, although in some cases it will be used to refer to complex numbers. In the first five chapters of this book, the term *scalars* will always refer to real numbers. Often the term *real vector space* is used to indicate that the set of scalars is the set of real numbers. The symbol $\mathbf{0}$ was used in Axiom 3 in order to distinguish the zero vector from the scalar 0.

An important component of the definition is the closure properties of the two operations. These properties can be summarized as follows:

C1. If $\mathbf{x} \in V$ and α is a scalar, then $\alpha\mathbf{x} \in V$.
C2. If $\mathbf{x}, \mathbf{y} \in V$, then $\mathbf{x} + \mathbf{y} \in V$.

To see the importance of the closure properties, consider the following example. Let

$$W = \{(a, 1) \mid a \text{ real}\}$$

with addition and scalar multiplication defined in the usual way. The elements $(3, 1)$ and $(5, 1)$ are in W, but the sum

$$(3, 1) + (5, 1) = (8, 2)$$

is not an element of W. The operation $+$ is not really an operation on the set W because property C2 fails to hold. Similarly, scalar multiplication is not defined on W because property C1 fails to hold. The set W together with the operations of addition and scalar multiplication is *not* a vector space.

On the other hand, if we are given a set U on which the operations of addition and scalar multiplication have been defined and satisfy properties C1 and C2, we must check to see if the eight axioms are valid in order to determine whether U is a vector space. We leave it to the reader to verify that R^n and $R^{m \times n}$, with the usual addition and scalar multiplication of matrices, are both vector spaces. There are a number of other important examples of vector spaces.

The Vector Space $C[a, b]$

Let $C[a, b]$ denote the set of all real-valued functions that are defined and continuous on the closed interval $[a, b]$. In this case our universal set is a set of functions. Thus our vectors are the functions in $C[a, b]$. The sum $f + g$ of two functions in $C[a, b]$ is defined by

$$(f + g)(x) = f(x) + g(x)$$

for all x in $[a, b]$. The new function $f + g$ is an element of $C[a, b]$, since the sum of two continuous functions is continuous. If f is a function in $C[a, b]$ and α is a real number, define αf by

$$(\alpha f)(x) = \alpha f(x)$$

for all x in $[a, b]$. Clearly, αf is in $C[a, b]$ since a constant times a continuous function is always continuous. Thus on $C[a, b]$ we have defined the operations of addition and scalar multiplication. To show that the first axiom, $f + g = g + f$, is satisfied, we must show that

$$(f + g)(x) = (g + f)(x) \qquad \text{for every } x \text{ in } [a, b]$$

This follows since

$$(f + g)(x) = f(x) + g(x) = g(x) + f(x) = (g + f)(x)$$

for every x in $[a, b]$. Axiom 3 is satisfied since the function

$$z(x) = 0 \quad \text{for all } x \text{ in } [a, b]$$

acts as the zero vector; that is,

$$f + z = f \quad \text{for all } f \text{ in } C[a, b]$$

We leave it to the reader to verify that the remaining vector space axioms are all satisfied.

The Vector Space P_n

Let P_n denote the set of all polynomials of degree less than n. Define $p + q$ and αp by

$$(p + q)(x) = p(x) + q(x)$$

and

$$(\alpha p)(x) = \alpha p(x)$$

for all real numbers x. In this case, the zero vector is the zero polynomial,

$$z(x) = 0x^{n-1} + 0x^{n-2} + \cdots + 0x + 0$$

It is easily verified that all the vector space axioms hold. Thus P_n, with the standard addition and scalar multiplication of functions, is a vector space.

Additional Properties of Vector Spaces

We close this section with a theorem that states three more fundamental properties of vector spaces. Other important properties are given in Exercises 7, 8, and 9.

▶ **THEOREM 3.1.1** *If V is a vector space and* $\mathbf{x}$ *is any element of V, then*

    (i) $0\mathbf{x} = \mathbf{0}$.

    (ii) $\mathbf{x} + \mathbf{y} = \mathbf{0}$ *implies that* $\mathbf{y} = -\mathbf{x}$ *(i.e., the additive inverse of* $\mathbf{x}$ *is unique).*

    (iii) $(-1)\mathbf{x} = -\mathbf{x}$.

▶ **Proof.** It follows from axioms A6 and A8 that

$$\mathbf{x} = 1\mathbf{x} = (1+0)\mathbf{x} = 1\mathbf{x} + 0\mathbf{x} = \mathbf{x} + 0\mathbf{x}$$

Thus,

$$-\mathbf{x} + \mathbf{x} = -\mathbf{x} + (\mathbf{x} + 0\mathbf{x}) = (-\mathbf{x} + \mathbf{x}) + 0\mathbf{x} \qquad\qquad \text{(A2)}$$

$$\mathbf{0} = \mathbf{0} + 0\mathbf{x} = 0\mathbf{x} \qquad\qquad \text{(A1, A3, and A4)}$$

To prove (ii), suppose that $\mathbf{x} + \mathbf{y} = \mathbf{0}$. Then

$$-\mathbf{x} = -\mathbf{x} + \mathbf{0} = -\mathbf{x} + (\mathbf{x} + \mathbf{y})$$

Therefore,

$$-\mathbf{x} = (-\mathbf{x} + \mathbf{x}) + \mathbf{y} = \mathbf{0} + \mathbf{y} = \mathbf{y} \qquad \text{(A1, A2, A3, and A4)}$$

Finally, to prove (iii), note that

$$\mathbf{0} = 0\mathbf{x} = (1 + (-1))\mathbf{x} = 1\mathbf{x} + (-1)\mathbf{x} \qquad \text{[(i) and A6]}$$

Thus

$$\mathbf{x} + (-1)\mathbf{x} = \mathbf{0} \qquad \text{(A8)}$$

and it follows from part (ii) that

$$(-1)\mathbf{x} = -\mathbf{x} \qquad\qquad\qquad ◀$$

EXERCISES

1. Consider the vectors $\mathbf{x}_1 = (8,\ 6)^T$ and $\mathbf{x}_2 = (4,\ -1)^T$ in R^2.

    (a) Determine the length of each vector.

    (b) Let $\mathbf{x}_3 = \mathbf{x}_1 + \mathbf{x}_2$. Determine the length of $\mathbf{x}_3$. How does its length compare to the sum of the lengths of $\mathbf{x}_1$ and $\mathbf{x}_2$?

    (c) Draw a graph illustrating how $\mathbf{x}_3$ can be constructed geometrically using $\mathbf{x}_1$ and $\mathbf{x}_2$. Use this graph to give a geometrical interpretation of your answer to the question in part (b).

2. Repeat Exercise 1 for the vectors $\mathbf{x}_1 = (2,\ 1)^T$ and $\mathbf{x}_2 = (6,\ 3)^T$.

3. Let C be the set of complex numbers. Define addition on C by
$$(a + bi) + (c + di) = (a + c) + (b + d)i$$
and define scalar multiplication by
$$\alpha(a + bi) = \alpha a + \alpha bi$$
for all real numbers α. Show that C is a vector space with these operations.

4. Show that $R^{m \times n}$, with the usual addition and scalar multiplication of matrices, satisfies the eight axioms of a vector space.

5. Show that $C[a, b]$, with the usual scalar multiplication and addition of functions, satisfies the eight axioms of a vector space.

6. Let P be the set of all polynomials. Show that P, with the usual addition and scalar multiplication of functions, forms a vector space.

7. Show that the element $\mathbf{0}$ in a vector space is unique.

8. Let $\mathbf{x}$, $\mathbf{y}$, and $\mathbf{z}$ be vectors in a vector space V. Prove that if
$$\mathbf{x} + \mathbf{y} = \mathbf{x} + \mathbf{z}$$
then $\mathbf{y} = \mathbf{z}$.

9. Let V be a vector space and let $\mathbf{x} \in V$. Show that:

(a) $\beta\mathbf{0} = \mathbf{0}$ for each scalar β.
(b) If $\alpha\mathbf{x} = \mathbf{0}$, then either $\alpha = 0$ or $\mathbf{x} = \mathbf{0}$.

10. Let S be the set of all ordered pairs of real numbers. Define scalar multiplication and addition on S by
$$\alpha(x_1, x_2) = (\alpha x_1, \alpha x_2)$$
$$(x_1, x_2) \oplus (y_1, y_2) = (x_1 + y_1, 0)$$
We use the symbol $\oplus$ to denote the addition operation for this system in order to avoid confusion with the usual addition $\mathbf{x} + \mathbf{y}$ of row vectors. Show that S, with the ordinary scalar multiplication and addition operation $\oplus$, is not a vector space. Which of the eight axioms fail to hold?

11. Let V be the set of all ordered pairs of real numbers with addition defined by
$$(x_1, x_2) + (y_1, y_2) = (x_1 + y_1, x_2 + y_2)$$
and scalar multiplication defined by
$$\alpha \circ (x_1, x_2) = (\alpha x_1, x_2)$$
The scalar multiplication for this system is defined in an unusual way, and consequently we use the symbol $\circ$ in order to avoid confusion with the ordinary scalar multiplication of row vectors. Is V a vector space with these operations? Justify your answer.

12. Let R^+ denote the set of positive real numbers. Define the operation of scalar multiplication, denoted $\circ$, by
$$\alpha \circ x = x^\alpha$$

for each $x \in R^+$ and for any real number α. Define the operation of addition, denoted $\oplus$, by

$$x \oplus y = x \cdot y \qquad \text{for all} \qquad x, y \in R^+$$

Thus for this system the scalar product of -3 times $\frac{1}{2}$ is given by

$$-3 \circ \frac{1}{2} = \left(\frac{1}{2}\right)^{-3} = 8$$

and the sum of 2 and 5 is given by

$$2 \oplus 5 = 2 \cdot 5 = 10$$

Is R^+ a vector space with these operations? Prove your answer.

13. Let R denote the set of real numbers. Define scalar multiplication by

$$\alpha x = \alpha \cdot x \qquad \text{(the usual multiplication of real numbers)}$$

and define addition, denoted $\oplus$, by

$$x \oplus y = \max(x, y) \qquad \text{(the maximum of the two numbers)}$$

Is R a vector space with these operations? Prove your answer.

14. Let Z denote the set of all integers with addition defined in the usual way and define the scalar multiplication, denoted $\circ$, by

$$\alpha \circ k = [[\alpha]] \cdot k \qquad \text{for all} \qquad k \in Z$$

where $[[\alpha]]$ denotes the greatest integer less than or equal to α. For example,

$$2.25 \circ 4 = [[2.25]] \cdot 4 = 2 \cdot 4 = 8$$

Show that Z, together with these operations, is not a vector space. Which axioms fail to hold?

15. Let S denote the set of all infinite sequences of real numbers with scalar multiplication and addition defined by

$$\alpha\{a_n\} = \{\alpha a_n\}$$

$$\{a_n\} + \{b_n\} = \{a_n + b_n\}$$

Show that S is a vector space.

16. We can define a one-to-one correspondence between the elements of P_n and R^n by

$$p(x) = a_1 + a_2 x + \cdots + a_n x^{n-1} \leftrightarrow (a_1, \ldots, a_n)^T = \mathbf{a}$$

Show that if $p \leftrightarrow \mathbf{a}$ and $q \leftrightarrow \mathbf{b}$ then

(a) $\alpha p \leftrightarrow \alpha \mathbf{a}$ for any scalar α.
(b) $p + q \leftrightarrow \mathbf{a} + \mathbf{b}$.

[In general, two vector spaces are said to be *isomorphic* if their elements can be put into a one-to-one correspondence that is preserved under scalar multiplication and addition as in (a) and (b).]

Given a vector space V, it is often possible to form another vector space by taking a subset S of V and using the operations of V. Since V is a vector space, the operations of addition and scalar multiplication always produce another vector in V. For a new system using a subset S of V as its universal set to be a vector space, the set S must be closed under the operations of addition and scalar multiplication. That is, the sum of two elements of S must always be an element of S, and the product of a scalar and an element of S must always be an element of S.

EXAMPLE 1. Let $S = \left\{ \begin{pmatrix} x_1 \\ x_2 \end{pmatrix} \,\middle|\, x_2 = 2x_1 \right\}$. S is a subset of R^2. If $\begin{pmatrix} c \\ 2c \end{pmatrix}$ is any element of S and α is any scalar, then

$$\alpha \begin{bmatrix} c \\ 2c \end{bmatrix} = \begin{bmatrix} \alpha c \\ 2\alpha c \end{bmatrix}$$

which is an element of S. If $\begin{bmatrix} a \\ 2a \end{bmatrix}$ and $\begin{bmatrix} b \\ 2b \end{bmatrix}$ are any two elements of S, their sum

$$\begin{bmatrix} a + b \\ 2a + 2b \end{bmatrix} = \begin{bmatrix} a + b \\ 2(a + b) \end{bmatrix}$$

is also an element of S. It is easily seen that the mathematical system consisting of the set S (instead of R^2), together with the operations from R^2, is itself a vector space. ◄

▶ **DEFINITION** If S is a nonempty subset of a vector space V, and S satisfies the following conditions:

(i) $\alpha \mathbf{x} \in S$ whenever $\mathbf{x} \in S$ for any scalar α

(ii) $\mathbf{x} + \mathbf{y} \in S$ whenever $\mathbf{x} \in S$ and $\mathbf{y} \in S$

then S is said to be a **subspace** of V. ◄

Condition (i) says that S is closed under scalar multiplication. That is, whenever an element of S is multiplied by a scalar, the result is an element of S. Condition (ii) says that S is closed under addition. That is, the sum of two elements of S is always an element of S. Thus, if we do arithmetic using the operations from V and the elements of S, we will always end up with elements of S. A subspace of V, then, is a subset S that is closed under the operations of V.

Let S be a subspace of a vector space V. Using the operations of addition and scalar multiplication as defined on V, we can form a new mathematical system with S as the universal set. It is easily seen that all eight axioms will remain valid for this new system. Axioms A3 and A4 follow from Theorem 3.1.1 and condition (i) of the definition of a subspace. The remaining six axioms are valid for any elements of V, so, in particular, they are valid for the elements of S. Thus every subspace is a vector space in its own right.

REMARK In a vector space V it can be readily verified that $\{\mathbf{0}\}$ and V are subspaces of V. All other subspaces are referred to as *proper subspaces*. ◄

EXAMPLE 2. Let $S = \{(x_1, x_2, x_3)^T \mid x_1 = x_2\}$. S is nonempty since $\mathbf{x} = (1, 1, 0)^T \in S$. To show that S is a subspace of R^3, we need to verify that the two closure properties hold.

(i) If $\mathbf{x} = (a, a, b)^T \in S$, then

$$\alpha\mathbf{x} = (\alpha a, \alpha a, \alpha b)^T \in S$$

(ii) If $(a, a, b)^T$ and $(c, c, d)^T$ are arbitrary elements of S, then

$$(a, a, b)^T + (c, c, d)^T = (a + c, a + c, b + d)^T \in S \tag{1}$$

Since S is nonempty and satisfies the two closure conditions, it follows that S is a subspace of R^3. ◀

EXAMPLE 3. Let $S = \left\{ \begin{pmatrix} x \\ 1 \end{pmatrix} \middle| \ x \text{ is a real number} \right\}$. S is not a subspace of R^2. In this case both conditions fail. S is not closed under scalar multiplication, since $\alpha \begin{pmatrix} x \\ 1 \end{pmatrix} \notin S$ unless $\alpha = 1$. S is not closed under addition, since

$$\begin{pmatrix} x \\ 1 \end{pmatrix} + \begin{pmatrix} y \\ 1 \end{pmatrix} = \begin{pmatrix} x + y \\ 2 \end{pmatrix} \notin S$$

◀

EXAMPLE 4. Let $S = \{A \in R^{2\times2} \mid a_{12} = -a_{21}\}$. The set S forms a subspace of $R^{2\times2}$, since

(i) If $A \in S$, then A must be of the form

$$A = \begin{pmatrix} a & b \\ -b & c \end{pmatrix}$$

and hence

$$\alpha A = \begin{pmatrix} \alpha a & \alpha b \\ -\alpha b & \alpha c \end{pmatrix}$$

Since the $(2, 1)$ entry of αA is the negative of the $(1, 2)$ entry, $\alpha A \in S$.

(ii) If $A, B \in S$, then they must be of the form

$$A = \begin{pmatrix} a & b \\ -b & c \end{pmatrix} \quad \text{and} \quad B = \begin{pmatrix} d & e \\ -e & f \end{pmatrix}$$

It follows that

$$A + B = \begin{pmatrix} a + d & b + e \\ -(b + e) & c + f \end{pmatrix}$$

Hence $A + B \in S$. ◀

EXAMPLE 5.   Let S be the set of all polynomials of degree less than n with the property $p(0) = 0$. The set S is nonempty since it contains the zero polynomial. We claim that S is a subspace of P_n. This follows, since:

(i) If $p(x) \in S$ and α is a scalar, then

$$\alpha p(0) = \alpha \cdot 0 = 0$$

and hence $\alpha p \in S$.

(ii) If $p(x)$ and $q(x)$ are elements of S, then

$$(p + q)(0) = p(0) + q(0) = 0 + 0 = 0$$

and hence $p + q \in S$.   ◀

EXAMPLE 6.   Let $C^n[a, b]$ be the set of all functions f that have a continuous nth derivative on $[a, b]$. We leave it to the reader to verify that $C^n[a, b]$ is a subspace of $C[a, b]$.   ◀

EXAMPLE 7.   The function $f(x) = |x|$ is in $C[-1, 1]$, but it is not differentiable at $x = 0$ and hence it is not in $C^1[-1, 1]$. This shows that $C^1[-1, 1]$ is a proper subspace of $C[-1, 1]$. The function $g(x) = x|x|$ is in $C^1[-1, 1]$ since it is differentiable at every point in $[-1, 1]$ and $g'(x) = 2|x|$ is continuous on $[-1, 1]$. However, $g \notin C^2[-1, 1]$, since $g''(x)$ is not defined when $x = 0$. Thus the vector space $C^2[-1, 1]$ is a proper subspace of both $C[-1, 1]$ and $C^1[-1, 1]$.   ◀

EXAMPLE 8.   Let S be the set of all f in $C^2[a, b]$ such that

$$f''(x) + f(x) = 0$$

for all x in $[a, b]$. The set S is nonempty since the zero function is in S. If $f \in S$ and α is any scalar, then for any x in $[a, b]$

$$(\alpha f)''(x) + (\alpha f)(x) = \alpha f''(x) + \alpha f(x)$$
$$= \alpha(f''(x) + f(x)) = \alpha \cdot 0 = 0$$

Thus $\alpha f \in S$. If f and g are both in S, then

$$(f + g)''(x) + (f + g)(x) = f''(x) + g''(x) + f(x) + g(x)$$
$$= [f''(x) + f(x)] + [g''(x) + g(x)]$$
$$= 0 + 0 = 0$$

Thus the set of all solutions on $[a, b]$ to the differential equation $y'' + y = 0$ forms a subspace of $C^2[a, b]$. If we note that $f(x) = \sin x$ and $g(x) = \cos x$ are both in S, it follows that any function of the form $c_1 \sin x + c_2 \cos x$ must also be in S. We can easily verify that functions of this form are solutions to $y'' + y = 0$.   ◀

The Nullspace of a Matrix

Let A be an $m \times n$ matrix. Let $N(A)$ denote the set of all solutions to the homogeneous system $A\mathbf{x} = \mathbf{0}$. Thus

$$N(A) = \{\mathbf{x} \in R^n \mid A\mathbf{x} = \mathbf{0}\}$$

We claim that $N(A)$ is a subspace of R^n. Clearly, $\mathbf{0} \in N(A)$, so $N(A)$ is nonempty. If $\mathbf{x} \in N(A)$ and α is a scalar, then

$$A(\alpha\mathbf{x}) = \alpha A\mathbf{x} = \alpha\mathbf{0} = \mathbf{0}$$

and hence $\alpha\mathbf{x} \in N(A)$. If $\mathbf{x}$ and $\mathbf{y}$ are elements of $N(A)$, then

$$A(\mathbf{x} + \mathbf{y}) = A\mathbf{x} + A\mathbf{y} = \mathbf{0} + \mathbf{0} = \mathbf{0}$$

Therefore, $\mathbf{x} + \mathbf{y} \in N(A)$. It follows then that $N(A)$ is a subspace of R^n. The set of all solutions to the homogeneous system $A\mathbf{x} = \mathbf{0}$ forms a subspace of R^n. The subspace $N(A)$ is called the *nullspace* of A.

EXAMPLE 9. Determine $N(A)$ if

$$A = \begin{bmatrix} 1 & 1 & 1 & 0 \\ 2 & 1 & 0 & 1 \end{bmatrix}$$

SOLUTION. Using Gauss–Jordan reduction to solve $A\mathbf{x} = \mathbf{0}$, we obtain

$$\begin{bmatrix} 1 & 1 & 1 & 0 & | & 0 \\ 2 & 1 & 0 & 1 & | & 0 \end{bmatrix} \rightarrow \begin{bmatrix} 1 & 1 & 1 & 0 & | & 0 \\ 0 & -1 & -2 & 1 & | & 0 \end{bmatrix}$$

$$\rightarrow \begin{bmatrix} 1 & 0 & -1 & 1 & | & 0 \\ 0 & -1 & -2 & 1 & | & 0 \end{bmatrix} \rightarrow \begin{bmatrix} 1 & 0 & -1 & 1 & | & 0 \\ 0 & 1 & 2 & -1 & | & 0 \end{bmatrix}$$

The reduced row echelon form involves two free variables, x_3 and x_4.

$$x_1 = x_3 - x_4$$
$$x_2 = -2x_3 + x_4$$

Thus, if we set $x_3 = \alpha$ and $x_4 = \beta$, then

$$\mathbf{x} = \begin{bmatrix} \alpha - \beta \\ -2\alpha + \beta \\ \alpha \\ \beta \end{bmatrix} = \alpha \begin{bmatrix} 1 \\ -2 \\ 1 \\ 0 \end{bmatrix} + \beta \begin{bmatrix} -1 \\ 1 \\ 0 \\ 1 \end{bmatrix}$$

is a solution to $A\mathbf{x} = \mathbf{0}$. The vector space $N(A)$ consists of all vectors of the form

$$\alpha \begin{bmatrix} 1 \\ -2 \\ 1 \\ 0 \end{bmatrix} + \beta \begin{bmatrix} -1 \\ 1 \\ 0 \\ 1 \end{bmatrix}$$

where α and β are scalars. ◄

Span and Spanning Sets

▶ **DEFINITION** Let $\mathbf{v}_1, \mathbf{v}_2, \ldots, \mathbf{v}_n$ be vectors in a vector space V. A sum of the form $\alpha_1 \mathbf{v}_1 + \alpha_2 \mathbf{v}_2 + \cdots + \alpha_n \mathbf{v}_n$, where $\alpha_1, \ldots, \alpha_n$ are scalars, is called a **linear combination** of $\mathbf{v}_1, \mathbf{v}_2, \ldots, \mathbf{v}_n$. The set of all linear combinations of $\mathbf{v}_1, \mathbf{v}_2, \ldots, \mathbf{v}_n$ is called the **span** of $\mathbf{v}_1, \ldots, \mathbf{v}_n$. The span of $\mathbf{v}_1, \ldots, \mathbf{v}_n$ will be denoted by $\text{Span}(\mathbf{v}_1, \ldots, \mathbf{v}_n)$. ◄

In Example 9 we saw that the nullspace of A was the span of the vectors $(1, -2, 1, 0)^T$ and $(-1, 1, 0, 1)^T$.

EXAMPLE 10. In R^3 the span of $\mathbf{e}_1$ and $\mathbf{e}_2$ is the set of all vectors of the form

$$\alpha \mathbf{e}_1 + \beta \mathbf{e}_2 = \begin{bmatrix} \alpha \\ \beta \\ 0 \end{bmatrix}$$

The reader may verify that $\text{Span}(\mathbf{e}_1, \mathbf{e}_2)$ is a subspace of R^3. The subspace can be interpreted geometrically as the set of all vectors in three space that lie in the $x_1 x_2$-plane. (See Figure 3.2.1). The span of $\mathbf{e}_1, \mathbf{e}_2, \mathbf{e}_3$ is the set of all vectors of

FIGURE 3.2.1

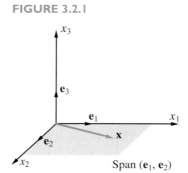

Span $(\mathbf{e}_1, \mathbf{e}_2)$

the form

$$\alpha_1 \mathbf{e}_1 + \alpha_2 \mathbf{e}_2 + \alpha_3 \mathbf{e}_3 = \begin{bmatrix} \alpha_1 \\ \alpha_2 \\ \alpha_3 \end{bmatrix}$$

Thus $\text{Span}(\mathbf{e}_1, \mathbf{e}_2, \mathbf{e}_3) = R^3$.   ◀

▶ **THEOREM 3.2.1** *If* $\mathbf{v}_1, \mathbf{v}_2, \dots, \mathbf{v}_n$ *are elements of a vector space* V, *then* $\text{Span}(\mathbf{v}_1, \mathbf{v}_2, \dots, \mathbf{v}_n)$ *is a subspace of* V.

▶ *Proof.* Let β be a scalar and let $\mathbf{v} = \alpha_1 \mathbf{v}_1 + \alpha_2 \mathbf{v}_2 + \cdots + \alpha_n \mathbf{v}_n$ be an arbitrary element of $\text{Span}(\mathbf{v}_1, \mathbf{v}_2, \dots, \mathbf{v}_n)$. Since

$$\beta \mathbf{v} = (\beta \alpha_1) \mathbf{v}_1 + (\beta \alpha_2) \mathbf{v}_2 + \cdots + (\beta \alpha_n) \mathbf{v}_n$$

it follows that $\beta \mathbf{v} \in \text{Span}(\mathbf{v}_1, \dots, \mathbf{v}_n)$. Next we must show that any sum of elements of $\text{Span}(\mathbf{v}_1, \dots, \mathbf{v}_n)$ is in $\text{Span}(\mathbf{v}_1, \dots, \mathbf{v}_n)$. Let $\mathbf{v} = \alpha_1 \mathbf{v}_1 + \cdots + \alpha_n \mathbf{v}_n$ and $\mathbf{w} = \beta_1 \mathbf{v}_1 + \cdots + \beta_n \mathbf{v}_n$.

$$\mathbf{v} + \mathbf{w} = (\alpha_1 + \beta_1) \mathbf{v}_1 + \cdots + (\alpha_n + \beta_n) \mathbf{v}_n \in \text{Span}(\mathbf{v}_1, \dots, \mathbf{v}_n)$$

Therefore, $\text{Span}(\mathbf{v}_1, \dots, \mathbf{v}_n)$ is a subspace of V.   ◀

A vector $\mathbf{x}$ in R^3 is in $\text{Span}(\mathbf{e}_1, \mathbf{e}_2)$ if and only if it lies in the $x_1 x_2$ plane in 3-space. Thus we can think of the $x_1 x_2$ plane as the geometrical representation of the subspace $\text{Span}(\mathbf{e}_1, \mathbf{e}_2)$ (see Figure 3.2.1). Similarly, given two vectors $\mathbf{x}$ and $\mathbf{y}$, if $(0, 0, 0)$, (x_1, x_2, x_3), and (y_1, y_2, y_3) are not collinear, these points determine a plane. If $\mathbf{z} = c_1 \mathbf{x} + c_2 \mathbf{y}$, then $\mathbf{z}$ is a sum of vectors parallel to $\mathbf{x}$ and $\mathbf{y}$ and hence must lie on the plane determined by the two vectors (see Figure 3.2.2). In general, if two vectors $\mathbf{x}$ and $\mathbf{y}$ can be used to determine a plane in 3-space, that plane is the geometrical representation of $\text{Span}(\mathbf{x}, \mathbf{y})$.

Let $\mathbf{v}_1, \mathbf{v}_2, \dots, \mathbf{v}_n$ be vectors in a vector space V. We will refer to $\text{Span}(\mathbf{v}_1, \dots, \mathbf{v}_n)$ as the subspace of V *spanned* by $\mathbf{v}_1, \mathbf{v}_2, \dots, \mathbf{v}_n$. It may happen that $\text{Span}(\mathbf{v}_1, \dots, \mathbf{v}_n) = V$, in which case we say that the vectors $\mathbf{v}_1, \dots, \mathbf{v}_n$ *span* V or that $\{\mathbf{v}_1, \dots, \mathbf{v}_n\}$ is a *spanning set* for V. Thus we have the following definition.

FIGURE 3.2.2

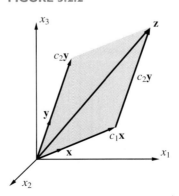

▶ **DEFINITION** The set $\{\mathbf{v}_1, \ldots, \mathbf{v}_n\}$ is a **spanning set** for V if and only if every vector in V can be written as a linear combination of $\mathbf{v}_1, \mathbf{v}_2, \ldots, \mathbf{v}_n$. ◀

EXAMPLE 11. Which of the following are spanning sets for R^3?

A. $\{\mathbf{e}_1, \mathbf{e}_2, \mathbf{e}_3, (1, 2, 3)^T\}$
B. $\{(1, 1, 1)^T, (1, 1, 0)^T, (1, 0, 0)^T\}$
C. $\{(1, 0, 1)^T, (0, 1, 0)^T\}$
D. $\{(1, 2, 4)^T, (2, 1, 3)^T, (4, -1, 1)^T\}$

SOLUTION. To determine whether a set spans R^3, we must determine whether an arbitrary vector $(a, b, c)^T$ in R^3 can be written as a linear combination of the vectors in the set. In part (A) it is easily seen that $(a, b, c)^T$ can be written as

$$(a, b, c)^T = a\mathbf{e}_1 + b\mathbf{e}_2 + c\mathbf{e}_3 + 0(1, 2, 3)^T$$

For part (B) we must determine whether it is possible to find constants $\alpha_1, \alpha_2, \alpha_3$ such that

$$\begin{bmatrix} a \\ b \\ c \end{bmatrix} = \alpha_1 \begin{bmatrix} 1 \\ 1 \\ 1 \end{bmatrix} + \alpha_2 \begin{bmatrix} 1 \\ 1 \\ 0 \end{bmatrix} + \alpha_3 \begin{bmatrix} 1 \\ 0 \\ 0 \end{bmatrix}$$

This leads to the system of equations

$$\alpha_1 + \alpha_2 + \alpha_3 = a$$
$$\alpha_1 + \alpha_2 \qquad = b$$
$$\alpha_1 \qquad\qquad = c$$

Since the coefficient matrix of the system is nonsingular, the system has a unique solution. In fact, we find that

$$\begin{bmatrix} \alpha_1 \\ \alpha_2 \\ \alpha_3 \end{bmatrix} = \begin{bmatrix} c \\ b - c \\ a - b \end{bmatrix}$$

Thus

$$\begin{bmatrix} a \\ b \\ c \end{bmatrix} = c \begin{bmatrix} 1 \\ 1 \\ 1 \end{bmatrix} + (b - c) \begin{bmatrix} 1 \\ 1 \\ 0 \end{bmatrix} + (a - b) \begin{bmatrix} 1 \\ 0 \\ 0 \end{bmatrix}$$

so the three vectors span R^3.

For part (C), we should note that linear combinations of $(1, 0, 1)^T$ and $(0, 1, 0)^T$ produce vectors of the form $(\alpha, \beta, \alpha)^T$. Thus any vector $(a, b, c)^T$ in R^3 where $a \neq c$ would not be in the span of these two vectors.

Part (D) can be done in the same manner as part (B). If

$$\begin{pmatrix} a \\ b \\ c \end{pmatrix} = \alpha_1 \begin{pmatrix} 1 \\ 2 \\ 4 \end{pmatrix} + \alpha_2 \begin{pmatrix} 2 \\ 1 \\ 3 \end{pmatrix} + \alpha_3 \begin{pmatrix} 4 \\ -1 \\ 1 \end{pmatrix}$$

then

$$\alpha_1 + 2\alpha_2 + 4\alpha_3 = a$$
$$2\alpha_1 + \alpha_2 - \alpha_3 = b$$
$$4\alpha_1 + 3\alpha_2 + \alpha_3 = c$$

In this case, however, the coefficient matrix is singular. Gaussian elimination will yield a system of the form

$$\alpha_1 + 2\alpha_2 + 4\alpha_3 = a$$
$$\alpha_2 + 3\alpha_3 = \frac{2a - b}{3}$$
$$0 = 2a - 3c + 5b$$

If

$$2a - 3c + 5b \neq 0$$

then the system is inconsistent. Hence, for most choices of a, b, c, it is impossible to express $(a, b, c)^T$ as a linear combination of $(1, 2, 4)^T$, $(2, 1, 3)^T$, $(4, -1, 1)^T$. The vectors do not span R^3. ◀

EXAMPLE 12. The vectors $1 - x^2$, $x + 2$, and x^2 span P_3. Thus, if $ax^2 + bx + c$ is any polynomial in P_3, it is possible to find scalars α_1, α_2, and α_3 such that

$$ax^2 + bx + c = \alpha_1(1 - x^2) + \alpha_2(x + 2) + \alpha_3 x^2$$

Indeed,

$$\alpha_1(1 - x^2) + \alpha_2(x + 2) + \alpha_3 x^2$$
$$= (\alpha_3 - \alpha_1)x^2 + \alpha_2 x + (\alpha_1 + 2\alpha_2)$$

Setting

$$\alpha_3 - \alpha_1 = a$$
$$\alpha_2 = b$$
$$\alpha_1 + 2\alpha_2 = c$$

and solving, we see that $\alpha_1 = c - 2b$, $\alpha_2 = b$, and $\alpha_3 = a + c - 2b$. ◀

In Example 11A we saw that the vectors $e_1, e_2, e_3, (1, 2, 3)^T$ span R^3. Clearly, R^3 could be spanned using only the vectors e_1, e_2, e_3. The vector $(1, 2, 3)^T$ is really not necessary. In the next section we consider the problem of finding minimal spanning sets for a vector space V (i.e., spanning sets that contain the smallest possible number of vectors).

EXERCISES

1. Determine whether the following sets form subspaces of R^2.

   (a) $\{(x_1, x_2)^T \mid x_1 + x_2 = 0\}$
   (b) $\{(x_1, x_2)^T \mid x_1 x_2 = 0\}$
   (c) $\{(x_1, x_2)^T \mid x_1 = 3x_2\}$
   (d) $\{(x_1, x_2)^T \mid |x_1| = |x_2|\}$
   (e) $\{(x_1, x_2)^T \mid x_1^2 = x_2^2\}$

2. Determine whether the following sets form subspaces of R^3.

   (a) $\{(x_1, x_2, x_3)^T \mid x_1 + x_3 = 1\}$
   (b) $\{(x_1, x_2, x_3)^T \mid x_1 = x_2 = x_3\}$
   (c) $\{(x_1, x_2, x_3)^T \mid x_3 = x_1 + x_2\}$
   (d) $\{(x_1, x_2, x_3)^T \mid x_3 = x_1 \text{ or } x_3 = x_2\}$

3. Determine whether the following are subspaces of $R^{2\times2}$.

   (a) The set of all 2×2 diagonal matrices
   (b) The set of all 2×2 triangular matrices
   (c) The set of all 2×2 matrices A such that $a_{12} = 1$
   (d) The set of all 2×2 matrices B such that $b_{11} = 0$
   (e) The set of all symmetric 2×2 matrices
   (f) The set of all singular 2×2 matrices

4. Determine the nullspace of each of the following matrices.

   (a) $\begin{bmatrix} 2 & 1 \\ 3 & 2 \end{bmatrix}$

   (b) $\begin{bmatrix} 1 & 2 & -3 & -1 \\ -2 & -4 & 6 & 3 \end{bmatrix}$

   (c) $\begin{bmatrix} 1 & 3 & -4 \\ 2 & -1 & -1 \\ -1 & -3 & 4 \end{bmatrix}$

   (d) $\begin{bmatrix} 1 & 1 & -1 & 2 \\ 2 & 2 & -3 & 1 \\ -1 & -1 & 0 & -5 \end{bmatrix}$

5. Determine whether the following are subspaces of P_4. (Be careful!)

   (a) The set of polynomials in P_4 of even degree
   (b) The set of all polynomials of degree 3
   (c) The set of all polynomials $p(x)$ in P_4 such that $p(0) = 0$
   (d) The set of all polynomials in P_4 having at least one real root

6. Determine whether the following are subspaces of $C[-1, 1]$.

(a) The set of functions f in $C[-1, 1]$ such that $f(-1) = f(1)$

(b) The set of odd functions in $C[-1, 1]$

(c) The set of continuous nondecreasing functions on $[-1, 1]$

(d) The set of functions f in $C[-1, 1]$ such that $f(-1) = 0$ and $f(1) = 0$

(e) The set of functions f in $C[-1, 1]$ such that $f(-1) = 0$ or $f(1) = 0$

7. Show that $C^n[a, b]$ is a subspace of $C[a, b]$.

8. Let A be a particular vector in $R^{2 \times 2}$. Determine whether the following are subspaces of $R^{2 \times 2}$.

(a) $S_1 = \{B \in R^{2 \times 2} \mid AB = BA\}$

(b) $S_2 = \{B \in R^{2 \times 2} \mid AB \neq BA\}$

(c) $S_3 = \{B \in R^{2 \times 2} \mid BA = O\}$

9. Determine whether the following are spanning sets for R^2.

(a) $\left\{ \begin{bmatrix} 2 \\ 1 \end{bmatrix}, \begin{bmatrix} 3 \\ 2 \end{bmatrix} \right\}$

(b) $\left\{ \begin{bmatrix} 2 \\ 3 \end{bmatrix}, \begin{bmatrix} 4 \\ 6 \end{bmatrix} \right\}$

(c) $\left\{ \begin{bmatrix} -2 \\ 1 \end{bmatrix}, \begin{bmatrix} 1 \\ 3 \end{bmatrix}, \begin{bmatrix} 2 \\ 4 \end{bmatrix} \right\}$

(d) $\left\{ \begin{bmatrix} -1 \\ 2 \end{bmatrix}, \begin{bmatrix} 1 \\ -2 \end{bmatrix}, \begin{bmatrix} 2 \\ -4 \end{bmatrix} \right\}$

(e) $\left\{ \begin{bmatrix} 1 \\ 2 \end{bmatrix}, \begin{bmatrix} -1 \\ 1 \end{bmatrix} \right\}$

10. Which of the following are spanning sets for R^3? Justify your answers.

(a) $\{(1, 0, 0)^T, (0, 1, 1)^T, (1, 0, 1)^T\}$

(b) $\{(1, 0, 0)^T, (0, 1, 1)^T, (1, 0, 1)^T, (1, 2, 3)^T\}$

(c) $\{(2, 1, -2)^T, (3, 2, -2)^T, (2, 2, 0)^T\}$

(d) $\{(2, 1, -2)^T, (-2, -1, 2)^T, (4, 2, -4)^T\}$

(e) $\{(1, 1, 3)^T, (0, 2, 1)^T\}$

11. Given

$$\mathbf{x}_1 = \begin{bmatrix} -1 \\ 2 \\ 3 \end{bmatrix}, \quad \mathbf{x}_2 = \begin{bmatrix} 3 \\ 4 \\ 2 \end{bmatrix}, \quad \mathbf{x} = \begin{bmatrix} 2 \\ 6 \\ 6 \end{bmatrix}, \quad \mathbf{y} = \begin{bmatrix} -9 \\ -2 \\ 5 \end{bmatrix}$$

(a) Is $\mathbf{x} \in \text{Span}(\mathbf{x}_1, \mathbf{x}_2)$?

(b) Is $\mathbf{y} \in \text{Span}(\mathbf{x}_1, \mathbf{x}_2)$?

Prove your answers.

12. Let $\{x_1, x_2, \ldots, x_k\}$ be a spanning set for a vector space V.

   (a) If we add an additional vector x_{k+1} to the set, will we still have a spanning set? Explain.

   (b) If we delete one of the vectors, say x_k, from the set, will we still have a spanning set? Explain.

13. In $R^{2\times2}$ let

$$E_{11} = \begin{bmatrix} 1 & 0 \\ 0 & 0 \end{bmatrix}, \qquad E_{12} = \begin{bmatrix} 0 & 1 \\ 0 & 0 \end{bmatrix}$$

$$E_{21} = \begin{bmatrix} 0 & 0 \\ 1 & 0 \end{bmatrix}, \qquad E_{22} = \begin{bmatrix} 0 & 0 \\ 0 & 1 \end{bmatrix}$$

Show that $E_{11}, E_{12}, E_{21}, E_{22}$ span $R^{2\times2}$.

14. Which of the following are spanning sets for P_3? Justify your answers.

   (a) $\{1, x^2, x^2 - 2\}$ (b) $\{2, x^2, x, 2x + 3\}$

   (c) $\{x + 2, x + 1, x^2 - 1\}$ (d) $\{x + 2, x^2 - 1\}$

15. Let S be the vector space of infinite sequences defined in Exercise 15 of Section 1. Let S_0 be the set of $\{a_n\}$ with the property $a_n \to 0$ as $n \to \infty$. Show that S_0 is a subspace of S.

16. Prove that if S is a subspace of R^1 either $S = \{0\}$ or $S = R^1$.

17. Let A be an $n \times n$ matrix. Prove that the following statements are equivalent.

   (a) $N(A) = \{0\}$.

   (b) A is nonsingular.

   (c) For each $b \in R^n$, the system $Ax = b$ has a unique solution.

18. Let U and V be subspaces of a vector space W. Prove that $U \cap V$ is also a subspace of W.

19. Let S be the subspace of R^2 spanned by e_1 and let T be the subspace of R^2 spanned by e_2. Is $S \cup T$ a subspace of R^2? Explain.

20. Let U and V be subspaces of a vector space W. Define

$$U + V = \{z \mid z = u + v \text{ where } u \in U \text{ and } v \in V\}$$

Show that $U + V$ is a subspace of W.

3   LINEAR INDEPENDENCE

In this section we look more closely at the structure of vector spaces. To begin with, we restrict ourselves to vector spaces that can be generated from a finite set of elements. Each vector in the vector space can be built up from the elements in this generating set using only the operations of addition and scalar multiplication. The generating set is usually referred to as a spanning set. In particular, it is desirable to find a "minimal" spanning set. By minimal we mean a spanning set with no unnecessary elements (i.e., all the elements in the set are needed in order to span the vector space). To see how to find a minimal spanning set, it is necessary to consider how the vectors in the collection "depend" on each other. Consequently, we introduce the concepts of *linear dependence* and *linear independence*. These simple concepts provide the keys to understanding the structure of vector spaces.

Consider the following vectors in R^3:

$$\mathbf{x}_1 = \begin{pmatrix} 1 \\ -1 \\ 2 \end{pmatrix}, \qquad \mathbf{x}_2 = \begin{pmatrix} -2 \\ 3 \\ 1 \end{pmatrix}, \qquad \mathbf{x}_3 = \begin{pmatrix} -1 \\ 3 \\ 8 \end{pmatrix}$$

Let S be the subspace of R^3 spanned by $\mathbf{x}_1, \mathbf{x}_2, \mathbf{x}_3$. Actually, S can be represented in terms of the two vectors $\mathbf{x}_1$ and $\mathbf{x}_2$, since the vector $\mathbf{x}_3$ is already in the span of $\mathbf{x}_1$ and $\mathbf{x}_2$.

$$(1) \qquad\qquad \mathbf{x}_3 = 3\mathbf{x}_1 + 2\mathbf{x}_2$$

Any linear combination of $\mathbf{x}_1, \mathbf{x}_2, \mathbf{x}_3$ can be reduced to a linear combination of $\mathbf{x}_1$ and $\mathbf{x}_2$:

$$\alpha_1\mathbf{x}_1 + \alpha_2\mathbf{x}_2 + \alpha_3\mathbf{x}_3 = \alpha_1\mathbf{x}_1 + \alpha_2\mathbf{x}_2 + \alpha_3(3\mathbf{x}_1 + 2\mathbf{x}_2)$$
$$= (\alpha_1 + 3\alpha_3)\mathbf{x}_1 + (\alpha_2 + 2\alpha_3)\mathbf{x}_2$$

Thus

$$S = \text{Span}(\mathbf{x}_1, \mathbf{x}_2, \mathbf{x}_3) = \text{Span}(\mathbf{x}_1, \mathbf{x}_2)$$

Equation (1) can be rewritten in the form

$$(2) \qquad\qquad 3\mathbf{x}_1 + 2\mathbf{x}_2 - 1\mathbf{x}_3 = \mathbf{0}$$

Since the three coefficients in (2) are nonzero, we could solve for any vector in terms of the other two.

$$\mathbf{x}_1 = -\frac{2}{3}\mathbf{x}_2 + \frac{1}{3}\mathbf{x}_3, \qquad \mathbf{x}_2 = -\frac{3}{2}\mathbf{x}_1 + \frac{1}{2}\mathbf{x}_3, \qquad \mathbf{x}_3 = 3\mathbf{x}_1 + 2\mathbf{x}_2$$

It follows that

$$\text{Span}(\mathbf{x}_1, \mathbf{x}_2, \mathbf{x}_3) = \text{Span}(\mathbf{x}_2, \mathbf{x}_3) = \text{Span}(\mathbf{x}_1, \mathbf{x}_3) = \text{Span}(\mathbf{x}_1, \mathbf{x}_2)$$

Because of the dependency relation (2), the subspace S can be represented as the span of any two of the given vectors.

On the other hand, no such dependency relationship exists between $\mathbf{x}_1$ and $\mathbf{x}_2$. Indeed, if there were scalars c_1 and c_2, not both 0, such that

$$(3) \qquad\qquad c_1\mathbf{x}_1 + c_2\mathbf{x}_2 = \mathbf{0}$$

then we could solve for one of the vectors in terms of the other.

$$\mathbf{x}_1 = -\frac{c_2}{c_1}\mathbf{x}_2 \quad (c_1 \neq 0) \qquad \text{or} \qquad \mathbf{x}_2 = -\frac{c_1}{c_2}\mathbf{x}_1 \quad (c_2 \neq 0)$$

However, neither of the two vectors in question is a multiple of the other. Therefore, $\text{Span}(\mathbf{x}_1)$ and $\text{Span}(\mathbf{x}_2)$ are both proper subspaces of $\text{Span}(\mathbf{x}_1, \mathbf{x}_2)$, and the only way that (3) can hold is if $c_1 = c_2 = 0$.

We can generalize this example by making the following observations.

(i) If $\mathbf{v}_1, \mathbf{v}_2, \ldots, \mathbf{v}_n$ span a vector space V and one of these vectors can be written as a linear combination of the other $n-1$ vectors, then those $n-1$ vectors span V.

(ii) Given n vectors $\mathbf{v}_1, \ldots, \mathbf{v}_n$, it is possible to write one of the vectors as a linear combination of the other $n-1$ vectors if and only if there exist scalars $c_1, \ldots, c_n$ not all zero such that

$$c_1 \mathbf{v}_1 + c_2 \mathbf{v}_2 + \cdots + c_n \mathbf{v}_n = \mathbf{0}$$

▶ **Proof of (i).** Suppose that $\mathbf{v}_n$ can be written as a linear combination of the vectors $\mathbf{v}_1, \mathbf{v}_2, \ldots, \mathbf{v}_{n-1}$.

$$\mathbf{v}_n = \beta_1 \mathbf{v}_1 + \beta_2 \mathbf{v}_2 + \cdots + \beta_{n-1} \mathbf{v}_{n-1}$$

Let $\mathbf{v}$ be any element of V. Since $\mathbf{v}_1, \ldots, \mathbf{v}_n$ span V, we can write

$$\mathbf{v} = \alpha_1 \mathbf{v}_1 + \alpha_2 \mathbf{v}_2 + \cdots + \alpha_{n-1} \mathbf{v}_{n-1} + \alpha_n \mathbf{v}_n$$

$$= \alpha_1 \mathbf{v}_1 + \alpha_2 \mathbf{v}_2 + \cdots + \alpha_{n-1} \mathbf{v}_{n-1} + \alpha_n (\beta_1 \mathbf{v}_1 + \cdots + \beta_{n-1} \mathbf{v}_{n-1})$$

$$= (\alpha_1 + \alpha_n \beta_1) \mathbf{v}_1 + (\alpha_2 + \alpha_n \beta_2) \mathbf{v}_2 + \cdots + (\alpha_{n-1} + \alpha_n \beta_{n-1}) \mathbf{v}_{n-1}$$

Thus any vector $\mathbf{v}$ in V can be written as a linear combination of $\mathbf{v}_1, \mathbf{v}_2, \ldots, \mathbf{v}_{n-1}$, and hence these vectors span V. ◀

▶ **Proof of (ii).** Suppose that one of the vectors $\mathbf{v}_1, \mathbf{v}_2, \ldots, \mathbf{v}_n$, say $\mathbf{v}_n$, can be written as a linear combination of the others.

$$\mathbf{v}_n = \alpha_1 \mathbf{v}_1 + \alpha_2 \mathbf{v}_2 + \cdots + \alpha_{n-1} \mathbf{v}_{n-1}$$

If we set $c_i = \alpha_i$ for $i = 1, \ldots, n-1$ and $c_n = -1$, it follows that

$$\sum_{i=1}^{n} c_i \mathbf{v}_i = \sum_{i=1}^{n-1} \alpha_i \mathbf{v}_i - \sum_{i=1}^{n-1} \alpha_i \mathbf{v}_i = \mathbf{0}$$

Conversely, if

$$c_1 \mathbf{v}_1 + c_2 \mathbf{v}_2 + \cdots + c_n \mathbf{v}_n = \mathbf{0}$$

and at least one of the c_i's, say c_n, is nonzero, then

$$\mathbf{v}_n = \frac{-c_1}{c_n} \mathbf{v}_1 + \frac{-c_2}{c_n} \mathbf{v}_2 + \cdots + \frac{-c_{n-1}}{c_n} \mathbf{v}_{n-1} \qquad ◀$$

▶ **DEFINITION** The vectors $\mathbf{v}_1, \mathbf{v}_2, \ldots, \mathbf{v}_n$ in a vector space V are said to be **linearly independent** if

$$c_1 \mathbf{v}_1 + c_2 \mathbf{v}_2 + \cdots + c_n \mathbf{v}_n = \mathbf{0}$$

implies that all the scalars $c_1, \ldots, c_n$ must equal 0. ◀

It follows from (i) and (ii) that, if $\{\mathbf{v}_1, \mathbf{v}_2, \ldots, \mathbf{v}_n\}$ is a minimal spanning set, then $\mathbf{v}_1, \mathbf{v}_2, \ldots, \mathbf{v}_n$ are linearly independent. Conversely, if $\mathbf{v}_1, \ldots, \mathbf{v}_n$ are linearly independent and span V, then $\{\mathbf{v}_1, \ldots, \mathbf{v}_n\}$ is a minimal spanning set for V (see

Exercise 17). A minimal spanning set is called a *basis*. The concept of a basis will be studied in more detail in the next section.

EXAMPLE I. The vectors $\begin{bmatrix} 1 \\ 1 \end{bmatrix}$ and $\begin{bmatrix} 1 \\ 2 \end{bmatrix}$ are linearly independent, since if

$$c_1 \begin{bmatrix} 1 \\ 1 \end{bmatrix} + c_2 \begin{bmatrix} 1 \\ 2 \end{bmatrix} = \begin{bmatrix} 0 \\ 0 \end{bmatrix}$$

then

$$c_1 + \ c_2 = 0$$
$$c_1 + 2c_2 = 0$$

and the only solution to this system is $c_1 = 0, c_2 = 0$. ◄

▶ **DEFINITION** The vectors $\mathbf{v}_1, \mathbf{v}_2, \ldots, \mathbf{v}_n$ in a vector space V are said to be **linearly dependent** if there exist scalars $c_1, c_2, \ldots, c_n$ not all zero such that

$$c_1 \mathbf{v}_1 + c_2 \mathbf{v}_2 + \cdots + c_n \mathbf{v}_n = \mathbf{0}$$

◄

EXAMPLE 2. Let $\mathbf{x} = (1, 2, 3)^T$. The vectors $\mathbf{e}_1, \mathbf{e}_2, \mathbf{e}_3, \mathbf{x}$ are linearly dependent, since

$$\mathbf{e}_1 + 2\mathbf{e}_2 + 3\mathbf{e}_3 - \mathbf{x} = \mathbf{0}$$

(In this case $c_1 = 1, c_2 = 2, c_3 = 3, c_4 = -1$.) ◄

Given a set of vectors $\{\mathbf{v}_1, \mathbf{v}_2, \ldots, \mathbf{v}_n\}$ in a vector space V, it is trivial to find scalars $c_1, c_2, \ldots, c_n$ such that

$$c_1 \mathbf{v}_1 + c_2 \mathbf{v}_2 + \cdots + c_n \mathbf{v}_n = \mathbf{0}$$

Just take

$$c_1 = c_2 = \cdots = c_n = 0$$

> If there are nontrivial choices of scalars for which the linear combination $c_1 \mathbf{v}_1 + \cdots + c_n \mathbf{v}_n$ equals the zero vector, then $\mathbf{v}_1, \ldots, \mathbf{v}_n$ are linearly dependent. If the *only* way the linear combination $c_1 \mathbf{v}_1 + \cdots + c_n \mathbf{v}_n$ can equal the zero vector is for all the scalars $c_1, \ldots, c_n$ to be 0, then $\mathbf{v}_1, \ldots, \mathbf{v}_n$ are linearly independent.

Geometric Interpretation

If $\mathbf{x}$ and $\mathbf{y}$ are linearly dependent in R^2, then

$$c_1 \mathbf{x} + c_2 \mathbf{y} = \mathbf{0}$$

where c_1 and c_2 are not both 0. If, say, $c_1 \neq 0$, we can write

$$\mathbf{x} = -\frac{c_2}{c_1}\mathbf{y}$$

If two vectors in R^2 are linearly dependent, one of the vectors can be written as a scalar multiple of the other. Thus, if both vectors are placed at the origin, they will lie along the same line (see Figure 3.3.1).

If

$$\mathbf{x} = \begin{bmatrix} x_1 \\ x_2 \\ x_3 \end{bmatrix} \quad \text{and} \quad \mathbf{y} = \begin{bmatrix} y_1 \\ y_2 \\ y_3 \end{bmatrix}$$

are linearly independent in R^3, then the two points (x_1, x_2, x_3) and (y_1, y_2, y_3) will not lie on the same line through the origin in 3-space. Since $(0, 0, 0)$, (x_1, x_2, x_3), and (y_1, y_2, y_3) are not collinear, they determine a plane. If (z_1, z_2, z_3) lies on this plane, the vector $\mathbf{z} = (z_1, z_2, z_3)^T$ can be written as a linear combination of $\mathbf{x}$ and $\mathbf{y}$, and hence $\mathbf{x}$, $\mathbf{y}$, and $\mathbf{z}$ are linearly dependent. If (z_1, z_2, z_3) does not lie on the plane, the three vectors will be linearly independent (see Figure 3.3.2).

FIGURE 3.3.1

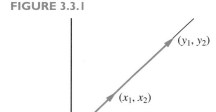

(a) $\mathbf{x}$ and $\mathbf{y}$ linearly dependent

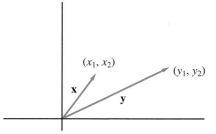

(b) $\mathbf{x}$ and $\mathbf{y}$ linearly independent

FIGURE 3.3.2

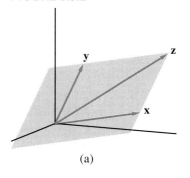

(a)

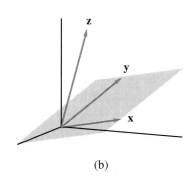

(b)

Theorems and Examples

EXAMPLE 3.   Which of the following collections of vectors are linearly independent in R^3?

A. $(1, 1, 1)^T$, $(1, 1, 0)^T$, $(1, 0, 0)^T$
B. $(1, 0, 1)^T$, $(0, 1, 0)^T$
C. $(1, 2, 4)^T$, $(2, 1, 3)^T$, $(4, -1, 1)^T$

SOLUTION.

A. These three vectors are linearly independent. To verify this, we must show that the only way for

$$(4) \qquad c_1(1, 1, 1)^T + c_2(1, 1, 0)^T + c_3(1, 0, 0)^T = (0, 0, 0)^T$$

is if the scalars c_1, c_2, c_3 are all zero. Equation (4) can be written as a linear system with unknowns c_1, c_2, c_3.

$$c_1 + c_2 + c_3 = 0$$
$$c_1 + c_2 \qquad = 0$$
$$c_1 \qquad\qquad = 0$$

The only solution to this system is $c_1 = 0$, $c_2 = 0$, $c_3 = 0$.

B. If

$$c_1(1, 0, 1)^T + c_2(0, 1, 0)^T = (0, 0, 0)^T$$

then

$$(c_1, c_2, c_1)^T = (0, 0, 0)^T$$

so $c_1 = c_2 = 0$. Therefore, the two vectors are linearly independent.

C. If

$$c_1(1, 2, 4)^T + c_2(2, 1, 3)^T + c_3(4, -1, 1)^T = (0, 0, 0)^T$$

then

$$c_1 + 2c_2 + 4c_3 = 0$$
$$2c_1 + c_2 - c_3 = 0$$
$$4c_1 + 3c_2 + c_3 = 0$$

The coefficient matrix of the system is singular. Hence the system has nontrivial solutions, and hence the vectors are linearly dependent.   ◀

Notice in Example 3, parts (A) and (C), that it was necessary to solve a 3 × 3 system to determine whether the three vectors were linearly independent. In part (A), where the coefficient matrix was nonsingular, the vectors were linearly independent, while in part (C), where the coefficient matrix was singular, the vectors were linearly dependent. This illustrates a special case of the following theorem.

▶ **THEOREM 3.3.1** *Let* $x_1, x_2, \ldots, x_n$ *be n vectors in* R^n *and let*

$$x_i = (x_{1i}, x_{2i}, \ldots, x_{ni})^T$$

for $i = 1, \ldots, n$. *If* $X = (x_1, x_2, \ldots, x_n)$, *then the vectors* $x_1, x_2, \ldots, x_n$ *will be linearly dependent if and only if X is singular.*

▶ *Proof.* The equation

$$c_1 x_1 + c_2 x_2 + \cdots + c_n x_n = 0$$

is equivalent to the system of equations

$$c_1 x_{11} + c_2 x_{12} + \cdots + c_n x_{1n} = 0$$
$$c_1 x_{21} + c_2 x_{22} + \cdots + c_n x_{2n} = 0$$
$$\vdots$$
$$c_1 x_{n1} + c_2 x_{n2} + \cdots + c_n x_{nn} = 0$$

If we let $c = (c_1, c_2, \ldots, c_n)^T$, the system can be written as a matrix equation

$$Xc = 0$$

This equation will have a nontrivial solution if and only if X is singular. Thus $x_1, \ldots, x_n$ will be linearly dependent if and only if X is singular.   ◀

We can use Theorem 3.3.1 to test whether n vectors are linearly independent in R^n. Simply form a matrix X whose columns are the vectors being tested. To determine whether X is singular, calculate the value of $\det(X)$. If $\det(X) = 0$, the vectors are linearly dependent. If $\det(X) \neq 0$, the vectors are linearly independent.

EXAMPLE 4. Determine whether the vectors $(4, 2, 3)^T$, $(2, 3, 1)^T$, and $(2, -5, 3)^T$ are linearly dependent.

SOLUTION. Since

$$\begin{vmatrix} 4 & 2 & 2 \\ 2 & 3 & -5 \\ 3 & 1 & 3 \end{vmatrix} = 0$$

the vectors are linearly dependent.   ◀

Next we consider a very important property of linearly independent vectors. Linear combinations of linearly independent vectors are unique. More precisely, we have the following theorem.

▶ **THEOREM 3.3.2** *Let* $v_1, \ldots, v_n$ *be vectors in a vector space V. A vector* $v \in$ *Span*$(v_1, \ldots, v_n)$ *can be written uniquely as a linear combination of* $v_1, \ldots, v_n$ *if and only if* $v_1, \ldots, v_n$ *are linearly independent.*

▶ *Proof.* If $\mathbf{v} \in \text{Span}(\mathbf{v}_1, \dots, \mathbf{v}_n)$, then $\mathbf{v}$ can be written as a linear combination

(5)
$$\mathbf{v} = \alpha_1 \mathbf{v}_1 + \alpha_2 \mathbf{v}_2 + \cdots + \alpha_n \mathbf{v}_n$$

Suppose that $\mathbf{v}$ can also be expressed as a linear combination

(6)
$$\mathbf{v} = \beta_1 \mathbf{v}_1 + \beta_2 \mathbf{v}_2 + \cdots + \beta_n \mathbf{v}_n$$

We will show that, if $\mathbf{v}_1, \dots, \mathbf{v}_n$ are linearly independent, then $\beta_i = \alpha_i, i = 1, \dots, n$, and if $\mathbf{v}_1, \dots, \mathbf{v}_n$ are linearly dependent, then it is possible to choose the β_i's different from the α_i's.

If $\mathbf{v}_1, \dots, \mathbf{v}_n$ are linearly independent, then subtracting (6) from (5) yields

(7)
$$(\alpha_1 - \beta_1)\mathbf{v}_1 + (\alpha_2 - \beta_2)\mathbf{v}_2 + \cdots + (\alpha_n - \beta_n)\mathbf{v}_n = \mathbf{0}$$

By the linear independence of $\mathbf{v}_1, \dots, \mathbf{v}_n$, the coefficients of (7) must all be 0. Hence

$$\alpha_1 = \beta_1, \quad \alpha_2 = \beta_2, \quad \dots, \quad \alpha_n = \beta_n$$

Thus the representation (5) is unique when $\mathbf{v}_1, \dots, \mathbf{v}_n$ are linearly independent.

On the other hand, if $\mathbf{v}_1, \dots, \mathbf{v}_n$ are linearly dependent, there exist $c_1, \dots, c_n$ not all 0 such that

(8)
$$\mathbf{0} = c_1 \mathbf{v}_1 + c_2 \mathbf{v}_2 + \cdots + c_n \mathbf{v}_n$$

Now if we set

$$\beta_1 = \alpha_1 + c_1, \quad \beta_2 = \alpha_2 + c_2, \quad \dots, \quad \beta_n = \alpha_n + c_n$$

then, adding (5) and (8), we get

$$\mathbf{v} = (\alpha_1 + c_1)\mathbf{v}_1 + (\alpha_2 + c_2)\mathbf{v}_2 + \cdots + (\alpha_n + c_n)\mathbf{v}_n$$
$$= \beta_1 \mathbf{v}_1 + \beta_2 \mathbf{v}_2 + \cdots + \beta_n \mathbf{v}_n$$

Since the c_i's are not all 0, $\beta_i \neq \alpha_i$ for at least one value of i. Thus, if $\mathbf{v}_1, \dots, \mathbf{v}_n$ are linearly dependent, the representation of a vector as a linear combination of $\mathbf{v}_1, \dots, \mathbf{v}_n$ is not unique.   ◀

Vector Spaces of Functions

To determine whether a set of vectors is linearly independent in R^n, we must solve a homogeneous linear system of equations. A similar situation holds for the vector space P_n.

The Vector Space P_n

To test whether the following polynomials $p_1, p_2, \dots, p_k$ are linearly independent in P_n, we set

(9)
$$c_1 p_1 + c_2 p_2 + \cdots + c_k p_k = z$$

where z represents the zero polynomial.

$$z(x) = 0x^{n-1} + 0x^{n-2} + \cdots + 0x + 0$$

If the polynomial on the left-hand side of equation (9) is rewritten in the form $a_1 x^{n-1} + a_2 x^{n-2} + \cdots + a_{n-1}x + a_n$, then, since two polynomials are equal if and only if their coefficients are equal, it follows that the coefficients a_i must all be 0. But each of the a_i's is a linear combination of the c_j's. This leads to a homogeneous linear system with unknowns $c_1, c_2, \dots, c_k$. If the system has only the trivial solution, the polynomials are linearly independent; otherwise, they are linearly dependent.

EXAMPLE 5. To test whether the vectors

$$p_1(x) = x^2 - 2x + 3 \quad p_2(x) = 2x^2 + x + 8 \quad p_3(x) = x^2 + 8x + 7$$

are linearly independent, set

$$c_1 p_1(x) + c_2 p_2(x) + c_3 p_3(x) = 0x^2 + 0x + 0$$

Grouping terms by powers of x, we get

$$(c_1 + 2c_2 + c_3)x^2 + (-2c_1 + c_2 + 8c_3)x + (3c_1 + 8c_2 + 7c_3) = 0x^2 + 0x + 0$$

Equating coefficients leads to the system

$$
\begin{aligned}
c_1 + 2c_2 + c_3 &= 0 \\
-2c_1 + c_2 + 8c_3 &= 0 \\
3c_1 + 8c_2 + 7c_3 &= 0
\end{aligned}
$$

The coefficient matrix to this system is singular and hence there are nontrivial solutions. Therefore, p_1, p_2, p_3 are linearly dependent.   ◀

The Vector Space $C^{(n-1)}[a, b]$

In Example 4 a determinant was used to test whether three vectors were linearly independent in R^3. Determinants can also be used to help to decide if a set of n vectors is linearly independent in $C^{(n-1)}[a, b]$. Indeed, let $f_1, f_2, \ldots, f_n$ be elements of $C^{(n-1)}[a, b]$. If these vectors are linearly dependent, then there exist scalars $c_1, c_2, \ldots, c_n$ not all zero such that

(10) $$c_1 f_1(x) + c_2 f_2(x) + \cdots + c_n f_n(x) = 0$$

for each x in $[a, b]$. Taking the derivative with respect to x of both sides of (10) yields

$$c_1 f_1'(x) + c_2 f_2'(x) + \cdots + c_n f_n'(x) = 0$$

If we continue taking derivatives of both sides, we end up with the system

$$
\begin{aligned}
c_1 f_1(x) &+ c_2 f_2(x) &+ \cdots + &c_n f_n(x) &= 0 \\
c_1 f_1'(x) &+ c_2 f_2'(x) &+ \cdots + &c_n f_n'(x) &= 0 \\
&\vdots \\
c_1 f_1^{(n-1)}(x) &+ c_2 f_2^{(n-1)}(x) &+ \cdots + &c_n f_n^{(n-1)}(x) &= 0
\end{aligned}
$$

For each fixed x in $[a, b]$, the matrix equation

(11)
$$\begin{bmatrix} f_1(x) & f_2(x) & \cdots & f_n(x) \\ f_1'(x) & f_2'(x) & \cdots & f_n'(x) \\ \vdots & & & \\ f_1^{(n-1)}(x) & f_2^{(n-1)}(x) & \cdots & f_n^{(n-1)}(x) \end{bmatrix} \begin{bmatrix} \alpha_1 \\ \alpha_2 \\ \vdots \\ \alpha_n \end{bmatrix} = \begin{bmatrix} 0 \\ 0 \\ \vdots \\ 0 \end{bmatrix}$$

will have the same nontrivial solution $(c_1, c_2, \ldots, c_n)^T$. Thus, if $f_1, \ldots, f_n$ are linearly dependent in $C^{(n-1)}[a, b]$, then, for each fixed x in $[a, b]$, the coefficient matrix of system (11) is singular. If the matrix is singular, its determinant is zero.

▶ **DEFINITION** Let $f_1, f_2, \ldots, f_n$ be functions in $C^{(n-1)}[a, b]$, and define the function $W[f_1, f_2, \ldots, f_n](x)$ on $[a, b]$ by

$$W[f_1, f_2, \ldots, f_n](x) = \begin{vmatrix} f_1(x) & f_2(x) & \cdots & f_n(x) \\ f_1'(x) & f_2'(x) & \cdots & f_n'(x) \\ \vdots & & & \\ f_1^{(n-1)}(x) & f_2^{(n-1)}(x) & \cdots & f_n^{(n-1)}(x) \end{vmatrix}$$

The function $W[f_1, f_2, \ldots, f_n]$ is called the **Wronskian** of $f_1, f_2, \ldots, f_n$. ◀

▶ **THEOREM 3.3.3** *Let $f_1, f_2, \ldots, f_n$ be elements of $C^{(n-1)}[a, b]$. If there exists a point x_0 in $[a, b]$ such that $W[f_1, f_2, \ldots, f_n](x_0) \neq 0$, then $f_1, f_2, \ldots, f_n$ are linearly independent.*

▶*Proof.* If $f_1, f_2, \ldots, f_n$ were linearly dependent, then by the preceding discussion the coefficient matrix in (11) would be singular for each x in $[a, b]$ and hence $W[f_1, f_2, \ldots, f_n](x)$ would be identically zero on $[a, b]$. ◀

If $f_1, f_2, \ldots, f_n$ are linearly independent in $C^{(n-1)}[a, b]$, they will also be linearly independent in $C[a, b]$.

EXAMPLE 6. Show that e^x and e^{-x} are linearly independent in $C(-\infty, \infty)$.

SOLUTION.

$$W[e^x, e^{-x}] = \begin{vmatrix} e^x & e^{-x} \\ e^x & -e^{-x} \end{vmatrix} = -2$$

Since $W[e^x, e^{-x}]$ is not identically zero, e^x and e^{-x} are linearly independent. ◀

EXAMPLE 7. Consider the functions x^2 and $x|x|$ in $C[-1, 1]$. Both functions are in the subspace $C^1[-1, 1]$ (see Example 7 of Section 2), so we can compute the Wronskian

$$W[x^2, x|x|] = \begin{vmatrix} x^2 & x|x| \\ 2x & 2|x| \end{vmatrix} \equiv 0$$

Since the Wronskian is identically zero, it gives no information as to whether the functions are linearly independent. To answer the question, suppose that

$$c_1 x^2 + c_2 x|x| = 0$$

for all x in $[-1, 1]$. Then, in particular for $x = 1$ and $x = -1$, we have

$$c_1 + c_2 = 0$$
$$c_1 - c_2 = 0$$

and the only solution to this system is $c_1 = c_2 = 0$. Thus the functions x^2 and $x|x|$ are linearly independent in $C[-1, 1]$ even though $W[x^2, x|x|] \equiv 0$.

This example shows that the converse to Theorem 3.3.3 is not valid.   ◄

EXAMPLE 8. Show that the vectors $1, x, x^2, x^3$ are linearly independent in P_4.

SOLUTION.

$$W[1, x, x^2, x^3] = \begin{vmatrix} 1 & x & x^2 & x^3 \\ 0 & 1 & 2x & 3x^2 \\ 0 & 0 & 2 & 6x \\ 0 & 0 & 0 & 6 \end{vmatrix} = 12$$

Since $W[1, x, x^2, x^3] \neq 0$, the vectors are linearly independent.   ◄

EXERCISES

1. Determine whether the following vectors are linearly independent in R^2.

(a) $\begin{bmatrix} 2 \\ 1 \end{bmatrix}, \begin{bmatrix} 3 \\ 2 \end{bmatrix}$

(b) $\begin{bmatrix} 2 \\ 3 \end{bmatrix}, \begin{bmatrix} 4 \\ 6 \end{bmatrix}$

(c) $\begin{bmatrix} -2 \\ 1 \end{bmatrix}, \begin{bmatrix} 1 \\ 3 \end{bmatrix}, \begin{bmatrix} 2 \\ 4 \end{bmatrix}$

(d) $\begin{bmatrix} -1 \\ 2 \end{bmatrix}, \begin{bmatrix} 1 \\ -2 \end{bmatrix}, \begin{bmatrix} 2 \\ -4 \end{bmatrix}$

(e) $\begin{bmatrix} 1 \\ 2 \end{bmatrix}, \begin{bmatrix} -1 \\ 1 \end{bmatrix}$

2. Determine whether the following vectors are linearly independent in R^3.

(a) $\begin{bmatrix} 1 \\ 0 \\ 0 \end{bmatrix}, \begin{bmatrix} 0 \\ 1 \\ 1 \end{bmatrix}, \begin{bmatrix} 1 \\ 0 \\ 1 \end{bmatrix}$

(b) $\begin{bmatrix} 1 \\ 0 \\ 0 \end{bmatrix}, \begin{bmatrix} 0 \\ 1 \\ 1 \end{bmatrix}, \begin{bmatrix} 1 \\ 0 \\ 1 \end{bmatrix}, \begin{bmatrix} 1 \\ 2 \\ 3 \end{bmatrix}$

(c) $\begin{bmatrix} 2 \\ 1 \\ -2 \end{bmatrix}, \begin{bmatrix} 3 \\ 2 \\ -2 \end{bmatrix}, \begin{bmatrix} 2 \\ 2 \\ 0 \end{bmatrix}$ (d) $\begin{bmatrix} 2 \\ 1 \\ -2 \end{bmatrix}, \begin{bmatrix} -2 \\ -1 \\ 2 \end{bmatrix}, \begin{bmatrix} 4 \\ 2 \\ -4 \end{bmatrix}$

(e) $\begin{bmatrix} 1 \\ 1 \\ 3 \end{bmatrix}, \begin{bmatrix} 0 \\ 2 \\ 1 \end{bmatrix}$

3. For each of the sets of vectors in Exercise 2, describe geometrically the span of the given vectors.

4. Determine whether the following vectors are linearly independent in $R^{2 \times 2}$.

(a) $\begin{bmatrix} 1 & 0 \\ 1 & 1 \end{bmatrix}, \begin{bmatrix} 0 & 1 \\ 0 & 0 \end{bmatrix}$ (b) $\begin{bmatrix} 1 & 0 \\ 0 & 1 \end{bmatrix}, \begin{bmatrix} 0 & 1 \\ 0 & 0 \end{bmatrix}, \begin{bmatrix} 0 & 0 \\ 1 & 0 \end{bmatrix}$

(c) $\begin{bmatrix} 1 & 0 \\ 0 & 1 \end{bmatrix}, \begin{bmatrix} 0 & 1 \\ 0 & 0 \end{bmatrix}, \begin{bmatrix} 2 & 3 \\ 0 & 2 \end{bmatrix}$

5. Let $\mathbf{x}_1, \mathbf{x}_2, \dots, \mathbf{x}_k$ be linearly independent vectors in a vector space V.

(a) If we add a vector $\mathbf{x}_{k+1}$ to the collection, will we still have a linearly independent collection of vectors? Explain.

(b) If we delete a vector, say $\mathbf{x}_k$, from the collection, will we still have a linearly independent collection of vectors? Explain.

6. Determine whether the following vectors are linearly independent in P_3.

(a) $1, x^2, x^2 - 2$ (b) $2, x^2, x, 2x + 3$

(c) $x + 2, x + 1, x^2 - 1$ (d) $x + 2, x^2 - 1$

7. For each of the following, show that the given vectors are linearly independent in $C[0, 1]$.

(a) $\cos \pi x, \ \sin \pi x$ (b) $x^{3/2}, \ x^{5/2}$

(c) $1, \ e^x + e^{-x}, \ e^x - e^{-x}$ (d) $e^x, \ e^{-x}, \ e^{2x}$

8. Determine whether the vectors $\cos x, 1, \sin^2(x/2)$ are linearly independent in $C[-\pi, \pi]$.

9. Consider the vectors $\cos(x + \alpha)$ and $\sin x$ in $C[-\pi, \pi]$. For what values of α will the two vectors be linearly dependent? Give a graphical interpretation of your answer.

10. Given the functions $2x$ and $|x|$, show that:

(a) These two vectors are linearly independent in $C[-1, 1]$.

(b) The vectors are linearly dependent in $C[0, 1]$.

11. Prove that any finite set of vectors that contains the zero vector must be linearly dependent.

12. Let $\mathbf{v}_1, \mathbf{v}_2$ be two vectors in a vector space V. Show that $\mathbf{v}_1$ and $\mathbf{v}_2$ are linearly dependent if and only if one of the vectors is a scalar multiple of the other.

13. Prove that any nonempty subset of a linearly independent set of vectors $\{\mathbf{v}_1, \ldots, \mathbf{v}_n\}$ is also linearly independent.

14. Let A be an $m \times n$ matrix. Show that if A has linearly independent column vectors then $N(A) = \{\mathbf{0}\}$. [*Hint:* For any $\mathbf{x} \in R^n$, $A\mathbf{x} = x_1\mathbf{a}_1 + x_2\mathbf{a}_2 + \cdots + x_n\mathbf{a}_n$.]

15. Let $\mathbf{x}_1, \ldots, \mathbf{x}_k$ be linearly independent vectors in R^n, and let A be a nonsingular $n \times n$ matrix. Define $\mathbf{y}_i = A\mathbf{x}_i$ for $i = 1, \ldots, k$. Show that $\mathbf{y}_1, \ldots, \mathbf{y}_k$ are linearly independent.

16. Let $\{\mathbf{v}_1, \ldots, \mathbf{v}_n\}$ be a spanning set for the vector space V, and let $\mathbf{v}$ be any other vector in V. Show that $\mathbf{v}, \mathbf{v}_1, \ldots, \mathbf{v}_n$ are linearly dependent.

17. Let $\mathbf{v}_1, \mathbf{v}_2, \ldots, \mathbf{v}_n$ be linearly independent vectors in a vector space V. Show that $\mathbf{v}_2, \ldots, \mathbf{v}_n$ cannot span V.

4 BASIS AND DIMENSION

In Section 3 we showed that a spanning set for a vector space is minimal if its elements are linearly independent. The elements of a minimal spanning set form the basic building blocks for the whole vector space and, consequently, we say that they form a "basis" for the vector space.

▶ **DEFINITION** The vectors $\mathbf{v}_1, \mathbf{v}_2, \ldots, \mathbf{v}_n$ form a **basis** for a vector space V if and only if

    (i) $\mathbf{v}_1, \ldots, \mathbf{v}_n$ are linearly independent.
    (ii) $\mathbf{v}_1, \ldots, \mathbf{v}_n$ span V.   ◀

EXAMPLE I.   The "standard basis" for R^3 is $\{\mathbf{e}_1, \mathbf{e}_2, \mathbf{e}_3\}$; however, there are many bases that we could choose for R^3. For example,

$$\left\{ \begin{bmatrix} 1 \\ 1 \\ 1 \end{bmatrix}, \begin{bmatrix} 0 \\ 1 \\ 1 \end{bmatrix}, \begin{bmatrix} 2 \\ 0 \\ 1 \end{bmatrix} \right\} \quad \text{and} \quad \left\{ \begin{bmatrix} 1 \\ 1 \\ 1 \end{bmatrix}, \begin{bmatrix} 1 \\ 1 \\ 0 \end{bmatrix}, \begin{bmatrix} 1 \\ 0 \\ 1 \end{bmatrix} \right\}$$

are both bases for R^3. We will see shortly that any basis for R^3 must have exactly three elements.   ◀

EXAMPLE 2.   In $R^{2 \times 2}$, consider the set $\{E_{11}, E_{12}, E_{21}, E_{22}\}$, where

$$E_{11} = \begin{bmatrix} 1 & 0 \\ 0 & 0 \end{bmatrix}, \qquad E_{12} = \begin{bmatrix} 0 & 1 \\ 0 & 0 \end{bmatrix},$$

$$E_{21} = \begin{bmatrix} 0 & 0 \\ 1 & 0 \end{bmatrix}, \qquad E_{22} = \begin{bmatrix} 0 & 0 \\ 0 & 1 \end{bmatrix}$$

If

$$c_1 E_{11} + c_2 E_{12} + c_3 E_{21} + c_4 E_{22} = O$$

then

$$\begin{bmatrix} c_1 & c_2 \\ c_3 & c_4 \end{bmatrix} = \begin{bmatrix} 0 & 0 \\ 0 & 0 \end{bmatrix}$$

so $c_1 = c_2 = c_3 = c_4 = 0$. Therefore, $E_{11}, E_{12}, E_{21}, E_{22}$ are linearly independent. If A is in $R^{2\times 2}$, then

$$A = a_{11} E_{11} + a_{12} E_{12} + a_{21} E_{21} + a_{22} E_{22}$$

Thus $E_{11}, E_{12}, E_{21}, E_{22}$ span $R^{2\times 2}$ and hence form a basis for $R^{2\times 2}$. ◀

In many applications it is necessary to find a particular subspace of a vector space V. This can be done by finding the basis elements of the subspace. For example, to find all solutions to the system

$$x_1 + x_2 + x_3 \qquad\quad = 0$$
$$2x_1 + x_2 \qquad\quad + x_4 = 0$$

we must find the nullspace of the matrix

$$A = \begin{bmatrix} 1 & 1 & 1 & 0 \\ 2 & 1 & 0 & 1 \end{bmatrix}$$

In Example 9 of Section 2 we saw that $N(A)$ is the subspace of R^4 spanned by the vectors

$$\begin{bmatrix} 1 \\ -2 \\ 1 \\ 0 \end{bmatrix} \quad \text{and} \quad \begin{bmatrix} -1 \\ 1 \\ 0 \\ 1 \end{bmatrix}$$

Since these two vectors are linearly independent, they form a basis for $N(A)$.

▶ **THEOREM 3.4.1** *If* $\{\mathbf{v}_1, \mathbf{v}_2, \dots, \mathbf{v}_n\}$ *is a spanning set for a vector space V, then any collection of m vectors in V, where $m > n$, is linearly dependent.*

▶*Proof.* Let $\mathbf{u}_1, \mathbf{u}_2, \dots, \mathbf{u}_m$ be m vectors in V where $m > n$. Then, since $\mathbf{v}_1, \mathbf{v}_2, \dots, \mathbf{v}_n$ span V, we have

$$\mathbf{u}_i = a_{i1}\mathbf{v}_1 + a_{i2}\mathbf{v}_2 + \cdots + a_{in}\mathbf{v}_n \qquad \text{for} \qquad i = 1, 2, \dots, m$$

A linear combination $c_1\mathbf{u}_1 + c_2\mathbf{u}_2 + \cdots + c_m\mathbf{u}_m$ can be written in the form

$$c_1 \sum_{j=1}^{n} a_{1j}\mathbf{v}_j + c_2 \sum_{j=1}^{n} a_{2j}\mathbf{v}_j + \cdots + c_m \sum_{j=1}^{n} a_{mj}\mathbf{v}_j$$

Rearranging the terms, we see that

$$c_1\mathbf{u}_1 + c_2\mathbf{u}_2 + \cdots + c_m\mathbf{u}_m = \sum_{i=1}^{m} \left[c_i \left(\sum_{j=1}^{n} a_{ij}\mathbf{v}_j \right) \right]$$

$$= \sum_{j=1}^{n} \left(\sum_{i=1}^{m} a_{ij}c_i \right) \mathbf{v}_j$$

Now consider the system of equations

$$\sum_{i=1}^{m} a_{ij}c_i = 0 \qquad j = 1, 2, \ldots, n$$

This is a homogeneous system with more unknowns than equations. Therefore, by Theorem 1.2.1, the system must have a nontrivial solution $(\hat{c}_1, \hat{c}_2, \ldots, \hat{c}_m)^T$. But then

$$\hat{c}_1\mathbf{u}_1 + \hat{c}_2\mathbf{u}_2 + \cdots + \hat{c}_m\mathbf{u}_m = \sum_{j=1}^{n} 0\mathbf{v}_j = \mathbf{0}$$

and hence $\mathbf{u}_1, \mathbf{u}_2, \ldots, \mathbf{u}_m$ are linearly dependent.   ◀

▶ **COROLLARY 3.4.2** *If* $\{\mathbf{v}_1, \ldots, \mathbf{v}_n\}$ *and* $\{\mathbf{u}_1, \ldots, \mathbf{u}_m\}$ *are both bases for a vector space V, then* $n = m$.

▶ *Proof.* Let $\mathbf{v}_1, \mathbf{v}_2, \ldots, \mathbf{v}_n$ and $\mathbf{u}_1, \mathbf{u}_2, \ldots, \mathbf{u}_m$ both be bases fosr V. Since $\mathbf{v}_1, \mathbf{v}_2, \ldots, \mathbf{v}_n$ span V and $\mathbf{u}_1, \mathbf{u}_2, \ldots, \mathbf{u}_m$ are linearly independent, it follows from Theorem 3.4.1 that $m \leq n$. By the same reasoning $\mathbf{u}_1, \mathbf{u}_2, \ldots, \mathbf{u}_m$ span V, and $\mathbf{v}_1, \mathbf{v}_2, \ldots, \mathbf{v}_n$ are linearly independent, so $n \leq m$.   ◀

In view of Corollary 3.4.2, we can now refer to the number of elements in any basis for a given vector space. This leads to the following definition.

▶ **DEFINITION** Let V be a vector space. If V has a basis consisting of n vectors, we say that V has **dimension** n. The subspace $\{\mathbf{0}\}$ of V is said to have dimension 0. V is said to be **finite-dimensional** if there is a finite set of vectors that spans V; otherwise, we say that V is **infinite-dimensional**.   ◀

If $\mathbf{x}$ is a nonzero vector in R^3, then $\mathbf{x}$ spans a one-dimensional subspace Span$(\mathbf{x}) = \{\alpha\mathbf{x} \mid \alpha \text{ is a scalar}\}$. A vector $(a, b, c)^T$ will be in Span$(\mathbf{x})$ if and only if the point (a, b, c) is on the line determined by $(0, 0, 0)$ and (x_1, x_2, x_3). Thus a one-dimensional subspace of R^3 can be represented geometrically by a line through the origin.

If $\mathbf{x}$ and $\mathbf{y}$ are linearly independent in R^3, then

$$\text{Span}(\mathbf{x}, \mathbf{y}) = \{\alpha\mathbf{x} + \beta\mathbf{y} \mid \alpha \text{ and } \beta \text{ are scalars}\}$$

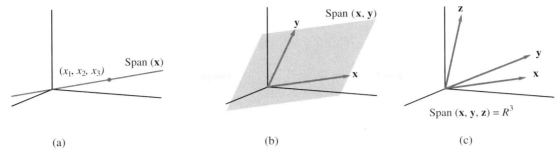

FIGURE 3.4.1

is a two-dimensional subspace of R^3. A vector $(a, b, c)^T$ will be in Span($\mathbf{x}, \mathbf{y}$) if and only if (a, b, c) lies on the plane determined by $(0, 0, 0)$, (x_1, x_2, x_3), and (y_1, y_2, y_3). Thus, we can think of a two-dimensional subspace of R^3 as a plane through the origin. If $\mathbf{x}$, $\mathbf{y}$, and $\mathbf{z}$ are linearly independent in R^3, they form a basis for R^3 and Span($\mathbf{x}, \mathbf{y}, \mathbf{z}$) $= R^3$. Thus any fourth point $(a, b, c)^T$ must lie in Span($\mathbf{x}, \mathbf{y}, \mathbf{z}$) (see Figure 3.4.1).

EXAMPLE 3. Let P be the vector space of all polynomials. We claim that P is infinite-dimensional. If P were finite-dimensional, say of dimension n, any set of $n + 1$ vectors would be linearly dependent. However, $1, x, x^2, \ldots, x^n$ are linearly independent, since $W[1, x, x^2, \ldots, x^n] > 0$. Therefore, P cannot be of dimension n. Since n was arbitrary, P must be infinite-dimensional. The same argument shows that $C[a, b]$ is infinite-dimensional. ◀

▶ **THEOREM 3.4.3** *If V is a vector space of dimension $n > 0$:*

    (i) *Any set of n linearly independent vectors spans V.*

    (ii) *Any n vectors that span V are linearly independent.*

▶ *Proof.* To prove (i), suppose that $\mathbf{v}_1, \ldots, \mathbf{v}_n$ are linearly independent and $\mathbf{v}$ is any other vector in V. Since V has dimension n, it has a basis consisting of n vectors and these vectors span V. It follows from Theorem 3.4.1 that $\mathbf{v}_1, \mathbf{v}_2, \ldots, \mathbf{v}_n, \mathbf{v}$ must be linearly dependent. Thus, there exist scalars $c_1, c_2, \ldots, c_n, c_{n+1}$ not all zero such that

$$(1) \qquad\qquad c_1\mathbf{v}_1 + c_2\mathbf{v}_2 + \cdots + c_n\mathbf{v}_n + c_{n+1}\mathbf{v} = \mathbf{0}$$

The scalar c_{n+1} cannot be zero, for then (1) would imply that $\mathbf{v}_1, \ldots, \mathbf{v}_n$ are linearly dependent. Thus (1) can be solved for $\mathbf{v}$.

$$\mathbf{v} = \alpha_1\mathbf{v}_1 + \alpha_2\mathbf{v}_2 + \cdots + \alpha_n\mathbf{v}_n$$

where $\alpha_i = -c_i/c_{n+1}$ for $i = 1, 2, \ldots, n$. Since $\mathbf{v}$ was an arbitrary vector in V, it follows that $\mathbf{v}_1, \mathbf{v}_2, \ldots, \mathbf{v}_n$ span V.

    To prove (ii), suppose that $\mathbf{v}_1, \ldots, \mathbf{v}_n$ span V. If $\mathbf{v}_1, \ldots, \mathbf{v}_n$ are linearly dependent, then one of the $\mathbf{v}_i$'s, say $\mathbf{v}_n$, can be written as a linear combination of the others. It follows that $\mathbf{v}_1, \ldots, \mathbf{v}_{n-1}$ will still span V. If $\mathbf{v}_1, \ldots, \mathbf{v}_{n-1}$ are linearly dependent, we can eliminate another vector and still have a spanning set. We can continue eliminating vectors in this way until we arrive at a linearly independent spanning set with $k < n$ elements. But this contradicts dim $V = n$. Therefore, $\mathbf{v}_1, \ldots, \mathbf{v}_n$ must be linearly independent. ◀

EXAMPLE 4. Show that $\left\{ \begin{bmatrix} 1 \\ 2 \\ 3 \end{bmatrix} , \begin{bmatrix} -2 \\ 1 \\ 0 \end{bmatrix} , \begin{bmatrix} 1 \\ 0 \\ 1 \end{bmatrix} \right\}$ is a basis for R^3.

SOLUTION. Since $\dim R^3 = 3$, we need only show that these three vectors are linearly independent. This follows since

$$\begin{vmatrix} 1 & -2 & 1 \\ 2 & 1 & 0 \\ 3 & 0 & 1 \end{vmatrix} = 2$$

◄

▶ **THEOREM 3.4.4** *If V is a vector space of dimension $n > 0$, then:*

   (i) *No set of less than n vectors can span V.*

   (ii) *Any subset of less than n linearly independent vectors can be extended to form a basis for V.*

   (iii) *Any spanning set containing more than n vectors can be pared down to form a basis for V.*

▶*Proof.* Observation (i) follows by the same reasoning that was used to prove (ii) of Theorem 3.4.3. To prove (ii), suppose that $\mathbf{v}_1, \ldots, \mathbf{v}_k$ are linearly independent and $k < n$. It follows from (i) that $\mathrm{Span}(\mathbf{v}_1, \ldots, \mathbf{v}_k)$ is a proper subspace of V and hence there exists a vector $\mathbf{v}_{k+1}$ that is in V but not in $\mathrm{Span}(\mathbf{v}_1, \ldots, \mathbf{v}_k)$. It follows then that $\mathbf{v}_1, \mathbf{v}_2, \ldots, \mathbf{v}_k, \mathbf{v}_{k+1}$ must be linearly independent. If $k + 1 < n$, then in the same manner $\{\mathbf{v}_1, \ldots, \mathbf{v}_k, \mathbf{v}_{k+1}\}$ can be extended to a set of $k + 2$ linearly independent vectors. This extension process may be continued until a set $\{\mathbf{v}_1, \mathbf{v}_2, \ldots, \mathbf{v}_k, \mathbf{v}_{k+1}, \ldots, \mathbf{v}_n\}$ of n linearly independent vectors is obtained.

To prove (iii), suppose that $\mathbf{v}_1, \ldots, \mathbf{v}_m$ span V and $m > n$. By Theorem 3.4.1, $\mathbf{v}_1, \ldots, \mathbf{v}_m$ must be linearly dependent. It follows that one of the vectors, say $\mathbf{v}_m$, can be written as a linear combination of the others. Hence, if $\mathbf{v}_m$ is eliminated from the set, the remaining $m - 1$ vectors will still span V. If $m - 1 > n$, we can continue to eliminate vectors in this manner until we arrive at a spanning set containing n elements.

◄

Standard Bases

In Example 1 we referred to the set $\{\mathbf{e}_1, \mathbf{e}_2, \mathbf{e}_3\}$ as the *standard basis* for R^3. We refer to this basis as the standard basis since it is the most natural one to use for representing vectors in R^3. More generally, the standard basis for R^n is the set $\{\mathbf{e}_1, \mathbf{e}_2, \ldots, \mathbf{e}_n\}$.

The most natural way to represent matrices in $R^{2 \times 2}$ is in terms of the basis $\{E_{11}, E_{12}, E_{21}, E_{22}\}$ given in Example 2. This then is the standard basis for $R^{2 \times 2}$.

The standard way to represent a polynomial in P_n is in terms of the functions 1, $x, x^2, \ldots, x^{n-1}$ and, consequently, the standard basis for P_n is $\{1, x, x^2, \ldots, x^{n-1}\}$.

Although these standard bases appear to be the simplest and most natural to use, they are not the most appropriate bases for many applied problems. (See, for example, the least squares problems in Chapter 5 or the eigenvalue applications in Chapter 6.) Indeed, the key to solving many applied problems is to switch from one of the standard bases to a basis that is in some sense natural for the particular application. Once the application is solved in terms of the new basis, it is a simple matter to switch back and represent the solution in terms of the standard basis. In the next section we will learn how to switch from one basis to another.

EXERCISES

1. In Exercise 1 of Section 3, indicate whether the given vectors form a basis for R^2.

2. In Exercise 2 of Section 3, indicate whether the given vectors form a basis for R^3.

3. Given the vectors

$$\mathbf{x}_1 = \begin{bmatrix} 2 \\ 1 \end{bmatrix}, \qquad \mathbf{x}_2 = \begin{bmatrix} 4 \\ 3 \end{bmatrix}, \qquad \mathbf{x}_3 = \begin{bmatrix} 7 \\ -3 \end{bmatrix}$$

(a) Show that $\mathbf{x}_1$ and $\mathbf{x}_2$ form a basis for R^2.
(b) Why must $\mathbf{x}_1, \mathbf{x}_2, \mathbf{x}_3$ be linearly dependent?
(c) What is the dimension of $\mathrm{Span}(\mathbf{x}_1, \mathbf{x}_2, \mathbf{x}_3)$?

4. Given the vectors

$$\mathbf{x}_1 = \begin{bmatrix} 3 \\ -2 \\ 4 \end{bmatrix}, \qquad \mathbf{x}_2 = \begin{bmatrix} -3 \\ 2 \\ -4 \end{bmatrix}, \qquad \mathbf{x}_3 = \begin{bmatrix} -6 \\ 4 \\ -8 \end{bmatrix}$$

What is the dimension of $\mathrm{Span}(\mathbf{x}_1, \mathbf{x}_2, \mathbf{x}_3)$?

5. Given

$$\mathbf{x}_1 = \begin{bmatrix} 2 \\ 1 \\ 3 \end{bmatrix}, \qquad \mathbf{x}_2 = \begin{bmatrix} 3 \\ -1 \\ 4 \end{bmatrix}, \qquad \mathbf{x}_3 = \begin{bmatrix} 2 \\ 6 \\ 4 \end{bmatrix}$$

(a) Show that $\mathbf{x}_1, \mathbf{x}_2, \mathbf{x}_3$ are linearly dependent.
(b) Show that $\mathbf{x}_1$ and $\mathbf{x}_2$ are linearly independent.
(c) What is the dimension of $\mathrm{Span}(\mathbf{x}_1, \mathbf{x}_2, \mathbf{x}_3)$?
(d) Give a geometric description of $\mathrm{Span}(\mathbf{x}_1, \mathbf{x}_2, \mathbf{x}_3)$.

6. In Exercise 2 of Section 2, some of the sets formed subspaces of R^3. In each of these cases, find a basis for the subspace and determine its dimension.

7. Find a basis for the subspace S of R^4 consisting of all vectors of the form $(a + b, a - b + 2c, b, c)^T$, where a, b, and c are all real numbers. What is the dimension of S?

8. Given $\mathbf{x}_1 = (1, \ 1, \ 1)^T$ and $\mathbf{x}_2 = (3, \ -1, \ 4)^T$:

    (a) Do $\mathbf{x}_1$ and $\mathbf{x}_2$ span R^3? Explain.
    (b) Let $\mathbf{x}_3$ be a third vector in R^3 and set $X = (\mathbf{x}_1 \ \ \mathbf{x}_2 \ \ \mathbf{x}_3)$. What condition(s) would X have to satisfy in order for $\mathbf{x}_1$, $\mathbf{x}_2$, $\mathbf{x}_3$ to form a basis for R^3?
    (c) Find a third vector $\mathbf{x}_3$ that will extend the set $\{\mathbf{x}_1, \mathbf{x}_2\}$ to a basis for R^3.

9. Let $\mathbf{a}_1$ and $\mathbf{a}_2$ be linearly independent vectors in R^3, and let $\mathbf{x}$ be a vector in R^2.

    (a) Describe geometrically Span$(\mathbf{a}_1, \mathbf{a}_2)$.
    (b) If $A = (\mathbf{a}_1, \mathbf{a}_2)$ and $\mathbf{b} = A\mathbf{x}$, then what is the dimension of Span$(\mathbf{a}_1, \mathbf{a}_2, \mathbf{b})$? Explain.

10. The vectors

$$\mathbf{x}_1 = \begin{pmatrix} 1 \\ 2 \\ 2 \end{pmatrix}, \quad \mathbf{x}_2 = \begin{pmatrix} 2 \\ 5 \\ 4 \end{pmatrix}, \quad \mathbf{x}_3 = \begin{pmatrix} 1 \\ 3 \\ 2 \end{pmatrix}, \quad \mathbf{x}_4 = \begin{pmatrix} 2 \\ 7 \\ 4 \end{pmatrix}, \quad \mathbf{x}_5 = \begin{pmatrix} 1 \\ 1 \\ 0 \end{pmatrix}$$

    span R^3. Pare down the set $\{\mathbf{x}_1, \mathbf{x}_2, \mathbf{x}_3, \mathbf{x}_4, \mathbf{x}_5\}$ to form a basis for R^3.

11. Let S be the subspace of P_3 consisting of all polynomials of the form $ax^2 + bx + 2a + 3b$. Find a basis for S.

12. In Exercise 3 of Section 2, some of the sets formed subspaces of $R^{2\times2}$. In each of these cases, find a basis for the subspace and determine its dimension.

13. In $C[-\pi, \pi]$, find the dimension of the subspace spanned by 1, $\cos 2x$, $\cos^2 x$.

14. In each of the following, find the dimension of the subspace of P_3 spanned by the given vectors.

    (a) $x, x - 1, x^2 + 1$
    (b) $x, x - 1, x^2 + 1, x^2 - 1$
    (c) $x^2, x^2 - x - 1, x + 1$
    (d) $2x, x - 2$

15. Let S be the subspace of P_3 consisting of all polynomials $p(x)$ such that $p(0) = 0$, and let T be the subspace of all polynomials $q(x)$ such that $q(1) = 0$. Find bases for

    (a) S
    (b) T
    (c) $S \cap T$

16. In R^4, let U be the subspace of all vectors of the form $(u_1, u_2, 0, 0)^T$, and let V be the subspace of all vectors of the form $(0, v_2, v_3, 0)^T$. What are the dimensions U, V, $U \cap V$, $U + V$? Find a basis for each of these four subspaces.

17. Is it possible to find a pair of two-dimensional subspaces U and V of R^3 such that $U \cap V = \{0\}$? Prove your answer. Give a geometrical interpretation of your conclusion. [*Hint:* Let $\{\mathbf{u}_1, \mathbf{u}_2\}$ and $\{\mathbf{v}_1, \mathbf{v}_2\}$ be bases for U and V, respectively. Show that $\mathbf{u}_1, \mathbf{u}_2, \mathbf{v}_1, \mathbf{v}_2$ are linearly dependent.]

5 · CHANGE OF BASIS

Many applied problems can be simplified by changing from one coordinate system to another. Changing coordinate systems in a vector space is essentially the same as changing from one basis to another. For example, in describing the motion of a particle in the plane at a particular time, it is often convenient to use a basis for R^2 consisting of a unit tangent vector $\mathbf{t}$ and a unit normal vector $\mathbf{n}$ instead of the standard basis $\{\mathbf{e}_1, \mathbf{e}_2\}$.

In this section we discuss the problem of switching from one coordinate system to another. We will show that this can be accomplished by multiplying a given coordinate vector $\mathbf{x}$ by a nonsingular matrix S. The product $\mathbf{y} = S\mathbf{x}$ will be the coordinate vector for the new coordinate system.

Changing Coordinates in R^2

The standard basis for R^2 is $\{\mathbf{e}_1, \mathbf{e}_2\}$. Any vector $\mathbf{x}$ in R^2 can be expressed as a linear combination

$$\mathbf{x} = x_1 \mathbf{e}_1 + x_2 \mathbf{e}_2$$

The scalars x_1 and x_2 can be thought of as the *coordinates* of $\mathbf{x}$ with respect to the standard basis. Actually, for any basis $\{\mathbf{y}, \mathbf{z}\}$ for R^2, it follows from Theorem 3.3.2 that a given vector $\mathbf{x}$ can be represented uniquely as a linear combination

$$\mathbf{x} = \alpha \mathbf{y} + \beta \mathbf{z}$$

The scalars α and β are the coordinates of $\mathbf{x}$ with respect to the basis $\{\mathbf{y}, \mathbf{z}\}$. Let us order the basis elements so that $\mathbf{y}$ is considered the first basis vector and $\mathbf{z}$ is considered the second, and denote the ordered basis by $[\mathbf{y}, \mathbf{z}]$. We can then refer to the vector $(\alpha, \beta)^T$ as the *coordinate vector* of $\mathbf{x}$ with respect to $[\mathbf{y}, \mathbf{z}]$. Note that, if we reverse the order of the basis vectors and take $[\mathbf{z}, \mathbf{y}]$, then we must also reorder the coordinate vector. The coordinate vector of $\mathbf{x}$ with respect to $[\mathbf{z}, \mathbf{y}]$ will be $(\beta, \alpha)^T$. When we refer to a basis using subscripts, such as $\{\mathbf{u}_1, \mathbf{u}_2\}$, the subscripts assign an ordering to the basis vectors.

EXAMPLE 1. Let $\mathbf{y} = (2, 1)^T$ and $\mathbf{z} = (1, 4)^T$. The vectors $\mathbf{y}$ and $\mathbf{z}$ are linearly independent and hence form a basis for R^2. The vector $\mathbf{x} = (7, 7)^T$ can be written as a linear combination

$$\mathbf{x} = 3\mathbf{y} + \mathbf{z}$$

Thus the coordinate vector of $\mathbf{x}$ with respect to $[\mathbf{y}, \mathbf{z}]$ is $(3, 1)^T$. Geometrically, the coordinate vector specifies how to get from the origin to the point $(7, 7)$ moving first in the direction of $\mathbf{y}$ and then in the direction of $\mathbf{z}$. The coordinate vector of $\mathbf{x}$ with respect to the ordered basis $[\mathbf{z}, \mathbf{y}]$ is $(1, 3)^T$. Geometrically, this vector tells us how to get from the origin to $(7, 7)$ moving first in the direction of $\mathbf{z}$ and then in the direction of $\mathbf{y}$ (see Figure 3.5.1). ◀

As an example of a problem for which it is helpful to change coordinates, consider the following application.

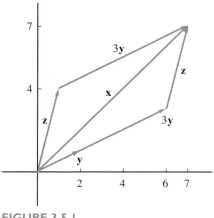

FIGURE 3.5.1

APPLICATION 1: POPULATION MIGRATION

Suppose that the total population of a large metropolitan area remains relatively fixed; however, each year 6% of the people living in the city move to the suburbs and 2% of the people living in the suburbs move to the city. If, initially, 30% of the population lives in the city and 70% lives in the suburbs, what will these percentages be in 10 years? 30 years? 50 years? What are the long-term implications?

The changes in population can be determined by matrix multiplications. If we set

$$A = \begin{bmatrix} 0.94 & 0.02 \\ 0.06 & 0.98 \end{bmatrix} \quad \text{and} \quad \mathbf{x}_0 = \begin{bmatrix} 0.30 \\ 0.70 \end{bmatrix}$$

then the percentages of people living in the city and suburbs after 1 year can be determined by setting $\mathbf{x}_1 = A\mathbf{x}_0$. The percentages after 2 years can be calculated by setting $\mathbf{x}_2 = A\mathbf{x}_1 = A^2\mathbf{x}_0$. In general, the percentages after n years will be given by $\mathbf{x}_n = A^n\mathbf{x}$. If we calculate these percentages for $n = 10, 30$, and 50 years and round to the nearest percent we get

$$\mathbf{x}_{10} = \begin{bmatrix} 0.27 \\ 0.73 \end{bmatrix} \qquad \mathbf{x}_{30} = \begin{bmatrix} 0.25 \\ 0.75 \end{bmatrix} \qquad \mathbf{x}_{50} = \begin{bmatrix} 0.25 \\ 0.75 \end{bmatrix}$$

In fact, as n increases, the sequence of vectors $\mathbf{x}_n = A^n\mathbf{x}_0$ converges to a limit $\mathbf{x} = (0.25, 0.75)^T$. The limit vector $\mathbf{x}$ is called a *steady-state vector* for the process.

To understand why the process approaches a steady state, it is helpful to switch to a different coordinate system. For the new coordinate system we will pick vectors $\mathbf{u}_1$ and $\mathbf{u}_2$, for which it is easy to see the effect of multiplication by the matrix A. In particular, if we pick $\mathbf{u}_1$ to be any multiple of the steady-state vector $\mathbf{x}$, then $A\mathbf{u}_1$ will equal $\mathbf{u}_1$. Let us choose $\mathbf{u}_1 = (1 \ \ 3)^T$ and $\mathbf{u}_2 = (-1 \ \ 1)^T$. The second vector was chosen because the effect of multiplying by A is just to scale the vector by a factor of 0.92. Thus, our new basis vectors satisfy

$$A\mathbf{u}_1 = \begin{bmatrix} 0.94 & 0.02 \\ 0.06 & 0.98 \end{bmatrix} \begin{bmatrix} 1 \\ 3 \end{bmatrix} = \begin{bmatrix} 1 \\ 3 \end{bmatrix} = \mathbf{u}_1$$

$$A\mathbf{u}_2 = \begin{bmatrix} 0.94 & 0.02 \\ 0.06 & 0.98 \end{bmatrix} \begin{bmatrix} -1 \\ 1 \end{bmatrix} = \begin{bmatrix} -0.92 \\ 0.92 \end{bmatrix} = 0.92\mathbf{u}_2$$

The initial vector $\mathbf{x}_0$ can be written as a linear combination of the new basis vectors.

$$\mathbf{x}_0 = \begin{bmatrix} 0.30 \\ 0.70 \end{bmatrix} = 0.25 \begin{bmatrix} 1 \\ 3 \end{bmatrix} - 0.05 \begin{bmatrix} -1 \\ 1 \end{bmatrix} = 0.25\mathbf{u}_1 - 0.05\mathbf{u}_2$$

It follows that

$$\mathbf{x}_n = A^n\mathbf{x}_0 = 0.25\mathbf{u}_1 - 0.05(0.92)^n\mathbf{u}_2$$

The entries of the second component approach 0 as n gets large. In fact, for $n > 27$ the entries will be small enough so that the rounded values of $\mathbf{x}_n$ are all equal to

$$0.25\mathbf{u}_1 = \begin{bmatrix} 0.25 \\ 0.75 \end{bmatrix}$$

This application is an example of a type of mathematical model called a *Markov process*. The sequence of vectors $\mathbf{x}_1, \mathbf{x}_2, \ldots$ is called a *Markov chain*. The matrix A has special structure in that its entries are nonnegative and its columns all add up to 1. Such matrices are called *stochastic matrices*. More precise definitions will be given later when we study these types of applications in Chapter 6. What we want to stress here is that the key to understanding these types of processes is to switch to a basis for which the effect of the matrix is quite simple. In particular, if A is $n \times n$, then we will want to choose basis vectors so that the effect of the matrix A on each basis vector $\mathbf{u}_j$ is simply to scale it by some factor λ_j, that is,

(1) $$A\mathbf{u}_j = \lambda_j\mathbf{u}_j \quad j = 1, 2, \ldots, n$$

In many applied problems involving an $n \times n$ matrix A, the key to solving the problem often is to find basis vectors $\mathbf{u}_1, \ldots, \mathbf{u}_n$ and scalars $\lambda_1, \ldots, \lambda_n$ such that (1) is satisfied. The new basis vectors can be thought of as a natural coordinate system to use with the matrix A and the scalars can be thought of as natural frequencies for the basis vectors. We will study these types of applications in more detail in Chapter 6.

Changing Coordinates

Once we have decided to work with a new basis, we have the problem of finding the coordinates with respect to that basis. Suppose, for example, that instead of using the standard basis $[\mathbf{e}_1, \mathbf{e}_2]$ for R^2 we wish to use a different basis, say

$$\mathbf{u}_1 = \begin{bmatrix} 3 \\ 2 \end{bmatrix}, \qquad \mathbf{u}_2 = \begin{bmatrix} 1 \\ 1 \end{bmatrix}$$

Indeed, we may want to switch back and forth between the two coordinate systems. Let us consider the following two problems.

I.   Given a vector $\mathbf{x} = (x_1, x_2)^T$, find its coordinates with respect to $\mathbf{u}_1$ and $\mathbf{u}_2$.

II. Given a vector $c_1\mathbf{u}_1 + c_2\mathbf{u}_2$, find its coordinates with respect to $\mathbf{e}_1$ and $\mathbf{e}_2$.

We will solve II first since it turns out to be the easier problem. To switch bases from $[\mathbf{u}_1, \mathbf{u}_2]$ to $[\mathbf{e}_1, \mathbf{e}_2]$, we must express the old basis elements $\mathbf{u}_1$ and $\mathbf{u}_2$ in terms of the new basis elements $\mathbf{e}_1$ and $\mathbf{e}_2$.

$$\mathbf{u}_1 = 3\mathbf{e}_1 + 2\mathbf{e}_2$$
$$\mathbf{u}_2 = \ \mathbf{e}_1 + \ \mathbf{e}_2$$

It follows then that

$$c_1\mathbf{u}_1 + c_2\mathbf{u}_2 = (3c_1\mathbf{e}_1 + 2c_1\mathbf{e}_2) + (c_2\mathbf{e}_1 + c_2\mathbf{e}_2)$$
$$= (3c_1 + c_2)\mathbf{e}_1 + (2c_1 + c_2)\mathbf{e}_2$$

Thus the coordinate vector of $c_1\mathbf{u}_1 + c_2\mathbf{u}_2$ with respect to $[\mathbf{e}_1, \mathbf{e}_2]$ is

$$\mathbf{x} = \begin{bmatrix} 3c_1 + c_2 \\ 2c_1 + c_2 \end{bmatrix} = \begin{bmatrix} 3 & 1 \\ 2 & 1 \end{bmatrix} \begin{bmatrix} c_1 \\ c_2 \end{bmatrix}$$

If we set

$$U = (\mathbf{u}_1, \mathbf{u}_2) = \begin{bmatrix} 3 & 1 \\ 2 & 1 \end{bmatrix}$$

then, given any coordinate vector $\mathbf{c}$ with respect to $[\mathbf{u}_1, \mathbf{u}_2]$, to find the corresponding coordinate vector $\mathbf{x}$ with respect to $[\mathbf{e}_1, \mathbf{e}_2]$, we simply multiply U times $\mathbf{c}$.

$$(2) \qquad\qquad\qquad\qquad \mathbf{x} = U\mathbf{c}$$

The matrix U is called the *transition matrix* from the ordered basis $[\mathbf{u}_1, \mathbf{u}_2]$ to the basis $[\mathbf{e}_1, \mathbf{e}_2]$.

To solve problem I, we must find the transition matrix from $[\mathbf{e}_1, \mathbf{e}_2]$ to $[\mathbf{u}_1, \mathbf{u}_2]$. The matrix U in (2) is nonsingular, since its column vectors, $\mathbf{u}_1$ and $\mathbf{u}_2$, are linearly independent. It follows from (2) that

$$\mathbf{c} = U^{-1}\mathbf{x}$$

Thus, given a vector

$$\mathbf{x} = (x_1, x_2)^T = x_1\mathbf{e}_1 + x_2\mathbf{e}_2$$

we need only multiply by U^{-1} to find its coordinate vector with respect to $[\mathbf{u}_1, \mathbf{u}_2]$. U^{-1} is the transition matrix from $[\mathbf{e}_1, \mathbf{e}_2]$ to $[\mathbf{u}_1, \mathbf{u}_2]$.

EXAMPLE 2.   Let $\mathbf{u}_1 = (3, 2)^T$, $\mathbf{u}_2 = (1, 1)^T$, and $\mathbf{x} = (7, 4)^T$. Find the coordinates of $\mathbf{x}$ with respect to $\mathbf{u}_1$ and $\mathbf{u}_2$.

SOLUTION.   By the preceding discussion, the transition matrix from $[\mathbf{e}_1, \mathbf{e}_2]$ to $[\mathbf{u}_1, \mathbf{u}_2]$ is the inverse of

$$
U = (\mathbf{u}_1, \mathbf{u}_2) = \begin{pmatrix} 3 & 1 \\ 2 & 1 \end{pmatrix}
$$

Thus

$$
\mathbf{c} = U^{-1}\mathbf{x} = \begin{pmatrix} 1 & -1 \\ -2 & 3 \end{pmatrix} \begin{pmatrix} 7 \\ 4 \end{pmatrix} = \begin{pmatrix} 3 \\ -2 \end{pmatrix}
$$

is the desired coordinate vector and

$$
\mathbf{x} = 3\mathbf{u}_1 - 2\mathbf{u}_2
$$

EXAMPLE 3.   Let $\mathbf{b}_1 = (1, -1)^T$ and $\mathbf{b}_2 = (-2, 3)^T$. Find the transition matrix from $[\mathbf{e}_1, \mathbf{e}_2]$ to $[\mathbf{b}_1, \mathbf{b}_2]$ and determine the coordinates of $\mathbf{x} = (1, 2)^T$ with respect to $[\mathbf{b}_1, \mathbf{b}_2]$.

SOLUTION.   The transition matrix from $[\mathbf{b}_1, \mathbf{b}_2]$ to $[\mathbf{e}_1, \mathbf{e}_2]$ is

$$
B = (\mathbf{b}_1, \mathbf{b}_2) = \begin{pmatrix} 1 & -2 \\ -1 & 3 \end{pmatrix}
$$

and hence the transition matrix from $[\mathbf{e}_1, \mathbf{e}_2]$ to $[\mathbf{b}_1, \mathbf{b}_2]$ is

$$
B^{-1} = \begin{pmatrix} 3 & 2 \\ 1 & 1 \end{pmatrix}
$$

The coordinate vector of $\mathbf{x}$ with respect to $[\mathbf{b}_1, \mathbf{b}_2]$ is

$$
\mathbf{c} = B^{-1}\mathbf{x} = \begin{pmatrix} 3 & 2 \\ 1 & 1 \end{pmatrix} \begin{pmatrix} 1 \\ 2 \end{pmatrix} = \begin{pmatrix} 7 \\ 3 \end{pmatrix}
$$

and hence

$$
\mathbf{x} = 7\mathbf{b}_1 + 3\mathbf{b}_2
$$

If

$$
S = \begin{pmatrix} s_{11} & s_{12} \\ s_{21} & s_{22} \end{pmatrix}
$$

is the transition matrix from an ordered basis $[\mathbf{v}_1, \mathbf{v}_2]$ of R^2 to an ordered basis $[\mathbf{u}_1, \mathbf{u}_2]$, then, since

$$\mathbf{v}_1 = 1\mathbf{v}_1 + 0\mathbf{v}_2$$

the coordinate vector of $\mathbf{v}_1$ with respect to $[\mathbf{u}_1, \mathbf{u}_2]$ is given by

$$\mathbf{s}_1 = \begin{bmatrix} s_{11} & s_{12} \\ s_{21} & s_{22} \end{bmatrix} \begin{bmatrix} 1 \\ 0 \end{bmatrix} = \begin{bmatrix} s_{11} \\ s_{21} \end{bmatrix}$$

Similarly,

$$\mathbf{v}_2 = 0\mathbf{v}_1 + 1\mathbf{v}_2$$

and hence its coordinate vector with respect to $[\mathbf{u}_1, \mathbf{u}_2]$ is

$$\mathbf{s}_2 = \begin{bmatrix} s_{11} & s_{12} \\ s_{21} & s_{22} \end{bmatrix} \begin{bmatrix} 0 \\ 1 \end{bmatrix} = \begin{bmatrix} s_{12} \\ s_{22} \end{bmatrix}$$

Thus,

(3)
$$\mathbf{v}_1 = s_{11}\mathbf{u}_1 + s_{21}\mathbf{u}_2$$
$$\mathbf{v}_2 = s_{12}\mathbf{u}_1 + s_{22}\mathbf{u}_2$$

In general, if the old basis elements $\mathbf{v}_1$ and $\mathbf{v}_2$ are written in terms of the new basis $[\mathbf{u}_1, \mathbf{u}_2]$, the coordinate vector $\mathbf{s}_1 = (s_{11}, s_{21})^T$ corresponding to $\mathbf{v}_1$ will be the first column of the transition matrix S and the coordinate vector $\mathbf{s}_2 = (s_{12}, s_{22})^T$, corresponding to $\mathbf{v}_2$, will be the second column of S. Thus S is the transpose of the coefficient array in (3).

EXAMPLE 4. Find the transition matrix corresponding to the change of basis from $[\mathbf{v}_1, \mathbf{v}_2]$ to $[\mathbf{u}_1, \mathbf{u}_2]$, where

$$\mathbf{v}_1 = \begin{bmatrix} 5 \\ 2 \end{bmatrix}, \quad \mathbf{v}_2 = \begin{bmatrix} 7 \\ 3 \end{bmatrix} \quad \text{and} \quad \mathbf{u}_1 = \begin{bmatrix} 3 \\ 2 \end{bmatrix}, \quad \mathbf{u}_2 = \begin{bmatrix} 1 \\ 1 \end{bmatrix}$$

SOLUTION. We must express $\mathbf{v}_1$ and $\mathbf{v}_2$ in terms of the new basis elements $\mathbf{u}_1$ and $\mathbf{u}_2$.

$$\mathbf{v}_1 = s_{11}\mathbf{u}_1 + s_{21}\mathbf{u}_2$$
$$\mathbf{v}_2 = s_{12}\mathbf{u}_1 + s_{22}\mathbf{u}_2$$

The first equation can be written

$$\begin{bmatrix} 5 \\ 2 \end{bmatrix} = \begin{bmatrix} 3s_{11} + s_{21} \\ 2s_{11} + s_{21} \end{bmatrix}$$

The solution to this system is $(s_{11}, s_{21})^T = (3, -4)^T$. Similarly, the second equation leads to the system

$$\begin{bmatrix} 7 \\ 3 \end{bmatrix} = \begin{bmatrix} 3s_{12} + s_{22} \\ 2s_{12} + s_{22} \end{bmatrix}$$

The solution to this system is $(s_{12}, s_{22})^T = (4, -5)^T$. Thus

$$S = \begin{bmatrix} 3 & 4 \\ -4 & -5 \end{bmatrix}$$

is the transition matrix from $[\mathbf{v}_1, \mathbf{v}_2]$ to $[\mathbf{u}_1, \mathbf{u}_2]$.   ◀

An alternative method to change from a basis $[\mathbf{v}_1, \mathbf{v}_2]$ to another basis $[\mathbf{u}_1, \mathbf{u}_2]$ is to first change from $[\mathbf{v}_1, \mathbf{v}_2]$ to the standard basis and then change to $[\mathbf{u}_1, \mathbf{u}_2]$. Given a vector $\mathbf{x}$ in R^2, if $\mathbf{c}$ is the coordinate vector of $\mathbf{x}$ with respect to $[\mathbf{v}_1, \mathbf{v}_2]$ and $\mathbf{d}$ is the coordinate vector of $\mathbf{x}$ with respect to $[\mathbf{u}_1, \mathbf{u}_2]$, then

$$c_1\mathbf{v}_1 + c_2\mathbf{v}_2 = x_1\mathbf{e}_1 + x_2\mathbf{e}_2 = d_1\mathbf{u}_1 + d_2\mathbf{u}_2$$

Since V is the transition matrix from E to $[\mathbf{e}_1, \mathbf{e}_2]$ and U^{-1} is the transition matrix from $[\mathbf{e}_1, \mathbf{e}_2]$ to F, it follows that

$$V\mathbf{c} = \mathbf{x} \qquad \text{and} \qquad U^{-1}\mathbf{x} = \mathbf{d}$$

and hence

$$U^{-1}V\mathbf{c} = U^{-1}\mathbf{x} = \mathbf{d}$$

Thus, $U^{-1}V$ is the transition matrix from $[\mathbf{v}_1, \mathbf{v}_2]$ to $[\mathbf{u}_1, \mathbf{u}_2]$ (see Figure 3.5.2).

EXAMPLE 5. Let $[\mathbf{v}_1, \mathbf{v}_2]$ and $[\mathbf{u}_1, \mathbf{u}_2]$ be the ordered bases given in Example 4. The transition matrix from $[\mathbf{v}_1, \mathbf{v}_2]$ to $[\mathbf{u}_1, \mathbf{u}_2]$ is given by

$$U^{-1}V = \begin{bmatrix} 1 & -1 \\ -2 & 3 \end{bmatrix} \begin{bmatrix} 5 & 7 \\ 2 & 3 \end{bmatrix} = \begin{bmatrix} 3 & 4 \\ -4 & -5 \end{bmatrix}$$ ◀

FIGURE 3.5.2

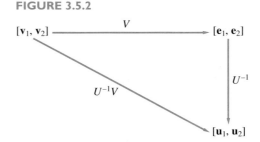

Change of Basis for a General Vector Space

Everything we have done so far can easily be generalized to apply to any finite-dimensional vector space. We begin by defining coordinate vectors for an n-dimensional vector space.

▶ **DEFINITION** Let V be a vector space and let $E = [\mathbf{v}_1, \mathbf{v}_2, \ldots, \mathbf{v}_n]$ be an ordered basis for V. If $\mathbf{v}$ is any element of V, then $\mathbf{v}$ can be written in the form

$$\mathbf{v} = c_1\mathbf{v}_1 + c_2\mathbf{v}_2 + \cdots + c_n\mathbf{v}_n$$

where $c_1, c_2, \ldots, c_n$ are scalars. Thus we can associate with each vector $\mathbf{v}$ a unique vector $\mathbf{c} = (c_1, c_2, \ldots, c_n)^T$ in R^n. The vector $\mathbf{c}$ defined in this way is called the **coordinate vector** of $\mathbf{v}$ with respect to the ordered basis E and is denoted $[\mathbf{v}]_E$. The c_i's are called the **coordinates** of $\mathbf{v}$ relative to E. ◀

The examples considered so far have all dealt with changing coordinates in R^2. Similar techniques could be used for R^n. In the case of R^n the transition matrices will be $n \times n$.

EXAMPLE 6. Let

$$E = [\mathbf{v}_1, \mathbf{v}_2, \mathbf{v}_3] = [(1, 1, 1)^T, (2, 3, 2)^T, (1, 5, 4)^T]$$
$$F = [\mathbf{u}_1, \mathbf{u}_2, \mathbf{u}_3] = [(1, 1, 0)^T, (1, 2, 0)^T, (1, 2, 1)^T]$$

Find the transition matrix from E to F. If

$$\mathbf{x} = 3\mathbf{v}_1 + 2\mathbf{v}_2 - \mathbf{v}_3 \qquad \text{and} \qquad \mathbf{y} = \mathbf{v}_1 - 3\mathbf{v}_2 + 2\mathbf{v}_3$$

find the coordinates of $\mathbf{x}$ and $\mathbf{y}$ with respect to the ordered basis F.

SOLUTION. As in Example 5, the transition matrix is given by

$$U^{-1}V = \begin{pmatrix} 2 & -1 & 0 \\ -1 & 1 & -1 \\ 0 & 0 & 1 \end{pmatrix} \begin{pmatrix} 1 & 2 & 1 \\ 1 & 3 & 5 \\ 1 & 2 & 4 \end{pmatrix} = \begin{pmatrix} 1 & 1 & -3 \\ -1 & -1 & 0 \\ 1 & 2 & 4 \end{pmatrix}$$

The coordinate vectors of $\mathbf{x}$ and $\mathbf{y}$ with respect to the ordered basis F are given by

$$[\mathbf{x}]_F = \begin{pmatrix} 1 & 1 & -3 \\ -1 & -1 & 0 \\ 1 & 2 & 4 \end{pmatrix} \begin{pmatrix} 3 \\ 2 \\ -1 \end{pmatrix} = \begin{pmatrix} 8 \\ -5 \\ 3 \end{pmatrix}$$

and

$$[\mathbf{y}]_F = \begin{pmatrix} 1 & 1 & -3 \\ -1 & -1 & 0 \\ 1 & 2 & 4 \end{pmatrix} \begin{pmatrix} 1 \\ -3 \\ 2 \end{pmatrix} = \begin{pmatrix} -8 \\ 2 \\ 3 \end{pmatrix}$$

The reader may verify that

$$U = \begin{pmatrix} 1 & 1 & 1 \\ 1 & 2 & 2 \\ 0 & 0 & 1 \end{pmatrix}$$

$$8\mathbf{u}_1 - 5\mathbf{u}_2 + 3\mathbf{u}_3 = 3\mathbf{v}_1 + 2\mathbf{v}_2 - \mathbf{v}_3$$

$$-8\mathbf{u}_1 + 2\mathbf{u}_2 + 3\mathbf{u}_3 = \mathbf{v}_1 - 3\mathbf{v}_2 + 2\mathbf{v}_3 \qquad \blacktriangleleft$$

$2 - 1(1) + 0$

$2 - 1$

If V is any n-dimensional vector space, it is possible to change from one basis to another by means of an $n \times n$ transition matrix. We will show that such a transition matrix is necessarily nonsingular. To see how this is done, let $E = [\mathbf{w}_1, \ldots, \mathbf{w}_n]$ and $F = [\mathbf{v}_1, \ldots, \mathbf{v}_n]$ be two ordered bases for V. Each vector $\mathbf{w}_j$ can then be expressed as a linear combination of the $\mathbf{v}_i$'s.

$$\mathbf{w}_1 = s_{11}\mathbf{v}_1 + s_{21}\mathbf{v}_2 + \cdots + s_{n1}\mathbf{v}_n$$

$$\mathbf{w}_2 = s_{12}\mathbf{v}_1 + s_{22}\mathbf{v}_2 + \cdots + s_{n2}\mathbf{v}_n$$

(4)

$$\vdots$$

$$\mathbf{w}_n = s_{1n}\mathbf{v}_1 + s_{2n}\mathbf{v}_2 + \cdots + s_{nn}\mathbf{v}_n$$

Let $\mathbf{v} \in V$. If $\mathbf{x} = [\mathbf{v}]_E$, it follows from (4) that

$$\mathbf{v} = x_1\mathbf{w}_1 + x_2\mathbf{w}_2 + \cdots + x_n\mathbf{w}_n$$

$$= \left(\sum_{j=1}^{n} s_{1j}x_j \right) \mathbf{v}_1 + \left(\sum_{j=1}^{n} s_{2j}x_j \right) \mathbf{v}_2 + \cdots + \left(\sum_{j=1}^{n} s_{nj}x_j \right) \mathbf{v}_n$$

Thus, if $\mathbf{y} = [\mathbf{v}]_F$, then

$$y_i = \sum_{j=1}^{n} s_{ij}x_j \qquad i = 1, \ldots, n$$

and hence

$$\mathbf{y} = S\mathbf{x}$$

The matrix S defined by (4) is referred to as the *transition matrix*. Once S has been determined, it is a simple matter to change coordinate systems. To find the coordinates of $\mathbf{v} = x_1\mathbf{w}_1 + \cdots + x_n\mathbf{w}_n$ with respect to $[\mathbf{v}_1, \ldots, \mathbf{v}_n]$, we need only calculate $\mathbf{y} = S\mathbf{x}$.

The transition matrix S corresponding to the change of basis from $[\mathbf{w}_1, \ldots, \mathbf{w}_n]$ to $[\mathbf{v}_1, \ldots, \mathbf{v}_n]$ can be characterized by the condition

(5) $\quad S\mathbf{x} = \mathbf{y} \quad$ if and only if $\quad x_1\mathbf{w}_1 + \cdots + x_n\mathbf{w}_n = y_1\mathbf{v}_1 + \cdots + y_n\mathbf{v}_n$

Taking $\mathbf{y} = \mathbf{0}$ in (5), we see that $S\mathbf{x} = \mathbf{0}$ implies that

$$x_1\mathbf{w}_1 + \cdots + x_n\mathbf{w}_n = \mathbf{0}$$

Since the $\mathbf{w}_i$'s are linearly independent, it follows that $\mathbf{x} = \mathbf{0}$. Thus the equation $S\mathbf{x} = \mathbf{0}$ has only the trivial solution and hence the matrix S is nonsingular. The inverse matrix is characterized by the condition

$$S^{-1}\mathbf{y} = \mathbf{x} \quad \text{if and only if} \quad y_1\mathbf{v}_1 + \cdots + y_n\mathbf{v}_n = x_1\mathbf{w}_1 + \cdots + x_n\mathbf{w}_n$$

Thus, S^{-1} is the transition matrix used to change basis from $[\mathbf{v}_1, \ldots, \mathbf{v}_n]$ to $[\mathbf{w}_1, \ldots, \mathbf{w}_n]$.

EXAMPLE 7. Suppose that in P_3 we want to change from the ordered basis $[1, x, x^2]$ to the ordered basis $[1, 2x, 4x^2 - 2]$. Since $[1, x, x^2]$ is the standard basis for P_3, it is easier to find the transition matrix from $[1, 2x, 4x^2 - 2]$ to $[1, x, x^2]$. Since

$$
\begin{aligned}
1 &= 1 \cdot 1 + 0x + 0x^2 \\
2x &= 0 \cdot 1 + 2x + 0x^2 \\
4x^2 - 2 &= -2 \cdot 1 + 0x + 4x^2
\end{aligned}
$$

the transition matrix is

$$
S = \begin{bmatrix} 1 & 0 & -2 \\ 0 & 2 & 0 \\ 0 & 0 & 4 \end{bmatrix}
$$

The inverse of S will be the transition matrix from $[1, x, x^2]$ to $[1, 2x, 4x^2 - 2]$:

$$
S^{-1} = \begin{bmatrix} 1 & 0 & \frac{1}{2} \\ 0 & \frac{1}{2} & 0 \\ 0 & 0 & \frac{1}{4} \end{bmatrix}
$$

Given any $p(x) = a + bx + cx^2$ in P_3, to find the coordinates of $p(x)$ with respect to $[1, 2x, 4x^2 - 2]$, we simply multiply

$$
\begin{bmatrix} 1 & 0 & \frac{1}{2} \\ 0 & \frac{1}{2} & 0 \\ 0 & 0 & \frac{1}{4} \end{bmatrix} \begin{bmatrix} a \\ b \\ c \end{bmatrix} = \begin{bmatrix} a + \frac{1}{2}c \\ \frac{1}{2}b \\ \frac{1}{4}c \end{bmatrix}
$$

Thus,

$$
p(x) = (a + \tfrac{1}{2}c) \cdot 1 + (\tfrac{1}{2}b) \cdot 2x + \tfrac{1}{4}c \cdot (4x^2 - 2) \qquad \blacktriangleleft
$$

We have seen that each transition matrix is nonsingular. Actually, any nonsingular matrix can be thought of as a transition matrix. If S is an $n \times n$ nonsingular matrix and $[\mathbf{v}_1, \ldots, \mathbf{v}_n]$ is an ordered basis for V, then define $[\mathbf{w}_1, \mathbf{w}_2, \ldots, \mathbf{w}_n]$ by (4). To see that the $\mathbf{w}_j$'s are linearly independent, suppose that

$$
\sum_{j=1}^{n} x_j \mathbf{w}_j = \mathbf{0}
$$

It follows from (4) that

$$\sum_{i=1}^{n} \left(\sum_{j=1}^{n} s_{ij} x_j \right) \mathbf{v}_j = \mathbf{0}$$

By the linear independence of the $\mathbf{v}_i$'s, it follows that

$$\sum_{j=1}^{n} s_{ij} x_j = 0 \qquad i = 1, \dots, n$$

or, equivalently,

$$S\mathbf{x} = \mathbf{0}$$

Since S is nonsingular, $\mathbf{x}$ must equal $\mathbf{0}$. Therefore, $\mathbf{w}_1, \dots, \mathbf{w}_n$ are linearly independent and hence they form a basis for V. The matrix S is the transition matrix corresponding to the change from the ordered basis $[\mathbf{w}_1, \dots, \mathbf{w}_n]$ to $[\mathbf{v}_1, \dots, \mathbf{v}_n]$.

In many applied problems it is important to use the right type of basis for the particular application. In Chapter 5 we will see that the key to solving least squares problems is to switch to a special type of basis called an *orthonormal* basis. In Chapter 6 we will consider a number of applications involving the *eigenvalues* and *eigenvectors* associated with an $n \times n$ matrix A. The key to solving these types of problems is to switch to a basis for R^n consisting of eigenvectors of A.

EXERCISES

1. For each of the following, find the transition matrix corresponding to the change of basis from $[\mathbf{u}_1, \mathbf{u}_2]$ to $[\mathbf{e}_1, \mathbf{e}_2]$.

   (a) $\mathbf{u}_1 = (1, 1)^T$, $\mathbf{u}_2 = (-1, 1)^T$
   (b) $\mathbf{u}_1 = (1, 2)^T$, $\mathbf{u}_2 = (2, 5)^T$
   (c) $\mathbf{u}_1 = (0, 1)^T$, $\mathbf{u}_2 = (1, 0)^T$

2. For each of the ordered bases $[\mathbf{u}_1, \mathbf{u}_2]$ in Exercise 1, find the transition matrix corresponding to the change of basis from $[\mathbf{e}_1, \mathbf{e}_2]$ to $[\mathbf{u}_1, \mathbf{u}_2]$.

3. Let $\mathbf{v}_1 = (3, 2)^T$ and $\mathbf{v}_2 = (4, 3)^T$. For each ordered basis $[\mathbf{u}_1, \mathbf{u}_2]$ given in Exercise 1, find the transition matrix from $[\mathbf{v}_1, \mathbf{v}_2]$ to $[\mathbf{u}_1, \mathbf{u}_2]$.

4. Let $E = [(5, 3)^T, (3, 2)^T]$ and let $\mathbf{x} = (1, 1)^T$, $\mathbf{y} = (1, -1)^T$, and $\mathbf{z} = (10, 7)^T$. Determine the values of $[\mathbf{x}]_E$, $[\mathbf{y}]_E$, and $[\mathbf{z}]_E$.

5. Let $\mathbf{u}_1 = (1, 1, 1)^T$, $\mathbf{u}_2 = (1, 2, 2)^T$, $\mathbf{u}_3 = (2, 3, 4)^T$.

   (a) Find the transition matrix corresponding to the change of basis from $[\mathbf{e}_1, \mathbf{e}_2, \mathbf{e}_3]$ to $[\mathbf{u}_1, \mathbf{u}_2, \mathbf{u}_3]$.
   (b) Find the coordinates of each of the following vectors with respect to $[\mathbf{u}_1, \mathbf{u}_2, \mathbf{u}_3]$.

   (i) $(3, 2, 5)^T$ (ii) $(1, 1, 2)^T$ (iii) $(2, 3, 2)^T$

6. Let $v_1 = (4, 6, 7)^T$, $v_2 = (0, 1, 1)^T$, and $v_3 = (0, 1, 2)^T$, and let u_1, u_2, u_3 be the vectors given in Exercise 5.

(a) Find the transition matrix from $[v_1, v_2, v_3]$ to $[u_1, u_2, u_3]$.

(b) If $x = 2v_1 + 3v_2 - 4v_3$, determine the coordinates of x with respect to $[u_1, u_2, u_3]$.

7. Given

$$v_1 = \begin{bmatrix} 1 \\ 2 \end{bmatrix}, \quad v_2 = \begin{bmatrix} 2 \\ 3 \end{bmatrix}, \quad S = \begin{bmatrix} 3 & 5 \\ 1 & -2 \end{bmatrix}$$

Find vectors w_1 and w_2 so that S will be the transition matrix from $[w_1, w_2]$ to $[v_1, v_2]$.

8. Given

$$v_1 = \begin{bmatrix} 2 \\ 6 \end{bmatrix}, \quad v_2 = \begin{bmatrix} 1 \\ 4 \end{bmatrix}, \quad S = \begin{bmatrix} 4 & 1 \\ 2 & 1 \end{bmatrix}$$

Find vectors u_1 and u_2 so that S will be the transition matrix from $[v_1, v_2]$ to $[u_1, u_2]$.

9. Let $[x, 1]$ and $[2x - 1, 2x + 1]$ be ordered bases for P_2.

(a) Find the transition matrix representing the change in coordinates from $[2x - 1, 2x + 1]$ to $[x, 1]$.

(b) Find the transition matrix representing the change in coordinates from $[x, 1]$ to $[2x - 1, 2x + 1]$.

10. Find the transition matrix representing the change of coordinates on P_3 from the ordered basis $[1, x, x^2]$ to the ordered basis

$$[1, 1 + x, 1 + x + x^2]$$

11. Let $E = [u_1, \ldots, u_n]$ and $F = [v_1, \ldots, v_n]$ be two ordered bases for R^n and set

$$U = (u_1, \ldots, u_n), \qquad V = (v_1, \ldots, v_n)$$

Show that the transition matrix from E to F can be determined by calculating the reduced row echelon form of $(V|U)$.

6 ROW SPACE AND COLUMN SPACE

If A is an $m \times n$ matrix, each row of A is an n-tuple of real numbers and hence can be considered as a vector in $R^{1 \times n}$. The m vectors corresponding to the rows of A will be referred to as the *row vectors* of A. Similarly, each column of A can be considered as a vector in R^m, and we can associate n *column vectors* with the matrix A.

▶ **DEFINITION** If A is an $m \times n$ matrix, the subspace of $R^{1 \times n}$ spanned by the row vectors of A is called the **row space** of A. The subspace of R^m spanned by the column vectors of A is called the **column space** of $A. ◀

EXAMPLE 1. Let

$$A = \begin{bmatrix} 1 & 0 & 0 \\ 0 & 1 & 0 \end{bmatrix}$$

The row space of A is the set of all 3-tuples of the form

$$\alpha(1, 0, 0) + \beta(0, 1, 0) = (\alpha, \beta, 0)$$

The column space of A is the set of all vectors of the form

$$\alpha \begin{bmatrix} 1 \\ 0 \end{bmatrix} + \beta \begin{bmatrix} 0 \\ 1 \end{bmatrix} + \gamma \begin{bmatrix} 0 \\ 0 \end{bmatrix} = \begin{bmatrix} \alpha \\ \beta \end{bmatrix}$$

Thus the row space of A is a two-dimensional subspace of $R^{1 \times 3}$, and the column space of A is R^2. ◀

▶ **THEOREM 3.6.1** *Two row equivalent matrices have the same row space.*

▶*Proof.* If B is row equivalent to A, then B can be formed from A by a finite sequence of row operations. Thus the row vectors of B must be linear combinations of the row vectors of A. Consequently, the row space of B must be a subspace of the row space of A. Since A is row equivalent to B, by the same reasoning, the row space of A is a subspace of the row space of B. ◀

▶ **DEFINITION** The **rank** of a matrix A is the dimension of the row space of A. ◀

To determine the rank of a matrix, we can reduce the matrix to row echelon form. The nonzero rows of the row echelon matrix will form a basis for the row space.

EXAMPLE 2. Let

$$A = \begin{bmatrix} 1 & -2 & 3 \\ 2 & -5 & 1 \\ 1 & -4 & -7 \end{bmatrix}$$

Reducing A to row echelon form, we obtain the matrix

$$U = \begin{bmatrix} 1 & -2 & 3 \\ 0 & 1 & 5 \\ 0 & 0 & 0 \end{bmatrix}$$

Clearly, $(1, -2, 3)$ and $(0, 1, 5)$ form a basis for the row space of U. Since U and A are row equivalent, they have the same row space, and hence the rank of A is 2. ◀

Linear Systems

The concepts of row space and column space are useful in the study of linear systems. A system $A\mathbf{x} = \mathbf{b}$ can be written in the form

$$(1) \qquad x_1 \begin{Bmatrix} a_{11} \\ a_{21} \\ \vdots \\ a_{m1} \end{Bmatrix} + x_2 \begin{Bmatrix} a_{12} \\ a_{22} \\ \vdots \\ a_{m2} \end{Bmatrix} + \cdots + x_n \begin{Bmatrix} a_{1n} \\ a_{2n} \\ \vdots \\ a_{mn} \end{Bmatrix} = \begin{Bmatrix} b_1 \\ b_2 \\ \vdots \\ b_m \end{Bmatrix}$$

In Chapter 1 we used this representation to characterize when a linear system will be consistent. The result, Theorem 1.3.1, can now be restated in terms of the column space of the matrix.

▶ **THEOREM 3.6.2 (Consistency Theorem for Linear Systems)** *A linear system $A\mathbf{x} = \mathbf{b}$ is consistent if and only if $\mathbf{b}$ is in the column space of A.*   ◀

If $\mathbf{b}$ is replaced by the zero vector, then (1) becomes

$$(2) \qquad\qquad x_1\mathbf{a}_1 + x_2\mathbf{a}_2 \cdots + x_n\mathbf{a}_n = \mathbf{0}$$

It follows from (2) that the system $A\mathbf{x} = \mathbf{0}$ will have only the trivial solution $\mathbf{x} = \mathbf{0}$ if and only if the column vectors of A are linearly independent.

▶ **THEOREM 3.6.3** *Let A be an $m \times n$ matrix. The linear system $A\mathbf{x} = \mathbf{b}$ is consistent for every $\mathbf{b} \in R^m$ if and only if the column vectors of A span R^m. The system $A\mathbf{x} = \mathbf{b}$ has at most one solution for every $\mathbf{b} \in R^m$ if and only if the column vectors of A are linearly independent.*

▶ *Proof.* We have seen that the system $A\mathbf{x} = \mathbf{b}$ is consistent if and only if $\mathbf{b}$ is in the column space of A. It follows that $A\mathbf{x} = \mathbf{b}$ will be consistent for every $\mathbf{b} \in R^m$ if and only if the column vectors of A span R^m. To prove the second statement, note that, if $A\mathbf{x} = \mathbf{b}$ has at most one solution for every $\mathbf{b}$, then in particular the system $A\mathbf{x} = \mathbf{0}$ can have only the trivial solution, and hence the column vectors of A must be linearly independent. Conversely, if the column vectors of A are linearly independent, $A\mathbf{x} = \mathbf{0}$ has only the trivial solution. Now if $\mathbf{x}_1$ and $\mathbf{x}_2$ were both solutions to $A\mathbf{x} = \mathbf{b}$, then $\mathbf{x}_1 - \mathbf{x}_2$ would be a solution to $A\mathbf{x} = \mathbf{0}$,

$$A(\mathbf{x}_1 - \mathbf{x}_2) = A\mathbf{x}_1 - A\mathbf{x}_2 = \mathbf{b} - \mathbf{b} = \mathbf{0}$$

It follows that $\mathbf{x}_1 - \mathbf{x}_2 = \mathbf{0}$, and hence $\mathbf{x}_1$ must equal $\mathbf{x}_2$.   ◀

Let A be an $m \times n$ matrix. If the column vectors of A span R^m, then n must be greater than or equal to m, since no set of less than m vectors could span R^m. If the columns of A are linearly independent, then n must be less than or equal to m, since every set of more than m vectors in R^m is linearly dependent. Thus, if the column vectors of A form a basis for R^m, then n must equal m.

▶ **COROLLARY 3.6.4** *An $n \times n$ matrix A is nonsingular if and only if the column vectors of A form a basis for R^n.*   ◀

In general, the rank and the dimension of the nullspace always add up to the number of columns of the matrix. The dimension of the nullspace of a matrix is called the *nullity* of the matrix.

▶ **THEOREM 3.6.5 (The Rank–Nullity Theorem)** *If A is an m × n matrix, then the rank of A plus the nullity of A equals n.*

▶ *Proof.* Let U be the reduced row echelon form of A. The system $A\mathbf{x} = \mathbf{0}$ is equivalent to the system $U\mathbf{x} = \mathbf{0}$. If A has rank r, then U will have r nonzero rows, and consequently the system $U\mathbf{x} = \mathbf{0}$ will involve r lead variables and $n - r$ free variables. The dimension of $N(A)$ will equal the number of free variables.   ◀

EXAMPLE 3. Let

$$A = \begin{bmatrix} 1 & 2 & -1 & 1 \\ 2 & 4 & -3 & 0 \\ 1 & 2 & 1 & 5 \end{bmatrix}$$

Find a basis for the row space of A and a basis for $N(A)$. Verify that $\dim N(A) = n - r$.

SOLUTION. The reduced row echelon form of A is given by

$$U = \begin{bmatrix} 1 & 2 & 0 & 3 \\ 0 & 0 & 1 & 2 \\ 0 & 0 & 0 & 0 \end{bmatrix}$$

Thus, $\{(1, 2, 0, 3), (0, 0, 1, 2)\}$ is a basis for the row space of A, and A has rank 2. Since the systems $A\mathbf{x} = \mathbf{0}$ and $U\mathbf{x} = \mathbf{0}$ are equivalent, it follows that $\mathbf{x}$ is in $N(A)$ if and only if

$$x_1 + 2x_2 + \qquad 3x_4 = 0$$
$$x_3 + 2x_4 = 0$$

The lead variables x_1 and x_3 can be solved for in terms of the free variables x_2 and x_4.

$$x_1 = -2x_2 - 3x_4$$
$$x_3 = -2x_4$$

Let $x_2 = \alpha$ and $x_4 = \beta$. It follows that $N(A)$ consists of all vectors of the form

$$\begin{bmatrix} x_1 \\ x_2 \\ x_3 \\ x_4 \end{bmatrix} = \begin{bmatrix} -2\alpha - 3\beta \\ \alpha \\ -2\beta \\ \beta \end{bmatrix} = \alpha \begin{bmatrix} -2 \\ 1 \\ 0 \\ 0 \end{bmatrix} + \beta \begin{bmatrix} -3 \\ 0 \\ -2 \\ 1 \end{bmatrix}$$

The vectors $(-2, 1, 0, 0)^T$ and $(-3, 0, -2, 1)^T$ form a basis for $N(A)$. Note that

$$n - r = 4 - 2 = 2 = \dim N(A) \qquad \blacktriangleleft$$

The Column Space

The matrices A and U in Example 3 have different column spaces; however, their column vectors satisfy the same dependency relations. For the matrix U, the column vectors $\mathbf{u}_1$ and $\mathbf{u}_3$ are linearly independent, while

$$\mathbf{u}_2 = 2\mathbf{u}_1$$

$$\mathbf{u}_4 = 3\mathbf{u}_1 + 2\mathbf{u}_3$$

The same relations hold for the columns of A. The vectors $\mathbf{a}_1$ and $\mathbf{a}_3$ are linearly independent, while

$$\mathbf{a}_2 = 2\mathbf{a}_1$$

$$\mathbf{a}_4 = 3\mathbf{a}_1 + 2\mathbf{a}_3$$

In general, if A is an $m \times n$ matrix and U is the row echelon form of A, then, since $A\mathbf{x} = \mathbf{0}$ if and only if $U\mathbf{x} = \mathbf{0}$, their column vectors satisfy the same dependency relations. We will use this to prove that the dimension of the column space of A is equal to the dimension of the row space of A.

▶ **THEOREM 3.6.6** *If A is an $m \times n$ matrix, the dimension of the row space of A equals the dimension of the column space of A.*

▶ **Proof.** If A is an $m \times n$ matrix of rank r, the row echelon form U of A will have r lead 1's. The columns of U corresponding to the lead 1's will be linearly independent. They do not, however, form a basis for the column space of A, since in general A and U will have different column spaces. Let U_L denote the matrix obtained from U by deleting all the columns corresponding to the free variables. Delete the same columns from A and denote the new matrix by A_L. The matrices A_L and U_L are row equivalent. Thus, if $\mathbf{x}$ is a solution to $A_L\mathbf{x} = \mathbf{0}$, then $\mathbf{x}$ must also be a solution to $U_L\mathbf{x} = \mathbf{0}$. Since the columns of U_L are linearly independent, $\mathbf{x}$ must equal $\mathbf{0}$. It follows from the remarks preceding Theorem 3.6.3 that the columns of A_L are linearly independent. Since A_L has r columns, the dimension of the column space of A is at least r.

We have proved that for any matrix the dimension of the column space is greater than or equal to the dimension of the row space. Applying this result to the matrix A^T, we see that

$$\dim(\text{row space of } A) = \dim(\text{column space of } A^T)$$

$$\geq \dim(\text{row space of } A^T)$$

$$= \dim(\text{column space of } A)$$

Thus, for any matrix A, the dimension of the row space must equal the dimension of the column space. ◀

We can use the row echelon form U of A to find a basis for the column space of A. We need only determine the columns of U that correspond to the lead 1's. Those same columns of A will be linearly independent and form a basis for the column space of A.

NOTE The row echelon form U tells us only which columns of A to use to form a basis. We cannot use the column vectors from U since, in general, U and A have different column spaces.

EXAMPLE 4. Let

$$
A = \begin{bmatrix} 1 & -2 & 1 & 1 & 2 \\ -1 & 3 & 0 & 2 & -2 \\ 0 & 1 & 1 & 3 & 4 \\ 1 & 2 & 5 & 13 & 5 \end{bmatrix}
$$

The row echelon form of A is given by

$$
U = \begin{bmatrix} 1 & -2 & 1 & 1 & 2 \\ 0 & 1 & 1 & 3 & 0 \\ 0 & 0 & 0 & 0 & 1 \\ 0 & 0 & 0 & 0 & 0 \end{bmatrix}
$$

The lead 1's occur in the first, second, and fifth columns. Thus

$$
\mathbf{a}_1 = \begin{bmatrix} 1 \\ -1 \\ 0 \\ 1 \end{bmatrix}, \qquad \mathbf{a}_2 = \begin{bmatrix} -2 \\ 3 \\ 1 \\ 2 \end{bmatrix}, \qquad \mathbf{a}_5 = \begin{bmatrix} 2 \\ -2 \\ 4 \\ 5 \end{bmatrix}
$$

form a basis for the column space of A. ◀

EXAMPLE 5. Find the dimension of the subspace of R^4 spanned by

$$
\mathbf{x}_1 = \begin{bmatrix} 1 \\ 2 \\ -1 \\ 0 \end{bmatrix}, \qquad \mathbf{x}_2 = \begin{bmatrix} 2 \\ 5 \\ -3 \\ 2 \end{bmatrix}, \qquad \mathbf{x}_3 = \begin{bmatrix} 2 \\ 4 \\ -2 \\ 0 \end{bmatrix}, \qquad \mathbf{x}_4 = \begin{bmatrix} 3 \\ 8 \\ -5 \\ 4 \end{bmatrix}
$$

SOLUTION. The subspace $\mathrm{Span}(\mathbf{x}_1, \mathbf{x}_2, \mathbf{x}_3, \mathbf{x}_4)$ is the same as the column space of the matrix

$$X = \begin{bmatrix} 1 & 2 & 2 & 3 \\ 2 & 5 & 4 & 8 \\ -1 & -3 & -2 & -5 \\ 0 & 2 & 0 & 4 \end{bmatrix}$$

The row echelon form of X is

$$\begin{bmatrix} 1 & 2 & 2 & 3 \\ 0 & 1 & 0 & 2 \\ 0 & 0 & 0 & 0 \\ 0 & 0 & 0 & 0 \end{bmatrix}$$

The first two columns $\mathbf{x}_1, \mathbf{x}_2$ of X will form a basis for the column space of X. Thus $\dim \mathrm{Span}(\mathbf{x}_1, \mathbf{x}_2, \mathbf{x}_3, \mathbf{x}_4) = 2$. ◀

EXERCISES

1. For each of the following matrices, find a basis for the row space, a basis for the column space, and a basis for the nullspace.

(a) $\begin{bmatrix} 1 & 3 & 2 \\ 2 & 1 & 4 \\ 4 & 7 & 8 \end{bmatrix}$

(b) $\begin{bmatrix} -3 & 1 & 3 & 4 \\ 1 & 2 & -1 & -2 \\ -3 & 8 & 4 & 2 \end{bmatrix}$

(c) $\begin{bmatrix} 1 & 3 & -2 & 1 \\ 2 & 1 & 3 & 2 \\ 3 & 4 & 5 & 6 \end{bmatrix}$

2. In each of the following, determine the dimension of the subspace of R^3 spanned by the given vectors.

(a) $\begin{bmatrix} 1 \\ -2 \\ 2 \end{bmatrix}, \begin{bmatrix} 2 \\ -2 \\ 4 \end{bmatrix}, \begin{bmatrix} -3 \\ 3 \\ 6 \end{bmatrix}$

(b) $\begin{bmatrix} 1 \\ 1 \\ 1 \end{bmatrix}, \begin{bmatrix} 1 \\ 2 \\ 3 \end{bmatrix}, \begin{bmatrix} 2 \\ 3 \\ 1 \end{bmatrix}$

(c) $\begin{bmatrix} 1 \\ -1 \\ 2 \end{bmatrix}, \begin{bmatrix} -2 \\ 2 \\ -4 \end{bmatrix}, \begin{bmatrix} 3 \\ -2 \\ 5 \end{bmatrix}, \begin{bmatrix} 2 \\ -1 \\ 3 \end{bmatrix}$

3. Given

$$A = \begin{bmatrix} 1 & 2 & 2 & 3 & 1 & 4 \\ 2 & 4 & 5 & 5 & 4 & 9 \\ 3 & 6 & 7 & 8 & 5 & 9 \end{bmatrix}$$

   (a) Compute the reduced row echelon form U of A. Which column vectors of U correspond to the free variables? Write each of these vectors as a linear combination of the column vectors corresponding to the lead variables.

   (b) Which column vectors of A correspond to the lead variables of U? These column vectors form a basis for the column space of A. Write each of the remaining column vectors of A as a linear combination of these basis vectors.

4. For each of the following choices of A and $\mathbf{b}$, determine if $\mathbf{b}$ is in the column space of A and state whether the system $A\mathbf{x} = \mathbf{b}$ is consistent.

   (a) $A = \begin{bmatrix} 1 & 2 \\ 2 & 4 \end{bmatrix}$, $\mathbf{b} = \begin{bmatrix} 4 \\ 8 \end{bmatrix}$ (b) $A = \begin{bmatrix} 3 & 6 \\ 1 & 2 \end{bmatrix}$, $\mathbf{b} = \begin{bmatrix} 1 \\ 1 \end{bmatrix}$

   (c) $A = \begin{bmatrix} 2 & 1 \\ 3 & 4 \end{bmatrix}$, $\mathbf{b} = \begin{bmatrix} 4 \\ 6 \end{bmatrix}$ (d) $A = \begin{bmatrix} 1 & 1 & 2 \\ 1 & 1 & 2 \\ 1 & 1 & 2 \end{bmatrix}$, $\mathbf{b} = \begin{bmatrix} 1 \\ 2 \\ 3 \end{bmatrix}$

   (e) $A = \begin{bmatrix} 0 & 1 \\ 1 & 0 \\ 0 & 1 \end{bmatrix}$, $\mathbf{b} = \begin{bmatrix} 2 \\ 5 \\ 2 \end{bmatrix}$ (f) $A = \begin{bmatrix} 1 & 2 \\ 2 & 4 \\ 1 & 2 \end{bmatrix}$, $\mathbf{b} = \begin{bmatrix} 5 \\ 10 \\ 5 \end{bmatrix}$

5. For each of the consistent systems in Exercise 4, determine whether there will be one or infinitely many solutions by examining the column vectors of the coefficient matrix A.

6. How many solutions will the linear system $A\mathbf{x} = \mathbf{b}$ have if $\mathbf{b}$ is in the column space of A and the column vectors of A are linearly dependent? Explain.

7. Let A be an $m \times n$ matrix with $m > n$. Let $\mathbf{b} \in R^m$ and suppose that $N(A) = \{\mathbf{0}\}$.

   (a) What can you conclude about the column vectors of A? Are they linearly independent? Do they span R^m? Explain.

   (b) How many solutions will the system $A\mathbf{x} = \mathbf{b}$ have if $\mathbf{b}$ is not in the column space of A? How many solutions will there be if $\mathbf{b}$ is in the column space of A? Explain.

8. Let A and B be 6×5 matrices. If dim $N(A) = 2$, what is the rank of A? If the rank of B is 4, what is the dimension of $N(B)$?

9. Let A and B be row equivalent matrices.

   (a) Show that the dimension of the column space of A equals the dimension of the column space of B.

(b) Are the column spaces of the two matrices necessarily the same? Justify your answer.

10. Let A be a 4×4 matrix with reduced row echelon form given by

$$U = \begin{bmatrix} 1 & 0 & 2 & 1 \\ 0 & 1 & 1 & 4 \\ 0 & 0 & 0 & 0 \\ 0 & 0 & 0 & 0 \end{bmatrix}$$

If

$$\mathbf{a}_1 = \begin{bmatrix} -3 \\ 5 \\ 2 \\ 1 \end{bmatrix} \quad \text{and} \quad \mathbf{a}_2 = \begin{bmatrix} 4 \\ -3 \\ 7 \\ -1 \end{bmatrix}$$

determine $\mathbf{a}_3$ and $\mathbf{a}_4$.

11. Let A be a 4×5 matrix. If $\mathbf{a}_1$, $\mathbf{a}_2$, $\mathbf{a}_4$ are linearly independent and

$$\mathbf{a}_3 = \mathbf{a}_1 + 2\mathbf{a}_2 \qquad \mathbf{a}_5 = 2\mathbf{a}_1 - \mathbf{a}_2 + 3\mathbf{a}_4$$

determine the reduced row echelon form of A.

12. Prove that a linear system $A\mathbf{x} = \mathbf{b}$ is consistent if and only if the rank of $(A \mid \mathbf{b})$ equals the rank of A.

13. Let A be an $m \times n$ matrix.

(a) If B is a nonsingular $m \times m$ matrix, show that BA and A have the same nullspace and hence the same rank.

(b) If C is a nonsingular $n \times n$ matrix, show that AC and A have the same rank.

14. Prove Corollary 3.6.4.

15. Suppose that A and B are $n \times n$ matrices with the property that $A\mathbf{x} = B\mathbf{x}$ for all $\mathbf{x} \in R^n$. Show that:

(a) $N(A - B) = R^n$.

(b) $A - B$ must have rank 0 and consequently $A = B$.

16. Let A and B be $n \times n$ matrices. Show that $AB = O$ if and only if the column space of B is a subspace of the nullspace of A.

17. Let $A \in R^{m \times n}$, $\mathbf{b} \in R^m$ and let $\mathbf{x}_0$ be a particular solution to the system $A\mathbf{x} = \mathbf{b}$. Prove the following.

(a) A vector $\mathbf{y}$ in R^n will be a solution to $A\mathbf{x} = \mathbf{b}$ if and only if $\mathbf{y} = \mathbf{x}_0 + \mathbf{z}$, where $\mathbf{z} \in N(A)$.

(b) If $N(A) = \{\mathbf{0}\}$, then the solution $\mathbf{x}_0$ is unique.

18. Let **x** and **y** be nonzero vectors in R^m and R^n, respectively, and let $A = \mathbf{xy}^T$.

    (a) Show that $\{\mathbf{x}\}$ is a basis for the column space of A and that $\{\mathbf{y}^T\}$ is a basis for the row space of A.

    (b) What is the dimension of $N(A)$?

19. Let $A \in R^{m \times n}$, $B \in R^{n \times r}$, and $C = AB$. Show that:

    (a) The column space of C is a subspace of the column space of A.

    (b) The row space of C is a subspace of the row space of B.

    (c) $\text{rank}(C) \leq \min\{\text{rank}(A), \ \text{rank}(B)\}$.

20. Let $A \in R^{m \times n}$, $B \in R^{n \times r}$, and $C = AB$. Show that:

    (a) If A and B both have linearly independent column vectors, then the column vectors of C will also be linearly independent.

    (b) If A and B both have linearly independent row vectors, then the row vectors of C will also be linearly independent.

[*Hint:* Apply part (a) to C^T.]

21. Let $A \in R^{m \times n}$, $B \in R^{n \times r}$, and $C = AB$. Show that:

    (a) If the column vectors of B are linearly dependent, then the column vectors of C must be linearly dependent.

    (b) If the row vectors of A are linearly dependent, then the row vectors of C are linearly dependent.

[*Hint:* Apply part (a) to C^T.]

22. An $m \times n$ matrix A is said to have a *right inverse* if there exists an $n \times m$ matrix C such that $AC = I_m$. A is said to have a *left inverse* if there exists an $n \times m$ matrix D such that $DA = I_n$.

    (a) If A has a right inverse, show that the column vectors of A span R^m.

    (b) Is it possible for an $m \times n$ matrix to have a right inverse if $n < m$? $n \geq m$? Explain.

23. Prove: If A is an $m \times n$ matrix and the column vectors of A span R^m, then A has a right inverse. [*Hint:* Let $\mathbf{e}_j$ denote the jth column of I_m and solve $A\mathbf{x} = \mathbf{e}_j$ for $j = 1, \ldots, m$.]

24. Show that a matrix B has a left inverse if and only if B^T has a right inverse.

25. Let B be an $n \times m$ matrix whose columns are linearly independent. Show that B has a left inverse.

26. Prove that if a matrix B has a left inverse the columns of B are linearly independent.

27. If a matrix U is in row echelon form, show that the nonzero row vectors of U form a basis for the row space of U.

MATLAB EXERCISES

1. (Change of Basis) Set

$$U = \text{round}(20 * \text{rand}(4)) - 10, \qquad V = \text{round}(10 * \text{rand}(4))$$

and set $\mathbf{b} = \text{ones}(4, 1)$.

(a) We can use the MATLAB function **rank** to determine whether the column vectors of a matrix are linearly independent. What should the rank be if the column vectors of U are linearly independent? Compute the rank of U and verify that its column vectors are linearly independent and hence form a basis for R^4. Compute the rank of V and verify that its column vectors also form a basis for R^4.

(b) Use MATLAB to compute the transition matrix from the standard basis for R^4 to the ordered basis $E = [\mathbf{u}_1, \mathbf{u}_2, \mathbf{u}_3, \mathbf{u}_4]$. [Note that in MATLAB the notation for the jth column vector $\mathbf{u}_j$ is $U(:, j)$.] Use this transition matrix to compute the coordinate vector $\mathbf{c}$ of $\mathbf{b}$ with respect to E. Verify that

$$\mathbf{b} = c_1\mathbf{u}_1 + c_2\mathbf{u}_2 + c_3\mathbf{u}_3 + c_4\mathbf{u}_4 = U\mathbf{c}$$

(c) Use MATLAB to compute the transition matrix from the standard basis to the basis $F = [\mathbf{v}_1, \mathbf{v}_2, \mathbf{v}_3, \mathbf{v}_4]$, and use this transition matrix to find the coordinate vector $\mathbf{d}$ of $\mathbf{b}$ with respect to F. Verify that

$$\mathbf{b} = d_1\mathbf{v}_1 + d_2\mathbf{v}_2 + d_3\mathbf{v}_3 + d_4\mathbf{v}_4 = V\mathbf{d}$$

(d) Use MATLAB to compute the transition matrix S from E to F and the transition matrix T from F to E. How are S and T related? Verify that $S\mathbf{c} = \mathbf{d}$ and $T\mathbf{d} = \mathbf{c}$.

2. (Rank-Deficient Matrices) In this exercise we consider how to generate matrices with specified ranks using MATLAB.

(a) In general, if A is an $m \times n$ matrix with rank r, then $r \leq \min(m, n)$. Why? Explain. If the entries of A are random numbers, we would expect that $r = \min(m, n)$. Why? Explain. Check this out in MATLAB by generating random 6×6, 8×6, 5×8 matrices and checking their ranks using the MATLAB command **rank**. Whenever the rank of an $m \times n$ matrix equals $\min(m, n)$, we say that the matrix has *full rank*. Otherwise, we say that the matrix is *rank deficient*.

(b) MATLAB's **rand** and **round** commands can be used to generate random $m \times n$ matrices with integer entries in a given range $[a, b]$. This can be done with a command of the form

$$A = \text{round}((b - a) * \text{rand}(m, n)) + a$$

For example, the command

$$A = \text{round}(4 * \text{rand}(6, 8)) + 3$$

will generate a 6×8 matrix whose entries are random integers in the range from 3 to 7. Using the range $[1, 10]$, create random integer 10×7, 8×12, and 10×15 matrices and in each case check the rank of the matrix. Do these integer matrices all have full rank?

(c) Suppose that we want to generate matrices with less than full rank using MATLAB. It is easy to generate rank 1 matrices. If **x** and **y** are nonzero vectors in R^m and R^n, respectively, then $A = \mathbf{xy}^T$ will be an $m \times n$ matrix with rank 1. Why? Explain. Verify this using MATLAB by setting

$$\mathbf{x} = \mathbf{round}(9 * \mathbf{rand}(8, 1)) + 1, \qquad \mathbf{y} = \mathbf{round}(9 * \mathbf{rand}(6, 1)) + 1$$

and using these vectors to construct an 8×6 matrix A. Check the rank of A using the MATLAB command **rank**.

(d) In general,

(3) $$\mathrm{rank}(AB) \le \min(\mathrm{rank}(A), \mathrm{rank}(B))$$

(See Exercise 19 in Section 6 of this chapter.) If A and B are noninteger random matrices, the relation (3) should be an equality. Generate an 8×6 matrix A by setting

$$X = \mathbf{rand}(8, 2), \qquad Y = \mathbf{rand}(6, 2), \qquad A = X * Y'$$

What would you expect the rank of A to be? Explain. Test the rank of A using MATLAB.

(e) Use MATLAB to generate matrices A, B, C such that

   (i) A is 8×8 with rank 3.

  (ii) B is 6×9 with rank 4.

 (iii) C is 10×7 with rank 5.

3. (Column Space and Reduced Row Echelon Form) Set $B = \mathbf{round}(10 * \mathbf{rand}(8, 4))$, $X = \mathbf{round}(10 * \mathbf{rand}(4, 3))$, $C = B * X$, and $A = [B \ \ C]$.

(a) How are the column spaces of B and C related? (See Exercise 17 in Section 6 of this chapter.) What would you expect the rank of A to be? Explain. Use MATLAB to check your answer.

(b) Which column vectors of A should form a basis for its column space? Explain. If U is the reduced row echelon form of A, what would you expect its first four columns to be? Explain. What would you expect its last four rows to be? Explain. Use MATLAB to verify your answers by computing U.

(c) Use MATLAB to construct another matrix $D = (E \ \ EY)$, where E is a random 6×4 matrix and Y is a random 4×2 matrix. What would you expect the reduced row echelon form of D to be? Compute it using MATLAB. Show that, in general, if B is an $m \times n$ matrix of rank n and X is an $n \times k$ matrix, the reduced row echelon form of $(B \ \ BX)$ will have block structure

$$(I \ \ X) \text{ if } m = n \qquad \text{or} \qquad \begin{bmatrix} I & X \\ O & O \end{bmatrix} \text{ if } m > n$$

4. (Rank 1 Updates of Linear Systems)

(a) Set $A = \mathbf{round}(10 * \mathbf{rand}(8))$, $\mathbf{b} = \mathbf{round}(10 * \mathbf{rand}(8, 1))$, and $M = \mathbf{inv}(A)$. Use the matrix M to solve the system $A\mathbf{y} = \mathbf{b}$ for $\mathbf{y}$.

(b) Consider now a new system $C\mathbf{x} = \mathbf{b}$, where the C is constructed as follows:

$$\mathbf{u} = \mathbf{round}(10 * \mathbf{rand}(8, 1)), \quad \mathbf{v} = \mathbf{round}(10 * \mathbf{rand}(8, 1))$$
$$E = \mathbf{u} * \mathbf{v}', \qquad\qquad\qquad C = A + E$$

The matrices C and A differ by the rank 1 matrix E. Use MATLAB to verify that the rank of E is 1. MATLAB has a counter **flops** that keeps track of the number of arithmetic operations performed in a MATLAB session. Enter the MATLAB command **flops**(0) to reset the flops counter to 0 and then solve the system $C\mathbf{x} = \mathbf{b}$ using the "\" operation. Type **flops** to check the value of flops in order to see how much work was required.

(c) Let us now solve $C\mathbf{x} = \mathbf{b}$ by a new method that takes advantage of the fact that A and C differ by a rank 1 matrix. This new procedure is called a *rank 1 update* method. Use the command **flops**(0) to reset the flops counter to 0 again and then set

$$\mathbf{z} = M * \mathbf{u}, \quad c = \mathbf{v}' * \mathbf{y}, \quad d = \mathbf{v}' * \mathbf{z}, \quad e = c/(1+d)$$

The solution $\mathbf{x}$ is given by

$$\mathbf{x} = \mathbf{y} - e * \mathbf{z}$$

Check the flop count for computing $\mathbf{x}$ by this second method. Is this method more efficient than using the "\" operation? Use MATLAB to compute the residual vector $\mathbf{b} - C\mathbf{x}$.

(d) To see why the rank 1 update method works, use MATLAB to compute and compare

$$C\mathbf{y} \qquad \text{and} \qquad \mathbf{b} + c\mathbf{u}$$

Prove that if all computations had been carried out in exact arithmetic these two vectors would be equal. Also compute

$$C\mathbf{z} \qquad \text{and} \qquad (1+d)\mathbf{u}$$

Prove that if all computations had been carried out in exact arithmetic these two vectors would be equal. Use these identities to prove that $C\mathbf{x} = \mathbf{b}$. Assuming that A is nonsingular, will the rank 1 update method always work? Under what conditions could it fail? Explain.

CHAPTER TEST

Answer each of the following *true* or *false*. In each case, explain or prove your answer.

1. If S is a subspace of a vector space V, then S is a vector space.

2. R^2 is a subspace of R^3.

3. It is possible to find a pair of two-dimensional subspaces S and T of R^3 such that $S \cap T = \{\mathbf{0}\}$.

4. If S and T are subspaces of a vector space V, then $S \cup T$ is a subspace of V.

5. If S and T are subspaces of a vector space V, then $S \cap T$ is a subspace of V.

6. If $\mathbf{x}_1, \mathbf{x}_2, \ldots, \mathbf{x}_n$ span R^n, then they are linearly independent.

7. If $\mathbf{x}_1, \mathbf{x}_2, \ldots, \mathbf{x}_n$ span a vector space V, then they are linearly independent.

8. If $\mathbf{x}_1, \mathbf{x}_2, \ldots, \mathbf{x}_k$ are vectors in a vector space V and

$$\text{Span}(\mathbf{x}_1, \mathbf{x}_2, \ldots, \mathbf{x}_k) = \text{Span}(\mathbf{x}_1, \mathbf{x}_2, \ldots, \mathbf{x}_{k-1})$$

then $\mathbf{x}_1, \mathbf{x}_2, \ldots, \mathbf{x}_k$ are linearly dependent.

9. If A is an $m \times n$ matrix, then A and A^T have the same rank.

10. If A is an $m \times n$ matrix, then A and A^T have the same nullity.

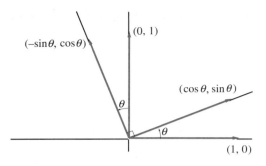

CHAPTER 4

LINEAR TRANSFORMATIONS

Linear mappings from one vector space to another play an important role in mathematics. This chapter provides an introduction to the theory of such mappings. In Section 1 the definition of a linear transformation is given and a number of examples are presented. In Section 2 it is shown that each linear transformation L mapping an n-dimensional vector space V into an m-dimensional vector space W can be represented by an $m \times n$ matrix A. Thus we can work with the matrix A in place of the operator L. In the case that the linear transformation L maps V into itself, the matrix representing L will depend on the ordered basis chosen for V. Thus L may be represented by a matrix A with respect to one ordered basis and by another matrix B with respect to another ordered basis. In Section 3 we consider the relationship between different matrices that represent the same linear transformation. In many applications it is desirable to choose the basis for V so that the matrix representing the linear transformation is either diagonal or in some other simple form.

▌ DEFINITION AND EXAMPLES

In the study of vector spaces the most important types of mappings are linear transformations.

▶ **DEFINITION** A mapping L from a vector space V into a vector space W is said to be a **linear transformation** or a **linear operator** if

(1) $$L(\alpha\mathbf{v}_1 + \beta\mathbf{v}_2) = \alpha L(\mathbf{v}_1) + \beta L(\mathbf{v}_2)$$

for all $\mathbf{v}_1, \mathbf{v}_2 \in V$ and for all scalars α and β. ◀

If L is a linear transformation mapping a vector space V into W, it follows from (1) that

(2) $L(\mathbf{v}_1 + \mathbf{v}_2) = L(\mathbf{v}_1) + L(\mathbf{v}_2)$ $(\alpha = \beta = 1)$

and

(3) $L(\alpha\mathbf{v}) = \alpha L(\mathbf{v})$ $(\mathbf{v} = \mathbf{v}_1, \beta = 0)$

Conversely, if L satisfies (2) and (3), then

$$L(\alpha\mathbf{v}_1 + \beta\mathbf{v}_2) = L(\alpha\mathbf{v}_1) + L(\beta\mathbf{v}_2)$$
$$= \alpha L(\mathbf{v}_1) + \beta L(\mathbf{v}_2)$$

Thus L is a linear operator if and only if L satisfies (2) and (3).

Notation   A mapping L from a vector space V into a vector space W will be denoted

$$L: V \rightarrow W$$

When the arrow notation is used, it will be assumed that V and W represent vector spaces.

Let us now consider some examples of linear transformations. We begin with linear operators that map R^2 into itself. In this case it is easier to see the geometric effect of the operator.

Linear Transformations on R^2

EXAMPLE I. Let L be the operator defined by

$$L(\mathbf{x}) = 3\mathbf{x}$$

for each $\mathbf{x} \in R^2$. Since

$$L(\alpha\mathbf{x}) = 3(\alpha\mathbf{x}) = \alpha(3\mathbf{x}) = \alpha L(\mathbf{x})$$

and

$$L(\mathbf{x} + \mathbf{y}) = 3(\mathbf{x} + \mathbf{y}) = (3\mathbf{x}) + (3\mathbf{y}) = L(\mathbf{x}) + L(\mathbf{y})$$

it follows that L is a linear transformation. We can think of L as a stretching by a factor of 3 (see Figure 4.1.1). In general, if α is a positive scalar, the linear transformation $F(\mathbf{x}) = \alpha\mathbf{x}$ can be thought of as a stretching or shrinking by a factor of α. ◀

EXAMPLE 2. Consider the mapping L defined by

$$L(\mathbf{x}) = x_1\mathbf{e}_1$$

for each $\mathbf{x} \in R^2$. Thus, if $\mathbf{x} = (x_1, x_2)^T$, then $L(\mathbf{x}) = (x_1, 0)^T$. If $\mathbf{y} = (y_1, y_2)^T$, then

$$\alpha\mathbf{x} + \beta\mathbf{y} = \begin{bmatrix} \alpha x_1 + \beta y_1 \\ \alpha x_2 + \beta y_2 \end{bmatrix}$$

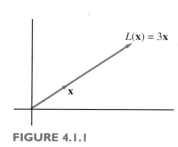

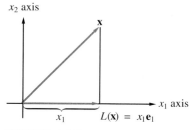

FIGURE 4.1.1   **FIGURE 4.1.2**

and it follows that

$$L(\alpha\mathbf{x} + \beta\mathbf{y}) = (\alpha x_1 + \beta y_1)\mathbf{e}_1 = \alpha(x_1\mathbf{e}_1) + \beta(y_1\mathbf{e}_1) = \alpha L(\mathbf{x}) + \beta L(\mathbf{y})$$

Thus L is a linear transformation. We can think of L as a projection onto the x_1 axis (see Figure 4.1.2). ◀

EXAMPLE 3. Let L be the operator defined by

$$L(\mathbf{x}) = (x_1, -x_2)^T$$

for each $\mathbf{x} = (x_1, x_2)^T$ in R^2. Since

$$L(\alpha\mathbf{x} + \beta\mathbf{y}) = \begin{bmatrix} \alpha x_1 + \beta y_1 \\ -(\alpha x_2 + \beta y_2) \end{bmatrix}$$

$$= \alpha \begin{bmatrix} x_1 \\ -x_2 \end{bmatrix} + \beta \begin{bmatrix} y_1 \\ -y_2 \end{bmatrix}$$

$$= \alpha L(\mathbf{x}) + \beta L(\mathbf{y})$$

it follows that L is a linear transformation. The operator L has the effect of reflecting vectors about the x_1 axis (see Figure 4.1.3). ◀

FIGURE 4.1.3

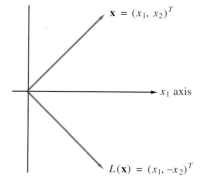

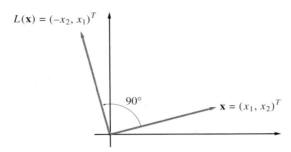

FIGURE 4.1.4

EXAMPLE 4. The operator L defined by

$$L(\mathbf{x}) = (-x_2, x_1)^T$$

is linear, since

$$L(\alpha \mathbf{x} + \beta \mathbf{y}) = \begin{bmatrix} -(\alpha x_2 + \beta y_2) \\ \alpha x_1 + \beta y_1 \end{bmatrix}$$

$$= \alpha \begin{bmatrix} -x_2 \\ x_1 \end{bmatrix} + \beta \begin{bmatrix} -y_2 \\ y_1 \end{bmatrix}$$

$$= \alpha L(\mathbf{x}) + \beta L(\mathbf{y})$$

The operator L has the effect of rotating each vector in R^2 by $90°$ in the counterclockwise direction (see Figure 4.1.4). ◀

EXAMPLE 5. Consider the mapping M defined by

$$M(\mathbf{x}) = (x_1^2 + x_2^2)^{1/2}$$

Since

$$M(\alpha \mathbf{x}) = (\alpha^2 x_1^2 + \alpha^2 x_2^2)^{1/2} = |\alpha| M(\mathbf{x})$$

it follows that

$$\alpha M(\mathbf{x}) \neq M(\alpha \mathbf{x})$$

whenever $\alpha < 0$ and $\mathbf{x} \neq \mathbf{0}$. Therefore, M is not a linear transformation. ◀

Linear Operators from R^n to R^m

EXAMPLE 6. The mapping $L: R^2 \rightarrow R^1$ defined by

$$L(\mathbf{x}) = x_1 + x_2$$

is a linear transformation since

$$L(\alpha\mathbf{x} + \beta\mathbf{y}) = (\alpha x_1 + \beta y_1) + (\alpha x_2 + \beta y_2)$$
$$= \alpha(x_1 + x_2) + \beta(y_1 + y_2)$$
$$= \alpha L(\mathbf{x}) + \beta L(\mathbf{y})$$ ◀

EXAMPLE 7. The mapping L from R^2 to R^3 defined by

$$L(\mathbf{x}) = (x_2, x_1, x_1 + x_2)^T$$

is linear, since

$$L(\alpha\mathbf{x}) = (\alpha x_2, \alpha x_1, \alpha x_1 + \alpha x_2)^T = \alpha L(\mathbf{x})$$

and

$$L(\mathbf{x} + \mathbf{y}) = (x_2 + y_2, x_1 + y_1, x_1 + y_1 + x_2 + y_2)^T$$
$$= (x_2, x_1, x_1 + x_2)^T + (y_2, y_1, y_1 + y_2)^T$$
$$= L(\mathbf{x}) + L(\mathbf{y})$$

Note that if we define the matrix A by

$$A = \begin{bmatrix} 0 & 1 \\ 1 & 0 \\ 1 & 1 \end{bmatrix}$$

then

$$L(\mathbf{x}) = \begin{bmatrix} x_2 \\ x_1 \\ x_1 + x_2 \end{bmatrix} = A\mathbf{x}$$

for each $\mathbf{x} \in R^2$.   ◀

In general, if A is any $m \times n$ matrix, we can define a linear operator L_A from R^n to R^m by

$$L_A(\mathbf{x}) = A\mathbf{x}$$

for each $\mathbf{x} \in R^n$. The operator L_A is linear since

$$L_A(\alpha\mathbf{x} + \beta\mathbf{y}) = A(\alpha\mathbf{x} + \beta\mathbf{y})$$
$$= \alpha A\mathbf{x} + \beta A\mathbf{y}$$
$$= \alpha L_A(\mathbf{x}) + \beta L_A(\mathbf{y})$$

Thus, we can think of each $m \times n$ matrix A as a linear operator from R^n to R^m.

In Example 7 we saw that the operator L could have been defined in terms of a matrix A. In the next section we will see that this is true for all linear operators from R^n to R^m.

Linear Operators from V to W

If L is a linear operator mapping a vector space V into a vector space W, then

(i) $L(\mathbf{0}_V) = \mathbf{0}_W$ (where $\mathbf{0}_V$ and $\mathbf{0}_W$ are the zero vectors in V and W, respectively).

(ii) If $\mathbf{v}_1, \ldots, \mathbf{v}_n$ are elements of V and $\alpha_1, \ldots, \alpha_n$ are scalars, then

$$L(\alpha_1\mathbf{v}_1 + \alpha_2\mathbf{v}_2 + \cdots + \alpha_n\mathbf{v}_n) = \alpha_1 L(\mathbf{v}_1) + \alpha_2 L(\mathbf{v}_2) + \cdots + \alpha_n L(\mathbf{v}_n)$$

(iii) $L(-\mathbf{v}) = -L(\mathbf{v})$ for all $\mathbf{v} \in V$.

Statement (i) follows from the condition $L(\alpha\mathbf{v}) = \alpha L(\mathbf{v})$ with $\alpha = 0$. Statement (ii) can easily be proved by mathematical induction. We leave this to the reader as an exercise. To prove (iii), note that

$$\mathbf{0}_W = L(\mathbf{0}_V) = L(\mathbf{v} + (-\mathbf{v})) = L(\mathbf{v}) + L(-\mathbf{v})$$

Therefore, $L(-\mathbf{v})$ is the additive inverse of $L(\mathbf{v})$, that is,

$$L(-\mathbf{v}) = -L(\mathbf{v})$$

EXAMPLE 8. If V is any vector space, then the identity operator $\mathcal{I}$ is defined by

$$\mathcal{I}(\mathbf{v}) = \mathbf{v}$$

for all $\mathbf{v} \in V$. Clearly, $\mathcal{I}$ is a linear transformation that maps V into itself.

$$\mathcal{I}(\alpha\mathbf{v}_1 + \beta\mathbf{v}_2) = \alpha\mathbf{v}_1 + \beta\mathbf{v}_2 = \alpha\mathcal{I}(\mathbf{v}_1) + \beta\mathcal{I}(\mathbf{v}_2) \qquad \blacktriangleleft$$

EXAMPLE 9. Let L be the mapping from $C[a, b]$ to R^1 defined by

$$L(f) = \int_a^b f(x)\, dx$$

If f and g are any vectors in $C[a, b]$, then

$$L(\alpha f + \beta g) = \int_a^b (\alpha f + \beta g)(x)\, dx$$

$$= \alpha \int_a^b f(x)\, dx + \beta \int_a^b g(x)\, dx$$

$$= \alpha L(f) + \beta L(g)$$

Therefore, L is a linear transformation. $\qquad \blacktriangleleft$

EXAMPLE 10. Let D be the operator mapping $C^1[a, b]$ into $C[a, b]$ defined by

$$D(f) = f' \qquad \text{(the derivative of } f\text{)}$$

D is a linear transformation, since

$$D(\alpha f + \beta g) = \alpha f' + \beta g' = \alpha D(f) + \beta D(g) \qquad \blacktriangleleft$$

The Image and Kernel

Let $L: V \rightarrow W$ be a linear transformation. We close this section by considering the effect that L has on subspaces of V. Of particular importance is the set of vectors in V that get mapped into the zero vector of W.

▶ **DEFINITION** Let $L: V \rightarrow W$ be a linear transformation. The **kernel** of L, denoted $\ker(L)$, is defined by

$$\ker(L) = \{\mathbf{v} \in V \,|\, L(\mathbf{v}) = \mathbf{0}_W\} \qquad \blacktriangleleft$$

▶ **DEFINITION** Let $L : V \rightarrow W$ be a linear transformation and let S be a subspace of V. The **image** of S, denoted $L(S)$, is defined by

$$L(S) = \{\mathbf{w} \in W \,|\, \mathbf{w} = L(\mathbf{v}) \quad \text{for some} \quad \mathbf{v} \in S\}$$

The image of the entire vector space, $L(V)$, is called the **range** of L. $\qquad \blacktriangleleft$

Let $L: V \rightarrow W$ be a linear transformation. It is easily seen that $\ker(L)$ is a subspace of V, and if S is any subspace of V, then $L(S)$ is a subspace of W. In particular, $L(V)$ is a subspace of W. Indeed, we have the following theorem.

▶ **THEOREM 4.1.1** *If $L: V \rightarrow W$ is a linear transformation and S is a subspace of V, then*

    (i) $\ker(L)$ *is a subspace of V.*

    (ii) $L(S)$ *is a subspace of W.*

▶ *Proof.* It is obvious that the $\ker(L)$ is nonempty since $\mathbf{0}_V$, the zero vector of V, is in $\ker(L)$. To prove (i), we must show that $\ker(L)$ is closed under scalar multiplication and addition of vectors. If $\mathbf{v} \in \ker(L)$ and α is a scalar, then

$$L(\alpha \mathbf{v}) = \alpha L(\mathbf{v}) = \alpha \mathbf{0}_W = \mathbf{0}_W$$

Therefore, $\alpha \mathbf{v} \in \ker(L)$.

    If $\mathbf{v}_1, \mathbf{v}_2 \in \ker(L)$, then

$$L(\mathbf{v}_1 + \mathbf{v}_2) = L(\mathbf{v}_1) + L(\mathbf{v}_2) = \mathbf{0}_W + \mathbf{0}_W = \mathbf{0}_W$$

Therefore, $\mathbf{v}_1 + \mathbf{v}_2 \in \ker(L)$ and hence $\ker(L)$ is a subspace of V.

    The proof of (ii) is similar. $L(S)$ is nonempty since $\mathbf{0}_W = L(\mathbf{0}_V) \in L(S)$. If $\mathbf{w} \in L(S)$, then $\mathbf{w} = L(\mathbf{v})$ for some $\mathbf{v} \in S$. For any scalar α,

$$\alpha \mathbf{w} = \alpha L(\mathbf{v}) = L(\alpha \mathbf{v})$$

Since $\alpha\mathbf{v} \in S$, it follows that $\alpha\mathbf{w} \in L(S)$, and hence $L(S)$ is closed under scalar multiplication. If $\mathbf{w}_1, \mathbf{w}_2 \in L(S)$, then there exist $\mathbf{v}_1, \mathbf{v}_2 \in S$ such that $L(\mathbf{v}_1) = \mathbf{w}_1$ and $L(\mathbf{v}_2) = \mathbf{w}_2$. Thus

$$\mathbf{w}_1 + \mathbf{w}_2 = L(\mathbf{v}_1) + L(\mathbf{v}_2) = L(\mathbf{v}_1 + \mathbf{v}_2)$$

and hence $L(S)$ is closed under addition. ◀

EXAMPLE 11. Let L be the linear transformation from R^2 into R^2 defined by

$$L(\mathbf{x}) = \begin{bmatrix} x_1 \\ 0 \end{bmatrix}$$

A vector $\mathbf{x}$ is in $\ker(L)$ if and only if $x_1 = 0$. Thus $\ker(L)$ is the one-dimensional subspace of R^2 spanned by $\mathbf{e}_2$. A vector $\mathbf{y}$ is in the range of L if and only if $\mathbf{y}$ is a multiple of $\mathbf{e}_1$. Thus $L(R^2)$ is the one-dimensional subspace of R^2 spanned by $\mathbf{e}_1$. ◀

EXAMPLE 12. Let $L: R^3 \rightarrow R^2$ be the linear transformation defined by

$$L(\mathbf{x}) = (x_1 + x_2,\ x_2 + x_3)^T$$

and let S be the subspace of R^3 spanned by $\mathbf{e}_1$ and $\mathbf{e}_3$.

If $\mathbf{x} \in \ker(L)$, then

$$x_1 + x_2 = 0 \qquad \text{and} \qquad x_2 + x_3 = 0$$

Setting the free variable $x_3 = a$, we get

$$x_2 = -a, \qquad x_1 = a$$

and hence $\ker(L)$ is the one-dimensional subspace of R^3 consisting of all vectors of the form $a(1, -1, 1)^T$.

If $\mathbf{x} \in S$, then $\mathbf{x}$ must be of the form $(a, 0, b)^T$, and hence $L(\mathbf{x}) = (a, b)^T$. Clearly, $L(S) = R^2$. Since the image of the subspace S is all of R^2, it follows that the entire range of L must be R^2 [i.e., $L(R^3) = R^2$]. ◀

EXAMPLE 13. Let $D: P_3 \rightarrow P_3$ be the differentiation operator defined by

$$D(p(x)) = p'(x)$$

The kernel of D consists of all polynomials of degree 0. Thus $\ker(D) = P_1$. The derivative of any polynomial in P_3 will be a polynomial of degree 1 or less. Conversely, any polynomial in P_2 will have antiderivatives in P_3, so each polynomial in P_2 will be the image of polynomials in P_3 under the operator D. It follows then that $D(P_3) = P_2$. ◀

EXERCISES

1. Show that each of the following are linear transformations from R^2 into R^2. Describe geometrically what each linear transformation accomplishes.

(a) $L(\mathbf{x}) = (-x_1, x_2)^T$ (b) $L(\mathbf{x}) = -\mathbf{x}$ (c) $L(\mathbf{x}) = (x_2, x_1)^T$

(d) $L(\mathbf{x}) = \frac{1}{2}\mathbf{x}$ (e) $L(\mathbf{x}) = x_2\mathbf{e}_2$

2. Let L be the linear transformation mapping R^2 into itself defined by

$$L(\mathbf{x}) = (x_1 \cos\alpha - x_2 \sin\alpha, \; x_1 \sin\alpha + x_2 \cos\alpha)^T$$

Express x_1, x_2, and $L(\mathbf{x})$ in terms of polar coordinates. Describe geometrically the effect of the linear transformation.

3. Let $\mathbf{a}$ be a fixed vector in R^2. A mapping of the form

$$L(\mathbf{x}) = \mathbf{x} + \mathbf{a}$$

is called a *translation*. Show that if $\mathbf{a} \neq \mathbf{0}$, then L is not a linear transformation. Illustrate geometrically the effect of a translation.

4. Let $L: R^2 \to R^2$ be a linear transformation. If

$$L((1, 2)^T) = (-2, 3)^T \qquad \text{and} \qquad L((1, -1)^T) = (5, 2)^T$$

determine the value of $L((7, 5)^T)$.

5. Determine whether the following are linear transformations from R^3 into R^2.

(a) $L(\mathbf{x}) = (x_2, x_3)^T$ (b) $L(\mathbf{x}) = (0, 0)^T$

(c) $L(\mathbf{x}) = (1 + x_1, x_2)^T$ (d) $L(\mathbf{x}) = (x_3, x_1 + x_2)^T$

6. Determine whether the following are linear transformations from R^2 into R^3.

(a) $L(\mathbf{x}) = (x_1, x_2, 1)^T$ (b) $L(\mathbf{x}) = (x_1, x_2, x_1 + 2x_2)^T$

(c) $L(\mathbf{x}) = (x_1, 0, 0)^T$ (d) $L(\mathbf{x}) = (x_1, x_2, x_1^2 + x_2^2)^T$

7. Determine whether the following are linear transformations from $R^{n \times n}$ into $R^{n \times n}$.

(a) $L(A) = 2A$ (b) $L(A) = A^T$

(c) $L(A) = A + I$ (d) $L(A) = A - A^T$

8. Determine whether the following are linear transformations from P_2 to P_3.

(a) $L(p(x)) = xp(x)$ (b) $L(p(x)) = x^2 + p(x)$

(c) $L(p(x)) = p(x) + xp(x) + x^2 p'(x)$

9. For each $f \in C[0, 1]$ define $L(f) = F$, where

$$F(x) = \int_0^x f(t)\, dt \qquad 0 \le x \le 1$$

Show that L is a linear transformation from $C[0, 1]$ to $C[0, 1]$. Then determine $L(e^x)$ and $L(x^2)$.

10. Determine whether the following are linear transformations from $C[0, 1]$ into R^1.

(a) $L(f) = f(0)$ (b) $L(f) = |f(0)|$

(c) $L(f) = [f(0) + f(1)]/2$ (d) $L(f) = \left\{ \int_0^1 [f(x)]^2\, dx \right\}^{1/2}$

11. If L is a linear transformation from V to W, use mathematical induction to prove that

$$L(\alpha_1 \mathbf{v}_1 + \alpha_2 \mathbf{v}_2 + \cdots + \alpha_n \mathbf{v}_n) = \alpha_1 L(\mathbf{v}_1) + \alpha_2 L(\mathbf{v}_2) + \cdots + \alpha_n L(\mathbf{v}_n)$$

12. Let $\{\mathbf{v}_1, \ldots, \mathbf{v}_n\}$ be a basis for a vector space V, and let L_1 and L_2 be two linear transformations mapping V into a vector space W. Show that if

$$L_1(\mathbf{v}_i) = L_2(\mathbf{v}_i)$$

for each $i = 1, \ldots, n$ then $L_1 = L_2$ [i.e., show that $L_1(\mathbf{v}) = L_2(\mathbf{v})$ for all $\mathbf{v} \in V$].

13. Let L be a linear transformation from R^1 into R^1 and let $a = L(1)$. Show that $L(x) = ax$ for all $x \in R^1$.

14. Let L be a linear operator mapping a vector space V into itself. Define L^n, $n \geq 1$, recursively by

$$L^1 = L$$
$$L^{k+1}(\mathbf{v}) = L(L^k(\mathbf{v})) \qquad \text{for all } \mathbf{v} \in V$$

Show that L^n is a linear operator on V for each $n \geq 1$.

15. Let $L_1 \colon U \to V$ and $L_2 \colon V \to W$ be linear transformations, and let $L = L_2 \circ L_1$ be the mapping defined by

$$L(\mathbf{u}) = L_2(L_1(\mathbf{u}))$$

for each $\mathbf{u} \in U$. Show that L is a linear transformation mapping U into W.

16. Determine the kernel and range of each of the following linear transformations from R^3 into R^3.

(a) $L(\mathbf{x}) = (x_3, x_2, x_1)^T$ (b) $L(\mathbf{x}) = (x_1, x_2, 0)^T$ (c) $L(\mathbf{x}) = (x_1, x_1, x_1)^T$

17. Let S be the subspace of R^3 spanned by $\mathbf{e}_1$ and $\mathbf{e}_2$. For each of the linear operators L in Exercise 16, determine $L(S)$.

18. Determine the kernel and range of each of the following linear transformations from P_3 into P_3.

(a) $L(p(x)) = xp'(x)$ (b) $L(p(x)) = p(x) - p'(x)$
(c) $L(p(x)) = p(0)x + p(1)$

19. Let $L \colon V \to W$ be a linear transformation, and let T be a subspace of W. The *inverse image* of T, denoted $L^{-1}(T)$, is defined by

$$L^{-1}(T) = \{\mathbf{v} \in V \,|\, L(\mathbf{v}) \in T\}$$

Show that $L^{-1}(T)$ is a subspace of V.

20. A linear transformation $L \colon V \to W$ is said to be *one-to-one* if $L(\mathbf{v}_1) = L(\mathbf{v}_2)$ implies that $\mathbf{v}_1 = \mathbf{v}_2$ (i.e., no two distinct vectors $\mathbf{v}_1, \mathbf{v}_2$ in V get mapped into the same vector $\mathbf{w} \in W$). Show that L is one-to-one if and only if $\ker(L) = \{\mathbf{0}_V\}$.

21. A linear operator $L : V \rightarrow W$ is said to map V *onto* W if $L(V) = W$. Show that the operator $L : R^3 \rightarrow R^3$ defined by

$$L(\mathbf{x}) = (x_1, x_1 + x_2, x_1 + x_2 + x_3)^T$$

maps R^3 onto R^3.

22. Which of the operators defined in Exercise 16 are one-to-one? Which map R^3 onto R^3?

23. Let A be a 2×2 matrix, and let L_A be the linear operator defined by

$$L_A(\mathbf{x}) = A\mathbf{x}$$

Show that:

(a) L_A maps R^2 onto the column space of A.

(b) If A is nonsingular, then L_A maps R^2 onto R^2.

24. Let D be the differentiation operator on P_3, and let

$$S = \{p \in P_3 \mid p(0) = 0\}$$

Show that:

(a) D maps P_3 onto P_2, but is not one-to-one.

(b) $D : S \rightarrow P_3$ is one-to-one but not onto.

2 MATRIX REPRESENTATIONS OF LINEAR TRANSFORMATIONS

In Section 1 it was shown that each $m \times n$ matrix A defines a linear transformation L_A from R^n to R^m, where

$$L_A(\mathbf{x}) = A\mathbf{x}$$

for each $\mathbf{x} \in R^n$. In this section we will see that for every linear transformation L mapping R^n into R^m there is an $m \times n$ matrix A such that

$$L(\mathbf{x}) = A\mathbf{x}$$

We will also see how any linear operator between finite-dimensional spaces can be represented by a matrix.

▶ **THEOREM 4.2.1** *If L is a linear operator mapping R^n into R^m, there is an $m \times n$ matrix A such that*

$$L(\mathbf{x}) = A\mathbf{x}$$

for each $\mathbf{x} \in R^n$. In fact, the jth column vector of A is given by

$$\mathbf{a}_j = L(\mathbf{e}_j) \qquad j = 1, 2, \dots, n$$

▶ *Proof.* For $j = 1, \dots, n$, define

$$\mathbf{a}_j = (a_{1j}, a_{2j}, \dots, a_{mj})^T = L(\mathbf{e}_j)$$

Let

$$A = (a_{ij}) = (\mathbf{a}_1, \mathbf{a}_2, \ldots, \mathbf{a}_n)$$

If

$$\mathbf{x} = x_1\mathbf{e}_1 + x_2\mathbf{e}_2 + \cdots + x_n\mathbf{e}_n$$

is an arbitrary element of R^n, then

$$L(\mathbf{x}) = x_1 L(\mathbf{e}_1) + x_2 L(\mathbf{e}_2) + \cdots + x_n L(\mathbf{e}_n)$$

$$= x_1\mathbf{a}_1 + x_2\mathbf{a}_2 + \cdots + x_n\mathbf{a}_n$$

$$= (\mathbf{a}_1, \mathbf{a}_2, \ldots, \mathbf{a}_n) \begin{bmatrix} x_1 \\ x_2 \\ \vdots \\ x_n \end{bmatrix}$$

$$= A\mathbf{x} \qquad \blacktriangleleft$$

We have established that each linear transformation from R^n into R^m can be represented in terms of an $m \times n$ matrix. Theorem 4.2.1 tells us how to construct the matrix A corresponding to a particular linear operator L. To get the first column of A, see what L does to the first basis element $\mathbf{e}_1$ of R^n. Set $\mathbf{a}_1 = L(\mathbf{e}_1)$. To get the second column of A, determine the effect of L on $\mathbf{e}_2$ and set $\mathbf{a}_2 = L(\mathbf{e}_2)$, and so on. Since the standard basis elements $\mathbf{e}_1, \mathbf{e}_2, \ldots, \mathbf{e}_n$ were used for R^n, we refer to A as the *standard matrix representation* of L. Later we will see how to represent a linear operator with respect to other bases.

EXAMPLE I. Define the operator $L: R^3 \to R^2$ by

$$L(\mathbf{x}) = (x_1 + x_2, x_2 + x_3)^T$$

for each $\mathbf{x} = (x_1, x_2, x_3)^T$ in R^3. It is easily verified that L is a linear operator. We wish to find a matrix A such that $L(\mathbf{x}) = A\mathbf{x}$ for each $\mathbf{x} \in R^3$. To do this, we must determine $L(\mathbf{e}_1)$, $L(\mathbf{e}_2)$, and $L(\mathbf{e}_3)$.

$$L(\mathbf{e}_1) = L((1, 0, 0)^T) = \begin{bmatrix} 1 \\ 0 \end{bmatrix}$$

$$L(\mathbf{e}_2) = L((0, 1, 0)^T) = \begin{bmatrix} 1 \\ 1 \end{bmatrix}$$

$$L(\mathbf{e}_3) = L((0, 0, 1)^T) = \begin{bmatrix} 0 \\ 1 \end{bmatrix}$$

We choose these vectors to be the columns of A,

$$A = \begin{bmatrix} 1 & 1 & 0 \\ 0 & 1 & 1 \end{bmatrix}$$

To check the result, compute $A\mathbf{x}$.

$$A\mathbf{x} = \begin{bmatrix} 1 & 1 & 0 \\ 0 & 1 & 1 \end{bmatrix} \begin{bmatrix} x_1 \\ x_2 \\ x_3 \end{bmatrix} = \begin{bmatrix} x_1 + x_2 \\ x_2 + x_3 \end{bmatrix}$$ ◄

EXAMPLE 2. Let L be the linear transformation mapping R^2 into R^2 that rotates each vector by an angle θ in the counterclockwise direction. We can see from Figure 4.2.1(a) that $\mathbf{e}_1$ is mapped into $(\cos\theta, \sin\theta)^T$ and the image of $\mathbf{e}_2$ is $(-\sin\theta, \cos\theta)^T$. The matrix A representing the transformation will have $(\cos\theta, \sin\theta)^T$ as its first column and $(-\sin\theta, \cos\theta)^T$ as its second column.

$$A = \begin{bmatrix} \cos\theta & -\sin\theta \\ \sin\theta & \cos\theta \end{bmatrix}$$

If $\mathbf{x}$ is any vector in R^2, then to rotate $\mathbf{x}$ counterclockwise by an angle θ we simply multiply by A [see Figure 4.2.1(b)]. ◄

Now that we have seen how matrices are used to represent linear operators from R^n to R^m, we may ask if it is possible to find a similar representation for linear operators from V into W, where V and W are vector spaces of dimension n and m, respectively. To see how this is done, let $E = [\mathbf{v}_1, \mathbf{v}_2, \dots, \mathbf{v}_n]$ be an ordered basis for V and $F = [\mathbf{w}_1, \mathbf{w}_2, \dots, \mathbf{w}_m]$ be an ordered basis for W. Let L be a linear operator mapping V into W. If $\mathbf{v}$ is any vector in V, then we can express $\mathbf{v}$ in terms of the basis E:

$$\mathbf{v} = x_1\mathbf{v}_1 + x_2\mathbf{v}_2 + \cdots + x_n\mathbf{v}_n$$

We will show that there exists an $m \times n$ matrix A representing the operator L in the sense that

$$A\mathbf{x} = \mathbf{y} \quad \text{if and only if} \quad L(\mathbf{v}) = y_1\mathbf{w}_1 + y_2\mathbf{w}_2 + \cdots + y_m\mathbf{w}_m$$

FIGURE 4.2.1

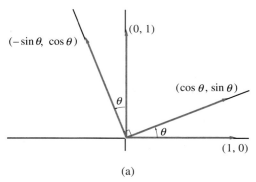

(a)

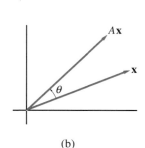

(b)

The matrix A characterizes the effect of the operator L. If $\mathbf{x}$ is the coordinate vector of $\mathbf{v}$ with respect to E, then the coordinate vector of $L(\mathbf{v})$ with respect to F is given by

$$[L(\mathbf{v})]_F = A\mathbf{x}$$

The procedure for determining the matrix representation A is essentially the same as before. For $j = 1, \ldots, n$, let $\mathbf{a}_j = (a_{1j}, a_{2j}, \ldots, a_{mj})^T$ be the coordinate vector of $L(\mathbf{v}_j)$ with respect to $[\mathbf{w}_1, \mathbf{w}_2, \ldots, \mathbf{w}_m]$.

$$L(\mathbf{v}_j) = a_{1j}\mathbf{w}_1 + a_{2j}\mathbf{w}_2 + \cdots + a_{mj}\mathbf{w}_m \qquad 1 \le j \le n$$

Let $A = (a_{ij}) = (\mathbf{a}_1, \ldots, \mathbf{a}_n)$. If

$$\mathbf{v} = x_1\mathbf{v}_1 + x_2\mathbf{v}_2 + \cdots + x_n\mathbf{v}_n$$

then

$$
\begin{aligned}
L(\mathbf{v}) &= L\left(\sum_{j=1}^{n} x_j\mathbf{v}_j\right) \\
&= \sum_{j=1}^{n} x_j L(\mathbf{v}_j) \\
&= \sum_{j=1}^{n} x_j \left(\sum_{i=1}^{m} a_{ij}\mathbf{w}_i\right) \\
&= \sum_{i=1}^{m} \left(\sum_{j=1}^{n} a_{ij}x_j\right) \mathbf{w}_i
\end{aligned}
$$

For $i = 1, \ldots, m$, let

$$y_i = \sum_{j=1}^{n} a_{ij}x_j$$

Thus,

$$\mathbf{y} = (y_1, y_2, \ldots, y_m)^T = A\mathbf{x}$$

is the coordinate vector of $L(\mathbf{v})$ with respect to $[\mathbf{w}_1, \mathbf{w}_2, \ldots, \mathbf{w}_m]$. We have established the following theorem.

▶ **THEOREM 4.2.2 (Matrix Representation Theorem)** *If $E = [\mathbf{v}_1, \mathbf{v}_2, \ldots, \mathbf{v}_n]$ and $F = [\mathbf{w}_1, \mathbf{w}_2, \ldots, \mathbf{w}_m]$ are ordered bases for vector spaces V and W, respectively, then corresponding to each linear transformation $L: V \to W$ there is an $m \times n$ matrix A such that*

$$[L(\mathbf{v})]_F = A[\mathbf{v}]_E \qquad \text{for each } \mathbf{v} \in V$$

A is the matrix representing L relative to the ordered bases E and F. In fact,

$$\mathbf{a}_j = \left[L(\mathbf{v}_j)\right]_F \qquad j = 1, 2, \ldots, n$$

◀

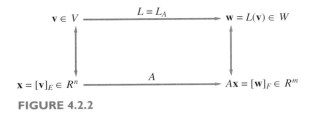

FIGURE 4.2.2

Theorem 4.2.2 is illustrated in Figure 4.2.2. If A is the matrix representing L with respect to the bases E and F and

$$\mathbf{x} = [\mathbf{v}]_E \quad \text{(the coordinate vector of } \mathbf{v} \text{ with respect to } E)$$

$$\mathbf{y} = [\mathbf{w}]_F \quad \text{(the coordinate vector of } \mathbf{w} \text{ with respect to } F)$$

then L maps $\mathbf{v}$ into $\mathbf{w}$ if and only if A maps $\mathbf{x}$ into $\mathbf{y}$.

EXAMPLE 3.   Let L be a linear transformation mapping R^3 into R^2 defined by

$$L(\mathbf{x}) = x_1\mathbf{b}_1 + (x_2 + x_3)\mathbf{b}_2$$

for each $\mathbf{x} \in R^3$, where

$$\mathbf{b}_1 = \begin{bmatrix} 1 \\ 1 \end{bmatrix} \quad \text{and} \quad \mathbf{b}_2 = \begin{bmatrix} -1 \\ 1 \end{bmatrix}$$

Find the matrix A representing L with respect to the ordered bases $[\mathbf{e}_1, \mathbf{e}_2, \mathbf{e}_3]$ and $[\mathbf{b}_1, \mathbf{b}_2]$.

SOLUTION.

$$L(\mathbf{e}_1) = 1\mathbf{b}_1 + 0\mathbf{b}_2$$

$$L(\mathbf{e}_2) = 0\mathbf{b}_1 + 1\mathbf{b}_2$$

$$L(\mathbf{e}_3) = 0\mathbf{b}_1 + 1\mathbf{b}_2$$

The ith column of A is determined by the coordinates of $L(\mathbf{e}_i)$ with respect to $[\mathbf{b}_1, \mathbf{b}_2]$ for $i = 1, 2, 3$. Thus

$$A = \begin{bmatrix} 1 & 0 & 0 \\ 0 & 1 & 1 \end{bmatrix} \qquad \blacktriangleleft$$

EXAMPLE 4.   Let L be a linear transformation mapping R^2 into itself defined by

$$L(\alpha\mathbf{b}_1 + \beta\mathbf{b}_2) = (\alpha + \beta)\mathbf{b}_1 + 2\beta\mathbf{b}_2$$

where $[\mathbf{b}_1, \mathbf{b}_2]$ is the ordered basis defined in Example 3. Find the matrix A representing L with respect to $[\mathbf{b}_1, \mathbf{b}_2]$.

SOLUTION.

$$L(\mathbf{b}_1) = 1\mathbf{b}_1 + 0\mathbf{b}_2$$

$$L(\mathbf{b}_2) = 1\mathbf{b}_1 + 2\mathbf{b}_2$$

Thus

$$A = \begin{pmatrix} 1 & 1 \\ 0 & 2 \end{pmatrix}$$ ◀

EXAMPLE 5. The linear operator D defined by $D(p) = p'$ maps P_3 into P_2. Given the ordered bases $[x^2, x, 1]$ and $[x, 1]$ for P_3 and P_2, respectively, we wish to determine a matrix representation for D. To do this, we apply D to each of the basis elements of P_3.

$$D(x^2) = 2x + 0 \cdot 1$$

$$D(x) = 0x + 1 \cdot 1$$

$$D(1) = 0x + 0 \cdot 1$$

In P_2, the coordinate vectors for $D(x^2)$, $D(x)$, $D(1)$ are $(2, 0)^T$, $(0, 1)^T$, $(0, 0)^T$, respectively. The matrix A is formed using these vectors as its columns.

$$A = \begin{pmatrix} 2 & 0 & 0 \\ 0 & 1 & 0 \end{pmatrix}$$

If $p(x) = ax^2 + bx + c$, then the coordinate vector of p with respect to the ordered basis of P_3 is $(a, b, c)^T$. To find the coordinate vector of $D(p)$ with respect to the ordered basis of P_2, we simply multiply

$$\begin{pmatrix} 2 & 0 & 0 \\ 0 & 1 & 0 \end{pmatrix} \begin{pmatrix} a \\ b \\ c \end{pmatrix} = \begin{pmatrix} 2a \\ b \end{pmatrix}$$

Thus, $D(ax^2 + bx + c) = 2ax + b$. ◀

In order to find the matrix representation A for a linear transformation $L: R^n \rightarrow R^m$ with respect to the ordered bases $E = [\mathbf{u}_1, \ldots, \mathbf{u}_n]$ and $F = [\mathbf{b}_1, \ldots, \mathbf{b}_m]$, we must represent each vector $L(\mathbf{u}_j)$ as a linear combination of $\mathbf{b}_1, \ldots, \mathbf{b}_m$. The following theorem shows that determining this representation of $L(\mathbf{u}_j)$ is equivalent to solving the linear system $B\mathbf{x} = L(\mathbf{u}_j)$.

▶ **THEOREM 4.2.3** *Let $E = [\mathbf{u}_1, \ldots, \mathbf{u}_n]$ and $F = [\mathbf{b}_1, \ldots, \mathbf{b}_m]$ be ordered bases for R^n and R^m, respectively. If $L: R^n \rightarrow R^m$ is a linear transformation and A is the matrix representing L with respect to E and F, then*

$$\mathbf{a}_j = B^{-1}L(\mathbf{u}_j) \quad \text{for } j = 1, \ldots, n$$

where $B = (\mathbf{b}_1, \ldots, \mathbf{b}_m)$.

▶*Proof.* If A is representing L with respect to E and F, then for $j = 1, \ldots, n$,

$$L(\mathbf{u}_j) = a_{1j}\mathbf{b}_1 + a_{2j}\mathbf{b}_2 + \cdots + a_{mj}\mathbf{b}_m$$
$$= B\mathbf{a}_j$$

The matrix B is nonsingular since its column vectors form a basis for R^m. Hence

$$\mathbf{a}_j = B^{-1}L(\mathbf{u}_j) \qquad j = 1, \ldots, n \qquad ◀$$

One consequence of this theorem is that we can determine the matrix representation of the operator by computing the reduced row echelon form of an augmented matrix. The following corollary shows how this is done.

▶ **COROLLARY 4.2.4** *If A is the matrix representing the linear operator $L\colon R^n \to R^m$ with respect to the bases*

$$E = [\mathbf{u}_1, \ldots, \mathbf{u}_n] \quad and \quad F = [\mathbf{b}_1, \ldots, \mathbf{b}_m]$$

then the reduced row echelon form of $(\mathbf{b}_1, \ldots, \mathbf{b}_m \mid L(\mathbf{u}_1), \ldots, L(\mathbf{u}_n))$ is $(I \mid A)$.

▶*Proof.* Let $B = (\mathbf{b}_1, \ldots, \mathbf{b}_m)$. The matrix $(B \mid L(\mathbf{u}_1), \ldots, L(\mathbf{u}_n))$ is row equivalent to

$$B^{-1}(B \mid L(\mathbf{u}_1), \ldots, L(\mathbf{u}_n)) = (I \mid B^{-1}L(\mathbf{u}_1), \ldots, B^{-1}L(\mathbf{u}_n))$$
$$= (I \mid \mathbf{a}_1, \ldots, \mathbf{a}_n)$$
$$= (I \mid A) \qquad ◀$$

EXAMPLE 6.   Let $L\colon R^2 \to R^3$ be the linear transformation defined by

$$L(\mathbf{x}) = (x_2, x_1 + x_2, x_1 - x_2)^T$$

Find the matrix representations of L with respect to the ordered bases $[\mathbf{u}_1, \mathbf{u}_2]$ and $[\mathbf{b}_1, \mathbf{b}_2, \mathbf{b}_3]$, where

$$\mathbf{u}_1 = (1, 2)^T, \qquad \mathbf{u}_2 = (3, 1)^T$$

and

$$\mathbf{b}_1 = (1, 0, 0)^T, \qquad \mathbf{b}_2 = (1, 1, 0)^T, \qquad \mathbf{b}_3 = (1, 1, 1)^T$$

SOLUTION.   We must compute $L(\mathbf{u}_1), L(\mathbf{u}_2)$ and then transform the matrix $(\mathbf{b}_1, \mathbf{b}_2, \mathbf{b}_3 \mid L(\mathbf{u}_1), L(\mathbf{u}_2))$ to reduced row echelon form.

$$L(\mathbf{u}_1) = (2, 3, -1)^T \qquad and \qquad L(\mathbf{u}_2) = (1, 4, 2)^T$$

$$\begin{pmatrix} 1 & 1 & 1 & 2 & 1 \\ 0 & 1 & 1 & 3 & 4 \\ 0 & 0 & 1 & -1 & 2 \end{pmatrix} \to \begin{pmatrix} 1 & 0 & 0 & -1 & -3 \\ 0 & 1 & 0 & 4 & 2 \\ 0 & 0 & 1 & -1 & 2 \end{pmatrix}$$

The matrix representing L with respect to the given ordered bases is

$$A = \begin{bmatrix} -1 & -3 \\ 4 & 2 \\ -1 & 2 \end{bmatrix}$$

The reader may verify that

$$L(\mathbf{u}_1) = -\mathbf{b}_1 + 4\mathbf{b}_2 - \mathbf{b}_3$$

$$L(\mathbf{u}_2) = -3\mathbf{b}_1 + 2\mathbf{b}_2 + 2\mathbf{b}_3 \qquad \blacktriangleleft$$

APPLICATION 1: COMPUTER GRAPHICS AND ANIMATION

A picture in the plane can be stored in the computer as a set of vertices. The vertices can then be plotted and connected by lines to produce the picture. If there are n vertices, they are stored in a $2 \times n$ matrix. The x coordinates of the vertices are stored in the first row and the y coordinates in the second. Each successive pair of points is connected by a straight line.

For example, to generate a triangle with vertices $(0, 0)$, $(1, 1)$, $(1, -1)$, we store the pairs as columns of a matrix:

$$T = \begin{bmatrix} 0 & 1 & 1 & 0 \\ 0 & 1 & -1 & 0 \end{bmatrix}$$

An additional copy of the vertex $(0, 0)$ is stored in the last column of T so that the previous point $(1, -1)$ will be connected back to $(0, 0)$. [See Figure 4.2.3(a).]

We can transform a figure by changing the positions of the vertices and then redrawing the figure. If the transformation is linear, it can be carried out as a matrix multiplication. Viewing a succession of such drawings will produce the effect of animation.

The four primary geometric transformations that are used in computer graphics are as follows:

1. *Dilations and contractions.* A linear transformation of the form

$$L(\mathbf{x}) = c\mathbf{x}$$

   is a *dilation* if $c > 1$ and a *contraction* if $0 < c < 1$. The transformation L is represented by the matrix cI, where I is the 2×2 identity matrix. A dilation increases the size of the figure by a factor $c > 1$, and a contraction shrinks the figure by a factor $c < 1$. Figure 4.2.3(b) shows a dilation by a factor of 1.5 of the triangle stored in the matrix T.

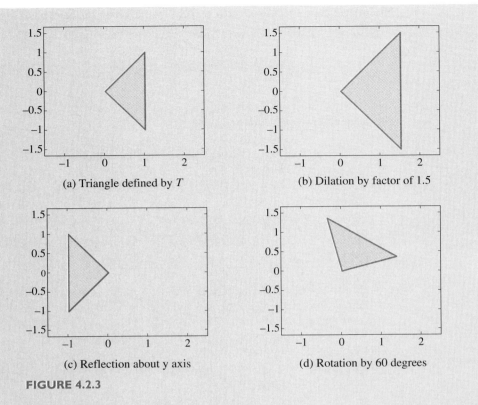

(a) Triangle defined by T

(b) Dilation by factor of 1.5

(c) Reflection about y axis

(d) Rotation by 60 degrees

FIGURE 4.2.3

2. *Reflection about an axis.* If L_x is an operator that reflects a vector $\mathbf{x}$ about the x axis, then L_x is a linear transformation and hence it can be represented by a 2×2 matrix A. Since

$$L_x(\mathbf{e}_1) = \mathbf{e}_1 \qquad \text{and} \qquad L_x(\mathbf{e}_2) = -\mathbf{e}_2$$

it follows that

$$A = \begin{bmatrix} 1 & 0 \\ 0 & -1 \end{bmatrix}$$

Similarly, if L_y is the linear transformation that reflects a vector about the y axis, then L_y is represented by the matrix

$$\begin{bmatrix} -1 & 0 \\ 0 & 1 \end{bmatrix}$$

Figure 4.2.3(c) shows the image of the triangle T after a reflection about the y-axis. In Chapter 7 we will learn a simple method for constructing reflection matrices that have the effect of reflecting a vector about any line through the origin.

3. *Rotations.* Let L be a transformation that rotates a vector about the origin by an angle θ in the counterclockwise direction. We saw in Example 2 that L is a linear transformation and that $L(\mathbf{x}) = A\mathbf{x}$, where

$$A = \begin{bmatrix} \cos\theta & -\sin\theta \\ \sin\theta & \cos\theta \end{bmatrix}$$

Figure 4.2.3(d) shows the result of rotating the triangle by $60°$ in the counterclockwise direction.

4. *Translations.* A *translation* by a vector $\mathbf{a}$ is a transformation of the form

$$L(\mathbf{x}) = \mathbf{x} + \mathbf{a}$$

If $\mathbf{a} \neq \mathbf{0}$, then L is not a linear transformation and hence L cannot be represented by a 2×2 matrix. However, in computer graphics it is desirable to do all transformations as matrix multiplications. The way around the problem is to introduce a new system of coordinates called *homogeneous coordinates*. *This new system will allow us to perform translations as linear transformations.*

Homogeneous Coordinates

The *homogeneous coordinate system* is formed by equating each vector in R^2 with a vector in R^3 having the same first two coordinates and having 1 as it third coordinate.

$$\begin{bmatrix} x_1 \\ x_2 \end{bmatrix} \leftrightarrow \begin{bmatrix} x_1 \\ x_2 \\ 1 \end{bmatrix}$$

When we want to plot a point represented by the homogeneous coordinate vector $(x_1, x_2, 1)^T$, we simply ignore the third coordinate and plot the ordered pair (x_1, x_2).

The linear transformations discussed earlier must now be represented by 3×3 matrices. To do this, we take the 2×2 matrix representation and augment it by attaching the third row and third column of the 3×3 identity matrix. For example, in place of the 2×2 dilation matrix

$$\begin{bmatrix} 3 & 0 \\ 0 & 3 \end{bmatrix}$$

we have the 3×3 matrix

$$\begin{bmatrix} 3 & 0 & 0 \\ 0 & 3 & 0 \\ 0 & 0 & 1 \end{bmatrix}$$

Note that

$$\begin{bmatrix} 3 & 0 & 0 \\ 0 & 3 & 0 \\ 0 & 0 & 1 \end{bmatrix} \begin{bmatrix} x_1 \\ x_2 \\ 1 \end{bmatrix} = \begin{bmatrix} 3x_1 \\ 3x_2 \\ 1 \end{bmatrix}$$

If L is a translation by a vector $\mathbf{a}$ in R^2, we can find a matrix representation for L with respect to the homogeneous coordinate system. We simply take the 3×3 identity matrix and replace the first two entries of its third column with the entries of $\mathbf{a}$. To see that this works, consider for example a translation corresponding to the vector $\mathbf{a} = (6, 2)^T$. In homogeneous coordinates this transformation is accomplished by the matrix multiplication

$$A\mathbf{x} = \begin{bmatrix} 1 & 0 & 6 \\ 0 & 1 & 2 \\ 0 & 0 & 1 \end{bmatrix} \begin{bmatrix} x_1 \\ x_2 \\ 1 \end{bmatrix} = \begin{bmatrix} x_1 + 6 \\ x_2 + 2 \\ 1 \end{bmatrix}$$

Figure 4.2.4(a) shows a stick figure generated from a 3×81 matrix S. If we multiply S by the translation matrix A, the graph of AS is the translated image given in Figure 4.2.4(b).

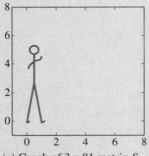

(a) Graph of 3 x 81 matrix S

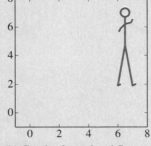

(b) Graph of translated figure AS

FIGURE 4.2.4

EXERCISES

1. Refer to Exercise 1 of Section 1. For each linear transformation L, find a matrix A representing L.

2. For each of the following linear transformations L mapping R^3 into R^2, find a matrix A such that $L(\mathbf{x}) = A\mathbf{x}$ for every $\mathbf{x}$ in R^3.

   (a) $L((x_1, x_2, x_3)^T) = (x_1 + x_2, 0)^T$ (b) $L((x_1, x_2, x_3)^T) = (x_1, x_2)^T$
   (c) $L((x_1, x_2, x_3)^T) = (x_2 - x_1, x_3 - x_2)^T$

3. For each of the following linear transformations L from R^3 into R^3, find a matrix A such that $L(\mathbf{x}) = A\mathbf{x}$ for every $\mathbf{x}$ in R^3.

   (a) $L((x_1, x_2, x_3)^T) = (x_3, x_2, x_1)^T$
   (b) $L((x_1, x_2, x_3)^T) = (x_1, x_1 + x_2, x_1 + x_2 + x_3)^T$
   (c) $L((x_1, x_2, x_3)^T) = (2x_3, x_2 + 3x_1, 2x_1 - x_3)^T$

4. Let L be the linear transformation mapping R^3 into R^3 defined by
   $$L(\mathbf{x}) = (2x_1 - x_2 - x_3, 2x_2 - x_1 - x_3, 2x_3 - x_1 - x_2)^T$$
   Determine the standard matrix representation A of L, and use A to find $L(\mathbf{x})$ for each of the following vectors $\mathbf{x}$.

   (a) $\mathbf{x} = (1, 1, 1)^T$ (b) $\mathbf{x} = (2, 1, 1)^T$ (c) $\mathbf{x} = (-5, 3, 2)^T$

5. Find the standard matrix representation for each of the following linear transformations.

   (a) L is the linear transformation that rotates each $\mathbf{x}$ in R^2 by $45°$ in the clockwise direction.
   (b) L is the linear transformation reflects each vector $\mathbf{x}$ in R^2 about the x_1 axis and then rotates it $90°$ in the counterclockwise direction.
   (c) L doubles the length of $\mathbf{x}$ and then rotates it $30°$ in the counterclockwise direction.
   (d) L reflects each vector $\mathbf{x}$ about the line $x_2 = x_1$ and then projects it onto the x_1 axis.

6. Let
   $$\mathbf{b}_1 = \begin{bmatrix} 1 \\ 1 \\ 0 \end{bmatrix}, \qquad \mathbf{b}_2 = \begin{bmatrix} 1 \\ 0 \\ 1 \end{bmatrix}, \qquad \mathbf{b}_3 = \begin{bmatrix} 0 \\ 1 \\ 1 \end{bmatrix}$$
   and let L be the linear transformation from R^2 into R^3 defined by
   $$L(\mathbf{x}) = x_1\mathbf{b}_1 + x_2\mathbf{b}_2 + (x_1 + x_2)\mathbf{b}_3$$
   Find the matrix A representing L with respect to the bases $[\mathbf{e}_1, \mathbf{e}_2]$ and $[\mathbf{b}_1, \mathbf{b}_2, \mathbf{b}_3]$.

7. Let
   $$\mathbf{y}_1 = \begin{bmatrix} 1 \\ 1 \\ 1 \end{bmatrix}, \qquad \mathbf{y}_2 = \begin{bmatrix} 1 \\ 1 \\ 0 \end{bmatrix}, \qquad \mathbf{y}_3 = \begin{bmatrix} 1 \\ 0 \\ 0 \end{bmatrix}$$
   and let $\mathcal{I}$ be the identity operator on R^3.

   (a) Find the coordinates of $\mathcal{I}(\mathbf{e}_1)$, $\mathcal{I}(\mathbf{e}_2)$, and $\mathcal{I}(\mathbf{e}_3)$ with respect to $[\mathbf{y}_1, \mathbf{y}_2, \mathbf{y}_3]$.
   (b) Find a matrix A such that $A\mathbf{x}$ is the coordinate vector of $\mathbf{x}$ with respect to $[\mathbf{y}_1, \mathbf{y}_2, \mathbf{y}_3]$.

8. Let y_1, y_2, y_3 be defined as in Exercise 7, and let L be the linear transformation from R^3 into R^3 defined by

$$L(c_1y_1 + c_2y_2 + c_3y_3) = (c_1 + c_2 + c_3)y_1 + (2c_1 + c_3)y_2 - (2c_2 + c_3)y_3$$

(a) Find a matrix representing L with respect to the ordered basis $[y_1, y_2, y_3]$.

(b) For each of the following, write the vector $\mathbf{x}$ as a linear combination of y_1, y_2, y_3 and use the matrix from part (a) to determine $L(\mathbf{x})$.

   (i) $\mathbf{x} = (7, 5, 2)^T$ (ii) $\mathbf{x} = (3, 2, 1)^T$ (iii) $\mathbf{x} = (1, 2, 3)^T$

9. Let

$$R = \begin{pmatrix} 0 & 0 & 1 & 1 & 0 \\ 0 & 1 & 1 & 0 & 0 \\ 1 & 1 & 1 & 1 & 1 \end{pmatrix}$$

The column vectors of R represent the homogeneous coordinates of points in the plane.

(a) Draw the figure whose vertices correspond to the column vectors of R. What type of figure is it?

(b) For each of the following choices of A, sketch the graph of the figure represented by AR and describe geometrically the effect of the linear transformation.

   (i) $A = \begin{pmatrix} \frac{1}{2} & 0 & 0 \\ 0 & \frac{1}{2} & 0 \\ 0 & 0 & 1 \end{pmatrix}$ (ii) $A = \begin{pmatrix} \frac{1}{\sqrt{2}} & \frac{1}{\sqrt{2}} & 0 \\ -\frac{1}{\sqrt{2}} & \frac{1}{\sqrt{2}} & 0 \\ 0 & 0 & 1 \end{pmatrix}$

   (iii) $A = \begin{pmatrix} 1 & 0 & 2 \\ 0 & 1 & -3 \\ 0 & 0 & 1 \end{pmatrix}$

10. For each of the following transformations from R^2 into R^2, find the matrix representation of the transformation with respect to the homogeneous coordinate system.

(a) The transformation L that rotates each vector by $120°$ in the counterclockwise direction

(b) The transformation L that translates each point 3 units to the left and 5 units up

(c) The transformation L that contracts each vector by a factor of one third

(d) The transformation that reflects a vector about the y-axis and then translates it up 2 units

11. Let L be the linear operator mapping P_2 into R^2 defined by

$$L(p(x)) = \begin{bmatrix} \int_0^1 p(x)\, dx \\ p(0) \end{bmatrix}$$

Find a matrix A such that

$$L(\alpha + \beta x) = A \begin{bmatrix} \alpha \\ \beta \end{bmatrix}$$

12. The linear operator L defined by

$$L(p(x)) = p'(x) + p(0)$$

maps P_3 into P_2. Find the matrix representation of L with respect to the ordered bases $[x^2, x, 1]$ and $[2, 1 - x]$. For each of the following vectors $p(x)$ in P_3, find the coordinates of $L(p(x))$ with respect to the ordered basis $[2, 1 - x]$.

(a) $x^2 + 2x - 3$ (b) $x^2 + 1$ (c) $3x$ (d) $4x^2 + 2x$

13. Let S be the subspace of $C[a, b]$ spanned by e^x, xe^x, and $x^2 e^x$. Let D be the differentiation operator of S. Find the matrix representing D with respect to $[e^x, xe^x, x^2 e^x]$.

14. Let L be a linear transformation from R^n into R^n. Suppose that $L(\mathbf{x}) = \mathbf{0}$ for some $\mathbf{x} \neq \mathbf{0}$. Let A be the matrix representing L with respect to the standard basis $[\mathbf{e}_1, \mathbf{e}_2, \dots, \mathbf{e}_n]$. Show that A is singular.

15. Let L be a linear operator mapping a vector space V into itself. Let A be the matrix representing L with respect to the ordered basis $[\mathbf{v}_1, \dots, \mathbf{v}_n]$ of V [i.e.,

$$L(\mathbf{v}_j) = \sum_{i=1}^n a_{ij} \mathbf{v}_i, \ j = 1, \dots, n].$$ Show that A^m is the matrix representing L^m
with respect to $[\mathbf{v}_1, \dots, \mathbf{v}_n]$.

16. Let $E = [\mathbf{u}_1, \mathbf{u}_2, \mathbf{u}_3]$ and $F = [\mathbf{b}_1, \mathbf{b}_2]$, where

$$\mathbf{u}_1 = (1, 0, -1)^T, \qquad \mathbf{u}_2 = (1, 2, 1)^T, \qquad \mathbf{u}_3 = (-1, 1, 1)^T$$

and

$$\mathbf{b}_1 = (1, -1)^T, \qquad \mathbf{b}_2 = (2, -1)^T$$

For each of the following linear transformations L from R^3 into R^2, find the matrix representing L with respect to the ordered bases E and F.

(a) $L(\mathbf{x}) = (x_3, x_1)^T$ (b) $L(\mathbf{x}) = (x_1 + x_2, x_1 - x_3)^T$
(c) $L(\mathbf{x}) = (2x_2, -x_1)^T$

17. Suppose that $L_1 : V \to W$ and $L_2 : W \to Z$ are linear transformations and E, F, and G are ordered bases for V, W, and Z, respectively. Show that, if A represents L_1 relative to E and F and B represents L_2 relative to F and

G, then the matrix $C = BA$ represents $L_2 \circ L_1 : V \to Z$ relative to E and G. [*Hint:* Show $BA[\mathbf{v}]_E = [(L_2 \circ L_1)(\mathbf{v})]_G$ for all $\mathbf{v} \in V$.]

18. Let V, W be vector spaces with ordered bases E and F, respectively. If $L : V \to W$ is a linear transformation and A is the matrix representing L relative to E and F, show that:

(a) $\mathbf{v} \in \ker(L)$ if and only if $[\mathbf{v}]_E \in N(A)$.
(b) $\mathbf{w} \in L(V)$ if and only if $[\mathbf{w}]_F$ is in the column space of A.

3   SIMILARITY

If L is a linear transformation mapping an n-dimensional vector space V into itself, the matrix representation of L will depend on the ordered basis chosen for V. By using different bases, it is possible to represent L by different $n \times n$ matrices. In this section we consider different matrix representations of linear operators and characterize the relationship between matrices representing the same linear operator.

Let us begin by considering an example in R^2. Let L be the linear transformation mapping R^2 into itself defined by

$$L(\mathbf{x}) = (2x_1, x_1 + x_2)^T$$

Since

$$L(\mathbf{e}_1) = \begin{bmatrix} 2 \\ 1 \end{bmatrix} \qquad \text{and} \qquad L(\mathbf{e}_2) = \begin{bmatrix} 0 \\ 1 \end{bmatrix}$$

it follows that the matrix representing L with respect to $[\mathbf{e}_1, \mathbf{e}_2]$ is

$$A = \begin{bmatrix} 2 & 0 \\ 1 & 1 \end{bmatrix}$$

If we use a different basis for R^2, the matrix representation of L will change. If, for example, we use

$$\mathbf{u}_1 = \begin{bmatrix} 1 \\ 1 \end{bmatrix} \qquad \text{and} \qquad \mathbf{u}_2 = \begin{bmatrix} -1 \\ 1 \end{bmatrix}$$

for a basis, then to determine the matrix representation of L with respect to $[\mathbf{u}_1, \mathbf{u}_2]$, we must determine $L(\mathbf{u}_1)$, $L(\mathbf{u}_2)$ and express these vectors as linear combinations of $\mathbf{u}_1$ and $\mathbf{u}_2$. We can use the matrix A to determine $L(\mathbf{u}_1)$ and $L(\mathbf{u}_2)$.

$$L(\mathbf{u}_1) = A\mathbf{u}_1 = \begin{bmatrix} 2 & 0 \\ 1 & 1 \end{bmatrix} \begin{bmatrix} 1 \\ 1 \end{bmatrix} = \begin{bmatrix} 2 \\ 2 \end{bmatrix}$$

$$L(\mathbf{u}_2) = A\mathbf{u}_2 = \begin{bmatrix} 2 & 0 \\ 1 & 1 \end{bmatrix} \begin{bmatrix} -1 \\ 1 \end{bmatrix} = \begin{bmatrix} -2 \\ 0 \end{bmatrix}$$

To express these vectors in terms of $\mathbf{u}_1$ and $\mathbf{u}_2$, we use a transition matrix to change from the ordered basis $[\mathbf{e}_1, \mathbf{e}_2]$ to $[\mathbf{u}_1, \mathbf{u}_2]$. Let us first compute the transition matrix from $[\mathbf{u}_1, \mathbf{u}_2]$ to $[\mathbf{e}_1, \mathbf{e}_2]$. This is simply

$$U = (\mathbf{u}_1, \mathbf{u}_2) = \begin{bmatrix} 1 & -1 \\ 1 & 1 \end{bmatrix}$$

The transition matrix from $[\mathbf{e}_1, \mathbf{e}_2]$ to $[\mathbf{u}_1, \mathbf{u}_2]$ will then be

$$U^{-1} = \begin{bmatrix} \frac{1}{2} & \frac{1}{2} \\ -\frac{1}{2} & \frac{1}{2} \end{bmatrix}$$

To determine the coordinates of $L(\mathbf{u}_1)$, $L(\mathbf{u}_2)$ with respect to $[\mathbf{u}_1, \mathbf{u}_2]$, we multiply the vectors by U^{-1}.

$$U^{-1}L(\mathbf{u}_1) = U^{-1}A\mathbf{u}_1 = \begin{bmatrix} \frac{1}{2} & \frac{1}{2} \\ -\frac{1}{2} & \frac{1}{2} \end{bmatrix} \begin{bmatrix} 2 \\ 2 \end{bmatrix} = \begin{bmatrix} 2 \\ 0 \end{bmatrix}$$

$$U^{-1}L(\mathbf{u}_2) = U^{-1}A\mathbf{u}_2 = \begin{bmatrix} \frac{1}{2} & \frac{1}{2} \\ -\frac{1}{2} & \frac{1}{2} \end{bmatrix} \begin{bmatrix} -2 \\ 0 \end{bmatrix} = \begin{bmatrix} -1 \\ 1 \end{bmatrix}$$

Thus,

$$L(\mathbf{u}_1) = 2\mathbf{u}_1 + 0\mathbf{u}_2$$
$$L(\mathbf{u}_2) = -1\mathbf{u}_1 + 1\mathbf{u}_2$$

and the matrix representing L with respect to $[\mathbf{u}_1, \mathbf{u}_2]$ is

$$B = \begin{bmatrix} 2 & -1 \\ 0 & 1 \end{bmatrix}$$

How are A and B related? Note that the columns of B are

$$\begin{bmatrix} 2 \\ 0 \end{bmatrix} = U^{-1}A\mathbf{u}_1 \qquad \text{and} \qquad \begin{bmatrix} -1 \\ 1 \end{bmatrix} = U^{-1}A\mathbf{u}_2$$

Hence,

$$B = (U^{-1}A\mathbf{u}_1, U^{-1}A\mathbf{u}_2) = U^{-1}A(\mathbf{u}_1, \mathbf{u}_2) = U^{-1}AU$$

Thus, if

(i) B is the matrix representing L with respect to $[\mathbf{u}_1, \mathbf{u}_2]$
(ii) A is the matrix representing L with respect to $[\mathbf{e}_1, \mathbf{e}_2]$
(iii) U is the transition matrix corresponding to the change of basis from $[\mathbf{u}_1, \mathbf{u}_2]$ to $[\mathbf{e}_1, \mathbf{e}_2]$

then

(1) $$B = U^{-1}AU$$

The results that we have established for this particular linear operator on R^2 are typical of what happens in a much more general setting. We will show next that the same sort of relationship as given in (1) will hold for any two matrix representations of a linear operator that maps an n-dimensional vector space into itself.

▶ **THEOREM 4.3.1** *Let $E = [\mathbf{v}_1, \ldots, \mathbf{v}_n]$ and $F = [\mathbf{w}_1, \ldots, \mathbf{w}_n]$ be two ordered bases for a vector space V, and let L be a linear operator mapping V into itself. Let S be the transition matrix representing the change from F to E. If A is the matrix representing L with respect to E and B is the matrix representing L with respect to F, then $B = S^{-1}AS$.*

▶ **Proof.** Let $\mathbf{x}$ be any vector in R^n and let
$$\mathbf{v} = x_1\mathbf{w}_1 + x_2\mathbf{w}_2 + \cdots + x_n\mathbf{w}_n$$

Let

(2) $$\mathbf{y} = S\mathbf{x}, \qquad \mathbf{t} = A\mathbf{y}, \qquad \mathbf{z} = B\mathbf{x}$$

It follows from the definition of S that $\mathbf{y} = [\mathbf{v}]_E$ and hence
$$\mathbf{v} = y_1\mathbf{v}_1 + \cdots + y_n\mathbf{v}_n$$

Since A represents L with respect to E and B represents L with respect to F, we have
$$\mathbf{t} = [L(\mathbf{v})]_E \qquad \text{and} \qquad \mathbf{z} = [L(\mathbf{v})]_F$$

The transition matrix from E to F is S^{-1}. Therefore,

(3) $$S^{-1}\mathbf{t} = \mathbf{z}$$

It follows from (2) and (3) that
$$S^{-1}AS\mathbf{x} = S^{-1}A\mathbf{y} = S^{-1}\mathbf{t} = \mathbf{z} = B\mathbf{x}$$

(see Figure 4.3.1). Thus
$$S^{-1}AS\mathbf{x} = B\mathbf{x}$$

for every $\mathbf{x} \in R^n$, and hence $S^{-1}AS = B$.   ◀

Another way of viewing Theorem 4.3.1 is to consider S as the matrix representing the identity transformation $\mathcal{I}$ with respect to the ordered bases $F = [\mathbf{w}_1, \ldots, \mathbf{w}_n]$ and $E = [\mathbf{v}_1, \ldots, \mathbf{v}_n]$. Thus, if

FIGURE 4.3.1

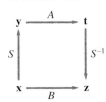

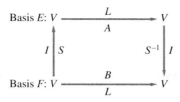

FIGURE 4.3.2

S represents $\mathcal{I}$ relative to F and E,

A represents L relative to E,

S^{-1} represents $\mathcal{I}$ relative to E and F,

then L can be expressed as a composite operator $\mathcal{I} \circ L \circ \mathcal{I}$, and the matrix representation of the composite will be the product of the matrix representations of the components. Thus the matrix representation of $\mathcal{I} \circ L \circ \mathcal{I}$ relative to F is $S^{-1}AS$. If B is the matrix representing L relative to F, then B must equal $S^{-1}AS$ (see Figure 4.3.2).

▶ **DEFINITION** Let A and B be $n \times n$ matrices. B is said to be **similar** to A if there exists a nonsingular matrix S such that $B = S^{-1}AS$.   ◀

Note that if B is similar to A, then $A = (S^{-1})^{-1}BS^{-1}$ is similar to B. Thus we may simply say that A and B are similar matrices.

It follows from Theorem 4.3.1 that, if A and B are $n \times n$ matrices representing the same operator L, then A and B are similar. Conversely, suppose that A represents L with respect to the ordered basis $[\mathbf{v}_1, \ldots , \mathbf{v}_n]$ and $B = S^{-1}AS$ for some nonsingular matrix S. If $\mathbf{w}_1, \ldots , \mathbf{w}_n$ are defined by

$$\mathbf{w}_1 = s_{11}\mathbf{v}_1 + s_{21}\mathbf{v}_2 + \cdots + s_{n1}\mathbf{v}_n$$
$$\mathbf{w}_2 = s_{12}\mathbf{v}_1 + s_{22}\mathbf{v}_2 + \cdots + s_{n2}\mathbf{v}_n$$
$$\vdots$$
$$\mathbf{w}_n = s_{1n}\mathbf{v}_1 + s_{2n}\mathbf{v}_2 + \cdots + s_{nn}\mathbf{v}_n$$

then $[\mathbf{w}_1, \ldots , \mathbf{w}_n]$ is an ordered basis for V, and B is the matrix representing L with respect to $[\mathbf{w}_1, \ldots , \mathbf{w}_n]$.

EXAMPLE I. Let D be the differentiation operator on P_3. Find the matrix B representing D with respect to $[1, x, x^2]$ and the matrix A representing D with respect to $[1, 2x, 4x^2 - 2]$.

SOLUTION.

$$D(1) = 0 \cdot 1 + 0 \cdot x + 0 \cdot x^2$$
$$D(x) = 1 \cdot 1 + 0 \cdot x + 0 \cdot x^2$$
$$D(x^2) = 0 \cdot 1 + 2 \cdot x + 0 \cdot x^2$$

The matrix B is then given by

$$B = \begin{bmatrix} 0 & 1 & 0 \\ 0 & 0 & 2 \\ 0 & 0 & 0 \end{bmatrix}$$

Applying D to 1, $2x$, and $4x^2 - 2$, we obtain

$$D(1) = 0 \cdot 1 + 0 \cdot 2x + 0 \cdot (4x^2 - 2)$$
$$D(2x) = 2 \cdot 1 + 0 \cdot 2x + 0 \cdot (4x^2 - 2)$$
$$D(4x^2 - 2) = 0 \cdot 1 + 4 \cdot 2x + 0 \cdot (4x^2 - 2)$$

Thus,

$$A = \begin{bmatrix} 0 & 2 & 0 \\ 0 & 0 & 4 \\ 0 & 0 & 0 \end{bmatrix}$$

The transition matrix S corresponding to the change of basis from $[1, 2x, 4x^2 - 2]$ to $[1, x, x^2]$ and its inverse are given by

$$S = \begin{bmatrix} 1 & 0 & -2 \\ 0 & 2 & 0 \\ 0 & 0 & 4 \end{bmatrix} \quad \text{and} \quad S^{-1} = \begin{bmatrix} 1 & 0 & \frac{1}{2} \\ 0 & \frac{1}{2} & 0 \\ 0 & 0 & \frac{1}{4} \end{bmatrix}$$

(See Example 7 from Chapter 3, Section 5.) The reader may verify that $A = S^{-1}BS$. ◀

EXAMPLE 2. Let L be the linear operator mapping R^3 into R^3 defined by $L(\mathbf{x}) = A\mathbf{x}$, where

$$A = \begin{bmatrix} 2 & 2 & 0 \\ 1 & 1 & 2 \\ 1 & 1 & 2 \end{bmatrix}$$

Thus the matrix A represents L with respect to $[\mathbf{e}_1, \mathbf{e}_2, \mathbf{e}_3]$. Find the matrix representing L with respect to $[\mathbf{y}_1, \mathbf{y}_2, \mathbf{y}_3]$, where

$$\mathbf{y}_1 = \begin{bmatrix} 1 \\ -1 \\ 0 \end{bmatrix}, \quad \mathbf{y}_2 = \begin{bmatrix} -2 \\ 1 \\ 1 \end{bmatrix}, \quad \mathbf{y}_3 = \begin{bmatrix} 1 \\ 1 \\ 1 \end{bmatrix}$$

SOLUTION.

$$L(\mathbf{y}_1) = A\mathbf{y}_1 = \mathbf{0} = 0\mathbf{y}_1 + 0\mathbf{y}_2 + 0\mathbf{y}_3$$
$$L(\mathbf{y}_2) = A\mathbf{y}_2 = \mathbf{y}_2 = 0\mathbf{y}_1 + 1\mathbf{y}_2 + 0\mathbf{y}_3$$
$$L(\mathbf{y}_3) = A\mathbf{y}_3 = 4\mathbf{y}_3 = 0\mathbf{y}_1 + 0\mathbf{y}_2 + 4\mathbf{y}_3$$

Thus the matrix representing L with respect to $[\mathbf{y}_1, \mathbf{y}_2, \mathbf{y}_3]$ is

$$D = \begin{bmatrix} 0 & 0 & 0 \\ 0 & 1 & 0 \\ 0 & 0 & 4 \end{bmatrix}$$

◄

We could have found D using the transition matrix $Y = (\mathbf{y}_1, \mathbf{y}_2, \mathbf{y}_3)$ and computing

$$D = Y^{-1}AY$$

This was unnecessary due to the simplicity of the action of L on the basis $[\mathbf{y}_1, \mathbf{y}_2, \mathbf{y}_3]$.

In Example 2 the linear operator L is represented by a diagonal matrix D with respect to the basis $[\mathbf{y}_1, \mathbf{y}_2, \mathbf{y}_3]$. It is much simpler to work with D than with A. For example, it is easier to compute $D\mathbf{x}$ and $D^n\mathbf{x}$ than $A\mathbf{x}$ and $A^n\mathbf{x}$. Generally, it is desirable to find as simple a representation as possible for a linear operator. In particular, if the operator can be represented by a diagonal matrix, that is usually the preferred representation. The problem of finding a diagonal representation for a linear operator will be studied in Chapter 6.

EXERCISES

1. For each of the following linear transformations L from R^2 into R^2, determine the matrix A representing L with respect to $[\mathbf{e}_1, \mathbf{e}_2]$ (see Exercise 1 of Section 2) and the matrix B representing L with respect to $[\mathbf{u}_1 = (1, 1)^T, \mathbf{u}_2 = (-1, 1)^T]$.

   (a) $L(\mathbf{x}) = (-x_1, x_2)^T$ (b) $L(\mathbf{x}) = -\mathbf{x}$ (c) $L(\mathbf{x}) = (x_2, x_1)^T$
   (d) $L(\mathbf{x}) = \frac{1}{2}\mathbf{x}$ (e) $L(\mathbf{x}) = x_2\mathbf{e}_2$

2. Let $[\mathbf{u}_1, \mathbf{u}_2]$ and $[\mathbf{v}_1, \mathbf{v}_2]$ be ordered bases for R^2, where

$$\mathbf{u}_1 = \begin{bmatrix} 1 \\ 1 \end{bmatrix}, \quad \mathbf{u}_2 = \begin{bmatrix} -1 \\ 1 \end{bmatrix} \quad \text{and} \quad \mathbf{v}_1 = \begin{bmatrix} 2 \\ 1 \end{bmatrix}, \quad \mathbf{v}_2 = \begin{bmatrix} 1 \\ 0 \end{bmatrix}$$

Let L be the linear transformation defined by

$$L(\mathbf{x}) = (-x_1, x_2)^T$$

and let B be the matrix representing L with respect to $[\mathbf{u}_1, \mathbf{u}_2]$ [from Exercise 1(a)].

   (a) Find the transition matrix S corresponding to the change of basis from $[\mathbf{u}_1, \mathbf{u}_2]$ to $[\mathbf{v}_1, \mathbf{v}_2]$.

(b) Find the matrix A representing L with respect to $[\mathbf{v}_1, \mathbf{v}_2]$ by computing SBS^{-1}.

(c) Verify that

$$L(\mathbf{v}_1) = a_{11}\mathbf{v}_1 + a_{21}\mathbf{v}_2$$

$$L(\mathbf{v}_2) = a_{12}\mathbf{v}_1 + a_{22}\mathbf{v}_2$$

3. Let L be the linear transformation on R^3 defined by

$$L(\mathbf{x}) = (2x_1 - x_2 - x_3, 2x_2 - x_1 - x_3, 2x_3 - x_1 - x_2)^T$$

and let A be the standard matrix representation of L (see Exercise 4 of Section 2). If $\mathbf{u}_1 = (1, 1, 0)^T$, $\mathbf{u}_2 = (1, 0, 1)^T$, and $\mathbf{u}_3 = (0, 1, 1)^T$, then $[\mathbf{u}_1, \mathbf{u}_2, \mathbf{u}_3]$ is an ordered basis for R^3 and $U = (\mathbf{u}_1, \mathbf{u}_2, \mathbf{u}_3)$ is the transition matrix corresponding to a change of basis from $[\mathbf{u}_1, \mathbf{u}_2, \mathbf{u}_3]$ to the standard basis $[\mathbf{e}_1, \mathbf{e}_2, \mathbf{e}_3]$. Determine the matrix B representing L with respect to the basis $[\mathbf{u}_1, \mathbf{u}_2, \mathbf{u}_3]$ by calculating $U^{-1}AU$.

4. Let L be the linear operator mapping R^3 into R^3 defined by $L(\mathbf{x}) = A\mathbf{x}$, where

$$A = \begin{pmatrix} 3 & -1 & -2 \\ 2 & 0 & -2 \\ 2 & -1 & -1 \end{pmatrix}$$

and let

$$\mathbf{v}_1 = \begin{pmatrix} 1 \\ 1 \\ 1 \end{pmatrix}, \qquad \mathbf{v}_2 = \begin{pmatrix} 1 \\ 2 \\ 0 \end{pmatrix}, \qquad \mathbf{v}_3 = \begin{pmatrix} 0 \\ -2 \\ 1 \end{pmatrix}$$

Find the transition matrix V corresponding to a change of basis from $[\mathbf{v}_1, \mathbf{v}_2, \mathbf{v}_3]$ to $[\mathbf{e}_1, \mathbf{e}_2, \mathbf{e}_3]$ and use it to determine the matrix B representing L with respect to $[\mathbf{v}_1, \mathbf{v}_2, \mathbf{v}_3]$.

5. Let L be the operator on P_3 defined by

$$L(p(x)) = xp'(x) + p''(x)$$

(a) Find the matrix A representing L with respect to $[1, x, x^2]$.

(b) Find the matrix B representing L with respect to $[1, x, 1 + x^2]$.

(c) Find the matrix S such that $B = S^{-1}AS$.

(d) If $p(x) = a_0 + a_1x + a_2(1 + x^2)$, calculate $L^n(p(x))$.

6. Let V be the subspace of $C[a, b]$ spanned by $1, e^x, e^{-x}$, and let D be the differentiation operator on V.

(a) Find the transition matrix S representing the change of coordinates from the ordered basis $[1, e^x, e^{-x}]$ to the ordered basis $[1, \cosh x, \sinh x]$. $[\cosh x = \frac{1}{2}(e^x + e^{-x}), \sinh x = \frac{1}{2}(e^x - e^{-x}).]$

(b) Find the matrix A representing D with respect to the ordered basis $[1, \cosh x, \sinh x]$.

(c) Find the matrix B representing D with respect to $[1, e^x, e^{-x}]$.

(d) Verify that $B = S^{-1}AS$.

7. Prove that if A is similar to B and B is similar to C, then A is similar to C.

8. Suppose that $A = S\Lambda S^{-1}$, where Λ is a diagonal matrix with diagonal elements $\lambda_1, \lambda_2, \ldots, \lambda_n$.

(a) Show that $A\mathbf{s}_i = \lambda_i \mathbf{s}_i, i = 1, \ldots, n$.

(b) Show that if $\mathbf{x} = \alpha_1 \mathbf{s}_1 + \alpha_2 \mathbf{s}_2 + \cdots + \alpha_n \mathbf{s}_n$, then

$$A^k \mathbf{x} = \alpha_1 \lambda_1^k \mathbf{s}_1 + \alpha_2 \lambda_2^k \mathbf{s}_2 + \cdots + \alpha_n \lambda_n^k \mathbf{s}_n$$

(c) Suppose that $|\lambda_i| < 1$ for $i = 1, \ldots, n$. What happens to $A^k \mathbf{x}$ as $k \to \infty$? Explain.

9. Suppose that $A = ST$, where S is nonsingular. Let $B = TS$. Show that B is similar to A.

10. Let A and B be $n \times n$ matrices. Show that if A is similar to B, then there exist $n \times n$ matrices S and T, with S nonsingular, such that

$$A = ST \qquad \text{and} \qquad B = TS$$

11. Show that if A and B are similar matrices, then $\det(A) = \det(B)$.

12. Let A and B be similar matrices. Show that:

(a) A^T and B^T are similar.

(b) A^k and B^k are similar for each positive integer k.

13. Show that if A is similar to B and A is nonsingular, then B must also be nonsingular and A^{-1} and B^{-1} are similar.

14. Let A and B be similar matrices and let λ be any scalar. Show that:

(a) $A - \lambda I$ and $B - \lambda I$ are similar.

(b) $\det(A - \lambda I) = \det(B - \lambda I)$.

15. The *trace* of an $n \times n$ matrix A, denoted $\operatorname{tr}(A)$, is the sum of its diagonal entries, that is,

$$\operatorname{tr}(A) = a_{11} + a_{22} + \cdots + a_{nn}$$

Show that:

(a) $\operatorname{tr}(AB) = \operatorname{tr}(BA)$

(b) If A is similar to B, then $\operatorname{tr}(A) = \operatorname{tr}(B)$.

MATLAB EXERCISES

1. Use MATLAB to generate a matrix W and a vector $\mathbf{x}$ by setting

$$W = \text{triu}(\text{ones}(5)) \quad \text{and} \quad \mathbf{x} = [1:5]'$$

The columns of W can be used to form an ordered basis

$$F = [\mathbf{w}_1, \mathbf{w}_2, \mathbf{w}_3, \mathbf{w}_4, \mathbf{w}_5]$$

Let $L: R^5 \to R^5$ be a linear operator such that

$$L(\mathbf{w}_1) = \mathbf{w}_2, \qquad L(\mathbf{w}_2) = \mathbf{w}_3, \qquad L(\mathbf{w}_3) = \mathbf{w}_4$$

and

$$L(\mathbf{w}_4) = 4\mathbf{w}_1 + 3\mathbf{w}_2 + 2\mathbf{w}_3 + \mathbf{w}_4$$
$$L(\mathbf{w}_5) = \mathbf{w}_1 + \mathbf{w}_2 + \mathbf{w}_3 + 3\mathbf{w}_4 + \mathbf{w}_5$$

(a) Determine the matrix A representing L with respect to F and enter it in MATLAB.
(b) Use MATLAB to compute the coordinate vector $\mathbf{y} = W^{-1}\mathbf{x}$ of $\mathbf{x}$ with respect to F.
(c) Use A to compute the coordinate vector $\mathbf{z}$ of $L(\mathbf{x})$ with respect to F.
(d) W is the transition matrix from F to the standard basis for R^5. Use W to compute the coordinate vector of $L(\mathbf{x})$ with respect to the standard basis.

2. Set $A = \text{triu}(\text{ones}(5)) * \text{tril}(\text{ones}(5))$. If L denotes the linear operator defined by $L(\mathbf{x}) = A\mathbf{x}$ for all $\mathbf{x}$ in R^n, then A is the matrix representing L with respect to the standard basis for R^5. Construct a 5×5 matrix U by setting

$$U = \text{hankel}(\text{ones}(5, 1), 1:5)$$

Use the MATLAB function **rank** to verify that the column vectors of U are linearly independent. Thus $E = [\mathbf{u}_1, \mathbf{u}_2, \mathbf{u}_3, \mathbf{u}_4, \mathbf{u}_5]$ is an ordered basis for R^5. The matrix U is the transition matrix from E to the standard basis.

(a) Use MATLAB to compute the matrix B representing L with respect to E. (The matrix B should be computed in terms of A, U, and U^{-1}.)
(b) Generate another matrix by setting

$$V = \text{toeplitz}([1, 0, 1, 1, 1])$$

Use MATLAB to check that V is nonsingular. It follows that the column vectors of V are linearly independent and hence form an ordered basis F for R^5. Use MATLAB to compute the matrix C, which represents L with respect to F. (The matrix C should be computed in terms of A, V, and V^{-1}.)
(c) The matrices B and C from parts (a) and (b) should be similar. Why? Explain. Use MATLAB to compute the transition matrix S from F to E. Compute the matrix C in terms of B, S, and S^{-1}. Compare your result with the result from part (b).

3. Let

$$A = \textbf{toeplitz}(1:7), \qquad S = \textbf{compan}(\textbf{ones}(8, 1))$$

and set $B = S^{-1} * A * S$. The matrices A and B are similar. Use MATLAB to verify that the following properties hold for these two matrices.

(a) $\det(B) = \det(A)$

(b) $B^T = S^T A^T (S^T)^{-1}$

(c) $B^{-1} = S^{-1} A^{-1} S$

(d) $B^9 = S^{-1} A^9 S$

(e) $B - 3I = S^{-1}(A - 3I)S$

(f) $\det(B - 3I) = \det(A - 3I)$

(g) $\text{tr}(B) = \text{tr}(A)$ (Note that the trace of a matrix A can be computed using the MATLAB command **trace**.)

These properties will hold in general for any pair of similar matrices. See Exercises 11–15 of Section 3.

CHAPTER TEST

In each of the following, answer *true* if the statement is always true and *false* otherwise. In the case of a true statement, explain or prove your answer. In the case of a false statement, give an example to show that the statement is not always true.

1. Let $L: R^n \rightarrow R^n$ be a linear transformation. If $L(\mathbf{x}_1) = L(\mathbf{x}_2)$, then the vectors $\mathbf{x}_1$ and $\mathbf{x}_2$ must be equal.

2. If L_1 and L_2 are both linear transformations mapping a vector space V into itself, then $L_1 + L_2$ is also a linear transformation, where $L_1 + L_2$ is the mapping defined by

$$(L_1 + L_2)(\mathbf{v}) = L_1(\mathbf{v}) + L_2(\mathbf{v}) \text{ for all } \mathbf{v} \in V$$

3. If $L: V \rightarrow V$ is a linear transformation and $\mathbf{x} \in \text{ker}(L)$, then $L(\mathbf{v} + \mathbf{x}) = L(\mathbf{v})$ for all $\mathbf{v} \in V$.

4. If L_1 rotates each vector $\mathbf{x}$ in R^2 by $60°$ and then reflects the resulting vector about the x axis and L_2 is a transformation that does the same two operations, but in the reverse order, then $L_1 = L_2$.

5. The set of all vectors $\mathbf{x}$ used in the homogeneous coordinate system (see the application on computer graphics and animation in Section 2) forms a subspace of R^3.

6. Let $L: R^2 \rightarrow R^2$ be a linear transformation, and let A be the standard matrix representation of L. If L^2 is defined by

$$L^2(\mathbf{x}) = L(L(\mathbf{x})) \text{ for all } \mathbf{x} \in R^2$$

then L^2 is a linear transformation and its standard matrix representation is A^2.

7. Let $E = [\mathbf{x}_1, \mathbf{x}_2, \dots, \mathbf{x}_n]$ be an ordered basis for R^n. If $L_1 : R^n \rightarrow R^n$ and $L_2 : R^n \rightarrow R^n$ have the same matrix representation with respect to E, then $L_1 = L_2$.

8. Let $L: R^n \rightarrow R^n$ be a linear transformation. If A is the standard matrix representation of L, then an $n \times n$ matrix B will also be a matrix representation of L if and only if B is similar to A.

9. Let A, B, C be $n \times n$ matrices. If A is similar to B and B is similar to C, then A is similar to C.

10. Any two matrices with the same trace are similar. (This statement is the converse of part (b) of Exercise 15 in Section 3.)

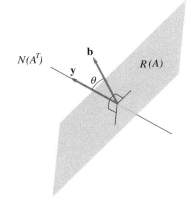

CHAPTER
5

ORTHOGONALITY

We can add to the structure of a vector space by defining a scalar or inner product. Such a product is not a true vector multiplication, since to every pair of vectors it associates a scalar rather than a third vector. For example, in R^2 we can define the scalar product of two vectors $\mathbf{x}$ and $\mathbf{y}$ to be $\mathbf{x}^T\mathbf{y}$. We can think of vectors in R^2 as directed line segments initiating at the origin. It is not difficult to show that the angle between two line segments will be a right angle if and only if the scalar product of the corresponding vectors is zero. In general, if V is a vector space with a scalar product, then two vectors in V are said to be *orthogonal* if their scalar product is zero.

We can think of orthogonality as a generalization of the concept of *perpendicularity* to any vector space with an inner product. To see the significance of this, consider the following problem. Let l be a line passing through the origin, and let Q be a point not on l. Find the point P on l that is closest to Q. The solution P to this problem is characterized by the condition that QP is perpendicular to OP (see Figure 5.0.1). If we think of the line l as corresponding to a subspace of R^2 and $\mathbf{v} = OQ$ as a vector in R^2, then the problem is to find a vector in the subspace that is "closest" to $\mathbf{v}$. The solution $\mathbf{p}$ will be characterized by the property that $\mathbf{p}$ is orthogonal to $\mathbf{v} - \mathbf{p}$ (see Figure 5.0.1). In the setting of a vector space with an inner product, we are able to consider general *least squares* problems. In these problems we are given a vector $\mathbf{v}$ in V and a subspace W. We wish to find a vector in W that is "closest" to $\mathbf{v}$. A solution $\mathbf{p}$ must be orthogonal to $\mathbf{v} - \mathbf{p}$. This orthogonality condition provides the key to solving the least squares problem. Least squares problems occur in many statistical applications involving data fitting.

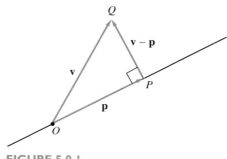

FIGURE 5.0.1

▌ THE SCALAR PRODUCT IN R^n

Two vectors $\mathbf{x}$ and $\mathbf{y}$ in R^n may be regarded as $n \times 1$ matrices. We can then form the matrix product $\mathbf{x}^T\mathbf{y}$. This product is a 1×1 matrix that may be regarded as a vector in R^1 or, more simply, as a real number. The product $\mathbf{x}^T\mathbf{y}$ is called the *scalar product* of $\mathbf{x}$ and $\mathbf{y}$. In particular, if $\mathbf{x} = (x_1, \dots, x_n)^T$ and $\mathbf{y} = (y_1, \dots, y_n)^T$, then

$$\mathbf{x}^T\mathbf{y} = x_1 y_1 + x_2 y_2 + \cdots + x_n y_n$$

EXAMPLE I.   If

$$\mathbf{x} = \begin{bmatrix} 3 \\ -2 \\ 1 \end{bmatrix} \quad \text{and} \quad \mathbf{y} = \begin{bmatrix} 4 \\ 3 \\ 2 \end{bmatrix}$$

then

$$\mathbf{x}^T\mathbf{y} = (3, -2, 1) \begin{bmatrix} 4 \\ 3 \\ 2 \end{bmatrix} = 3 \cdot 4 - 2 \cdot 3 + 1 \cdot 2 = 8 \qquad ◄$$

The Scalar Product in R^2 and R^3

In order to see the geometric significance of the scalar product, let us begin by restricting our attention to R^2 and R^3. Vectors in R^2 and R^3 can be represented by directed line segments. Given a vector $\mathbf{x}$ in either R^2 or R^3, its *Euclidean length* can be defined in terms of the scalar product.

$$\|\mathbf{x}\| = (\mathbf{x}^T\mathbf{x})^{1/2} = \begin{cases} \sqrt{x_1^2 + x_2^2} & \text{if } \mathbf{x} \in R^2 \\ \sqrt{x_1^2 + x_2^2 + x_3^2} & \text{if } \mathbf{x} \in R^3 \end{cases}$$

Given two nonzero vectors $\mathbf{x}$ and $\mathbf{y}$, we can think of them as directed line segments initiating at the same point. The angle between the two vectors is then defined as the angle θ between the line segments. We can measure the distance between

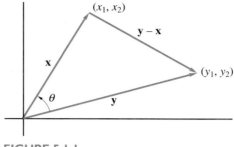

FIGURE 5.1.1

the vectors by measuring the length of the vector joining the terminal point of **x** to the terminal point of **y** (see Figure 5.1.1). Thus, we have the following definition.

▶ **DEFINITION** Let **x** and **y** be vectors in either R^2 or R^3. The distance between **x** and **y** is defined to be the number $\|\mathbf{x} - \mathbf{y}\|$. ◀

EXAMPLE 2. If $\mathbf{x} = (3, 4)^T$ and $\mathbf{y} = (-1, 7)^T$, then the distance between **x** and **y** given by

$$\|\mathbf{y} - \mathbf{x}\| = \sqrt{(-1-3)^2 + (7-4)^2} = 5$$ ◀

The angle between two vectors can be computed using the following theorem.

▶ **THEOREM 5.1.1** *If **x** and **y** are two nonzero vectors in either R^2 or R^3 and θ is the angle between them, then*

(1) $$\mathbf{x}^T\mathbf{y} = \|\mathbf{x}\|\,\|\mathbf{y}\|\cos\theta$$

▶*Proof.* We will prove the result for R^2. The proof for R^3 is similar. The vectors **x**, **y**, and $\mathbf{y} - \mathbf{x}$ may be used to form a triangle as in Figure 5.1.1. By the law of cosines,

$$\|\mathbf{y} - \mathbf{x}\|^2 = \|\mathbf{x}\|^2 + \|\mathbf{y}\|^2 - 2\|\mathbf{x}\|\,\|\mathbf{y}\|\cos\theta$$

or

$$\begin{aligned}
\|\mathbf{x}\|\,\|\mathbf{y}\|\cos\theta &= \tfrac{1}{2}(\|\mathbf{x}\|^2 + \|\mathbf{y}\|^2 - \|\mathbf{y} - \mathbf{x}\|^2) \\
&= \tfrac{1}{2}\left[x_1^2 + x_2^2 + y_1^2 + y_2^2 - (x_1 - y_1)^2 - (x_2 - y_2)^2\right] \\
&= x_1 y_1 + x_2 y_2 \\
&= \mathbf{x}^T\mathbf{y}
\end{aligned}$$ ◀

If **x** and **y** are nonzero vectors, then we can specify their directions by forming unit vectors

$$\mathbf{u} = \frac{1}{\|\mathbf{x}\|}\mathbf{x} \quad \text{and} \quad \mathbf{v} = \frac{1}{\|\mathbf{y}\|}\mathbf{y}$$

If θ is the angle between $\mathbf{x}$ and $\mathbf{y}$, then

$$\cos\theta = \frac{\mathbf{x}^T\mathbf{y}}{\|\mathbf{x}\|\|\mathbf{y}\|} = \mathbf{u}^T\mathbf{v}$$

The cosine of the angle between the vectors $\mathbf{x}$ and $\mathbf{y}$ is simply the scalar product of the corresponding direction vectors $\mathbf{u}$ and $\mathbf{v}$.

EXAMPLE 3. Let $\mathbf{x}$ and $\mathbf{y}$ be the vectors in Example 2. The directions of these vectors are given by the unit vectors

$$(3,4)^T \qquad (-1,7)^T$$

$$\mathbf{u} = \frac{1}{\|\mathbf{x}\|}\mathbf{x} = \begin{pmatrix} \frac{3}{5} \\ \frac{4}{5} \end{pmatrix} \qquad \text{and} \qquad \mathbf{v} = \frac{1}{\|\mathbf{y}\|}\mathbf{y} = \begin{pmatrix} -\frac{1}{5\sqrt{2}} \\ \frac{7}{5\sqrt{2}} \end{pmatrix}$$

The cosine of the angle θ between the two vectors is

$$\cos\theta = \mathbf{u}^T\mathbf{v} = \frac{1}{\sqrt{2}}$$

and hence $\theta = \frac{\pi}{4}$. ◀

▶ **COROLLARY 5.1.2 (Cauchy–Schwarz Inequality)** *If $\mathbf{x}$ and $\mathbf{y}$ are vectors in either R^2 or R^3, then*

(2) $$|\mathbf{x}^T\mathbf{y}| \le \|\mathbf{x}\|\,\|\mathbf{y}\|$$

with equality holding if and only if one of the vectors is $\mathbf{0}$ or one vector is a multiple of the other.

▶ *Proof.* The inequality follows from (1). If one of the vectors is $\mathbf{0}$, then both sides of (2) are 0. If both vectors are nonzero, it follows from (1) that equality can hold in (2) if and only if $\cos\theta = \pm1$. But this would imply that the vectors are either in the same or opposite directions and hence that one vector must be a multiple of the other. ◀

If $\mathbf{x}^T\mathbf{y} = 0$, it follows from Theorem 5.1.1 that either one of the vectors is the zero vector or $\cos\theta = 0$. If $\cos\theta = 0$, the angle between the vectors is a right angle.

▶ **DEFINITION** The vectors $\mathbf{x}$ and $\mathbf{y}$ in R^2 (or R^3) are said to be **orthogonal** if $\mathbf{x}^T\mathbf{y} = 0$. ◀

EXAMPLE 4.

A. The vector $\mathbf{0}$ is orthogonal to every vector in R^2.

B. The vectors $\begin{pmatrix} 3 \\ 2 \end{pmatrix}$ and $\begin{pmatrix} -4 \\ 6 \end{pmatrix}$ are orthogonal in R^2.

C. The vectors $\begin{pmatrix} 2 \\ -3 \\ 1 \end{pmatrix}$ and $\begin{pmatrix} 1 \\ 1 \\ 1 \end{pmatrix}$ are orthogonal in R^3. ◀

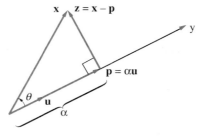

FIGURE 5.1.2

Scalar and Vector Projections

The scalar product can be used to find the component of one vector in the direction of another. Let **x** and **y** be nonzero vectors in either R^2 or R^3. We would like to write **x** as a sum of the form $\mathbf{p} + \mathbf{z}$, where **p** is in the direction of **y** and **z** is orthogonal to **p** (see Figure 5.1.2). To do this, let $\mathbf{u} = (1/\|\mathbf{y}\|)\mathbf{y}$. Thus **u** is a unit vector (length 1) in the direction of **y**. We wish to find α such that $\mathbf{p} = \alpha\mathbf{u}$ is orthogonal to $\mathbf{z} = \mathbf{x} - \alpha\mathbf{u}$. In order for **p** and **z** to be orthogonal, the scalar α must satisfy

$$\alpha = \|\mathbf{x}\| \cos\theta$$

$$= \frac{\|\mathbf{x}\|\,\|\mathbf{y}\| \cos\theta}{\|\mathbf{y}\|}$$

$$= \frac{\mathbf{x}^T \mathbf{y}}{\|\mathbf{y}\|}$$

The scalar α is called the *scalar projection* of **x** onto **y**, and the vector **p** is called the *vector projection* of **x** onto **y**.

> Scalar projection of **x** onto **y**:
>
> $$\alpha = \frac{\mathbf{x}^T \mathbf{y}}{\|\mathbf{y}\|}$$
>
> Vector projection of **x** onto **y**:
>
> $$\mathbf{p} = \alpha\mathbf{u} = \alpha\frac{1}{\|\mathbf{y}\|}\mathbf{y} = \frac{\mathbf{x}^T \mathbf{y}}{\mathbf{y}^T \mathbf{y}}\mathbf{y}$$

EXAMPLE 5. The point Q in Figure 5.1.3 is the point on the line $y = \frac{1}{3}x$ that is closest to the point $(1, 4)$. Determine the coordinates of Q.

SOLUTION. The vector $\mathbf{w} = (3, 1)^T$ is a vector in the direction of the line $y = \frac{1}{3}x$. Let $\mathbf{v} = (1, 4)^T$. If Q is the desired point, then Q^T is the vector projection of **v** onto **w**.

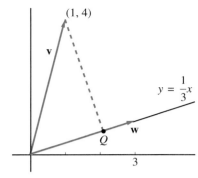

FIGURE 5.1.3

$$Q^T = \left(\frac{\mathbf{v}^T\mathbf{w}}{\mathbf{w}^T\mathbf{w}}\right)\mathbf{w} = \frac{7}{10}\begin{bmatrix} 3 \\ 1 \end{bmatrix} = \begin{bmatrix} 2.1 \\ 0.7 \end{bmatrix}$$

Thus $Q = (2.1, 0.7)$ is the closest point.   ◀

Notation   If P_1 and P_2 are two points in 3-space, we will denote the vector from P_1 to P_2 by $\overline{P_1 P_2}$.

If $\mathbf{N}$ is a nonzero vector and P_0 is a fixed point, the set of points P such that $\overline{P_0 P}$ is orthogonal to $\mathbf{N}$ forms a plane π in 3-space that passes through P_0. The vector $\mathbf{N}$ and the plane π are said to be *normal* to each other. A point $P = (x, y, z)$ will lie on π if and only if

$$(\overline{P_0 P})^T \mathbf{N} = 0$$

If $\mathbf{N} = (a, b, c)^T$ and $P_0 = (x_0, y_0, z_0)$, this equation can be written in the form

$$a(x - x_0) + b(y - y_0) + c(z - z_0) = 0$$

EXAMPLE 6.   Find the equation of the plane passing through the point $(2, -1, 3)$ and normal to the vector $\mathbf{N} = (2, 3, 4)^T$.

SOLUTION. $\overline{P_0 P} = (x - 2, y + 1, z - 3)^T$. The equation is $(\overline{P_0 P})^T\mathbf{N} = 0$, or

$$2(x - 2) + 3(y + 1) + 4(z - 3) = 0$$ ◀

EXAMPLE 7.   Find the distance from the point $(2, 0, 0)$ to the plane $x + 2y + 2z = 0$.

SOLUTION.   The vector $\mathbf{N} = (1, 2, 2)^T$ is normal to the plane and the plane passes through the origin. Let $\mathbf{v} = (2, 0, 0)^T$. The distance d from $(2, 0, 0)$ to the plane is simply the absolute value of the scalar projection of $\mathbf{v}$ onto $\mathbf{N}$. Thus

$$d = \frac{|\mathbf{v}^T\mathbf{N}|}{\|\mathbf{N}\|} = \frac{2}{3}$$ ◀

Orthogonality in R^n

The definitions that have been given for R^2 and R^3 can all be generalized to R^n. Indeed, if $\mathbf{x} \in R^n$, then the *Euclidean length* of $\mathbf{x}$ is defined by

$$\|\mathbf{x}\| = (\mathbf{x}^T \mathbf{x})^{1/2} = (x_1^2 + x_2^2 + \cdots + x_n^2)^{1/2}$$

If $\mathbf{x}$ and $\mathbf{y}$ are two vectors in R^n, then the distance between the vectors is $\|\mathbf{y} - \mathbf{x}\|$.

The Cauchy–Schwarz inequality holds in R^n. (We will prove this in Section 4.) Consequently

$$(3) \qquad\qquad -1 \leq \frac{\mathbf{x}^T \mathbf{y}}{\|\mathbf{x}\| \, \|\mathbf{y}\|} \leq 1$$

for any nonzero vectors $\mathbf{x}$ and $\mathbf{y}$ in R^n. In view of (3) the definition of the angle between two vectors that was used for R^2 can be generalized to R^n. Thus the angle θ between two nonzero vectors $\mathbf{x}$ and $\mathbf{y}$ in R^n is given by

$$\cos \theta = \frac{\mathbf{x}^T \mathbf{y}}{\|\mathbf{x}\| \, \|\mathbf{y}\|}, \qquad 0 \leq \theta \leq \pi$$

In talking about angles between vectors it is usually more convenient to scale the vectors so as to make them unit vectors. If we set

$$\mathbf{u} = \frac{1}{\|\mathbf{x}\|}\mathbf{x} \qquad \text{and} \qquad \mathbf{v} = \frac{1}{\|\mathbf{y}\|}\mathbf{y}$$

then the angle θ between $\mathbf{u}$ and $\mathbf{v}$ is clearly the same as the angle between $\mathbf{x}$ and $\mathbf{y}$, and its cosine can be computed by simply taking the scalar product of the two unit vectors.

$$\cos \theta = \frac{\mathbf{x}^T \mathbf{y}}{\|\mathbf{x}\| \, \|\mathbf{y}\|} = \mathbf{u}^T \mathbf{v}$$

The vectors $\mathbf{x}$ and $\mathbf{y}$ are said to be *orthogonal* if $\mathbf{x}^T \mathbf{y} = 0$. Often the symbol $\perp$ is used to indicate orthogonality. Thus, if $\mathbf{x}$ and $\mathbf{y}$ are orthogonal, we will write $\mathbf{x} \perp \mathbf{y}$. Vector and scalar projections are defined in R^n in the same way that they were defined for R^2.

If $\mathbf{x}$ and $\mathbf{y}$ are vectors in R^n, then

$$(4) \qquad\qquad \|\mathbf{x} + \mathbf{y}\|^2 = (\mathbf{x} + \mathbf{y})^T (\mathbf{x} + \mathbf{y}) = \|\mathbf{x}\|^2 + 2\mathbf{x}^T \mathbf{y} + \|\mathbf{y}\|^2$$

In the case that $\mathbf{x}$ and $\mathbf{y}$ are orthogonal, equation (4) becomes the *Pythagorean Law*

$$\|\mathbf{x} + \mathbf{y}\|^2 = \|\mathbf{x}\|^2 + \|\mathbf{y}\|^2$$

The Pythagorean Law is a generalization of the Pythagorean Theorem. When $\mathbf{x}$ and $\mathbf{y}$ are orthogonal vectors in R^2, we can use these vectors and their sum $\mathbf{x} + \mathbf{y}$

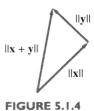

FIGURE 5.1.4

to form a right triangle as in Figure 5.1.4. The Pythagorean Law relates the lengths of the sides of the triangle. Indeed, if we set

$$a = \|\mathbf{x}\|, \quad b = \|\mathbf{y}\|, \quad c = \|\mathbf{x} + \mathbf{y}\|$$

then

$$c^2 = a^2 + b^2 \quad \text{(the famous Pythagorean Theorem)}$$

In many applications the cosine of the angle between two nonzero vectors is used as a measure of how closely the directions of the vectors match up. If $\cos\theta$ is near 1, then the angle between the vectors is small and hence the vectors are in nearly the same direction. A cosine value near zero would indicate that the angle between the vectors is nearly a right angle.

APPLICATION 1: INFORMATION RETRIEVAL REVISITED

In Section 3 of Chapter 1 we considered the problem of searching a database for documents that contain certain key words. If there are m possible key search words and a total of n documents in the collection, then the database can be represented by an $m \times n$ matrix A. Each column of A represents a document in the database. The entries of the jth column correspond to the relative frequencies of the keywords in the jth document.

Refined search techniques must deal with vocabulary disparities and the complexities of language. Two of the main problems are *polysemy* (words having multiple meanings) and *synonymy* (multiple words having the same meaning). Some of the words that you are searching for may have multiple meanings and could appear in contexts that are completely irrelevant to your particular search. For example, the word *calculus* would occur frequently in both mathematical papers and in dentistry papers. On the other hand, most words have synonyms, and it is possible that many of the documents may use the synonyms rather than the specified search words. For example, you could search for an article on rabies using the key word *dogs*; however, the author of the article may have preferred to use the word *canines* throughout his paper. In order to handle these problems, we need a technique to find the documents that best match the list of search words without necessarily matching every word on the list. We want to pick out the column vectors of the database matrix that most closely match a given search vector. To do this, we use the cosine of the angle between two vectors as a measure of how closely the vectors match up.

In practice, both m and n are quite large as there are many possible key words and many documents to search. For simplicity, let us consider an example where $m = 10$ and $n = 8$. Suppose that a Web site has eight modules for learning linear algebra and each module is located on a separate Web page. Our list of possible search words consists of

> *determinants, eigenvalues, linear, matrices, numerical,*
> *orthogonality, spaces, systems, transformations, vector*

(This key word list was compiled from the chapter headings for this book.) Table 1 shows the frequencies of the key words in each of the modules. The $(2, 6)$ entry of the table is 5, which indicates that the key word *eigenvalues* appears five times in the sixth module.

The database matrix is formed by scaling each column of the table so that all column vectors are unit vectors. Thus, if A is the matrix corresponding to Table 1, then the columns of the database matrix Q are determined by setting

$$\mathbf{q}_j = \frac{1}{\|\mathbf{a}_j\|}\mathbf{a}_j \quad j = 1, \dots, 8$$

To do a search for the key words *orthogonality, spaces, vector*, we form a search vector $\mathbf{x}$ whose entries are all 0 except for the three rows corresponding to the search rows. In order to obtain a unit search vector, we put $\frac{1}{\sqrt{3}}$ in each of the rows corresponding to the search words. For this example, the database matrix Q and search vector $\mathbf{x}$ (with entries rounded to three decimal places) are given by

TABLE I

FREQUENCY OF KEY WORDS

Key words	Modules							
	M1	M2	M3	M4	M5	M6	M7	M8
determinants	0	6	3	0	1	0	1	1
eigenvalues	0	0	0	0	0	5	3	2
linear	5	4	4	5	4	0	3	3
matrices	6	5	3	3	4	4	3	2
numerical	0	0	0	0	3	0	4	3
orthogonality	0	0	0	0	4	6	0	2
spaces	0	0	5	2	3	3	0	1
systems	5	3	3	2	4	2	1	1
transformations	0	0	0	5	1	3	1	0
vector	0	4	4	3	4	2	0	3

$$Q = \begin{bmatrix} 0.000 & 0.594 & 0.327 & 0.000 & 0.100 & 0.000 & 0.147 & 0.154 \\ 0.000 & 0.000 & 0.000 & 0.000 & 0.000 & 0.500 & 0.442 & 0.309 \\ 0.539 & 0.396 & 0.436 & 0.574 & 0.400 & 0.000 & 0.442 & 0.463 \\ 0.647 & 0.495 & 0.327 & 0.344 & 0.400 & 0.400 & 0.442 & 0.309 \\ 0.000 & 0.000 & 0.000 & 0.000 & 0.300 & 0.000 & 0.590 & 0.463 \\ 0.000 & 0.000 & 0.000 & 0.000 & 0.400 & 0.600 & 0.000 & 0.309 \\ 0.000 & 0.000 & 0.546 & 0.229 & 0.300 & 0.300 & 0.000 & 0.154 \\ 0.539 & 0.297 & 0.327 & 0.229 & 0.400 & 0.200 & 0.147 & 0.154 \\ 0.000 & 0.000 & 0.000 & 0.574 & 0.100 & 0.300 & 0.147 & 0.000 \\ 0.000 & 0.396 & 0.436 & 0.344 & 0.400 & 0.200 & 0.000 & 0.463 \end{bmatrix} \quad \mathbf{x} = \begin{bmatrix} 0.000 \\ 0.000 \\ 0.000 \\ 0.000 \\ 0.000 \\ 0.577 \\ 0.577 \\ 0.000 \\ 0.000 \\ 0.577 \end{bmatrix}$$

If we set $\mathbf{y} = Q^T \mathbf{x}$ then

$$y_i = \mathbf{q}_i^T \mathbf{x} = \cos \theta_i$$

where θ_i is the angle between the unit vectors $\mathbf{x}$ and $\mathbf{q}_i$. For our example,

$$\mathbf{y} = (0.000, 0.229, 0.567, 0.310, 0.635, 0.577, 0.000, 0.535)^T$$

Since $y_5 = 0.635$ is the entry of $\mathbf{y}$ that is closest to 1, this indicates that the direction of the search vector $\mathbf{x}$ is closest to the direction of $\mathbf{q}_5$ and hence that Module 5 is the one that best matches our search criteria. The next best matches come from modules 6 ($y_6 = 0.577$) and 3 ($y_3 = 0.567$). If a document doesn't contain any of the search words, then the corresponding column vector of the database matrix will be orthogonal to the search vector. Note that modules 1 and 7 do not have any of the three search words and consequently

$$y_1 = \mathbf{q}_1^T \mathbf{x} = 0 \qquad \text{and} \qquad y_7 = \mathbf{q}_7^T \mathbf{x} = 0$$

If a document doesn't satisfy any of the search criteria, then the column vector representing the document will be orthogonal to the search vector.

This example illustrates some of the basic ideas behind database searches. Using modern matrix techniques, it is possible to improve the search process significantly. We can speed up searches and at the same time correct for errors due to polysemy and synonomy. These advanced techniques are referred to as *latent semantic indexing* (LSI) and depend on a matrix factorization, the *singular value decomposition*, which we will discuss in Section 5 of Chapter 6.

There are many other important applications involving angles between vectors. In particular, statisticians use the cosine of angle between two vectors as a measure of how closely the two vectors are correlated.

APPLICATION 2: STATISTICS — CORRELATION AND COVARIANCE MATRICES

Suppose that we wanted to compare how closely exam scores for a class correlate to scores on homework assignments. As an example, we consider the total scores on assignments and tests of a mathematics class at the University of Massachusetts Dartmouth. The total scores for homework assignments during the semester for the class are given in the second column of Table 2. The third column represents the total scores for the two exams given during the semester, and the last column contains the scores on the final exam. In each case a perfect score would be 200 points. The last row of the table summarizes the class averages.

We would like to measure how student performance compares between each set of exam or assignment scores. In order to see how closely the two sets of scores are correlated and allow for any differences in difficulty, we need to adjust the scores so that each test has a mean of 0. If in each column we subtract the average score from each of the test scores, then the translated scores will each have an average of 0. Let us store these translated scores in a matrix:

$$X = \begin{bmatrix} 37 & 37 & 65 \\ -1 & 2 & 34 \\ -3 & -5 & 2 \\ -11 & 2 & -40 \\ 14 & 19 & 20 \\ -27 & -28 & -30 \\ -9 & -27 & -51 \end{bmatrix}$$

TABLE 2

MATH SCORES FALL 1996

Student	Scores		
	Assignments	Exams	Final
S1	198	200	196
S2	160	165	165
S3	158	158	133
S4	150	165	91
S5	175	182	151
S6	134	135	101
S7	152	136	80
Average	161	163	131

The column vectors of X represent the deviations from the mean for each of the three sets of scores. The three sets of translated data specified by the column vectors of X all have mean 0 and all sum to 0. To compare two sets of scores, we compute the cosine of the angle between the corresponding column vectors of X. A cosine value near 1 indicates that the two sets of scores are highly correlated. For example, correlation between the assignment scores and the exam scores is given by

$$\cos\theta = \frac{\mathbf{x}_1^T\mathbf{x}_2}{\|\mathbf{x}_1\|\,\|\mathbf{x}_2\|} \approx 0.92$$

A perfect correlation of 1 would correspond to the case where the two sets of translated scores are proportional. Thus, for a perfect correlation the translated vectors would satisfy

$$\mathbf{x}_2 = \alpha\mathbf{x}_1 \qquad (\alpha > 0)$$

and if the corresponding coordinates of the $\mathbf{x}_1$ and $\mathbf{x}_2$ were paired off, then each ordered pair would lie on the line $y = \alpha x$. Although the vectors $\mathbf{x}_1$ and $\mathbf{x}_2$ in our example are not perfectly correlated, the coefficient of 0.92 does indicate that the two sets of scores are highly correlated. Figure 5.1.5 shows how close the actual pairs are to lying on a line $y = \alpha x$. The slope of the line in the figure was determined by setting

$$\alpha = \frac{\mathbf{x}_1^T\mathbf{x}_2}{\mathbf{x}_1^T\mathbf{x}_1} = \frac{375}{258} \approx 1.05$$

This choice of slope yields an optimal *least squares* fit to the data points. (See Exercise 7 of Section 3.)

FIGURE 5.1.5

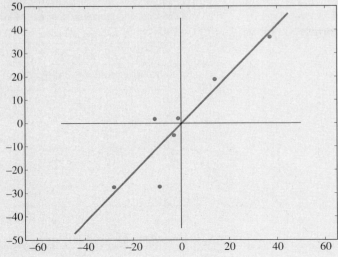

If we scale $\mathbf{x}_1$ and $\mathbf{x}_2$ to make them unit vectors

$$\mathbf{u}_1 = \frac{1}{\|\mathbf{x}_1\|}\mathbf{x}_1 \quad \text{and} \quad \mathbf{u}_2 = \frac{1}{\|\mathbf{x}_2\|}\mathbf{x}_2$$

then the cosine of the angle between the vectors will remain unchanged, and it can be computed simply by taking the scalar product $\mathbf{u}_1^T \mathbf{u}_2$. Let us scale all three sets of translated scores in this way and store the results in a matrix:

$$U = \begin{bmatrix} 0.74 & 0.65 & 0.62 \\ -0.02 & 0.03 & 0.33 \\ -0.06 & -0.09 & 0.02 \\ -0.22 & 0.03 & -0.38 \\ 0.28 & 0.33 & 0.19 \\ -0.54 & -0.49 & -0.29 \\ -0.18 & -0.47 & -0.49 \end{bmatrix}$$

If we set $C = U^T U$, then

$$C = \begin{bmatrix} 1 & 0.92 & 0.83 \\ 0.92 & 1 & 0.83 \\ 0.83 & 0.83 & 1 \end{bmatrix}$$

and the (i, j) entry of C represents the correlation between the ith and jth sets of scores. The matrix C is referred to as a *correlation matrix*.

The three sets of scores in our example are all *positively correlated* since the correlation coefficients are all positive. A negative coefficient would indicate that two data sets were *negatively correlated* and a coefficient of 0 would indicate that they were *uncorrelated*. Thus two sets of test scores would be uncorrelated if the corresponding vectors of deviations from the mean were orthogonal.

Another statistically important quantity that is closely related to the correlation matrix is the *covariance matrix*. Given a collection of n data points representing values of some variable x, we compute the mean $\overline{x}$ of the data points and form a vector $\mathbf{x}$ of the deviations from the mean. The *variance*, s^2, is defined by

$$s^2 = \frac{1}{n-1}\sum_1^n x_i^2 = \frac{\mathbf{x}^T \mathbf{x}}{n-1}$$

and the standard deviation s is the square root of the variance. If we have two data sets X_1 and X_2 each containing n values of a variable, we can form vectors $\mathbf{x}_1$ and $\mathbf{x}_2$ of deviations from the mean for both sets. The *covariance* is defined by

$$\text{cov}(X_1, X_2) = \frac{\mathbf{x}_1^T \mathbf{x}_2}{n-1}$$

If we have more than two data sets, we can form a matrix X whose columns represent the deviations from the mean for each data set and then form a *covariance matrix* S by setting

$$S = \frac{1}{n-1}X^T X$$

The covariance matrix for the three sets of mathematics scores is

$$S = \frac{1}{6}\begin{bmatrix} 37 & -1 & -3 & -11 & 14 & -27 & -9 \\ 37 & 2 & -5 & 2 & 19 & -28 & -27 \\ 65 & 34 & 2 & -40 & 20 & -30 & -51 \end{bmatrix}\begin{bmatrix} 37 & 37 & 65 \\ -1 & 2 & 34 \\ -3 & -5 & 2 \\ -11 & 2 & -40 \\ 14 & 19 & 20 \\ -27 & -28 & -30 \\ -9 & -27 & -51 \end{bmatrix}$$

$$= \begin{bmatrix} 417.7 & 437.5 & 725.7 \\ 437.5 & 546.0 & 830.0 \\ 725.7 & 830.0 & 1814.3 \end{bmatrix}$$

The diagonal entries of S are the variances for the three sets of scores, and the off-diagonal entries are the covariances.

To illustrate the importance of the correlation and covariance matrices, we will consider an application to the field of psychology.

APPLICATION 3: PSYCHOLOGY — FACTOR ANALYSIS AND PRINCIPAL COMPONENT ANALYSIS

Factor analysis had its start at the beginning of the twentieth century with the efforts of psychologists to identify the factor or factors that make up intelligence. The person most responsible for pioneering this field was the psychologist Charles Spearman. In a 1904 paper, Spearman analyzed a series of exam scores at a preparatory school. The exams were taken by a class of 23 pupils in a number of standard subject areas and also in pitch discrimination. The correlation matrix reported by Spearman is summarized in the following table.

Using this and other sets of data, Spearman observed a hierarchy of correlations among the test scores for the various disciplines. This led him to conclude that "All branches of intellectual activity have in common one fundamental function (or group of fundamental functions), ..." Although Spearman did not assign names to these functions, others have used terms such as verbal comprehension, spatial, perceptual, associative memory, and so on, to describe the hypothetical factors.

TABLE 3
SPEARMAN'S CORRELATION MATRIX

	Classics	French	English	Math	Discrim.	Music
Classics	1	0.83	0.78	0.70	0.66	0.63
French	0.83	1	0.67	0.67	0.65	0.57
English	0.78	0.67	1	0.64	0.54	0.51
Math	0.70	0.67	0.64	1	0.45	0.51
Discrim.	0.66	0.65	0.54	0.45	1	0.40
Music	0.63	0.57	0.51	0.51	0.40	1

The hypothetical factors can be isolated mathematically using a method known as *principal component analysis*. The basic idea is to form a matrix X of deviations from the mean and then factor it into a product UW, where the columns of U correspond to the hypothetical factors. While in practice the columns of X are positively correlated, the hypothetical factors should be uncorrelated. Thus the column vectors of U should be mutually orthogonal (i.e., $\mathbf{u}_i^T \mathbf{u}_j = 0$ whenever $i \neq j$). The entries in each column of U measure how well the individual students exhibit the particular intellectual ability represented by that column. The matrix W measures to what extent each test depends on the hypothetical factors.

The construction of the principal component vectors relies on the covariance matrix $S = \frac{1}{n-1} X^T X$. Since it depends on the *eigenvalues* and *eigenvectors* of the S, we will defer the details of the method until Chapter 6. In Section 5 of Chapter 6 we will revisit this application and learn an important factorization called the *singular value decomposition*, which is the main tool of principal component analysis.

REFERENCES

1. Spearman, C., 'General Intelligence', objectively determined and measured, *American Journal of Psychology*, **15**, 1904.
2. Hotelling, H., Analysis of a complex of statistical variables in principal components, *Journal of Educational Psychology*, **26**, 1933.
3. Maxwell, A. E., *Multivariate Analysis in Behavioral Research*, Chapman and Hall, London, 1977.

EXERCISES

1. Find the angle between the vectors $\mathbf{v}$ and $\mathbf{w}$ in each of the following.
   (a) $\mathbf{v} = (2, 1, 3)^T$, $\mathbf{w} = (6, 3, 9)^T$
   (b) $\mathbf{v} = (2, -3)^T$, $\mathbf{w} = (3, 2)^T$
   (c) $\mathbf{v} = (4, 1)^T$, $\mathbf{w} = (3, 2)^T$
   (d) $\mathbf{v} = (-2, 3, 1)^T$, $\mathbf{w} = (1, 2, 4)^T$

2. For each of the pairs of vectors in Exercise 1, find the scalar projection of **v** onto **w**. Also find the vector projection of **v** onto **w**.

3. For each of the following pairs of vectors **x** and **y**, find the vector projection **p** of **x** onto **y** and verify that **p** and **x** − **p** are orthogonal.

   (a) $\mathbf{x} = (3, 4)^T$, $\mathbf{y} = (1, 0)^T$ (b) $\mathbf{x} = (3, 5)^T$, $\mathbf{y} = (1, 1)^T$
   (c) $\mathbf{x} = (2, 4, 3)^T$, $\mathbf{y} = (1, 1, 1)^T$ (d) $\mathbf{x} = (2, -5, 4)^T$, $\mathbf{y} = (1, 2, -1)^T$

4. Let **x** and **y** be linearly independent vectors in R^2. If $\|\mathbf{x}\| = 2$ and $\|\mathbf{y}\| = 3$, what, if anything, can we conclude about the possible values of $|\mathbf{x}^T\mathbf{y}|$?

5. Find the point on the line $y = 2x$ that is closest to the point $(5, 2)$.

6. Find the point on the line $y = 2x + 1$ that is closest to the point $(5, 2)$.

7. Find the distance from the point $(1, 2)$ to the line $4x - 3y = 0$.

8. In each of the following, find the equation of the plane normal to the given vector **N** and passing through the point P_0.

   (a) $\mathbf{N} = (2, 4, 3)^T$, $P_0 = (0, 0, 0)$ (b) $\mathbf{N} = (-3, 6, 2)^T$, $P_0 = (4, 2, -5)$
   (c) $\mathbf{N} = (0, 0, 1)^T$, $P_0 = (3, 2, 4)$

9. Find the distance from the point $(1, 1, 1)$ to the plane $2x + 2y + z = 0$.

10. Find the distance from the point $(2, 1, -2)$ to the plane $6(x - 1) + 2(y - 3) + 3(z + 4) = 0$.

11. If $\mathbf{x} = (x_1, x_2)^T$, $\mathbf{y} = (y_1, y_2)^T$, and $\mathbf{z} = (z_1, z_2)^T$ are arbitrary vectors in R^2, prove:

    (a) $\mathbf{x}^T\mathbf{x} \geq 0$ (b) $\mathbf{x}^T\mathbf{y} = \mathbf{y}^T\mathbf{x}$ (c) $\mathbf{x}^T(\mathbf{y} + \mathbf{z}) = \mathbf{x}^T\mathbf{y} + \mathbf{x}^T\mathbf{z}$

12. If **u** and **v** are any vectors in R^2, show that $\|\mathbf{u} + \mathbf{v}\|^2 \leq (\|\mathbf{u}\| + \|\mathbf{v}\|)^2$ and hence $\|\mathbf{u} + \mathbf{v}\| \leq \|\mathbf{u}\| + \|\mathbf{v}\|$. When does equality hold? Give a geometric interpretation of the inequality.

13. Let $\mathbf{x}_1, \mathbf{x}_2, \mathbf{x}_3$ be vectors in R^3. If $\mathbf{x}_1 \perp \mathbf{x}_2$ and $\mathbf{x}_2 \perp \mathbf{x}_3$, is it necessarily true that $\mathbf{x}_1 \perp \mathbf{x}_3$? Prove your answer.

14. Let A be a 2×2 matrix with linearly independent column vectors $\mathbf{a}_1$ and $\mathbf{a}_2$. If $\mathbf{a}_1$ and $\mathbf{a}_2$ are used to form a parallelogram P with altitude h (see the figure), show that:

    (a) $h^2\|\mathbf{a}_2\|^2 = \|\mathbf{a}_1\|^2\|\mathbf{a}_2\|^2 - (\mathbf{a}_1^T\mathbf{a}_2)^2$ (b) Area of $P = |\det(A)|$

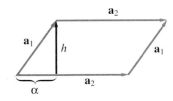

15. Given

$$\mathbf{x} = \begin{pmatrix} 4 \\ 4 \\ -4 \\ 4 \end{pmatrix} \quad \text{and} \quad \mathbf{y} = \begin{pmatrix} 4 \\ 2 \\ 2 \\ 1 \end{pmatrix}$$

(a) Determine the angle between $\mathbf{x}$ and $\mathbf{y}$.

(b) Determine the distance between $\mathbf{x}$ and $\mathbf{y}$.

16. Let $\mathbf{x}$ and $\mathbf{y}$ be vectors in R^n and define

$$\mathbf{p} = \frac{\mathbf{x}^T \mathbf{y}}{\mathbf{y}^T \mathbf{y}} \mathbf{y} \quad \text{and} \quad \mathbf{z} = \mathbf{x} - \mathbf{p}$$

(a) Show that $\mathbf{p} \perp \mathbf{z}$. Thus $\mathbf{p}$ is the *vector projection* of $\mathbf{x}$ onto $\mathbf{y}$; that is, $\mathbf{x} = \mathbf{p} + \mathbf{z}$, where $\mathbf{p}$ and $\mathbf{z}$ are orthogonal components of $\mathbf{x}$, and $\mathbf{p}$ is a scalar multiple of $\mathbf{y}$.

(b) If $\|\mathbf{p}\| = 6$ and $\|\mathbf{z}\| = 8$, determine the value of $\|\mathbf{x}\|$.

17. Use the database matrix U from Application 1 and search for the key words *orthogonality, spaces, vector*, only this time give the key word *orthogonality* twice the weight of the other two key words. Which of the eight modules best matches the search criteria? [*Hint:* Form the search vector using the weights 2, 1, 1 in the rows corresponding to the search words and then scale the vector to make it a unit vector.]

18. Five students in an elementary school take aptitude tests in English, Mathematics, and Science. Their scores are given in the table. Determine the correlation matrix and describe how the three sets of scores are correlated.

	Scores		
Student	English	Mathematics	Science
S1	61	53	53
S2	63	73	78
S3	78	61	82
S4	65	84	96
S5	63	59	71
Average	66	66	76

2 ORTHOGONAL SUBSPACES

Let A be an $m \times n$ matrix and let $\mathbf{x} \in N(A)$, the nullspace of A. Since $A\mathbf{x} = \mathbf{0}$, we have

(1) $$a_{i1}x_1 + a_{i2}x_2 + \cdots + a_{in}x_n = 0$$

for $i = 1, \ldots, m$. Equation (1) says that $\mathbf{x}$ is orthogonal to the ith column vector of A^T for $i = 1, \ldots, m$. Since $\mathbf{x}$ is orthogonal to each column vector of A^T, it is orthogonal to any linear combination of the column vectors of A^T. So if $\mathbf{y}$ is any vector in the column space of A^T, then $\mathbf{x}^T \mathbf{y} = 0$. Thus each vector in $N(A)$ is orthogonal to every vector in the column space of A^T. When two subspaces of R^n have this property, we say that they are orthogonal.

▶ **DEFINITION** Two subspaces X and Y of R^n are said to be **orthogonal** if $\mathbf{x}^T \mathbf{y} = 0$ for every $\mathbf{x} \in X$ and every $\mathbf{y} \in Y$. If X and Y are orthogonal, we write $X \perp Y$.   ◀

EXAMPLE I. Let X be the subspace of R^3 spanned by $\mathbf{e}_1$, and let Y be the subspace spanned by $\mathbf{e}_2$. If $\mathbf{x} \in X$ and $\mathbf{y} \in Y$, these vectors must be of the form

$$\mathbf{x} = \begin{bmatrix} x_1 \\ 0 \\ 0 \end{bmatrix} \quad \text{and} \quad \mathbf{y} = \begin{bmatrix} 0 \\ y_2 \\ 0 \end{bmatrix}$$

Thus,

$$\mathbf{x}^T \mathbf{y} = x_1 \cdot 0 + 0 \cdot y_2 + 0 \cdot 0 = 0$$

Therefore, $X \perp Y$.   ◀

The concept of orthogonal subspaces does not always agree with our intuitive idea of perpendicularity. For example, the floor and wall of the classroom "look" orthogonal, but the xy plane and the yz plane are not orthogonal subspaces. Indeed, we can think of the vectors $\mathbf{x}_1 = (1, 1, 0)^T$ and $\mathbf{x}_2 = (0, 1, 1)^T$ as lying in the xy and yz planes, respectively. Since

$$\mathbf{x}_1^T \mathbf{x}_2 = 1 \cdot 0 + 1 \cdot 1 + 0 \cdot 1 = 1$$

the subspaces are not orthogonal. The next example shows that the subspace corresponding to the z axis is orthogonal to the subspace corresponding to the xy plane.

EXAMPLE 2. Let X be the subspace of R^3 spanned by $\mathbf{e}_1$ and $\mathbf{e}_2$, and let Y be the subspace spanned by $\mathbf{e}_3$. If $\mathbf{x} \in X$ and $\mathbf{y} \in Y$, then

$$\mathbf{x}^T \mathbf{y} = x_1 \cdot 0 + x_2 \cdot 0 + 0 \cdot y_3 = 0$$

Thus, $X \perp Y$. Furthermore, if $\mathbf{z}$ is any vector in R^3 that is orthogonal to every vector in Y, then $\mathbf{z} \perp \mathbf{e}_3$, and hence

$$z_3 = \mathbf{z}^T \mathbf{e}_3 = 0$$

But if $z_3 = 0$, then $\mathbf{z} \in X$. Thus X is the set of all vectors in R^3 that are orthogonal to every vector in Y (see Figure 5.2.1).   ◀

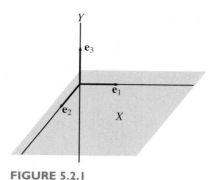

FIGURE 5.2.1

▶ **DEFINITION** Let Y be a subspace of R^n. The set of all vectors in R^n that are orthogonal to every vector in Y will be denoted $Y^\perp$. Thus

$$Y^\perp = \left\{ \mathbf{x} \in R^n \mid \mathbf{x}^T \mathbf{y} = 0 \quad \text{for every} \quad \mathbf{y} \in Y \right\}$$

The set $Y^\perp$ is called the **orthogonal complement** of Y.   ◀

NOTE The subspaces $X = \text{Span}(\mathbf{e}_1)$ and $Y = \text{Span}(\mathbf{e}_2)$ of R^3 given in Example 1 are orthogonal, but they are not orthogonal complements. Indeed,

$$X^\perp = \text{Span}(\mathbf{e}_2, \mathbf{e}_3) \qquad \text{and} \qquad Y^\perp = \text{Span}(\mathbf{e}_1, \mathbf{e}_3)$$

Remarks

1. If X and Y are orthogonal subspaces of R^n, then $X \cap Y = \{\mathbf{0}\}$.
2. If Y is a subspace of R^n, then $Y^\perp$ is also a subspace of R^n.

▶ *Proof of (1).* If $\mathbf{x} \in X \cap Y$ and $X \perp Y$, then $\|\mathbf{x}\|^2 = \mathbf{x}^T \mathbf{x} = 0$ and hence $\mathbf{x} = \mathbf{0}$.   ◀

▶ *Proof of (2).* If $\mathbf{x} \in Y^\perp$ and α is a scalar, then for any $\mathbf{y} \in Y$,
$$(\alpha \mathbf{x})^T \mathbf{y} = \alpha(\mathbf{x}^T \mathbf{y}) = \alpha \cdot 0 = 0$$
Therefore, $\alpha \mathbf{x} \in Y^\perp$. If $\mathbf{x}_1$ and $\mathbf{x}_2$ are elements of $Y^\perp$, then
$$(\mathbf{x}_1 + \mathbf{x}_2)^T \mathbf{y} = \mathbf{x}_1^T \mathbf{y} + \mathbf{x}_2^T \mathbf{y} = 0 + 0 = 0$$
for each $\mathbf{y} \in Y$. Hence $\mathbf{x}_1 + \mathbf{x}_2 \in Y^\perp$. Therefore, $Y^\perp$ is a subspace of R^n.   ◀

Fundamental Subspaces

Let A be an $m \times n$ matrix. We saw in Chapter 3 that a vector $\mathbf{b} \in R^m$ is in the column space of A if and only if $\mathbf{b} = A\mathbf{x}$ for some $\mathbf{x} \in R^n$. If we think of A as an operator mapping R^n into R^m, then the column space of A is the same as the range of A. Let us denote the range of A by $R(A)$. Thus

$$R(A) = \{\mathbf{b} \in R^m \mid \mathbf{b} = A\mathbf{x} \quad \text{for some} \quad \mathbf{x} \in R^n\}$$
$$= \text{the column space of } A$$

The column space of A^T, $R(A^T)$, is a subspace of R^n:

$$R(A^T) = \left\{ \mathbf{y} \in R^n \mid \mathbf{y} = A^T\mathbf{x} \quad \text{for some} \quad \mathbf{x} \in R^m \right\}$$

The column space of $R(A^T)$ is essentially the same as the row space of A except that it consists of vectors in R^n ($n \times 1$ matrices) rather than n-tuples. Thus $\mathbf{y} \in R(A^T)$ if and only if $\mathbf{y}^T$ is in the row space of A. We have seen that $R(A^T) \perp N(A)$. The following theorem shows that $N(A)$ is actually the orthogonal complement of $R(A^T)$.

▶ **THEOREM 5.2.1 (Fundamental Subspaces Theorem)** *If A is an $m \times n$ matrix, then $N(A) = R(A^T)^\perp$ and $N(A^T) = R(A)^\perp$.*

▶ **Proof.** We have already seen that $N(A) \perp R(A^T)$, and this implies that $N(A) \subset R(A^T)^\perp$. On the other hand, if $\mathbf{x}$ is any vector in $R(A^T)^\perp$, then $\mathbf{x}$ is orthogonal to each of the column vectors of A^T and, consequently, $A\mathbf{x} = \mathbf{0}$. Thus $\mathbf{x}$ must be an element of $N(A)$ and hence $N(A) = R(A^T)^\perp$. This proof does not depend on the dimensions of A. In particular, the result will also hold for the matrix $B = A^T$. Thus
$$N(A^T) = N(B) = R(B^T)^\perp = R(A)^\perp \qquad \blacktriangleleft$$

EXAMPLE 3. Let

$$A = \begin{bmatrix} 1 & 0 \\ 2 & 0 \end{bmatrix}$$

The column space of A consists of all vectors of the form

$$\begin{bmatrix} \alpha \\ 2\alpha \end{bmatrix} = \alpha \begin{bmatrix} 1 \\ 2 \end{bmatrix}$$

Note that if $\mathbf{x}$ is any vector in R^n and $\mathbf{b} = A\mathbf{x}$ then

$$\mathbf{b} = \begin{bmatrix} 1 & 0 \\ 2 & 0 \end{bmatrix} \begin{bmatrix} x_1 \\ x_2 \end{bmatrix} = \begin{bmatrix} 1x_1 \\ 2x_1 \end{bmatrix} = x_1 \begin{bmatrix} 1 \\ 2 \end{bmatrix}$$

The nullspace of A^T consists of all vectors of the form $\beta(-2, 1)^T$. Since $(1, 2)^T$ and $(-2, 1)^T$ are orthogonal, it follows that every vector in $R(A)$ will be orthogonal to every vector in $N(A^T)$. The same relationship holds between $R(A^T)$ and $N(A)$. $R(A^T)$ consists of vectors of the form $\alpha\mathbf{e}_1$, and $N(A)$ consists of all vectors of the form $\beta\mathbf{e}_2$. Since $\mathbf{e}_1$ and $\mathbf{e}_2$ are orthogonal, it follows that each vector in $R(A^T)$ is orthogonal to every vector in $N(A)$. $\qquad \blacktriangleleft$

Theorem 5.2.1 is one of the most important theorems in this chapter. In Section 3 we will see that the result $N(A^T) = R(A)^{\perp}$ provides a key to solving least squares problems. For the present we will use Theorem 5.2.1 to prove the following theorem, which, in turn, will be used to establish two more important results about orthogonal subspaces.

▶ **THEOREM 5.2.2** *If S is a subspace of R^n, then $\dim S + \dim S^{\perp} = n$. Furthermore, if $\{\mathbf{x}_1, \ldots, \mathbf{x}_r\}$ is a basis for S and $\{\mathbf{x}_{r+1}, \ldots, \mathbf{x}_n\}$ is a basis for $S^{\perp}$, then $\{\mathbf{x}_1, \ldots, \mathbf{x}_r, \mathbf{x}_{r+1}, \ldots, \mathbf{x}_n\}$ is a basis for R^n.*

▶ *Proof.* If $S = \{\mathbf{0}\}$, then $S^{\perp} = R^n$ and

$$\dim S + \dim S^{\perp} = 0 + n = n$$

If $S \neq \{\mathbf{0}\}$, then let $\{\mathbf{x}_1, \ldots, \mathbf{x}_r\}$ be a basis for S and define X to be an $r \times n$ matrix whose ith row is $\mathbf{x}_i^T$ for each i. By construction, the matrix X has rank r and $R(X^T) = S$. By Theorem 5.2.1,

$$S^{\perp} = R(X^T)^{\perp} = N(X)$$

It follows from Theorem 3.6.5 that

$$\dim S^{\perp} = \dim N(X) = n - r$$

To show that $\{\mathbf{x}_1, \ldots, \mathbf{x}_r, \mathbf{x}_{r+1}, \ldots, \mathbf{x}_n\}$ is a basis for R^n, it suffices to show that the n vectors are linearly independent. Suppose that

$$c_1 \mathbf{x}_1 + \cdots + c_r \mathbf{x}_r + c_{r+1} \mathbf{x}_{r+1} + \cdots + c_n \mathbf{x}_n = \mathbf{0}$$

Let $\mathbf{y} = c_1 \mathbf{x}_1 + \cdots + c_r \mathbf{x}_r$ and $\mathbf{z} = c_{r+1} \mathbf{x}_{r+1} + \cdots + c_n \mathbf{x}_n$. We then have

$$\mathbf{y} + \mathbf{z} = \mathbf{0}$$

$$\mathbf{y} = -\mathbf{z}$$

Thus, $\mathbf{y}$ and $\mathbf{z}$ are both elements of $S \cap S^{\perp}$. But $S \cap S^{\perp} = \{\mathbf{0}\}$. Therefore,

$$c_1 \mathbf{x}_1 + \cdots + c_r \mathbf{x}_r = \mathbf{0}$$

$$c_{r+1} \mathbf{x}_{r+1} + \cdots + c_n \mathbf{x}_n = \mathbf{0}$$

Since $\mathbf{x}_1, \ldots, \mathbf{x}_r$ are linearly independent

$$c_1 = c_2 = \cdots = c_r = 0$$

Similarly, $\mathbf{x}_{r+1}, \ldots, \mathbf{x}_n$ are linearly independent and hence

$$c_{r+1} = c_{r+2} = \cdots = c_n = 0$$

Therefore, $\mathbf{x}_1, \mathbf{x}_2, \ldots, \mathbf{x}_n$ are linearly independent and hence form a basis for R^n. ◀

Given a subspace S of R^n, we will use Theorem 5.2.2 to prove that each $\mathbf{x} \in R^n$ can be expressed uniquely as a sum $\mathbf{y} + \mathbf{z}$, where $\mathbf{y} \in S$ and $\mathbf{z} \in S^{\perp}$.

▶ **DEFINITION** If U and V are subspaces of a vector space W and each $\mathbf{w} \in W$ can be written uniquely as a sum $\mathbf{u} + \mathbf{v}$, where $\mathbf{u} \in U$ and $\mathbf{v} \in V$, then we say that W is a **direct sum** of U and V, and we write $W = U \oplus V$. ◀

▶ **THEOREM 5.2.3** *If S is a subspace of R^n, then*

$$R^n = S \oplus S^{\perp}$$

▶*Proof.* The result is trivial if either $S = \{0\}$ or $S = R^n$. In the case where $\dim S = r, 0 < r < n$, it follows from Theorem 5.2.2 that each vector $\mathbf{x} \in R^n$ can be represented in the form

$$\mathbf{x} = c_1\mathbf{x}_1 + \cdots + c_r\mathbf{x}_r + c_{r+1}\mathbf{x}_{r+1} + \cdots + c_n\mathbf{x}_n$$

where $\{\mathbf{x}_1, \ldots, \mathbf{x}_r\}$ is a basis for S and $\{\mathbf{x}_{r+1}, \ldots, \mathbf{x}_n\}$ is a basis for $S^\perp$. If we let

$$\mathbf{u} = c_1\mathbf{x}_1 + \cdots + c_r\mathbf{x}_r \qquad \text{and} \qquad \mathbf{v} = c_{r+1}\mathbf{x}_{r+1} + \cdots + c_n\mathbf{x}_n$$

then $\mathbf{u} \in S$, $\mathbf{v} \in S^\perp$, and $\mathbf{x} = \mathbf{u} + \mathbf{v}$. To show uniqueness, suppose that $\mathbf{x}$ can also be written as a sum $\mathbf{y} + \mathbf{z}$, where $\mathbf{y} \in S$ and $\mathbf{z} \in S^\perp$. Thus

$$\mathbf{u} + \mathbf{v} = \mathbf{x} = \mathbf{y} + \mathbf{z}$$

$$\mathbf{u} - \mathbf{y} = \mathbf{z} - \mathbf{v}$$

But $\mathbf{u} - \mathbf{y} \in S$ and $\mathbf{z} - \mathbf{v} \in S^\perp$, so each is in $S \cap S^\perp$. Since

$$S \cap S^\perp = \{0\}$$

it follows that

$$\mathbf{u} = \mathbf{y} \qquad \text{and} \qquad \mathbf{v} = \mathbf{z} \qquad ◀$$

▶ **THEOREM 5.2.4** *If S is a subspace of R^n, then $(S^\perp)^\perp = S$.*

▶*Proof.* If $\mathbf{x} \in S$, then $\mathbf{x}$ is orthogonal to each $\mathbf{y}$ in $S^\perp$. Therefore, $\mathbf{x} \in (S^\perp)^\perp$ and hence $S \subset (S^\perp)^\perp$. On the other hand, suppose that $\mathbf{z}$ is an arbitrary element of $(S^\perp)^\perp$. By Theorem 5.2.3, we can write $\mathbf{z}$ as a sum $\mathbf{u} + \mathbf{v}$, where $\mathbf{u} \in S$ and $\mathbf{v} \in S^\perp$. Since $\mathbf{v} \in S^\perp$, it is orthogonal to both $\mathbf{u}$ and $\mathbf{z}$. It follows then that

$$0 = \mathbf{v}^T\mathbf{z} = \mathbf{v}^T\mathbf{u} + \mathbf{v}^T\mathbf{v} = \mathbf{v}^T\mathbf{v}$$

and, consequently, $\mathbf{v} = \mathbf{0}$. Therefore, $\mathbf{z} = \mathbf{u} \in S$ and hence $S = (S^\perp)^\perp$. ◀

It follows from Theorem 5.2.4 that, if T is the orthogonal complement of a subspace S, then S is the orthogonal complement of T, and we may say simply that S and T are orthogonal complements. In particular, it follows from Theorem 5.2.1 that $N(A)$ and $R(A^T)$ are orthogonal complements of each other and $N(A^T)$ and $R(A)$ are orthogonal complements. Hence we may write

$$N(A)^\perp = R(A^T) \qquad \text{and} \qquad N(A^T)^\perp = R(A)$$

Recall that the system $A\mathbf{x} = \mathbf{b}$ is consistent if and only if $\mathbf{b} \in R(A)$. Since $R(A) = N(A^T)^\perp$, we have the following result, which may be considered a corollary to Theorem 5.2.1.

▶ **COROLLARY 5.2.5** *If A is an $m \times n$ matrix and $\mathbf{b} \in R^m$, then either there is a vector $\mathbf{x} \in R^n$ such that $A\mathbf{x} = \mathbf{b}$ or there is a vector $\mathbf{y} \in R^m$ such that $A^T\mathbf{y} = \mathbf{0}$ and $\mathbf{y}^T\mathbf{b} \neq 0$.* ◀

Corollary 5.2.5 is illustrated in Figure 5.2.2 for the case where $R(A)$ is a two-dimensional subspace of R^3. The angle θ in the figure will be a right angle if and only if $\mathbf{b} \in R(A)$.

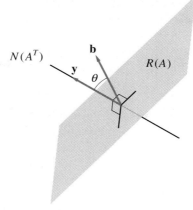

FIGURE 5.2.2

EXAMPLE 4. Let

$$A = \begin{bmatrix} 1 & 1 & 2 \\ 0 & 1 & 1 \\ 1 & 3 & 4 \end{bmatrix}$$

Find the bases for $N(A)$, $R(A^T)$, $N(A^T)$, and $R(A)$.

SOLUTION. We can find bases for $N(A)$ and $R(A^T)$ by transforming A into reduced row echelon form.

$$\begin{bmatrix} 1 & 1 & 2 \\ 0 & 1 & 1 \\ 1 & 3 & 4 \end{bmatrix} \rightarrow \begin{bmatrix} 1 & 1 & 2 \\ 0 & 1 & 1 \\ 0 & 2 & 2 \end{bmatrix} \rightarrow \begin{bmatrix} 1 & 0 & 1 \\ 0 & 1 & 1 \\ 0 & 0 & 0 \end{bmatrix}$$

Since $(1, 0, 1)$ and $(0, 1, 1)$ form a basis for the row space of A, it follows that $(1, 0, 1)^T$ and $(0, 1, 1)^T$ form a basis for $R(A^T)$. If $\mathbf{x} \in N(A)$, it follows from the reduced row echelon form of A that

$$x_1 + x_3 = 0$$
$$x_2 + x_3 = 0$$

Thus

$$x_1 = x_2 = -x_3$$

Setting $x_3 = \alpha$, we see that $N(A)$ consists of all vectors of the form $\alpha(-1, -1, 1)^T$. Note that $(-1, -1, 1)^T$ is orthogonal to $(1, 0, 1)^T$ and $(0, 1, 1)^T$.

To find bases for $R(A)$ and $N(A^T)$, reduce A^T to reduced row echelon form.

$$\begin{bmatrix} 1 & 0 & 1 \\ 1 & 1 & 3 \\ 2 & 1 & 4 \end{bmatrix} \rightarrow \begin{bmatrix} 1 & 0 & 1 \\ 0 & 1 & 2 \\ 0 & 1 & 2 \end{bmatrix} \rightarrow \begin{bmatrix} 1 & 0 & 1 \\ 0 & 1 & 2 \\ 0 & 0 & 0 \end{bmatrix}$$

Thus $(1, 0, 1)^T$ and $(0, 1, 2)^T$ form a basis for $R(A)$. If $\mathbf{x} \in N(A^T)$, then $x_1 = -x_3$, $x_2 = -2x_3$. Thus $N(A^T)$ is the subspace of R^3 spanned by $(-1, -2, 1)^T$. Note that $(-1, -2, 1)^T$ is orthogonal to $(1, 0, 1)^T$ and $(0, 1, 2)^T$.   ◀

We saw in Chapter 3 that the row space and the column space have the same dimension. If A has rank r, then

$$\dim R(A) = \dim R(A^T) = r$$

Actually, A can be used to establish a one-to-one correspondence between $R(A^T)$ and $R(A)$.

We can think of an $m \times n$ matrix A as a linear transformation from R^n to R^m.

$$\mathbf{x} \in R^n \rightarrow A\mathbf{x} \in R^m$$

Since $R(A^T)$ and $N(A)$ are orthogonal complements in R^n,

$$R^n = R(A^T) \oplus N(A)$$

Each vector $\mathbf{x} \in R^n$ can be written as a sum

$$\mathbf{x} = \mathbf{y} + \mathbf{z}, \qquad \mathbf{y} \in R(A^T), \qquad \mathbf{z} \in N(A)$$

It follows that

$$A\mathbf{x} = A\mathbf{y} + A\mathbf{z} = A\mathbf{y} \qquad \text{for each } \mathbf{x} \in R^n$$

and hence

$$R(A) = \left\{ A\mathbf{x} \mid \mathbf{x} \in R^n \right\} = \left\{ A\mathbf{y} \mid \mathbf{y} \in R(A^T) \right\}$$

Thus, if we restrict the domain of A to $R(A^T)$, then A maps $R(A^T)$ onto $R(A)$. Furthermore, the mapping is one-to-one. Indeed, if $\mathbf{x}_1, \mathbf{x}_2 \in R(A^T)$ and

$$A\mathbf{x}_1 = A\mathbf{x}_2$$

then

$$A(\mathbf{x}_1 - \mathbf{x}_2) = \mathbf{0}$$

and hence

$$\mathbf{x}_1 - \mathbf{x}_2 \in R(A^T) \cap N(A)$$

Since $R(A^T) \cap N(A) = \{\mathbf{0}\}$, it follows that $\mathbf{x}_1 = \mathbf{x}_2$. Thus we can think of A as determining a one-to-one correspondence between $R(A^T)$ and $R(A)$. Since each $\mathbf{b} \in R(A)$ corresponds to exactly one $\mathbf{y} \in R(A^T)$, we can define an inverse transformation from $R(A)$ to $R(A^T)$. Indeed, every $m \times n$ matrix A is invertible when viewed as a linear transformation from $R(A^T)$ to $R(A)$.

EXAMPLE 5. Let $A = \begin{bmatrix} 2 & 0 & 0 \\ 0 & 3 & 0 \end{bmatrix}$. $R(A^T)$ is spanned by $\mathbf{e}_1$ and $\mathbf{e}_2$, and $N(A)$ is spanned by $\mathbf{e}_3$. Any vector $\mathbf{x} \in R^3$ can be written as a sum

$$\mathbf{x} = \mathbf{y} + \mathbf{z}$$

where

$$\mathbf{y} = (x_1, x_2, 0)^T \in R(A^T) \quad \text{and} \quad \mathbf{z} = (0, 0, x_3)^T \in N(A)$$

If we restrict ourselves to vectors $\mathbf{y} \in R(A^T)$, then

$$\mathbf{y} = \begin{bmatrix} x_1 \\ x_2 \\ 0 \end{bmatrix} \to A\mathbf{y} = \begin{bmatrix} 2x_1 \\ 3x_2 \end{bmatrix}$$

In this case $R(A) = R^2$, and the inverse transformation from $R(A)$ to $R(A^T)$ is defined by

$$\mathbf{b} = \begin{bmatrix} b_1 \\ b_2 \end{bmatrix} \to \begin{bmatrix} \frac{1}{2}b_1 \\ \frac{1}{3}b_2 \\ 0 \end{bmatrix} \quad \blacktriangleleft$$

EXERCISES

1. For each of the following matrices, determine a basis for each of the subspaces $R(A^T)$, $N(A)$, $R(A)$, and $N(A^T)$.

(a) $A = \begin{bmatrix} 3 & 4 \\ 6 & 8 \end{bmatrix}$ (b) $A = \begin{bmatrix} 1 & 3 & 1 \\ 2 & 4 & 0 \end{bmatrix}$

(c) $A = \begin{bmatrix} 4 & -2 \\ 1 & 3 \\ 2 & 1 \\ 3 & 4 \end{bmatrix}$ (d) $A = \begin{bmatrix} 1 & 0 & 0 & 0 \\ 0 & 1 & 1 & 1 \\ 0 & 0 & 1 & 1 \\ 1 & 1 & 2 & 2 \end{bmatrix}$

2. Let S be the subspace of R^3 spanned by $\mathbf{x} = (1, -1, 1)^T$.

(a) Find a basis for $S^\perp$.

(b) Give a geometrical description of S and $S^\perp$.

3. (a) Let S be the subspace of R^3 spanned by the vectors $\mathbf{x} = (x_1, x_2, x_3)^T$ and $\mathbf{y} = (y_1, y_2, y_3)^T$. Let

$$A = \begin{bmatrix} x_1 & x_2 & x_3 \\ y_1 & y_2 & y_3 \end{bmatrix}$$

Show that $S^\perp = N(A)$.

(b) Find the orthogonal complement of the subspace of R^3 spanned by $(1, 2, 1)^T$ and $(1, -1, 2)^T$.

4. Let S be the subspace of R^4 spanned by $\mathbf{x}_1 = (1, 0, -2, 1)^T$ and $\mathbf{x}_2 = (0, 1, 3, -2)^T$. Find a basis for $S^\perp$.

5. Let $P_1 = (1, 1, 1)$, $P_2 = (2, 4, -1)$, and $P_3 = (0, -1, 5)$.

(a) Find a nonzero vector $\mathbf{N}$ that is orthogonal to $\overline{P_1 P_2}$ and $\overline{P_1 P_3}$.

(b) Find the equation of the plane determined by the three points.

6. Is it possible for a matrix to have the vector $(3, 1, 2)$ in its row space and $(2, 1, 1)^T$ in its nullspace? Explain.

7. Let $\mathbf{a}_j$ be a nonzero column vector of an $m \times n$ matrix A. Is it possible for $\mathbf{a}_j$ to be in $N(A^T)$? Explain.

8. Let S be the subspace of R^n spanned by the vectors $\mathbf{x}_1, \mathbf{x}_2, \dots, \mathbf{x}_k$. Show that $\mathbf{y} \in S^\perp$ if and only if $\mathbf{y} \perp \mathbf{x}_i$ for $i = 1, \dots, k$.

9. If A is an $m \times n$ matrix of rank r, what are the dimensions of $N(A)$ and $N(A^T)$? Explain.

10. Prove Corollary 5.2.5.

11. Prove: If A is an $m \times n$ matrix and $\mathbf{x} \in R^n$, then either $A\mathbf{x} = \mathbf{0}$ or there exists $\mathbf{y} \in R(A^T)$ such that $\mathbf{x}^T \mathbf{y} \neq 0$. Draw a picture similar to Figure 5.2.2 to illustrate this result geometrically for the case where $N(A)$ is a two-dimensional subspace of R^3.

12. Let A be an $m \times n$ matrix. Explain why the following are true.

(a) Any vector $\mathbf{x}$ in R^n can be uniquely written as a sum $\mathbf{y} + \mathbf{z}$, where $\mathbf{y} \in N(A)$ and $\mathbf{z} \in R(A^T)$.

(b) Any vector $\mathbf{b} \in R^m$ can be uniquely written as a sum $\mathbf{u} + \mathbf{v}$, where $\mathbf{u} \in N(A^T)$ and $\mathbf{v} \in R(A)$.

13. Let A be an $m \times n$ matrix. Show that:

(a) If $\mathbf{x} \in N(A^T A)$, then $A\mathbf{x}$ is in both $R(A)$ and $N(A^T)$.

(b) $N(A^T A) = N(A)$.

(c) A and $A^T A$ have the same rank.

(d) If A has linearly independent columns, then $A^T A$ is nonsingular.

14. Let A be an $m \times n$ matrix, B an $n \times r$ matrix, and $C = AB$. Show that:

(a) $N(B)$ is a subspace of $N(C)$.

(b) $N(C)^\perp$ is a subspace of $N(B)^\perp$ and, consequently, $R(C^T)$ is a subspace of $R(B^T)$.

15. Let U and V be subspaces of a vector space W. If $W = U \oplus V$, show that $U \cap V = \{\mathbf{0}\}$.

16. Let A be an $m \times n$ matrix of rank r and let $\{\mathbf{x}_1, \dots, \mathbf{x}_r\}$ be a basis for $R(A^T)$. Show that $\{A\mathbf{x}_1, \dots, A\mathbf{x}_r\}$ is a basis for $R(A)$.

3   LEAST SQUARES PROBLEMS

A standard technique in mathematical and statistical modeling is to find a *least squares* fit to a set of data points in the plane. The least squares curve is usually the graph of a standard type of function such as a linear function, a polynomial, or a trigonometric polynomial. Since the data may include errors in measurement or experiment inaccuracies, we do not require the curve to pass through all the data points. Instead we require the curve to provide an optimal approximation in the sense that the sum of squares of errors between the y-values of the data points and the corresponding y-values of the approximating curve are minimized.

The technique of least squares was developed independently by A. M. Legendre and Carl Friedrich Gauss. The first paper on the subject was published by Legendre in 1806, although there is clear evidence that Gauss had discovered it as a student nine years prior to Legendre's paper and had used the method to do astronomical calculations. Figure 5.3.1 shows a portrait of Gauss.

FIGURE 5.3.1

APPLICATION 1: ASTRONOMY–THE CERES ORBIT OF GAUSS

On January 1, 1801, the Italian astronomer G. Piazzi discovered the asteroid Ceres. He was able to track the asteroid for six weeks, but it was lost due to interference caused by the sun. A number of leading astronomers published papers predicting the orbit of the asteroid. Gauss also published a forecast, but his predicted orbit differed considerably from the others. Ceres was relocated by one observer on December 7 and by another on January 1, 1802. In both cases the position was very close to that predicted by Gauss. Gauss won instant fame in astronomical circles and for a time was more well known as an astronomer than as a mathematician. The key to his success was the use of the method of least squares.

Least Squares Solutions to Overdetermined Systems

A least squares problem can generally be formulated as an overdetermined linear system of equations. Recall that an overdetermined system is one involving more equations than unknowns. Such systems are usually inconsistent. Thus, given an $m \times n$ system $A\mathbf{x} = \mathbf{b}$ with $m > n$, we cannot expect in general to find a vector $\mathbf{x} \in R^n$ for which $A\mathbf{x}$ equals $\mathbf{b}$. Instead, we can look for a vector $\mathbf{x}$ for which $A\mathbf{x}$ is "closest" to $\mathbf{b}$. As you might expect, orthogonality plays an important role in finding such an $\mathbf{x}$.

Given a system of equations $A\mathbf{x} = \mathbf{b}$, where A is an $m \times n$ matrix with $m > n$ and $\mathbf{b} \in R^m$, then for each $\mathbf{x} \in R^n$ we can form a *residual*

$$r(\mathbf{x}) = \mathbf{b} - A\mathbf{x}$$

The distance between $\mathbf{b}$ and $A\mathbf{x}$ is given by

$$\|\mathbf{b} - A\mathbf{x}\| = \|r(\mathbf{x})\|$$

We wish to find a vector $\mathbf{x} \in R^n$ for which $\|r(\mathbf{x})\|$ will be a minimum. Minimizing $\|r(\mathbf{x})\|$ is equivalent to minimizing $\|r(\mathbf{x})\|^2$. A vector $\hat{\mathbf{x}}$ that accomplishes this is said to be a *least squares solution* to the system $A\mathbf{x} = \mathbf{b}$.

If $\hat{\mathbf{x}}$ is a least squares solution to the system $A\mathbf{x} = \mathbf{b}$ and $\mathbf{p} = A\hat{\mathbf{x}}$, then $\mathbf{p}$ is a vector in the column space of A that is closest to $\mathbf{b}$. The following theorem guarantees that such a closest vector $\mathbf{p}$ not only exists, but is unique. Additionally, it provides an important characterization of the closest vector.

▶ **THEOREM 5.3.1** *Let S be a subspace of R^m. For each $\mathbf{b} \in R^m$ there is a unique element $\mathbf{p}$ of S that is closest to $\mathbf{b}$, that is,*

$$\|\mathbf{b} - \mathbf{y}\| > \|\mathbf{b} - \mathbf{p}\|$$

for any $\mathbf{y} \neq \mathbf{p}$ in S. Furthermore, a given vector $\mathbf{p}$ in S will be closest to a given vector $\mathbf{b} \in R^m$ if and only if $\mathbf{b} - \mathbf{p} \in S^\perp$.

▶*Proof.* Since $R^m = S \oplus S^\perp$, each element $\mathbf{b}$ in R^m can be expressed uniquely as a sum

$$\mathbf{b} = \mathbf{p} + \mathbf{z}$$

where $\mathbf{p} \in S$ and $\mathbf{z} \in S^\perp$. If $\mathbf{y}$ is any other element of S, then

$$\|\mathbf{b} - \mathbf{y}\|^2 = \|(\mathbf{b} - \mathbf{p}) + (\mathbf{p} - \mathbf{y})\|^2$$

Since $\mathbf{p} - \mathbf{y} \in S$ and $\mathbf{b} - \mathbf{p} = \mathbf{z} \in S^\perp$, it follows from the Pythagorean Law that

$$\|\mathbf{b} - \mathbf{y}\|^2 = \|\mathbf{b} - \mathbf{p}\|^2 + \|\mathbf{p} - \mathbf{y}\|^2$$

Therefore,

$$\|\mathbf{b} - \mathbf{y}\| > \|\mathbf{b} - \mathbf{p}\|$$

Thus, if $\mathbf{p} \in S$ and $\mathbf{b} - \mathbf{p} \in S^{\perp}$, then $\mathbf{p}$ is the element of S that is closest to $\mathbf{b}$. Conversely, if $\mathbf{q} \in S$ and $\mathbf{b} - \mathbf{q} \notin S^{\perp}$, then $\mathbf{q} \neq \mathbf{p}$, and it follows from the preceding argument (with $\mathbf{y} = \mathbf{q}$) that

$$\|\mathbf{b} - \mathbf{q}\| > \|\mathbf{b} - \mathbf{p}\| \qquad \blacktriangleleft$$

In the special case that $\mathbf{b}$ is in the subspace S to begin with, we have

$$\mathbf{b} = \mathbf{p} + \mathbf{z}, \qquad \mathbf{p} \in S, \quad \mathbf{z} \in S^{\perp}$$

and

$$\mathbf{b} = \mathbf{b} + \mathbf{0}$$

By the uniqueness of the direct sum representation,

$$\mathbf{p} = \mathbf{b} \qquad \text{and} \qquad \mathbf{z} = \mathbf{0}$$

A vector $\hat{\mathbf{x}}$ will be a solution to the least squares problem $A\mathbf{x} = \mathbf{b}$ if and only if $\mathbf{p} = A\hat{\mathbf{x}}$ is the vector in $R(A)$ that is closest to $\mathbf{b}$. The vector $\mathbf{p}$ is said to be the *projection of $\mathbf{b}$ onto $R(A)$*. It follows from Theorem 5.3.1 that

$$\mathbf{b} - \mathbf{p} = \mathbf{b} - A\hat{\mathbf{x}} = r(\hat{\mathbf{x}})$$

must be an element of $R(A)^{\perp}$. Thus $\hat{\mathbf{x}}$ is a solution to the least squares problem if and only if

$$(1) \qquad\qquad\qquad r(\hat{\mathbf{x}}) \in R(A)^{\perp}$$

(see Figure 5.3.2).

How do we find a vector $\hat{\mathbf{x}}$ satisfying (1)? The key to solving the least squares problem is provided by Theorem 5.2.1, which states that

$$R(A)^{\perp} = N(A^{T})$$

A vector $\hat{\mathbf{x}}$ will be a least squares solution to the system $A\mathbf{x} = \mathbf{b}$ if and only if

$$r(\hat{\mathbf{x}}) \in N(A^{T})$$

FIGURE 5.3.2

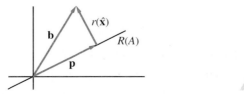

(a) $\mathbf{b} \in R^{2}$ and A is a 2 x 1 matrix of rank 1.   (b) $\mathbf{b} \in R^{2}$ and A is a 3 x 2 matrix of rank 2.

or, equivalently,

$$0 = A^T r(\hat{\mathbf{x}}) = A^T(\mathbf{b} - A\hat{\mathbf{x}})$$

Thus, to solve the least squares problem $A\mathbf{x} = \mathbf{b}$, we must solve

$$(2) \qquad\qquad A^T A\mathbf{x} = A^T \mathbf{b}$$

Equation (2) represents an $n \times n$ system of linear equations. These equations are called the *normal equations*. In general, it is possible to have more than one solution to the normal equations; however, if $\hat{\mathbf{x}}$ and $\hat{\mathbf{y}}$ are both solutions, then, since the projection $\mathbf{p}$ of $\mathbf{b}$ onto $R(A)$ is unique,

$$A\hat{\mathbf{x}} = A\hat{\mathbf{y}} = \mathbf{p}$$

The following theorem characterizes the conditions under which the least squares problem $A\mathbf{x} = \mathbf{b}$ will have a unique solution.

▶ **THEOREM 5.3.2** *If A is an $m \times n$ matrix of rank n, the normal equations*
$$A^T A\mathbf{x} = A^T \mathbf{b}$$
have a unique solution
$$\hat{\mathbf{x}} = (A^T A)^{-1} A^T \mathbf{b}$$
and $\hat{\mathbf{x}}$ is the unique least squares solution to the system $A\mathbf{x} = \mathbf{b}$.

▶ *Proof.* We will first show that $A^T A$ is nonsingular. To prove this, let $\mathbf{z}$ be a solution to

$$(3) \qquad\qquad A^T A\mathbf{x} = \mathbf{0}$$

Then $A\mathbf{z} \in N(A^T)$. Clearly, $A\mathbf{z} \in R(A) = N(A^T)^\perp$. Since $N(A^T) \cap N(A^T)^\perp = \{\mathbf{0}\}$, it follows that $A\mathbf{z} = \mathbf{0}$. If A has rank n, the column vectors of A are linearly independent and, consequently, $A\mathbf{x} = \mathbf{0}$ has only the trivial solution. Thus $\mathbf{z} = \mathbf{0}$ and (3) has only the trivial solution. Therefore, by Theorem 1.4.3, $A^T A$ is nonsingular. It follows that $\hat{\mathbf{x}} = (A^T A)^{-1} A^T \mathbf{b}$ is the unique solution to the normal equations and, consequently, the unique least squares solution to the system $A\mathbf{x} = \mathbf{b}$. ◀

The projection vector

$$\mathbf{p} = A\hat{\mathbf{x}} = A(A^T A)^{-1} A^T \mathbf{b}$$

is the element of $R(A)$ that is closest to $\mathbf{b}$ in the least squares sense. The matrix $P = A(A^T A)^{-1} A^T$ is called the *projection matrix*.

APPLICATION 2: SPRING CONSTANTS

Hooke's law states that the force applied to a spring is proportional to the distance that the spring is stretched. Thus, if F is the force applied and x is the distance that the spring has been stretched, then $F = kx$. The proportionality constant k is called the *spring constant*.

Some physics students want to determine the spring constant for a given spring. They apply forces of 3, 5, and 8 pounds, which have the effect of stretching the spring 4, 7, and 11 inches, respectively. Using Hooke's law, they derive the following system of equations:

$$4k = 3$$
$$7k = 5$$
$$11k = 8$$

The system is clearly inconsistent since each of the equations yields a different value of k. Rather than use any one of these values, the students decide to compute the least squares solution to the system.

$$(4, 7, 11) \begin{bmatrix} 4 \\ 7 \\ 11 \end{bmatrix} (k) = (4, 7, 11) \begin{bmatrix} 3 \\ 5 \\ 8 \end{bmatrix}$$

$$186k = 135$$
$$k \approx 0.726$$

EXAMPLE I. Find the least squares solution to the system

$$x_1 + x_2 = 3$$
$$-2x_1 + 3x_2 = 1$$
$$2x_1 - x_2 = 2$$

SOLUTION. The normal equations for this system are

$$\begin{bmatrix} 1 & -2 & 2 \\ 1 & 3 & -1 \end{bmatrix} \begin{bmatrix} 1 & 1 \\ -2 & 3 \\ 2 & -1 \end{bmatrix} \begin{bmatrix} x_1 \\ x_2 \end{bmatrix} = \begin{bmatrix} 1 & -2 & 2 \\ 1 & 3 & -1 \end{bmatrix} \begin{bmatrix} 3 \\ 1 \\ 2 \end{bmatrix}$$

This simplifies to the 2×2 system

$$\begin{bmatrix} 9 & -7 \\ -7 & 11 \end{bmatrix} \begin{bmatrix} x_1 \\ x_2 \end{bmatrix} = \begin{bmatrix} 5 \\ 4 \end{bmatrix}$$

The solution to the 2×2 system is $\left(\frac{83}{50}, \frac{71}{50} \right)^T$. ◄

Scientists often collect data and try to find a functional relationship among the variables. For example, the data may involve temperatures $T_0, T_1, \ldots, T_n$ of a liquid measured at times $t_0, t_1, \ldots, t_n$, respectively. If the temperature T can be represented as a function of the time t, this function can be used to predict the temperatures at future times. If the data consist of $n + 1$ points in the plane, it is possible to find a polynomial of degree n or less passing through all the points. Such a polynomial is called an *interpolating polynomial*. Actually, since the data usually involve experimental error, there is no reason to require that the function pass through all the points. Indeed, lower-degree polynomials that do not pass through the points exactly usually give a truer description of the relationship between the variables. If, for example, the relationship between the variables is actually linear and the data involve slight errors, it would be disastrous to use an interpolating polynomial (see Figure 5.3.3).

Given a table of data

x	x_1	x_2	$\cdots$	x_m
y	y_1	y_2	$\cdots$	y_m

we wish to find a linear function

$$y = c_0 + c_1 x$$

that best fits the data in the least squares sense. If we require that

$$y_i = c_0 + c_1 x_i \qquad \text{for} \qquad i = 1, \ldots, m$$

we get a system of m equations in two unknowns.

(4)
$$\begin{pmatrix} 1 & x_1 \\ 1 & x_2 \\ \vdots & \vdots \\ 1 & x_m \end{pmatrix} \begin{pmatrix} c_0 \\ c_1 \end{pmatrix} = \begin{pmatrix} y_1 \\ y_2 \\ \vdots \\ y_m \end{pmatrix}$$

FIGURE 5.3.3

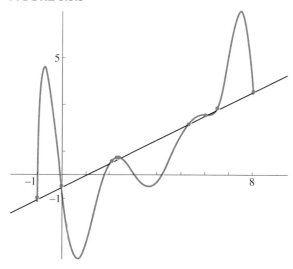

x	−1.00	0.00	2.10	2.30	2.40	5.30	6.00	6.50	8.00
y	−1.02	−0.52	0.55	0.70	0.70	2.13	2.52	2.82	3.54

The linear function whose coefficients are the least squares solution to (4) is said to be the best least squares fit to the data by a linear function.

EXAMPLE 2. Given the data

$$\begin{array}{c|ccc} x & 0 & 3 & 6 \\ \hline y & 1 & 4 & 5 \end{array}$$

Find the best least squares fit by a linear function.

SOLUTION. For this example the system (4) becomes

$$A\mathbf{c} = \mathbf{y}$$

where

$$A = \begin{bmatrix} 1 & 0 \\ 1 & 3 \\ 1 & 6 \end{bmatrix}, \qquad \mathbf{c} = \begin{bmatrix} c_0 \\ c_1 \end{bmatrix}, \qquad \text{and} \qquad \mathbf{y} = \begin{bmatrix} 1 \\ 4 \\ 5 \end{bmatrix}$$

The normal equations

$$A^T A \mathbf{c} = A^T \mathbf{y}$$

simplify to

(5)
$$\begin{bmatrix} 3 & 9 \\ 9 & 45 \end{bmatrix} \begin{bmatrix} c_0 \\ c_1 \end{bmatrix} = \begin{bmatrix} 10 \\ 42 \end{bmatrix}$$

The solution of this system is $\left(\frac{4}{3}, \frac{2}{3}\right)$. Thus the best linear least squares fit is given by

$$y = \tfrac{4}{3} + \tfrac{2}{3}x \qquad \blacktriangleleft$$

Example 2 could also have been solved using calculus. The residual $r(\mathbf{c})$ is given by

$$r(\mathbf{c}) = \mathbf{y} - A\mathbf{c}$$

and

$$\|r(\mathbf{c})\|^2 = \|\mathbf{y} - A\mathbf{c}\|^2$$
$$= [1 - (c_0 + 0c_1)]^2 + [4 - (c_0 + 3c_1)]^2 + [5 - (c_0 + 6c_1)]^2$$
$$= f(c_0, c_1)$$

Thus, $\|r(\mathbf{c})\|^2$ can be thought of as a function of two variables, $f(c_0, c_1)$. The minimum of this function will occur when its partials are zero.

$$\frac{\partial f}{\partial c_0} = -2(10 - 3c_0 - 9c_1) = 0$$

$$\frac{\partial f}{\partial c_1} = -6(14 - 3c_0 - 15c_1) = 0$$

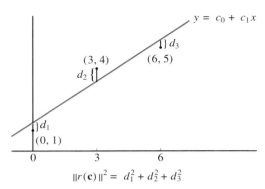

$$\|r(\mathbf{c})\|^2 = d_1^2 + d_2^2 + d_3^2$$

FIGURE 5.3.4

Dividing both equations through by 2 gives the same system as in (5) (see Figure 5.3.4).

If the data do not resemble a linear function, we could use a higher-degree polynomial. To find the coefficients $c_0, c_1, \ldots, c_n$ of the best least squares fit to the data

$$
\begin{array}{c|ccccc}
x & x_1 & x_2 & \cdots & x_m \\
\hline
y & y_1 & y_2 & \cdots & y_m
\end{array}
$$

by a polynomial of degree n, we must find the least squares solution to the system

(6)
$$
\begin{pmatrix}
1 & x_1 & x_1^2 & \cdots & x_1^n \\
1 & x_2 & x_2^2 & \cdots & x_2^n \\
\vdots & & & & \\
1 & x_m & x_m^2 & \cdots & x_m^n
\end{pmatrix}
\begin{pmatrix}
c_0 \\ c_1 \\ \vdots \\ c_n
\end{pmatrix}
=
\begin{pmatrix}
y_1 \\ y_2 \\ \vdots \\ y_m
\end{pmatrix}
$$

EXAMPLE 3. Find the best quadratic least squares fit to the data

$$
\begin{array}{c|cccc}
x & 0 & 1 & 2 & 3 \\
\hline
y & 3 & 2 & 4 & 4
\end{array}
$$

SOLUTION. For this example the system (6) becomes

$$
\begin{pmatrix}
1 & 0 & 0 \\
1 & 1 & 1 \\
1 & 2 & 4 \\
1 & 3 & 9
\end{pmatrix}
\begin{pmatrix}
c_0 \\ c_1 \\ c_2
\end{pmatrix}
=
\begin{pmatrix}
3 \\ 2 \\ 4 \\ 4
\end{pmatrix}
$$

Thus the normal equations are

$$
\begin{pmatrix}
1 & 1 & 1 & 1 \\
0 & 1 & 2 & 3 \\
0 & 1 & 4 & 9
\end{pmatrix}
\begin{pmatrix}
1 & 0 & 0 \\
1 & 1 & 1 \\
1 & 2 & 4 \\
1 & 3 & 9
\end{pmatrix}
\begin{pmatrix}
c_0 \\ c_1 \\ c_2
\end{pmatrix}
=
\begin{pmatrix}
1 & 1 & 1 & 1 \\
0 & 1 & 2 & 3 \\
0 & 1 & 4 & 9
\end{pmatrix}
\begin{pmatrix}
3 \\ 2 \\ 4 \\ 4
\end{pmatrix}
$$

These simplify to

$$\begin{bmatrix} 4 & 6 & 14 \\ 6 & 14 & 36 \\ 14 & 36 & 98 \end{bmatrix} \begin{bmatrix} c_0 \\ c_1 \\ c_2 \end{bmatrix} = \begin{bmatrix} 13 \\ 22 \\ 54 \end{bmatrix}$$

The solution to this system is $(2.75, -0.25, 0.25)$. The quadratic polynomial that gives the best least squares fit to the data is

$$p(x) = 2.75 - 0.25x + 0.25x^2 \qquad \blacktriangleleft$$

APPLICATION 3: COORDINATE METROLOGY

Many manufactured goods, such as rods, disks, and pipes, are circular in shape. A company will often employ quality control engineers to test whether items produced on the production line are meeting industrial standards. Sensing machines are used to record the coordinates of points on the perimeter of the manufactured products. To determine how close these points are to being circular, we can fit a least squares circle to the data and check to see how close the measured points are to the circle. (See Figure 5.3.5.)

In order to fit a circle

(7) $$(x - c_1)^2 + (y - c_2)^2 = r^2$$

to n sample pairs of coordinates $(x_1, y_1), (x_2, y_2), \ldots, (x_n, y_n)$, we must determine the center (c_1, c_2) and the radius r. Rewriting equation (7), we get

$$2xc_1 + 2yc_2 + (r^2 - c_1^2 - c_2^2) = x^2 + y^2$$

FIGURE 5.3.5

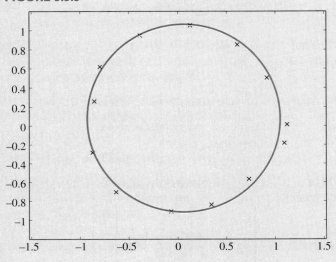

If we set $c_3 = r^2 - c_1^2 - c_2^2$, then the equation, takes the form

$$2xc_1 + 2yc_2 + c_3 = x^2 + y^2$$

Substituting each of the data points into this equation, we obtain the overdetermined system

$$
\begin{bmatrix}
2x_1 & 2y_1 & 1 \\
2x_2 & 2y_2 & 1 \\
\vdots & \vdots & \vdots \\
2x_n & 2y_n & 1
\end{bmatrix}
\begin{bmatrix}
c_1 \\
c_2 \\
c_3
\end{bmatrix}
=
\begin{bmatrix}
x_1^2 + y_1^2 \\
x_2^2 + y_2^2 \\
\vdots \\
x_n^2 + y_n^2
\end{bmatrix}
$$

Once we find the least squares solution $\mathbf{c}$, then the center of the least squares circle is (c_1, c_2) and the radius is determined by setting

$$r = \sqrt{c_3 + c_1^2 + c_2^2}$$

To measure how close the sampled points are to the circle, we can form a residual vector $\mathbf{r}$ by setting

$$r_i = r^2 - (x_i - c_1)^2 - (y_i - c_2)^2 \quad i = 1, \dots, n$$

We can then use $\|\mathbf{r}\|$ as a measure of how close the points are to the circle.

EXERCISES

1. Find the least squares solution to each of the following systems.

   (a) $\begin{aligned} x_1 + x_2 &= 3 \\ 2x_1 - 3x_2 &= 1 \\ 0x_1 + 0x_2 &= 2 \end{aligned}$

   (b) $\begin{aligned} -x_1 + x_2 &= 10 \\ 2x_1 + x_2 &= 5 \\ x_1 - 2x_2 &= 20 \end{aligned}$

   (c) $\begin{aligned} x_1 + x_2 + x_3 &= 4 \\ -x_1 + x_2 + x_3 &= 0 \\ - x_2 + x_3 &= 1 \\ x_1 \qquad + x_3 &= 2 \end{aligned}$

2. For each of your solutions $\hat{\mathbf{x}}$ in Exercise 1:

   (a) Determine the projection $\mathbf{p} = A\hat{\mathbf{x}}$.
   (b) Calculate the residual $r(\hat{\mathbf{x}})$.
   (c) Verify that $r(\hat{\mathbf{x}}) \in N(A^T)$.

3. For each of the following systems $A\mathbf{x} = \mathbf{b}$, find all least squares solutions.

   (a) $A = \begin{bmatrix} 1 & 2 \\ 2 & 4 \\ -1 & -2 \end{bmatrix}, \qquad \mathbf{b} = \begin{bmatrix} 3 \\ 2 \\ 1 \end{bmatrix}$

(b) $A = \begin{bmatrix} 1 & 1 & 3 \\ -1 & 3 & 1 \\ 1 & 2 & 4 \end{bmatrix}$, $\quad b = \begin{bmatrix} -2 \\ 0 \\ 8 \end{bmatrix}$

4. For each of the systems in Exercise 3, determine the projection **p** of **b** onto $R(A)$ and verify that $b - p$ is orthogonal to each of the column vectors of A.

5. (a) Find the best least squares fit by a linear function to the data

x	-1	0	1	2
y	0	1	3	9

   (b) Plot your linear function from part (a) along with the data on a coordinate system.

6. Find the best least squares fit to the data in Exercise 5 by a quadratic polynomial. Plot the points $x = -1, 0, 1, 2$ for your function and sketch the graph.

7. Given a collection of points $(x_1, y_1), (x_2, y_2), \ldots, (x_n, y_n)$, let

$$\mathbf{x} = (x_1, x_2, \ldots, x_n)^T \qquad \mathbf{y} = (y_1, y_2, \ldots, y_n)^T$$

$$\overline{x} = \frac{1}{n} \sum_{i=1}^{n} x_i \qquad \overline{y} = \frac{1}{n} \sum_{i=1}^{n} y_i$$

   and let $y = c_0 + c_1 x$ be the linear function that gives the best least squares fit to the points. Show that if $\overline{x} = 0$, then

$$c_0 = \overline{y} \qquad \text{and} \qquad c_1 = \frac{\mathbf{x}^T \mathbf{y}}{\mathbf{x}^T \mathbf{x}}$$

8. The point $(\overline{x}, \overline{y})$ is the *center of mass* for the collection of points in Exercise 7. Show that the least squares line must pass through the center of mass. [*Hint:* Use a change of variables $z = x - \overline{x}$ to translate the problem so that the new independent variable has mean 0.]

9. Let A be an $m \times n$ matrix of rank n and let $P = A(A^T A)^{-1} A^T$.

   (a) Show that $P\mathbf{b} = \mathbf{b}$ for every $\mathbf{b} \in R(A)$. Explain this in terms of projections.

   (b) If $\mathbf{b} \in R(A)^{\perp}$, show that $P\mathbf{b} = \mathbf{0}$.

   (c) Give a geometric illustration of parts (a) and (b) if $R(A)$ is a plane through the origin in R^3.

10. Let $P = A(A^T A)^{-1} A^T$, where A is an $m \times n$ matrix of rank n.

   (a) Show that $P^2 = P$.

   (b) Prove $P^k = P$ for $k = 1, 2, \ldots$.

   (c) Show that P is symmetric. [*Hint:* If B is nonsingular, then $(B^{-1})^T = (B^T)^{-1}$.]

11. Show that if

$$\begin{bmatrix} A & I \\ O & A^T \end{bmatrix} \begin{bmatrix} \hat{\mathbf{x}} \\ \mathbf{r} \end{bmatrix} = \begin{bmatrix} \mathbf{b} \\ \mathbf{0} \end{bmatrix}$$

then $\hat{\mathbf{x}}$ is a least squares solution to the system $A\mathbf{x} = \mathbf{b}$ and $\mathbf{r}$ is the residual vector.

12. Let $A \in R^{m \times n}$ and let $\hat{\mathbf{x}}$ be a solution to the least squares problem $A\mathbf{x} = \mathbf{b}$. Show that a vector $\mathbf{y} \in R^n$ will also be a solution if and only if $\mathbf{y} = \hat{\mathbf{x}} + \mathbf{z}$, for some vector $\mathbf{z} \in N(A)$. [*Hint:* $N(A^TA) = N(A)$.]

13. Find the equation of the circle that gives the best least squares circle fit to the points $(-1, -2)$, $(0, 2.4)$, $(1.1, -4)$, $(2.4, -1.6)$.

4 INNER PRODUCT SPACES

Scalar products are useful not only in R^n but in a wide variety of contexts. To generalize this concept to other vector spaces, we introduce the following definition.

Definition and Examples

▶ **DEFINITION** An **inner product** on a vector space V is an operation on V that assigns to each pair of vectors $\mathbf{x}$ and $\mathbf{y}$ in V a real number $\langle \mathbf{x}, \mathbf{y} \rangle$ satisfying the following conditions:

(i) $\langle \mathbf{x}, \mathbf{x} \rangle \geq 0$ with equality if and only if $\mathbf{x} = \mathbf{0}$.

(ii) $\langle \mathbf{x}, \mathbf{y} \rangle = \langle \mathbf{y}, \mathbf{x} \rangle$ for all $\mathbf{x}$ and $\mathbf{y}$ in V.

(iii) $\langle \alpha\mathbf{x} + \beta\mathbf{y}, \mathbf{z} \rangle = \alpha \langle \mathbf{x}, \mathbf{z} \rangle + \beta \langle \mathbf{y}, \mathbf{z} \rangle$ for all $\mathbf{x}, \mathbf{y}, \mathbf{z}$ in V and all scalars α and β. ◀

A vector space V with an inner product is called an **inner product space**.

The Vector Space R^n

The standard inner product for R^n is the scalar product

$$\langle \mathbf{x}, \mathbf{y} \rangle = \mathbf{x}^T \mathbf{y}$$

Given a vector $\mathbf{w}$ with positive entries, we could also define an inner product on R^n by

(1)
$$\langle \mathbf{x}, \mathbf{y} \rangle = \sum_{i=1}^{n} x_i y_i w_i$$

The entries w_i are referred to as *weights*.

The Vector Space $R^{m \times n}$

Given A and B in $R^{m \times n}$, we can define an inner product by

(2)
$$\langle A, B \rangle = \sum_{i=1}^{m} \sum_{j=1}^{n} a_{ij} b_{ij}$$

We leave it to the reader to verify that (2) does indeed define an inner product on $R^{m \times n}$.

The Vector Space C[a, b]

In $C[a, b]$ we may define an inner product by

$$(3) \qquad \langle f, g \rangle = \int_a^b f(x)g(x)\,dx$$

Note that

$$\langle f, f \rangle = \int_a^b (f(x))^2\,dx \geq 0$$

If $f(x_0) \neq 0$ for some x_0 in $[a, b]$, then, since $(f(x))^2$ is continuous, there exists a subinterval I of $[a, b]$ containing x_0 such that $(f(x))^2 \geq (f(x_0))^2/2$ for all x in I. If we let p represent the length of I, then it follows that

$$\langle f, f \rangle = \int_a^b (f(x))^2\,dx \geq \int_I (f(x))^2\,dx \geq \frac{(f(x_0))^2 p}{2} > 0$$

So if $\langle f, f \rangle = 0$, then $f(x)$ must be identically zero on $[a, b]$. We leave it to the reader to verify that (3) satisfies the other two conditions specified in the definition of an inner product.

If $w(x)$ is a positive continuous function on $[a, b]$, then

$$(4) \qquad \langle f, g \rangle = \int_a^b f(x)g(x)w(x)\,dx$$

also defines an inner product on $C[a, b]$. The function $w(x)$ is called a *weight function*. Thus it is possible to define many different inner products on $C[a, b]$.

The Vector Space P_n

Let $x_1, x_2, \ldots, x_n$ be distinct real numbers. For each pair of polynomials in P_n, define

$$(5) \qquad \langle p, q \rangle = \sum_{i=1}^n p(x_i)q(x_i)$$

It is easily seen that (5) satisfies conditions (ii) and (iii) of the definition of an inner product. To show that (i) holds, note that

$$\langle p, p \rangle = \sum_{i=1}^n (p(x_i))^2 \geq 0$$

If $\langle p, p \rangle = 0$, then $x_1, x_2, \ldots, x_n$ must be roots of $p(x) = 0$. Since $p(x)$ is of degree less than n, it must be the zero polynomial.

If $w(x)$ is a positive function, then

$$\langle p, q \rangle = \sum_{i=1}^n p(x_i)q(x_i)w(x_i)$$

also defines an inner product on P_n.

Basic Properties of Inner Product Spaces

The results presented in Section 1 for scalar products in R^n all generalize to inner product spaces. In particular, if $\mathbf{v}$ is a vector in an inner product space V, the *length* or *norm* of $\mathbf{v}$ is given by

$$\|\mathbf{v}\| = \sqrt{\langle \mathbf{v}, \mathbf{v}\rangle}$$

Two vectors $\mathbf{u}$ and $\mathbf{v}$ are said to be *orthogonal* if $\langle \mathbf{u}, \mathbf{v}\rangle = 0$. As in R^n, a pair of orthogonal vectors will satisfy the Pythagorean Law.

▶ **THEOREM 5.4.1 (The Pythagorean Law)** *If* $\mathbf{u}$ *and* $\mathbf{v}$ *are orthogonal vectors in an inner product space* V, *then*

$$\|\mathbf{u} + \mathbf{v}\|^2 = \|\mathbf{u}\|^2 + \|\mathbf{v}\|^2$$

▶ *Proof.*

$$\|\mathbf{u} + \mathbf{v}\|^2 = \langle \mathbf{u} + \mathbf{v}, \mathbf{u} + \mathbf{v}\rangle$$
$$= \langle \mathbf{u}, \mathbf{u}\rangle + 2\langle \mathbf{u}, \mathbf{v}\rangle + \langle \mathbf{v}, \mathbf{v}\rangle$$
$$= \|\mathbf{u}\|^2 + \|\mathbf{v}\|^2 \qquad ◀$$

Interpreted in R^2, this is just the familiar Pythagorean theorem as shown in Figure 5.4.1.

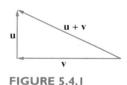

FIGURE 5.4.1

EXAMPLE 1. Consider the vector space $C[-1, 1]$ with inner product defined by (3). The vectors 1 and x are orthogonal since

$$\langle 1, x\rangle = \int_{-1}^{1} 1 \cdot x \, dx = 0$$

To determine the lengths of these vectors, we compute

$$\langle 1, 1\rangle = \int_{-1}^{1} 1 \cdot 1 \, dx = 2$$
$$\langle x, x\rangle = \int_{-1}^{1} x^2 \, dx = \frac{2}{3}$$

It follows that

$$\|1\| = (\langle 1, 1\rangle)^{1/2} = \sqrt{2}$$
$$\|x\| = (\langle x, x\rangle)^{1/2} = \frac{\sqrt{6}}{3}$$

Since 1 and x are orthogonal, they satisfy the Pythagorean Law:

$$\|1 + x\|^2 = \|1\|^2 + \|x\|^2 = 2 + \frac{2}{3} = \frac{8}{3}$$

The reader may verify that

$$\|1 + x\|^2 = \langle 1 + x, 1 + x \rangle = \int_{-1}^{1} (1 + x)^2 \, dx = \frac{8}{3} \qquad \blacktriangleleft$$

EXAMPLE 2. For the vector space $C[-\pi, \pi]$, if we use a constant weight function $w(x) = 1/\pi$ to define an inner product

(6) $$\langle f, g \rangle = \frac{1}{\pi} \int_{-\pi}^{\pi} f(x)g(x) \, dx$$

then it follows that

$$\langle \cos x, \sin x \rangle = \frac{1}{\pi} \int_{-\pi}^{\pi} \cos x \sin x \, dx = 0$$

$$\langle \cos x, \cos x \rangle = \frac{1}{\pi} \int_{-\pi}^{\pi} \cos x \cos x \, dx = 1$$

$$\langle \sin x, \sin x \rangle = \frac{1}{\pi} \int_{-\pi}^{\pi} \sin x \sin x \, dx = 1$$

Thus $\cos x$ and $\sin x$ are orthogonal unit vectors with respect to this inner product. It follows from the Pythagorean Law that

$$\| \cos x + \sin x \| = \sqrt{2} \qquad \blacktriangleleft$$

The inner product (6) plays a key role in Fourier analysis applications involving trigonometric approximation of functions. We will look at some of these applications in Section 5.

For the vector space $R^{m \times n}$ the norm derived from the inner product (2) is called the *Frobenius norm* and is denoted by $\| \cdot \|_F$. Thus, if $A \in R^{m \times n}$, then

$$\|A\|_F = (\langle A, A \rangle)^{1/2} = \left(\sum_{i=1}^{m} \sum_{j=1}^{n} a_{ij}^2 \right)^{1/2}$$

EXAMPLE 3. If

$$A = \begin{bmatrix} 1 & 1 \\ 1 & 2 \\ 3 & 3 \end{bmatrix} \qquad \text{and} \qquad B = \begin{bmatrix} -1 & 1 \\ 3 & 0 \\ -3 & 4 \end{bmatrix}$$

then

$$\langle A, B \rangle = 1 \cdot -1 + 1 \cdot 1 + 1 \cdot 3 + 2 \cdot 0 + 3 \cdot -3 + 3 \cdot 4 = 6$$

Thus, A is not orthogonal to B. The norms of these matrices are given by

$$\|A\|_F = (1 + 1 + 1 + 4 + 9 + 9)^{1/2} = 5$$
$$\|B\|_F = (1 + 1 + 9 + 0 + 9 + 16)^{1/2} = 6 \quad \blacktriangleleft$$

EXAMPLE 4.   In P_5, define an inner product by (5) with $x_i = (i - 1)/4$ for $i = 1, 2, \ldots , 5$. The length of the function $p(x) = 4x$ is given by

$$\|4x\| = (\langle 4x, 4x \rangle)^{1/2} = \left(\sum_{i=1}^{5} 16x_i^2 \right)^{1/2} = \left(\sum_{i=1}^{5} (i - 1)^2 \right)^{1/2} = \sqrt{30} \quad \blacktriangleleft$$

▶ **DEFINITION**   If $\mathbf{u}$ and $\mathbf{v}$ are vectors in an inner product space V and $\mathbf{v} \neq \mathbf{0}$, then the **scalar projection** of $\mathbf{u}$ onto $\mathbf{v}$ is given by

$$\alpha = \frac{\langle \mathbf{u}, \mathbf{v} \rangle}{\|\mathbf{v}\|}$$

and the **vector projection** of $\mathbf{u}$ onto $\mathbf{v}$ is given by

$$(7) \qquad \mathbf{p} = \alpha \left(\frac{1}{\|\mathbf{v}\|} \mathbf{v} \right) = \frac{\langle \mathbf{u}, \mathbf{v} \rangle}{\langle \mathbf{v}, \mathbf{v} \rangle} \mathbf{v} \quad \blacktriangleleft$$

Observations   If $\mathbf{v} \neq \mathbf{0}$ and $\mathbf{p}$ is the vector projection of $\mathbf{u}$ onto $\mathbf{v}$, then

    I. $\mathbf{u} - \mathbf{p}$ and $\mathbf{p}$ are orthogonal.
    II. $\mathbf{u} = \mathbf{p}$ if and only if $\mathbf{u}$ is a scalar multiple of $\mathbf{v}$.

▶ *Proof.* [Observation I] Since

$$\langle \mathbf{p}, \mathbf{p} \rangle = \left\langle \frac{\alpha}{\|\mathbf{v}\|} \mathbf{v}, \frac{\alpha}{\|\mathbf{v}\|} \mathbf{v} \right\rangle = \left(\frac{\alpha}{\|\mathbf{v}\|} \right)^2 \langle \mathbf{v}, \mathbf{v} \rangle = \alpha^2$$

and

$$\langle \mathbf{u}, \mathbf{p} \rangle = \frac{(\langle \mathbf{u}, \mathbf{v} \rangle)^2}{\langle \mathbf{v}, \mathbf{v} \rangle} = \alpha^2$$

it follows that

$$\langle \mathbf{u} - \mathbf{p}, \mathbf{p} \rangle = \langle \mathbf{u}, \mathbf{p} \rangle - \langle \mathbf{p}, \mathbf{p} \rangle = \alpha^2 - \alpha^2 = 0$$

Therefore, $\mathbf{u} - \mathbf{p}$ and $\mathbf{p}$ are orthogonal. $\quad \blacktriangleleft$

▶ *Proof.* [Observation II] If $\mathbf{u} = \beta\mathbf{v}$, then the vector projection of $\mathbf{u}$ onto $\mathbf{v}$ is given by

$$\mathbf{p} = \frac{\langle \beta\mathbf{v}, \mathbf{v} \rangle}{\langle \mathbf{v}, \mathbf{v} \rangle} \mathbf{v} = \beta\mathbf{v} = \mathbf{u}$$

Conversely, if $\mathbf{u} = \mathbf{p}$, it follows from (7) that

$$\mathbf{u} = \beta\mathbf{v} \qquad \text{where} \quad \beta = \frac{\alpha}{\|\mathbf{v}\|} \quad \blacktriangleleft$$

Observations I and II are useful for establishing the following theorem.

▶ **THEOREM 5.4.2 (The Cauchy–Schwarz Inequality)** *If* **u** *and* **v** *are any two vectors in an inner product space* V, *then*

(8) $$|\langle \mathbf{u}, \mathbf{v} \rangle| \le \|\mathbf{u}\| \, \|\mathbf{v}\|$$

Equality holds if and only if **u** *and* **v** *are linearly dependent.*

▶ *Proof.* If $\mathbf{v} = \mathbf{0}$, then

$$|\langle \mathbf{u}, \mathbf{v} \rangle| = 0 = \|\mathbf{u}\| \, \|\mathbf{v}\|$$

If $\mathbf{v} \ne \mathbf{0}$, then let **p** be the vector projection of **u** onto **v**. Since **p** is orthogonal to $\mathbf{u} - \mathbf{p}$, it follows from the Pythagorean Law that

$$\|\mathbf{p}\|^2 + \|\mathbf{u} - \mathbf{p}\|^2 = \|\mathbf{u}\|^2$$

Thus,

$$\frac{(\langle \mathbf{u}, \mathbf{v} \rangle)^2}{\|\mathbf{v}\|^2} = \|\mathbf{p}\|^2 = \|\mathbf{u}\|^2 - \|\mathbf{u} - \mathbf{p}\|^2$$

and hence

(9) $$(\langle \mathbf{u}, \mathbf{v} \rangle)^2 = \|\mathbf{u}\|^2 \|\mathbf{v}\|^2 - \|\mathbf{u} - \mathbf{p}\|^2 \|\mathbf{v}\|^2 \le \|\mathbf{u}\|^2 \|\mathbf{v}\|^2$$

Therefore,

$$|\langle \mathbf{u}, \mathbf{v} \rangle| \le \|\mathbf{u}\| \, \|\mathbf{v}\|$$

Equality holds in (9) if and only if $\mathbf{u} = \mathbf{p}$. It follows from observation II that equality will hold in (8) if and only if $\mathbf{v} = \mathbf{0}$ or if **u** is a multiple of **v**. More simply stated, equality will hold if and only if **u** and **v** are linearly dependent. ◀

One consequence of the Cauchy–Schwarz inequality is that if **u** and **v** are nonzero vectors, then

$$-1 \le \frac{\langle \mathbf{u}, \mathbf{v} \rangle}{\|\mathbf{u}\| \|\mathbf{v}\|} \le 1$$

and hence there is a unique angle θ in $[0, \pi]$ such that

(10) $$\cos \theta = \frac{\langle \mathbf{u}, \mathbf{v} \rangle}{\|\mathbf{u}\| \|\mathbf{v}\|}$$

Thus, equation (10) can be used to define the angle θ between two nonzero vectors **u** and **v**.

Norms

The word *norm* in mathematics has its own meaning independent of an inner product and its use here should be justified.

▶ **DEFINITION** A vector space V is said to be a **normed linear space** if to each vector $\mathbf{v} \in V$ there is associated a real number $\|\mathbf{v}\|$, called the **norm** of **v**, satisfying:

(i) $\|\mathbf{v}\| \ge 0$ with equality if and only if $\mathbf{v} = \mathbf{0}$.
(ii) $\|\alpha \mathbf{v}\| = |\alpha| \, \|\mathbf{v}\|$ for any scalar α.
(iii) $\|\mathbf{v} + \mathbf{w}\| \le \|\mathbf{v}\| + \|\mathbf{w}\|$ for all $\mathbf{v}, \mathbf{w} \in V$. ◀

The third condition is called the *triangle inequality* (see Figure 5.4.2).

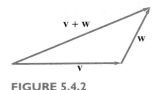

FIGURE 5.4.2

▶ **THEOREM 5.4.3** *If V is an inner product space, then the equation*

$$\|\mathbf{v}\| = \sqrt{\langle \mathbf{v}, \mathbf{v} \rangle} \qquad \text{for all} \quad \mathbf{v} \in V$$

defines a norm on V.

▶ *Proof.* It is easily seen that conditions (i) and (ii) of the definition are satisfied. We leave this for the reader to verify and proceed to show that condition (iii) is satisfied.

$$\begin{aligned}
\|\mathbf{u} + \mathbf{v}\|^2 &= \langle \mathbf{u} + \mathbf{v}, \mathbf{u} + \mathbf{v} \rangle \\
&= \langle \mathbf{u}, \mathbf{u} \rangle + 2\langle \mathbf{u}, \mathbf{v} \rangle + \langle \mathbf{v}, \mathbf{v} \rangle \\
&\leq \|\mathbf{u}\|^2 + 2\|\mathbf{u}\|\,\|\mathbf{v}\| + \|\mathbf{v}\|^2 \quad \text{(Cauchy–Schwarz)} \\
&= (\|\mathbf{u}\| + \|\mathbf{v}\|)^2
\end{aligned}$$

Thus

$$\|\mathbf{u} + \mathbf{v}\| \leq \|\mathbf{u}\| + \|\mathbf{v}\| \qquad\qquad ◀$$

It is possible to define many different norms on a given vector space. For example, in R^n we could define

$$\|\mathbf{x}\|_1 = \sum_{i=1}^{n} |x_i|$$

for every $\mathbf{x} = (x_1, x_2, \ldots, x_n)^T$. It is easily verified that $\|\cdot\|_1$ defines a norm on R^n. Another important norm on R^n is the *uniform norm* or *infinity norm*, which is defined by

$$\|\mathbf{x}\|_\infty = \max_{1 \leq i \leq n} |x_i|$$

More generally, we could define a norm on R^n by

$$\|\mathbf{x}\|_p = \left(\sum_{i=1}^{n} |x_i|^p \right)^{1/p}$$

for any real number $p \geq 1$. In particular, if $p = 2$, then

$$\|\mathbf{x}\|_2 = \left(\sum_{i=1}^{n} |x_i|^2 \right)^{1/2} = \sqrt{\langle \mathbf{x}, \mathbf{x} \rangle}$$

The norm $\|\cdot\|_2$ is the norm on R^n derived from the inner product. If $p \neq 2$, $\|\cdot\|_p$ does not correspond to any inner product. In the case of a norm that is not derived from an inner product, the Pythagorean Law will not hold. For example,

$$\mathbf{x}_1 = \begin{pmatrix} 1 \\ 2 \end{pmatrix} \qquad \text{and} \qquad \mathbf{x}_2 = \begin{pmatrix} -4 \\ 2 \end{pmatrix}$$

are orthogonal; however,

$$\|\mathbf{x}_1\|_\infty^2 + \|\mathbf{x}_2\|_\infty^2 = 4 + 16 = 20$$

while

$$\|\mathbf{x}_1 + \mathbf{x}_2\|_\infty^2 = 16$$

On the other hand, if $\|\cdot\|_2$ is used, then

$$\|\mathbf{x}_1\|_2^2 + \|\mathbf{x}_2\|_2^2 = 5 + 20 = 25 = \|\mathbf{x}_1 + \mathbf{x}_2\|_2^2$$

EXAMPLE 5. Let $\mathbf{x}$ be the vector $(4, -5, 3)^T$ in R^3. Compute $\|\mathbf{x}\|_1$, $\|\mathbf{x}\|_2$, and $\|\mathbf{x}\|_\infty$.

$$\|\mathbf{x}\|_1 = |4| + |-5| + |3| = 12$$
$$\|\mathbf{x}\|_2 = \sqrt{16 + 25 + 9} = 5\sqrt{2}$$
$$\|\mathbf{x}\|_\infty = \max(|4|, |-5|, |3|) = 5 \qquad \blacktriangleleft$$

It is also possible to define different matrix norms for $R^{m \times n}$. In Chapter 7 we will study other types of matrix norms that are useful in determining the sensitivity of linear systems.

In general, a norm provides a way of measuring the distance between vectors.

▶ **DEFINITION** Let $\mathbf{x}$ and $\mathbf{y}$ be vectors in a normed linear space. The distance between $\mathbf{x}$ and $\mathbf{y}$ is defined to be the number $\|\mathbf{y} - \mathbf{x}\|$. ◀

Many applications involve finding a unique closest vector in a subspace S to a given vector $\mathbf{v}$ in a vector space V. If the norm used for V is derived from an inner product, then the closest vector can be computed as a vector projection of $\mathbf{v}$ onto the subspace S. This type of approximation problem is discussed further in the next section.

EXERCISES

1. Let $\mathbf{x} = (-1, -1, 1, 1)^T$ and $\mathbf{y} = (1, 1, 5, -3)^T$. Show that $\mathbf{x} \perp \mathbf{y}$. Calculate $\|\mathbf{x}\|_2$, $\|\mathbf{y}\|_2$, $\|\mathbf{x} + \mathbf{y}\|_2$ and verify that the Pythagorean Law holds.

2. Given $\mathbf{x} = (1, 1, 1, 1)^T$ and $\mathbf{y} = (8, 2, 2, 0)^T$:

   (a) Determine the angle θ between $\mathbf{x}$ and $\mathbf{y}$.
   (b) Find the vector projection $\mathbf{p}$ of $\mathbf{x}$ onto $\mathbf{y}$.
   (c) Verify that $\mathbf{x} - \mathbf{p}$ is orthogonal to $\mathbf{p}$.
   (d) Compute $\|\mathbf{x} - \mathbf{p}\|_2$, $\|\mathbf{p}\|_2$, $\|\mathbf{x}\|_2$ and verify that the Pythagorean Law is satisfied.

3. Use equation (1) with weight vector $\mathbf{w} = \left(\frac{1}{4}, \frac{1}{2}, \frac{1}{4} \right)^T$ to define an inner product for R^3 and let $\mathbf{x} = (1, 1, 1)^T$ and $\mathbf{y} = (-5, 1, 3)^T$.

   (a) Show that $\mathbf{x}$ and $\mathbf{y}$ are orthogonal with respect to this inner product.
   (b) Compute the values of $\|\mathbf{x}\|$ and $\|\mathbf{y}\|$ with respect to this inner product.

4. Given

$$
A = \begin{bmatrix} 1 & 2 & 2 \\ 1 & 0 & 2 \\ 3 & 1 & 1 \end{bmatrix} \quad \text{and} \quad B = \begin{bmatrix} -4 & 1 & 1 \\ -3 & 3 & 2 \\ 1 & -2 & -2 \end{bmatrix}
$$

   Determine the value of each of the following.
   (a) $\langle A, B \rangle$ (b) $\|A\|_F$ (c) $\|B\|_F$ (d) $\|A + B\|_F$

5. Show that equation (2) defines an inner product on $R^{m \times n}$.

6. Show that the inner product defined by equation (3) satisfies the last two conditions of the definition of an inner product.

7. In $C[0, 1]$ with inner product defined by (3), compute:
   (a) $\langle e^x, e^{-x} \rangle$ (b) $\langle x, \sin \pi x \rangle$ (c) $\langle x^2, x^3 \rangle$

8. In $C[0, 1]$, with inner product defined by (3), consider the vectors 1 and x.

   (a) Find the angle θ between 1 and x.
   (b) Determine the vector projection $\mathbf{p}$ of 1 onto x and verify that $1 - \mathbf{p}$ is orthogonal to $\mathbf{p}$.
   (c) Compute $\|1 - \mathbf{p}\|$, $\|\mathbf{p}\|$, $\|1\|$ and verify that the Pythagorean Law holds.

9. In $C[-\pi, \pi]$ with inner product defined by (6), show that $\cos mx$ and $\sin nx$ are orthogonal and that both are unit vectors. Determine the distance between the two vectors.

10. Show that the functions x and x^2 are orthogonal in P_5 with inner product defined by (5), where $x_i = (i - 3)/2$ for $i = 1, \ldots, 5$.

11. In P_5 with inner product as in Exercise 10 and norm defined by

$$
\|p\| = \sqrt{\langle p, p \rangle} = \left\{ \sum_{i=1}^{5} [p(x_i)]^2 \right\}^{1/2}
$$

   compute:
   (a) $\|x\|$ (b) $\|x^2\|$ (c) The distance between x and x^2

12. If V is an inner product space, show that

$$
\|\mathbf{v}\| = \sqrt{\langle \mathbf{v}, \mathbf{v} \rangle}
$$

   satisfies the first two conditions in the definition of a norm.

13. Show that

$$\|\mathbf{x}\|_1 = \sum_{i=1}^{n} |x_i|$$

defines a norm on R^n.

14. Show that

$$\|\mathbf{x}\|_\infty = \max_{1 \le i \le n} |x_i|$$

defines a norm on R^n.

15. Compute $\|\mathbf{x}\|_1$, $\|\mathbf{x}\|_2$, and $\|\mathbf{x}\|_\infty$ for each of the following vectors in R^3.

   (a) $\mathbf{x} = (-3, 4, 0)^T$ (b) $\mathbf{x} = (-1, -1, 2)^T$ (c) $\mathbf{x} = (1, 1, 1)^T$

16. Let $\mathbf{x} = (5, 2, 4)^T$ and $\mathbf{y} = (3, 3, 2)^T$. Compute $\|\mathbf{x}-\mathbf{y}\|_1$, $\|\mathbf{x}-\mathbf{y}\|_2$, and $\|\mathbf{x}-\mathbf{y}\|_\infty$. Under which norm are the two vectors closest together? Under which norm are they farthest apart?

17. Let $\mathbf{x}$ and $\mathbf{y}$ be vectors in an inner product space. Show that if $\mathbf{x} \perp \mathbf{y}$, then the distance between $\mathbf{x}$ and $\mathbf{y}$ is

$$\left(\|\mathbf{x}\|^2 + \|\mathbf{y}\|^2 \right)^{1/2}$$

18. In R^n with inner product

$$\langle \mathbf{x}, \mathbf{y} \rangle = \mathbf{x}^T \mathbf{y}$$

derive a formula for the distance between two vectors $\mathbf{x} = (x_1, \dots, x_n)^T$ and $\mathbf{y} = (y_1, \dots, y_n)^T$.

19. Let $\mathbf{x} \in R^n$. Show that $\|\mathbf{x}\|_\infty \le \|\mathbf{x}\|_2$.

20. Let $\mathbf{x} \in R^2$. Show that $\|\mathbf{x}\|_2 \le \|\mathbf{x}\|_1$. [*Hint:* Write $\mathbf{x}$ in the form $x_1\mathbf{e}_1 + x_2\mathbf{e}_2$ and use the triangle inequality.]

21. Give an example of a nonzero vector $\mathbf{x} \in R^2$ for which

$$\|\mathbf{x}\|_\infty = \|\mathbf{x}\|_2 = \|\mathbf{x}\|_1$$

22. Show that in any vector space with a norm

$$\|-\mathbf{v}\| = \|\mathbf{v}\|$$

23. Show that for any $\mathbf{u}$ and $\mathbf{v}$ in a normed vector space

$$\|\mathbf{u} + \mathbf{v}\| \ge |\, \|\mathbf{u}\| - \|\mathbf{v}\| \,|$$

24. Prove that for any $\mathbf{u}$ and $\mathbf{v}$ in an inner product space V

$$\|\mathbf{u} + \mathbf{v}\|^2 + \|\mathbf{u} - \mathbf{v}\|^2 = 2\|\mathbf{u}\|^2 + 2\|\mathbf{v}\|^2$$

Give a geometric interpretation of this result for the vector space R^2.

25. The result of Exercise 24 is not valid for norms other than the norm derived from the inner product. Give an example of this in R^2 using $\|\cdot\|_1$.

26. Determine whether the following define norms on $C[a, b]$.

(a) $\|f\| = |f(a)| + |f(b)|$ (b) $\|f\| = \int_a^b |f(x)|\, dx$

(c) $\|f\| = \max_{a \le x \le b} |f(x)|$

27. Let $\mathbf{x} \in R^n$ and show that:

(a) $\|\mathbf{x}\|_1 \le n\|\mathbf{x}\|_\infty$ (b) $\|\mathbf{x}\|_2 \le \sqrt{n}\,\|\mathbf{x}\|_\infty$

Give examples of vectors in R^n for which equality holds in parts (a) and (b).

28. Sketch the set of points $(x_1, x_2) = \mathbf{x}^T$ in R^2 such that:

(a) $\|\mathbf{x}\|_2 = 1$ (b) $\|\mathbf{x}\|_1 = 1$ (c) $\|\mathbf{x}\|_\infty = 1$

29. Consider the vector space R^n with inner product $\langle \mathbf{x}, \mathbf{y} \rangle = \mathbf{x}^T \mathbf{y}$. For any $n \times n$ matrix A, show that:

(a) $\langle A\mathbf{x}, \mathbf{y} \rangle = \langle \mathbf{x}, A^T \mathbf{y} \rangle$ (b) $\langle A^T A \mathbf{x}, \mathbf{x} \rangle = \|A\mathbf{x}\|^2$

5   ORTHONORMAL SETS

In R^2 it is generally more convenient to use the standard basis $\{\mathbf{e}_1, \mathbf{e}_2\}$ than to use some other basis, such as $\{(2, 1)^T, (3, 5)^T\}$. For example, it would be easier to find the coordinates of $(x_1, x_2)^T$ with respect to the standard basis. The elements of the standard basis are orthogonal unit vectors. In working with an inner product space V, it is generally desirable to have a basis of mutually orthogonal unit vectors. This is convenient not only in finding coordinates of vectors, but also in solving least squares problems.

▶ **DEFINITION** Let $\mathbf{v}_1, \mathbf{v}_2, \ldots, \mathbf{v}_n$ be vectors in an inner product space V. If $\langle \mathbf{v}_i, \mathbf{v}_j \rangle = 0$ whenever $i \ne j$, then $\{\mathbf{v}_1, \mathbf{v}_2, \ldots, \mathbf{v}_n\}$ is said to be an **orthogonal set** of vectors. ◀

EXAMPLE I. The set $\{(1, 1, 1)^T, (2, 1, -3)^T, (4, -5, 1)^T\}$ is an orthogonal set in R^3, since

$$(1, 1, 1)(2, 1, -3)^T = 0$$
$$(1, 1, 1)(4, -5, 1)^T = 0$$
$$(2, 1, -3)(4, -5, 1)^T = 0$$
◀

▶ **THEOREM 5.5.1** *If $\{\mathbf{v}_1, \mathbf{v}_2, \ldots, \mathbf{v}_n\}$ is an orthogonal set of nonzero vectors in an inner product space V, then $\mathbf{v}_1, \mathbf{v}_2, \ldots, \mathbf{v}_n$ are linearly independent.*

▶ *Proof.* Suppose that $\mathbf{v}_1, \mathbf{v}_2, \ldots, \mathbf{v}_n$ are mutually orthogonal nonzero vectors and

(1) $$c_1\mathbf{v}_1 + c_2\mathbf{v}_2 + \cdots + c_n\mathbf{v}_n = \mathbf{0}$$

If $1 \le j \le n$, then, taking the inner product of $\mathbf{v}_j$ with both sides of equation (1), we see that

$$c_1 \langle \mathbf{v}_j, \mathbf{v}_1 \rangle + c_2 \langle \mathbf{v}_j, \mathbf{v}_2 \rangle + \cdots + c_n \langle \mathbf{v}_j, \mathbf{v}_n \rangle = 0$$

$$c_j \|\mathbf{v}_j\|^2 = 0$$

and hence all the scalars $c_1, c_2, \ldots, c_n$ must be 0.   ◀

▶ **DEFINITION** An **orthonormal** set of vectors is an orthogonal set of unit vectors.   ◀

The set $\{\mathbf{u}_1, \mathbf{u}_2, \ldots, \mathbf{u}_n\}$ will be orthonormal if and only if

$$\langle \mathbf{u}_i, \mathbf{u}_j \rangle = \delta_{ij}$$

where

$$\delta_{ij} = \begin{cases} 1 & \text{if } i = j \\ 0 & \text{if } i \neq j \end{cases}$$

Given any orthogonal set of nonzero vectors $\{\mathbf{v}_1, \mathbf{v}_2, \ldots, \mathbf{v}_n\}$, it is possible to form an orthonormal set by defining

$$\mathbf{u}_i = \left(\frac{1}{\|\mathbf{v}_i\|} \right) \mathbf{v}_i \qquad \text{for} \quad i = 1, 2, \ldots, n$$

The reader may verify that $\{\mathbf{u}_1, \mathbf{u}_2, \ldots, \mathbf{u}_n\}$ will be an orthonormal set.

EXAMPLE 2. We saw in Example 1 that, if $\mathbf{v}_1 = (1, 1, 1)^T$, $\mathbf{v}_2 = (2, 1, -3)^T$, and $\mathbf{v}_3 = (4, -5, 1)^T$, then $\{\mathbf{v}_1, \mathbf{v}_2, \mathbf{v}_3\}$ is an orthogonal set in R^3. To form an orthonormal set, let

$$\mathbf{u}_1 = \left(\frac{1}{\|\mathbf{v}_1\|} \right) \mathbf{v}_1 = \frac{1}{\sqrt{3}}(1, 1, 1)^T$$

$$\mathbf{u}_2 = \left(\frac{1}{\|\mathbf{v}_2\|} \right) \mathbf{v}_2 = \frac{1}{\sqrt{14}}(2, 1, -3)^T$$

$$\mathbf{u}_3 = \left(\frac{1}{\|\mathbf{v}_3\|} \right) \mathbf{v}_3 = \frac{1}{\sqrt{42}}(4, -5, 1)^T \qquad ◀$$

EXAMPLE 3. In $C[-\pi, \pi]$ with inner product

(2) $$\langle f, g \rangle = \frac{1}{\pi} \int_{-\pi}^{\pi} f(x)g(x)\,dx$$

the set $\{1, \cos x, \cos 2x, \ldots, \cos nx\}$ is an orthogonal set of vectors, since for any positive integers j and k

$$\langle 1, \cos kx \rangle = \frac{1}{\pi} \int_{-\pi}^{\pi} \cos kx\,dx = 0$$

$$\langle \cos jx, \cos kx \rangle = \frac{1}{\pi} \int_{-\pi}^{\pi} \cos jx \cos kx\,dx = 0 \qquad (j \neq k)$$

The functions $\cos x, \cos 2x, \ldots, \cos nx$ are already unit vectors since

$$\langle \cos kx, \cos kx \rangle = \frac{1}{\pi} \int_{-\pi}^{\pi} \cos^2 kx\,dx = 1 \qquad \text{for} \quad k = 1, 2, \ldots, n$$

To form an orthonormal set, we need only find a unit vector in the direction of 1.

$$\|1\|^2 = \langle 1, 1 \rangle = \frac{1}{\pi} \int_{-\pi}^{\pi} 1 \, dx = 2$$

Thus $1/\sqrt{2}$ is a unit vector and hence $\{1/\sqrt{2}, \cos x, \cos 2x, \ldots, \cos nx\}$ is an orthonormal set of vectors.   ◀

     It follows from Theorem 5.5.1 that, if $B = \{\mathbf{u}_1, \mathbf{u}_2, \ldots, \mathbf{u}_k\}$ is an orthonormal set in an inner product space V, then B is a basis for a subspace S of V. We say that B is an *orthonormal basis* for S. It is generally much easier to work with an orthonormal basis than with an ordinary basis. In particular, it is much easier to calculate the coordinates of a given vector $\mathbf{v}$ with respect to an orthonormal basis. Once these coordinates have been determined, they can be used to compute $\|\mathbf{v}\|$.

▶ **THEOREM 5.5.2** *Let* $\{\mathbf{u}_1, \mathbf{u}_2, \ldots, \mathbf{u}_n\}$ *be an orthonormal basis for an inner product space* V. *If* $\mathbf{v} = \sum_{i=1}^{n} c_i \mathbf{u}_i$, *then* $c_i = \langle \mathbf{u}_i, \mathbf{v} \rangle$.

▶ *Proof.*

$$\langle \mathbf{u}_i, \mathbf{v} \rangle = \left\langle \mathbf{u}_i, \sum_{j=1}^{n} c_j \mathbf{u}_j \right\rangle = \sum_{j=1}^{n} c_j \langle \mathbf{u}_i, \mathbf{u}_j \rangle = \sum_{j=1}^{n} c_j \delta_{ij} = c_i$$

◀

     As a consequence of Theorem 5.5.2 we can state two more important results.

▶ **COROLLARY 5.5.3** *Let* $\{\mathbf{u}_1, \mathbf{u}_2, \ldots, \mathbf{u}_n\}$ *be an orthonormal basis for an inner product space* V. *If* $\mathbf{u} = \sum_{i=1}^{n} a_i \mathbf{u}_i$ *and* $\mathbf{v} = \sum_{i=1}^{n} b_i \mathbf{u}_i$, *then*

$$\langle \mathbf{u}, \mathbf{v} \rangle = \sum_{i=1}^{n} a_i b_i$$

▶ *Proof.* By Theorem 5.5.2

$$\langle \mathbf{u}_i, \mathbf{v} \rangle = b_i \qquad i = 1, \ldots, n$$

Therefore,

$$\langle \mathbf{u}, \mathbf{v} \rangle = \left\langle \sum_{i=1}^{n} a_i \mathbf{u}_i, \mathbf{v} \right\rangle = \sum_{i=1}^{n} a_i \langle \mathbf{u}_i, \mathbf{v} \rangle = \sum_{i=1}^{n} a_i b_i$$

◀

▶ **COROLLARY 5.5.4 (Parseval's Formula)** *If* $\{\mathbf{u}_1, \ldots, \mathbf{u}_n\}$ *is an orthonormal basis for an inner product space* V *and* $\mathbf{v} = \sum_{i=1}^{n} c_i \mathbf{u}_i$, *then*

$$\|\mathbf{v}\|^2 = \sum_{i=1}^{n} c_i^2$$

▶ *Proof.* If $\mathbf{v} = \sum_{i=1}^{n} c_i \mathbf{u}_i$, then, by Corollary 5.5.3,

$$\|\mathbf{v}\|^2 = \langle \mathbf{v}, \mathbf{v} \rangle = \sum_{i=1}^{n} c_i^2 \qquad ◀$$

EXAMPLE 4.   The vectors

$$\mathbf{u}_1 = \left(\frac{1}{\sqrt{2}}, \frac{1}{\sqrt{2}} \right)^T \qquad \text{and} \qquad \mathbf{u}_2 = \left(\frac{1}{\sqrt{2}}, -\frac{1}{\sqrt{2}} \right)^T$$

form an orthonormal basis for R^2. If $\mathbf{x} \in R^2$, then

$$\mathbf{x}^T \mathbf{u}_1 = \frac{x_1 + x_2}{\sqrt{2}} \qquad \text{and} \qquad \mathbf{x}^T \mathbf{u}_2 = \frac{x_1 - x_2}{\sqrt{2}}$$

It follows from Theorem 5.5.2 that

$$\mathbf{x} = \frac{x_1 + x_2}{\sqrt{2}} \mathbf{u}_1 + \frac{x_1 - x_2}{\sqrt{2}} \mathbf{u}_2$$

and it follows from Corollary 5.5.4 that

$$\|\mathbf{x}\|^2 = \left(\frac{x_1 + x_2}{\sqrt{2}} \right)^2 + \left(\frac{x_1 - x_2}{\sqrt{2}} \right)^2 = x_1^2 + x_2^2 \qquad ◀$$

EXAMPLE 5.   Given that $\{1/\sqrt{2}, \cos 2x\}$ is an orthonormal set in $C[-\pi, \pi]$
(with inner product as in Example 3), determine the value of $\int_{-\pi}^{\pi} \sin^4 x \, dx$ without
computing antiderivatives.

SOLUTION.   Since

$$\sin^2 x = \frac{1 - \cos 2x}{2} = \frac{1}{\sqrt{2}} \frac{1}{\sqrt{2}} + \left(-\frac{1}{2} \right) \cos 2x$$

it follows from Parseval's formula that

$$\int_{-\pi}^{\pi} \sin^4 x \, dx = \pi \| \sin^2 x \|^2 = \pi \left(\frac{1}{2} + \frac{1}{4} \right) = \frac{3\pi}{4} \qquad ◀$$

Orthogonal Matrices

Of particular importance are $n \times n$ matrices whose column vectors form an
orthonormal set in R^n.

▶ **DEFINITION** An $n \times n$ matrix Q is said to be an **orthogonal matrix** if the
column vectors of Q form an orthonormal set in R^n. ◀

▶ **THEOREM 5.5.5** *An $n \times n$ matrix Q is orthogonal if and only if $Q^T Q = I$.*

▶ *Proof.* It follows from the definition that an $n \times n$ matrix Q is orthogonal if and
only if its column vectors satisfy

$$\mathbf{q}_i^T \mathbf{q}_j = \delta_{ij}$$

However, $\mathbf{q}_i^T \mathbf{q}_j$ is the (i, j) entry of the matrix $Q^T Q$. Thus Q is orthogonal if and only if $Q^T Q = I$. ◀

It follows from the theorem that if Q is an orthogonal matrix, then Q is invertible and $Q^{-1} = Q^T$.

EXAMPLE 6.   For any fixed θ, the matrix

$$Q = \begin{pmatrix} \cos\theta & -\sin\theta \\ \sin\theta & \cos\theta \end{pmatrix}$$

is orthogonal and

$$Q^{-1} = Q^T = \begin{pmatrix} \cos\theta & \sin\theta \\ -\sin\theta & \cos\theta \end{pmatrix}$$

◀

The matrix Q in Example 6 can be thought of as a linear transformation from R^2 onto R^2 that has the effect of rotating each vector by an angle θ while leaving the length of the vector unchanged (see Example 2 in Section 2 of Chapter 4). Similarly, Q^{-1} can be thought of as a rotation by the angle $-\theta$ (see Figure 5.5.1).

In general, inner products are preserved under multiplication by an orthogonal matrix [i.e., $\langle \mathbf{x}, \mathbf{y} \rangle = \langle Q\mathbf{x}, Q\mathbf{y} \rangle$]. This is true, since

$$\langle Q\mathbf{x}, Q\mathbf{y} \rangle = (Q\mathbf{y})^T Q\mathbf{x} = \mathbf{y}^T Q^T Q\mathbf{x} = \mathbf{y}^T \mathbf{x} = \langle \mathbf{x}, \mathbf{y} \rangle$$

In particular, if $\mathbf{x} = \mathbf{y}$, then $\|Q\mathbf{x}\|^2 = \|\mathbf{x}\|^2$ and hence $\|Q\mathbf{x}\| = \|\mathbf{x}\|$. Multiplication by an orthogonal matrix preserves the lengths of vectors.

Properties of Orthogonal Matrices

If Q is an $n \times n$ orthogonal matrix, then:

(a) The column vectors of Q form an orthonormal basis for R^n.

(b) $Q^T Q = I$

(c) $Q^T = Q^{-1}$

(d) $\langle Q\mathbf{x}, Q\mathbf{y} \rangle = \langle \mathbf{x}, \mathbf{y} \rangle$

(e) $\|Q\mathbf{x}\|_2 = \|\mathbf{x}\|_2$

FIGURE 5.5.1

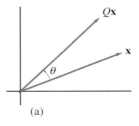

(a)

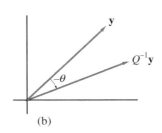

(b)

Permutation Matrices

A *permutation matrix* is a matrix formed from the identity matrix by reordering its columns. Clearly, then, permutation matrices are orthogonal matrices. If P is the permutation matrix formed by reordering the columns of I in the order $(k_1, \ldots, k_n)$, then $P = (\mathbf{e}_{k_1}, \ldots, \mathbf{e}_{k_n})$. If A is an $m \times n$ matrix, then

$$AP = (A\mathbf{e}_{k_1}, \ldots, A\mathbf{e}_{k_n}) = (\mathbf{a}_{k_1}, \ldots, \mathbf{a}_{k_n})$$

Postmultiplication of A by P reorders the columns of A in the order $(k_1, \ldots, k_n)$. For example, if

$$A = \begin{bmatrix} 1 & 2 & 3 \\ 1 & 2 & 3 \end{bmatrix} \qquad \text{and} \qquad P = \begin{bmatrix} 0 & 1 & 0 \\ 0 & 0 & 1 \\ 1 & 0 & 0 \end{bmatrix}$$

then

$$AP = \begin{bmatrix} 3 & 1 & 2 \\ 3 & 1 & 2 \end{bmatrix}$$

Since $P = (\mathbf{e}_{k_1}, \ldots, \mathbf{e}_{k_n})$ is orthogonal, it follows that

$$P^{-1} = P^T = \begin{bmatrix} \mathbf{e}_{k_1}^T \\ \vdots \\ \mathbf{e}_{k_n}^T \end{bmatrix}$$

The k_1 column of P^T will be $\mathbf{e}_1$, the k_2 column will be $\mathbf{e}_2$, and so on. Thus P^T is a permutation matrix. The matrix P^T can be formed directly from I by reordering its rows in the order $(k_1, k_2, \ldots, k_n)$. In general, a permutation matrix can be formed from I by either reordering its rows or its columns.

If Q is the permutation matrix formed by reordering the rows of I in the order $(k_1, k_2, \ldots, k_n)$ and B is an $n \times r$ matrix, then

$$QB = \begin{bmatrix} \mathbf{e}_{k_1}^T \\ \vdots \\ \mathbf{e}_{k_n}^T \end{bmatrix} B = \begin{bmatrix} \mathbf{e}_{k_1}^T B \\ \vdots \\ \mathbf{e}_{k_n}^T B \end{bmatrix} = \begin{bmatrix} \mathbf{b}(k_1, :) \\ \vdots \\ \mathbf{b}(k_n, :) \end{bmatrix}$$

Thus, QB is the matrix formed by reordering the rows of B in the order $(k_1, k_2, \ldots, k_n)$. For example, if

$$Q = \begin{bmatrix} 0 & 0 & 1 \\ 1 & 0 & 0 \\ 0 & 1 & 0 \end{bmatrix} \qquad \text{and} \qquad B = \begin{bmatrix} 1 & 1 \\ 2 & 2 \\ 3 & 3 \end{bmatrix}$$

then

$$QB = \begin{bmatrix} 3 & 3 \\ 1 & 1 \\ 2 & 2 \end{bmatrix}$$

In general, if P is an $n \times n$ permutation matrix, premultiplication of an $n \times r$ matrix B by P reorders the rows of B, and postmultiplication of an $m \times n$ matrix A by P reorders the columns of A.

Orthonormal Sets and Least Squares

Orthogonality plays an important role in solving least squares problems. Recall that if A is an $m \times n$ matrix of rank n, then the least squares problem $A\mathbf{x} = \mathbf{b}$ has a unique solution $\hat{\mathbf{x}}$ that is determined by solving the normal equations $A^T A\mathbf{x} = A^T \mathbf{b}$. The projection $\mathbf{p} = A\hat{\mathbf{x}}$ is the vector in $R(A)$ that is closest to $\mathbf{b}$. The least squares problem is especially easy to solve in the case where the column vectors of A form an orthonormal set in R^m.

▶ **THEOREM 5.5.6** *If the column vectors of A form an orthonormal set of vectors in R^m, then $A^T A = I$ and the solution to the least squares problem is*

$$\hat{\mathbf{x}} = A^T \mathbf{b}$$

▶ *Proof.* The ijth entry of $A^T A$ is formed from the ith row of A^T and the jth column of A. Thus the ijth entry is actually the scalar product of the ith and jth columns of A. Since the column vectors of A are orthonormal, it follows that

$$A^T A = (\delta_{ij}) = I$$

Consequently, the normal equations simplify to

$$\mathbf{x} = A^T \mathbf{b} \qquad\qquad ◀$$

What if the columns of A are not orthonormal? In the next section we will learn a method for finding an orthonormal basis for $R(A)$. From this method we will obtain a factorization of A into a product QR, where Q has an orthonormal set of column vectors and R is upper triangular. With this factorization, the least squares problem can be solved quickly and accurately.

If we have an orthonormal basis for $R(A)$, the projection $\mathbf{p} = A\hat{\mathbf{x}}$ can be determined in terms of the basis elements. Indeed, this is a special case of the more general least squares problem of finding the element $\mathbf{p}$ in a subspace S of an inner product space V that is closest to a given element $\mathbf{x}$ in V. This problem is easily solved if S has an orthonormal basis. We first prove the following theorem.

▶ **THEOREM 5.5.7** *Let S be a subspace of an inner product space V and let $\mathbf{x} \in V$. Let $\{\mathbf{x}_1, \mathbf{x}_2, \dots, \mathbf{x}_n\}$ be an orthonormal basis for S. If*

(3)
$$\mathbf{p} = \sum_{i=1}^{n} c_i \mathbf{x}_i$$

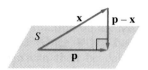

FIGURE 5.5.2

where

(4) $$c_i = \langle \mathbf{x}, \mathbf{x}_i \rangle \qquad \text{for each } i$$

then $\mathbf{p} - \mathbf{x} \in S^\perp$ *(see Figure 5.5.2).*

▶ *Proof.* We will show first that $(\mathbf{p} - \mathbf{x}) \perp \mathbf{x}_i$ for each i.

$$\begin{aligned}
\langle \mathbf{x}_i, \mathbf{p} - \mathbf{x} \rangle &= \langle \mathbf{x}_i, \mathbf{p} \rangle - \langle \mathbf{x}_i, \mathbf{x} \rangle \\
&= \left\langle \mathbf{x}_i, \sum_{j=1}^{n} c_j \mathbf{x}_j \right\rangle - c_i \\
&= \sum_{j=1}^{n} c_j \langle \mathbf{x}_i, \mathbf{x}_j \rangle - c_i \\
&= 0
\end{aligned}$$

So $\mathbf{p} - \mathbf{x}$ is orthogonal to all the $\mathbf{x}_i$'s. If $\mathbf{y} \in S$, then

$$\mathbf{y} = \sum_{i=1}^{n} \alpha_i \mathbf{x}_i$$

and hence

$$\langle \mathbf{p} - \mathbf{x}, \mathbf{y} \rangle = \left\langle \mathbf{p} - \mathbf{x}, \sum_{i=1}^{n} \alpha_i \mathbf{x}_i \right\rangle = \sum_{i=1}^{n} \alpha_i \langle \mathbf{p} - \mathbf{x}, \mathbf{x}_i \rangle = 0 \qquad \blacktriangleleft$$

If $\mathbf{x} \in S$, the preceding result is trivial, since by Theorem 5.5.2, $\mathbf{p} - \mathbf{x} = \mathbf{0}$. If $\mathbf{x} \notin S$, then $\mathbf{p}$ is the element in S closest to $\mathbf{x}$.

▶ **THEOREM 5.5.8** *Under the hypothesis of Theorem 5.5.7, $\mathbf{p}$ is the element of S that is closest to $\mathbf{x}$, that is,*

$$\|\mathbf{y} - \mathbf{x}\| > \|\mathbf{p} - \mathbf{x}\|$$

for any $\mathbf{y} \neq \mathbf{p}$ in S.

▶ *Proof.* If $\mathbf{y} \in S$ and $\mathbf{y} \neq \mathbf{p}$, then

$$\|\mathbf{y} - \mathbf{x}\|^2 = \|(\mathbf{y} - \mathbf{p}) + (\mathbf{p} - \mathbf{x})\|^2$$

Since $\mathbf{y} - \mathbf{p} \in S$, it follows from Theorem 5.5.7 and the Pythagorean Law that

$$\|\mathbf{y} - \mathbf{x}\|^2 = \|\mathbf{y} - \mathbf{p}\|^2 + \|\mathbf{p} - \mathbf{x}\|^2 > \|\mathbf{p} - \mathbf{x}\|^2$$

Therefore, $\|\mathbf{y} - \mathbf{x}\| > \|\mathbf{p} - \mathbf{x}\|$. $\qquad \blacktriangleleft$

The vector $\mathbf{p}$ defined by (3) and (4) is said to be the *projection of $\mathbf{x}$ onto S.*

▶ **COROLLARY 5.5.9** *Let S be a nonzero subspace of R^m and let $\mathbf{b} \in R^m$. If $\{\mathbf{u}_1, \mathbf{u}_2, \dots, \mathbf{u}_k\}$ is an orthonormal basis for S and $U = (\mathbf{u}_1, \mathbf{u}_2, \dots, \mathbf{u}_k)$, then the projection $\mathbf{p}$ of $\mathbf{b}$ onto S is given by*

$$\mathbf{p} = UU^T\mathbf{b}$$

▶**Proof.** It follows from Theorem 5.5.8 that the projection $\mathbf{p}$ of $\mathbf{b}$ onto S is given by

$$\mathbf{p} = c_1\mathbf{u}_1 + c_2\mathbf{u}_2 + \cdots + c_k\mathbf{u}_k = U\mathbf{c}$$

where

$$\mathbf{c} = \begin{pmatrix} c_1 \\ c_2 \\ \vdots \\ c_k \end{pmatrix} = \begin{pmatrix} \mathbf{u}_1^T\mathbf{b} \\ \mathbf{u}_2^T\mathbf{b} \\ \vdots \\ \mathbf{u}_k^T\mathbf{b} \end{pmatrix} = U^T\mathbf{b}$$

Therefore,

$$\mathbf{p} = UU^T\mathbf{b} \qquad\qquad ◀$$

The matrix UU^T in Corollary 5.5.9 is the projection matrix corresponding to the subspace S of R^m. To project any vector $\mathbf{b} \in R^m$ onto S, we need only find an orthonormal basis $\{\mathbf{u}_1, \mathbf{u}_2, \dots, \mathbf{u}_k\}$ for S, form the matrix UU^T, and then multiply UU^T times $\mathbf{b}$.

If P is a projection matrix corresponding to a subspace S of R^m, then for any $\mathbf{b} \in R^m$ the projection $\mathbf{p}$ of $\mathbf{b}$ onto S is unique. If Q is also a projection matrix corresponding to S, then

$$Q\mathbf{b} = \mathbf{p} = P\mathbf{b}$$

It follows then that

$$\mathbf{q}_j = Q\mathbf{e}_j = P\mathbf{e}_j = \mathbf{p}_j \qquad \text{for} \quad j = 1, \dots, m$$

and hence $Q = P$. Thus the projection matrix for a subspace S of R^m is unique.

EXAMPLE 7. Let S be the set of all vectors in R^3 of the form $(x, y, 0)^T$. Find the vector $\mathbf{p}$ in S that is closest to $\mathbf{w} = (5, 3, 4)^T$ (see Figure 5.5.3).

FIGURE 5.5.3

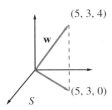

SOLUTION.   Let $\mathbf{u}_1 = (1, 0, 0)^T$ and $\mathbf{u}_2 = (0, 1, 0)^T$. Clearly, $\mathbf{u}_1$ and $\mathbf{u}_2$ form an orthonormal basis for S. Now

$$c_1 = \mathbf{w}^T \mathbf{u}_1 = 5$$

$$c_2 = \mathbf{w}^T \mathbf{u}_2 = 3$$

The vector $\mathbf{p}$ turns out to be exactly what we would expect:

$$\mathbf{p} = 5\mathbf{u}_1 + 3\mathbf{u}_2 = (5, 3, 0)^T$$

Alternatively, $\mathbf{p}$ could have been calculated using the projection matrix UU^T.

$$\mathbf{p} = UU^T \mathbf{w} = \begin{bmatrix} 1 & 0 & 0 \\ 0 & 1 & 0 \\ 0 & 0 & 0 \end{bmatrix} \begin{bmatrix} 5 \\ 3 \\ 4 \end{bmatrix} = \begin{bmatrix} 5 \\ 3 \\ 0 \end{bmatrix} \qquad \blacktriangleleft$$

Approximation of Functions

In many applications it is necessary to approximate a continuous function in terms of functions from some special type of approximating set. Most commonly we approximate by a polynomial of degree n or less. We can use Theorem 5.5.8 to obtain the best least squares approximation.

EXAMPLE 8.   Find the best least squares approximation to e^x on the interval $[0, 1]$ by a linear function.

SOLUTION.   Let S be the subspace of all linear functions in $C[0, 1]$. Although the functions 1 and x span S, they are not orthogonal. We seek a function of the form $x - a$ that is orthogonal to 1.

$$\langle 1, x - a \rangle = \int_0^1 (x - a)\, dx = \frac{1}{2} - a$$

Thus $a = \frac{1}{2}$. Since $\|x - \frac{1}{2}\| = 1/\sqrt{12}$, it follows that

$$u_1(x) = 1 \quad \text{and} \quad u_2(x) = \sqrt{12}\left(x - \tfrac{1}{2}\right)$$

form an orthonormal basis for S.
    Let

$$c_1 = \int_0^1 u_1(x)\, e^x\, dx = e - 1$$

$$c_2 = \int_0^1 u_2(x)\, e^x\, dx = \sqrt{3}\,(3 - e)$$

The projection

$$p(x) = c_1 u_1(x) + c_2 u_2(x)$$
$$= (e - 1) \cdot 1 + \sqrt{3}(3 - e)\left[\sqrt{12}\left(x - \tfrac{1}{2}\right)\right]$$
$$= (4e - 10) + 6(3 - e)x$$

is the best linear least squares approximation to e^x on $[0, 1]$ (see Figure 5.5.4). $\blacktriangleleft$

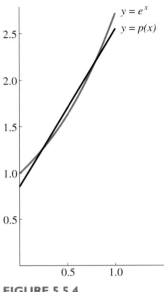

FIGURE 5.5.4

Approximation by Trigonometric Polynomials

Trigonometric polynomials are used to approximate periodic functions. By a *trigonometric polynomial* of degree n we mean a function of the form

$$t_n(x) = \frac{a_0}{2} + \sum_{k=1}^{n}(a_k \cos kx + b_k \sin kx)$$

We have already seen that the collection of functions

$$\frac{1}{\sqrt{2}}, \cos x, \cos 2x, \ldots, \cos nx$$

forms an orthonormal set with respect to the inner product (2). We leave it to the reader to verify that if the functions

$$\sin x, \sin 2x, \ldots, \sin nx$$

are added to the collection, it will still be an orthonormal set. Thus we can use Theorem 5.5.8 to find the best least squares approximation to a continuous 2π periodic function $f(x)$ by trigonometric polynomial of degree n or less. Note that

$$\left\langle f, \frac{1}{\sqrt{2}} \right\rangle \frac{1}{\sqrt{2}} = \langle f, 1 \rangle \frac{1}{2}$$

so that if

$$a_0 = \langle f, 1 \rangle = \frac{1}{\pi} \int_{-\pi}^{\pi} f(x)\,dx$$

and

$$a_k = \langle f, \cos kx \rangle = \frac{1}{\pi} \int_{-\pi}^{\pi} f(x) \cos kx \, dx$$

$$b_k = \langle f, \sin kx \rangle = \frac{1}{\pi} \int_{-\pi}^{\pi} f(x) \sin kx \, dx$$

for $k = 1, 2, \ldots, n$, then these coefficients determine the best least squares approximation to f. The a_k's and the b_k's turn out to be the well-known *Fourier coefficients* that occur in many applications involving trigonometric series approximations of functions.

Let us think of $f(x)$ as representing the position at time x of an object moving along a line, and let t_n be the Fourier approximation of degree n to f. If we set

$$r_k = \sqrt{a_k^2 + b_k^2} \qquad \text{and} \qquad \theta_k = \mathrm{Tan}^{-1}\left(\frac{b_k}{a_k}\right)$$

then

$$a_k \cos kx + b_k \sin kx = r_k \left(\frac{a_k}{r_k} \cos kx + \frac{b_k}{r_k} \sin kx \right)$$

$$= r_k \cos(kx - \theta_k)$$

Thus, the motion $f(x)$ is being represented as a sum of simple harmonic motions.

For signal processing applications it is useful to express the trigonometric approximation in complex form. To this end we define complex Fourier coefficients c_k in terms of the real Fourier coefficients a_k and b_k:

$$c_k = \frac{1}{2}(a_k - ib_k) = \frac{1}{2\pi} \int_{-\pi}^{\pi} f(x)(\cos kx - i \sin kx) \, dx$$

$$= \frac{1}{2\pi} \int_{-\pi}^{\pi} f(x) e^{-ikx} \, dx \qquad (k \geq 0)$$

The latter equality follows from the identity

$$e^{i\theta} = \cos \theta + i \sin \theta$$

We also define the coefficient c_{-k} to be the complex conjugate of c_k. Thus

$$c_{-k} = \overline{c_k} = \frac{1}{2}(a_k + ib_k) \qquad (k \geq 0)$$

Alternatively, if we solve for a_k and b_k, then

$$a_k = c_k + c_{-k} \qquad \text{and} \qquad b_k = i(c_k - c_{-k})$$

Using these identities, it follows that

$$c_k e^{ikx} + c_{-k} e^{-ikx} = (c_k + c_{-k}) \cos kx + i(c_k - c_{-k}) \sin kx$$

$$= a_k \cos kx + b_k \sin kx$$

and hence the trigonometric polynomial

$$t_n(x) = \frac{a_0}{2} + \sum_{k=1}^{n}(a_k \cos kx + b_k \sin kx)$$

can be rewritten in complex form as

$$t_n(x) = \sum_{k=-n}^{n} c_k e^{ikx}$$

APPLICATION 1: SIGNAL PROCESSING

The Discrete Fourier Transform

The function $f(x)$ pictured in Figure 5.5.5(a) corresponds to a noisy signal. Here the independent variable x represents time and the signal values are plotted as a function of time. In this context it is convenient to start with time 0. Thus we will choose $[0, 2\pi]$ as the interval for our inner product rather than $[-\pi, \pi]$.

Let us approximate $f(x)$ by a trigonometric polynomial

$$t_n(x) = \sum_{k=-n}^{n} c_k e^{ikx}$$

FIGURE 5.5.5

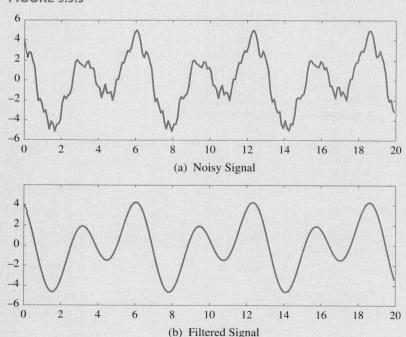

(a) Noisy Signal

(b) Filtered Signal

As noted in the previous discussion, the trigonometric approximation allows us to represent the function as a sum of simple harmonics. The kth harmonic can be written as $r_k \cos(kx - \theta_k)$. It is said to have *angular frequency* k. A signal is *smooth* if the coefficients c_k approach 0 rapidly as k increases. If some of the coefficients corresponding to larger frequencies are not small, the graph will appear to be noisy as in Figure 5.5.5(a). We can filter the signal by setting these coefficients equal to 0. Figure 5.5.5(b) shows the smooth function obtained by suppressing some of the higher frequencies from the original signal.

In actual signal processing applications, we do not have a mathematical formula for the signal function $f(x)$; rather, the signal is sampled over a sequence of times $x_0, x_1, \ldots, x_N$ where $x_j = \frac{2j\pi}{N}$. The function f is represented by the N sample values

$$y_0 = f(x_0), \ \ y_1 = f(x_1), \ldots, \ y_{N-1} = f(x_{N-1})$$

[Note: $y_N = f(2\pi) = f(0) = y_0$.] In this case it is not possible to compute the Fourier coefficients as integrals. Instead of using

$$c_k = \frac{1}{2\pi} \int_0^{2\pi} f(x)e^{-ikx}\,dx$$

we use a numerical integration method, the trapezoid rule, to approximate the integral. The approximation is given by

$$(5) \qquad d_k = \frac{1}{N} \sum_{j=0}^{N-1} f(x_j)e^{-ikx_j}$$

The d_k coefficients are approximations to the Fourier coefficients. The larger the sample size N, the closer d_k will be to c_k.

If we set

$$\omega_N = e^{-\frac{2\pi i}{N}} = \cos\frac{2\pi}{N} - i\sin\frac{2\pi}{N}$$

then equation (5) can be rewritten in the form

$$d_k = \frac{1}{N} \sum_{j=0}^{N-1} y_j \omega_N^{jk}$$

The finite sequence $\{d_0, d_1, \ldots, d_{N-1}\}$ is said to be the *discrete Fourier transform* of $\{y_0, y_1, \ldots, y_{N-1}\}$. The discrete Fourier transform can be determined by a single matrix vector multiplication. For example, if $N = 4$, the coefficients are given by

$$d_0 = \frac{1}{4}(y_0 + y_1 + y_2 + y_3)$$

$$d_1 = \frac{1}{4}(y_0 + \omega_4 y_1 + \omega_4^2 y_2 + \omega_4^3 y_3)$$

$$d_2 = \frac{1}{4}(y_0 + \omega_4^2 y_1 + \omega_4^4 y_2 + \omega_4^6 y_3)$$

$$d_3 = \frac{1}{4}(y_0 + \omega_4^3 y_1 + \omega_4^6 y_2 + \omega_4^9 y_3)$$

If we set

$$\mathbf{z} = \frac{1}{4}\mathbf{y} = \frac{1}{4}(y_0, y_1, y_2, y_3)^T$$

then the vector $\mathbf{d} = (d_0, d_1, d_2, d_3)^T$ is determined by multiplying $\mathbf{z}$ by the matrix

$$F_4 = \begin{bmatrix} 1 & 1 & 1 & 1 \\ 1 & \omega_4 & \omega_4^2 & \omega_4^3 \\ 1 & \omega_4^2 & \omega_4^4 & \omega_4^6 \\ 1 & \omega_4^3 & \omega_4^6 & \omega_4^9 \end{bmatrix} = \begin{bmatrix} 1 & 1 & 1 & 1 \\ 1 & -i & -1 & i \\ 1 & -1 & 1 & -1 \\ 1 & i & -1 & -i \end{bmatrix}$$

The matrix F_4 is called a *Fourier matrix*.

In the case of N sample values, $y_0, y_1, \ldots, y_{N-1}$, the coefficients are computed by setting

$$\mathbf{z} = \frac{1}{N}\mathbf{y} \qquad \text{and} \qquad \mathbf{d} = F_N \mathbf{z}$$

where $\mathbf{y} = (y_0, y_1, \ldots, y_{N-1})^T$ and F_N is the $N \times N$ matrix whose (j, k) entry is given by $f_{j,k} = \omega_N^{(j-1)(k-1)}$. The multiplication of F_N times $\mathbf{z}$ requires a multiple of N^2 arithmetic operations (roughly $8N^2$, since complex arithmetic is used).

In signal processing applications, N is generally very large and consequently the computation of the discrete Fourier transform can be prohibitively slow and costly even on modern high-powered computers. A revolution in signal processing occurred in 1965 with the introduction by J. W. Cooley and J. W. Tukey of a dramatically more efficient method for computing the discrete Fourier transform.

The Fast Fourier Transform

The method of Cooley and Tukey, known as the *fast Fourier transform* or simply the *FFT*, takes advantage of the special structure of the Fourier matrices. We illustrate this method in the case $N = 4$. To see the special structure, we rearrange the columns of F_4 so that its odd-numbered columns all come before the even-numbered columns. This rearrangement is equivalent to postmultiplying F_4 by the permutation matrix

$$P_4 = \begin{bmatrix} 1 & 0 & 0 & 0 \\ 0 & 0 & 1 & 0 \\ 0 & 1 & 0 & 0 \\ 0 & 0 & 0 & 1 \end{bmatrix}$$

If we set $\mathbf{w} = P_4^T \mathbf{z}$, then

$$F_4 \mathbf{z} = F_4 P_4 P_4^T \mathbf{z} = F_4 P_4 \mathbf{w}$$

Partitioning $F_4 P_4$ into 2×2 blocks, we get

$$F_4 P_4 = \begin{bmatrix} 1 & 1 & 1 & 1 \\ 1 & -1 & -i & i \\ 1 & 1 & -1 & -1 \\ 1 & -1 & i & -i \end{bmatrix}$$

The (1,1) and (2,1) blocks are both equal to the Fourier matrix F_2, and if we set

$$D_2 = \begin{bmatrix} 1 & 0 \\ 0 & -i \end{bmatrix}$$

then the (1,2) and (2,2) blocks are $D_2 F_2$ and $-D_2 F_2$, respectively. The computation of the Fourier transform can now be carried out as a block multiplication.

$$\mathbf{d}_4 = \begin{bmatrix} F_2 & D_2 F_2 \\ F_2 & -D_2 F_2 \end{bmatrix} \begin{bmatrix} \mathbf{w}_1 \\ \mathbf{w}_2 \end{bmatrix} = \begin{bmatrix} F_2 \mathbf{w}_1 + D_2 F_2 \mathbf{w}_2 \\ F_2 \mathbf{w}_1 - D_2 F_2 \mathbf{w}_2 \end{bmatrix}$$

The computation reduces to computing two Fourier transforms of length 2. If we set $\mathbf{q}_1 = F_2 \mathbf{w}_1$ and $\mathbf{q}_2 = D_2(F_2 \mathbf{w}_2)$, then

$$\mathbf{d}_4 = \begin{bmatrix} \mathbf{q}_1 + \mathbf{q}_2 \\ \mathbf{q}_1 - \mathbf{q}_2 \end{bmatrix}$$

The procedure we have just described will work in general whenever the number of sample points is even. If, say, $N = 2m$, and we permute the columns of F_{2m} so that the odd columns are first, then the reordered Fourier matrix $F_{2m} P_{2m}$ can be partitioned into $m \times m$ blocks

$$F_{2m} P_{2m} = \begin{bmatrix} F_m & D_m F_m \\ F_m & -D_m F_m \end{bmatrix}$$

where D_m is a diagonal matrix whose (j, j) entry is ω_m^{j-1}. The discrete Fourier transform can then be computed in terms of two transforms of length m. Furthermore, if m is even, then each length m transform can be computed in terms of two transforms of length $\frac{m}{2}$, and so on.

If, initially, N is a power of 2, say $N = 2^k$, then we can apply this procedure recursively through k levels of recursion. The amount of arithmetic required to compute the FFT is proportional to $Nk = N \log_2 N$. In fact, the actual amount of arithmetic operations required for the FFT is approximately $5N \log_2 N$. How dramatic of a speedup is this? If we consider for example the case where $N = 2^{20} = 1,048,576$, then the discrete Fourier transform requires $8N^2 = 8 \cdot 2^{40}$ operations, that is, approximately 8.8 trillion operations. On the other hand, the FFT requires only $100N = 100 \cdot 2^{20}$ or approximately 100 million operations. The ratio of these two operations counts is

$$r = \frac{8N^2}{5N \log_2 N} = 0.08 \cdot 1{,}048{,}576 = 83{,}886$$

In this case the FFT is approximately 84,000 times faster than the discrete Fourier transform.

EXERCISES

1. Which of the following sets of vectors form an orthonormal basis for R^2?

   (a) $\{(1, 0)^T, (0, 1)^T\}$

   (b) $\left\{ \left(\frac{3}{5}, \frac{4}{5}\right)^T, \left(\frac{5}{13}, \frac{12}{13}\right)^T \right\}$

   (c) $\{(1, -1)^T, (1, 1)^T\}$

   (d) $\left\{ \left(\frac{\sqrt{3}}{2}, \frac{1}{2}\right)^T, \left(-\frac{1}{2}, \frac{\sqrt{3}}{2}\right)^T \right\}$

2. Let

$$\mathbf{u}_1 = \begin{bmatrix} \frac{1}{3\sqrt{2}} \\ \frac{1}{3\sqrt{2}} \\ -\frac{4}{3\sqrt{2}} \end{bmatrix}, \quad \mathbf{u}_2 = \begin{bmatrix} \frac{2}{3} \\ \frac{2}{3} \\ \frac{1}{3} \end{bmatrix}, \quad \mathbf{u}_3 = \begin{bmatrix} \frac{1}{\sqrt{2}} \\ -\frac{1}{\sqrt{2}} \\ 0 \end{bmatrix}$$

   (a) Show that $\{\mathbf{u}_1, \mathbf{u}_2, \mathbf{u}_3\}$ is an orthonormal basis for R^3.
   (b) Let $\mathbf{x} = (1, 1, 1)^T$. Write $\mathbf{x}$ as a linear combination of $\mathbf{u}_1$, $\mathbf{u}_2$, and $\mathbf{u}_3$ using Theorem 5.5.2 and use Parseval's formula to compute $\|\mathbf{x}\|$.

3. Let S be the subspace of R^3 spanned by the vectors $\mathbf{u}_2$ and $\mathbf{u}_3$ of Exercise 2. Let $\mathbf{x} = (1, 2, 2)^T$. Find the projection $\mathbf{p}$ of $\mathbf{x}$ onto S. Show that $(\mathbf{p} - \mathbf{x}) \perp \mathbf{u}_2$ and $(\mathbf{p} - \mathbf{x}) \perp \mathbf{u}_3$.

4. Let θ be a fixed real number and let

$$\mathbf{x}_1 = \begin{bmatrix} \cos\theta \\ \sin\theta \end{bmatrix} \quad \text{and} \quad \mathbf{x}_2 = \begin{bmatrix} -\sin\theta \\ \cos\theta \end{bmatrix}$$

   (a) Show that $\{\mathbf{x}_1, \mathbf{x}_2\}$ is an orthonormal basis for R^2.
   (b) Given a vector $\mathbf{y}$ in R^2, write it as a linear combination $c_1\mathbf{x}_1 + c_2\mathbf{x}_2$.
   (c) Verify that

$$c_1^2 + c_2^2 = \|\mathbf{y}\|^2 = y_1^2 + y_2^2$$

5. Let $\mathbf{u}_1$ and $\mathbf{u}_2$ form an orthonormal basis for R^2 and let $\mathbf{u}$ be a unit vector in R^2. If $\mathbf{u}^T\mathbf{u}_1 = \frac{1}{2}$, determine the value of $|\mathbf{u}^T\mathbf{u}_2|$.

6. Let $\{\mathbf{u}_1, \mathbf{u}_2, \mathbf{u}_3\}$ be an orthonormal basis for an inner product space V and let

$$\mathbf{u} = \mathbf{u}_1 + 2\mathbf{u}_2 + 2\mathbf{u}_3 \qquad \text{and} \qquad \mathbf{v} = \mathbf{u}_1 + 7\mathbf{u}_3$$

Determine the value of each of the following.

(a) $\langle \mathbf{u}, \mathbf{v} \rangle$

(b) $\|\mathbf{u}\|$ and $\|\mathbf{v}\|$

(c) The angle θ between $\mathbf{u}$ and $\mathbf{v}$

7. Let $\{\mathbf{u}_1, \mathbf{u}_2, \mathbf{u}_3\}$ be an orthonormal basis for an inner product space V. If $\mathbf{x} = c_1\mathbf{u}_1 + c_2\mathbf{u}_2 + c_3\mathbf{u}_3$ is a vector with the properties $\|\mathbf{x}\| = 5$, $\langle \mathbf{u}_1, \mathbf{x} \rangle = 4$, and $\mathbf{x} \perp \mathbf{u}_2$, then what are the possible values of c_1, c_2, c_3?

8. The functions $\cos x$ and $\sin x$ form an orthonormal set in $C[-\pi, \pi]$. If

$$f(x) = 3\cos x + 2\sin x \quad \text{and} \quad g(x) = \cos x - \sin x$$

use Corollary 5.5.3 to determine the value of

$$\langle f, g \rangle = \frac{1}{\pi} \int_{-\pi}^{\pi} f(x)g(x)\, dx$$

9. The set

$$S = \left\{ \frac{1}{\sqrt{2}}, \cos x, \cos 2x, \cos 3x, \cos 4x \right\}$$

is an orthonormal set of vectors in $C[-\pi, \pi]$ with inner product defined by (2).

(a) Use trigonometric identities to write the function $\sin^4 x$ as a linear combination of elements of S.

(b) Use part (a) and Theorem 5.5.2 to find the values of the following integrals.

(i) $\int_{-\pi}^{\pi} \sin^4 x \cos x\, dx$ (ii) $\int_{-\pi}^{\pi} \sin^4 x \cos 2x\, dx$

(iii) $\int_{-\pi}^{\pi} \sin^4 x \cos 3x\, dx$ (iv) $\int_{-\pi}^{\pi} \sin^4 x \cos 4x\, dx$

10. Write out the Fourier matrix F_8. Show that $F_8 P_8$ can be partitioned into block form

$$\begin{bmatrix} F_4 & D_4 F_4 \\ F_4 & -D_4 F_4 \end{bmatrix}$$

11. Prove that the transpose of an orthogonal matrix is an orthogonal matrix.

12. If Q is an $n \times n$ orthogonal matrix and $\mathbf{x}$ and $\mathbf{y}$ are nonzero vectors in R^n, then how does the angle between $Q\mathbf{x}$ and $Q\mathbf{y}$ compare to the angle between $\mathbf{x}$ and $\mathbf{y}$? Prove your answer.

13. Let Q be an $n \times n$ orthogonal matrix. Use mathematical induction to prove each of the following.

(a) $(Q^m)^{-1} = (Q^T)^m = (Q^m)^T$ for any positive integer m.

(b) $\|Q^m \mathbf{x}\| = \|\mathbf{x}\|$ for any $\mathbf{x} \in R^n$.

14. Let **u** be a unit vector in R^n and let $H = I - 2\mathbf{u}\mathbf{u}^T$. Show that H is both orthogonal and symmetric and hence is its own inverse.

15. Let Q be an orthogonal matrix and let $d = \det(Q)$. Show that $|d| = 1$.

16. Show that the product of two orthogonal matrices is also an orthogonal matrix. Is the product of two permutation matrices a permutation matrix? Explain.

17. Show that if U is an $n \times n$ orthogonal matrix, then

$$\mathbf{u}_1\mathbf{u}_1^T + \mathbf{u}_2\mathbf{u}_2^T + \cdots + \mathbf{u}_n\mathbf{u}_n^T = I$$

18. Use mathematical induction to show that if $Q \in R^{n \times n}$ is both upper triangular and orthogonal, then $\mathbf{q}_j = \pm\mathbf{e}_j$, $j = 1, \ldots, n$.

19. Let

$$A = \begin{pmatrix} \frac{1}{2} & -\frac{1}{2} \\ \frac{1}{2} & -\frac{1}{2} \\ \frac{1}{2} & \frac{1}{2} \\ \frac{1}{2} & \frac{1}{2} \end{pmatrix}$$

(a) Show that the column vectors of A form an orthonormal set in R^4.
(b) Solve the least squares problem $A\mathbf{x} = \mathbf{b}$ for each of the following choices of **b**.
    (i) $\mathbf{b} = (4, 0, 0, 0)^T$ (ii) $\mathbf{b} = (1, 2, 3, 4)^T$ (iii) $\mathbf{b} = (1, 1, 2, 2)^T$

20. Let A be the matrix given in Exercise 19.

(a) Find the projection matrix P that projects vectors in R^4 onto $R(A)$.
(b) For each of your solutions **x** to Exercise 19(b), compute $A\mathbf{x}$ and compare it to $P\mathbf{b}$.

21. Let A be the matrix given in Exercise 19.

(a) Find an orthonormal basis for $N(A^T)$.
(b) Determine the projection matrix Q that projects vectors in R^4 onto $N(A^T)$.

22. Let A be an $m \times n$ matrix, let P be the projection matrix that projects vectors in R^m onto $R(A)$, and let Q be the projection matrix that projects vectors in R^n onto $R(A^T)$. Show that:

(a) $I - P$ is the projection matrix from R^m onto $N(A^T)$.
(b) $I - Q$ is the projection matrix from R^n onto $N(A)$.

23. Let P be the projection matrix corresponding to a subspace S of R^m. Show that:
(a) $P^2 = P$ (b) $P^T = P$

24. Let A be an $m \times n$ matrix whose column vectors are mutually orthogonal and let $\mathbf{b} \in R^m$. Show that if **y** is the least squares solution to the system $A\mathbf{x} = \mathbf{b}$ then

$$y_i = \frac{\mathbf{b}^T \mathbf{a}_i}{\mathbf{a}_i^T \mathbf{a}_i} \qquad i = 1, \ldots, n$$

25. Given the vector space $C[-1, 1]$ with inner product

$$\langle f, g \rangle = \int_{-1}^{1} f(x)g(x)\, dx$$

and norm

$$\|f\| = (\langle f, f \rangle)^{1/2}$$

(a) Show that the vectors 1 and x are orthogonal.
(b) Compute $\|1\|$ and $\|x\|$.
(c) Find the best least squares approximation to $x^{1/3}$ on $[-1, 1]$ by a linear function $l(x) = c_1 1 + c_2 x$.
(d) Sketch the graphs of $x^{1/3}$ and $l(x)$ on $[-1, 1]$.

26. Consider the inner product space $C[0, 1]$ with inner product defined by

$$\langle f, g \rangle = \int_{0}^{1} f(x)g(x)\, dx$$

Let S be the subspace spanned by the vectors 1 and $2x - 1$.

(a) Show that 1 and $2x - 1$ are orthogonal.
(b) Determine $\|1\|$ and $\|2x - 1\|$.
(c) Find the best least squares approximation to $\sqrt{x}$ by a function from the subspace S.

27. Let

$$S = \{1/\sqrt{2}, \cos x, \cos 2x, \ldots, \cos nx, \sin x, \sin 2x, \ldots, \sin nx\}$$

Show that S is an orthonormal set in $C[-\pi, \pi]$ with inner product defined by (2).

28. Find the best least squares approximation to $f(x) = |x|$ on $[-\pi, \pi]$ by a trigonometric polynomial of degree less than or equal to 2.

29. Let $\{\mathbf{x}_1, \mathbf{x}_2, \ldots, \mathbf{x}_k, \mathbf{x}_{k+1}, \ldots, \mathbf{x}_n\}$ be an orthonormal basis for an inner product space V. Let S_1 be the subspace of V spanned by $\mathbf{x}_1, \ldots, \mathbf{x}_k$, and let S_2 be the subspace spanned by $\mathbf{x}_{k+1}, \mathbf{x}_{k+2}, \ldots, \mathbf{x}_n$. Show that $S_1 \perp S_2$.

30. Let $\mathbf{x}$ be an element of the inner product space V in Exercise 29, and let $\mathbf{p}_1$ and $\mathbf{p}_2$ be the projections of $\mathbf{x}$ onto S_1 and S_2, respectively. Show that:

(a) $\mathbf{x} = \mathbf{p}_1 + \mathbf{p}_2$.
(b) If $\mathbf{x} \in S_1^{\perp}$, then $\mathbf{p}_1 = \mathbf{0}$ and hence $S^{\perp} = S_2$.

31. Let S be a subspace of an inner product space V. Let $\{\mathbf{x}_1, \ldots, \mathbf{x}_n\}$ be an orthogonal basis for S and let $\mathbf{x} \in V$. Show that the best least squares approximation to $\mathbf{x}$ by elements of S is given by

$$\mathbf{p} = \sum_{i=1}^{n} \frac{\langle \mathbf{x}, \mathbf{x}_i \rangle}{\langle \mathbf{x}_i, \mathbf{x}_i \rangle} \mathbf{x}_i$$

6 THE GRAM–SCHMIDT ORTHOGONALIZATION PROCESS

In this section we learn a process for constructing an orthonormal basis for an n-dimensional inner product space V. The method involves using projections transform an ordinary basis $\{\mathbf{x}_1, \mathbf{x}_2, \dots, \mathbf{x}_n\}$ into an orthonormal basis $\{\mathbf{u}_1, \mathbf{u}_2, \dots, \mathbf{u}_n\}$.

We will construct the $\mathbf{u}_i$'s so that

$$\text{Span}(\mathbf{u}_1, \dots, \mathbf{u}_k) = \text{Span}(\mathbf{x}_1, \dots, \mathbf{x}_k)$$

for $k = 1, \dots, n$. To begin the process, let

$$(1) \qquad \mathbf{u}_1 = \left(\frac{1}{\|\mathbf{x}_1\|}\right)\mathbf{x}_1$$

$\text{Span}(\mathbf{u}_1) = \text{Span}(\mathbf{x}_1)$, since $\mathbf{u}_1$ is a unit vector in the direction of $\mathbf{x}_1$. Let $\mathbf{p}_1$ denote the projection of $\mathbf{x}_2$ onto $\text{Span}(\mathbf{x}_1) = \text{Span}(\mathbf{u}_1)$.

$$\mathbf{p}_1 = \langle \mathbf{x}_2, \mathbf{u}_1 \rangle \, \mathbf{u}_1$$

By Theorem 5.5.7,

$$(\mathbf{x}_2 - \mathbf{p}_1) \perp \mathbf{u}_1$$

Note that $\mathbf{x}_2 - \mathbf{p}_1 \neq \mathbf{0}$, since

$$(2) \qquad \mathbf{x}_2 - \mathbf{p}_1 = \frac{-\langle \mathbf{x}_2, \mathbf{u}_1 \rangle}{\|\mathbf{x}_1\|}\mathbf{x}_1 + \mathbf{x}_2$$

and $\mathbf{x}_1$ and $\mathbf{x}_2$ are linearly independent. If we set

$$(3) \qquad \mathbf{u}_2 = \frac{1}{\|\mathbf{x}_2 - \mathbf{p}_1\|}(\mathbf{x}_2 - \mathbf{p}_1)$$

then $\mathbf{u}_2$ is a unit vector orthogonal to $\mathbf{u}_1$. It follows from (1), (2), and (3) that $\text{Span}(\mathbf{u}_1, \mathbf{u}_2) \subset \text{Span}(\mathbf{x}_1, \mathbf{x}_2)$. Since $\mathbf{u}_1$ and $\mathbf{u}_2$ are linearly independent, it follows that $\{\mathbf{u}_1, \mathbf{u}_2\}$ is an orthonormal basis for $\text{Span}(\mathbf{x}_1, \mathbf{x}_2)$, and hence $\text{Span}(\mathbf{x}_1, \mathbf{x}_2) = \text{Span}(\mathbf{u}_1, \mathbf{u}_2)$.

To construct $\mathbf{u}_3$, continue in the same manner. Let $\mathbf{p}_2$ be the projection of $\mathbf{x}_3$ onto $\text{Span}(\mathbf{x}_1, \mathbf{x}_2) = \text{Span}(\mathbf{u}_1, \mathbf{u}_2)$,

$$\mathbf{p}_2 = \langle \mathbf{x}_3, \mathbf{u}_1 \rangle \, \mathbf{u}_1 + \langle \mathbf{x}_3, \mathbf{u}_2 \rangle \, \mathbf{u}_2$$

and set

$$\mathbf{u}_3 = \frac{1}{\|\mathbf{x}_3 - \mathbf{p}_2\|}(\mathbf{x}_3 - \mathbf{p}_2)$$

and so on (see Figure 5.6.1).

▶ **THEOREM 5.6.1 (The Gram–Schmidt Process)** *Let $\{\mathbf{x}_1, \mathbf{x}_2, \dots, \mathbf{x}_n\}$ be a basis for the inner product space V. Let*

$$\mathbf{u}_1 = \left(\frac{1}{\|\mathbf{x}_1\|}\right)\mathbf{x}_1$$

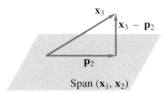

FIGURE 5.6.1

and define $\mathbf{u}_2, \ldots, \mathbf{u}_n$ *recursively by*

$$\mathbf{u}_{k+1} = \frac{1}{\|\mathbf{x}_{k+1} - \mathbf{p}_k\|}(\mathbf{x}_{k+1} - \mathbf{p}_k) \quad for \; k = 1, \ldots, n-1$$

where

$$\mathbf{p}_k = \langle \mathbf{x}_{k+1}, \mathbf{u}_1 \rangle \mathbf{u}_1 + \langle \mathbf{x}_{k+1}, \mathbf{u}_2 \rangle \mathbf{u}_2 + \cdots + \langle \mathbf{x}_{k+1}, \mathbf{u}_k \rangle \mathbf{u}_k$$

is the projection of $\mathbf{x}_{k+1}$ *onto* $\mathrm{Span}(\mathbf{u}_1, \mathbf{u}_2, \ldots, \mathbf{u}_k)$. *The set*

$$\{\mathbf{u}_1, \mathbf{u}_2, \ldots, \mathbf{u}_n\}$$

is an orthonormal basis for V.

▶*Proof.* We will argue inductively. Clearly, $\mathrm{Span}(\mathbf{u}_1) = \mathrm{Span}(\mathbf{x}_1)$. Suppose that $\mathbf{u}_1, \mathbf{u}_2, \ldots, \mathbf{u}_k$ have been constructed so that $\{\mathbf{u}_1, \mathbf{u}_2, \ldots, \mathbf{u}_k\}$ is an orthonormal set and

$$\mathrm{Span}(\mathbf{u}_1, \mathbf{u}_2, \ldots, \mathbf{u}_k) = \mathrm{Span}(\mathbf{x}_1, \mathbf{x}_2, \ldots, \mathbf{x}_k)$$

Since $\mathbf{p}_k$ is a linear combination of the $\mathbf{u}_i$'s, $1 \leq i \leq k$, it follows that $\mathbf{p}_k \in \mathrm{Span}(\mathbf{x}_1, \ldots, \mathbf{x}_k)$ and $\mathbf{x}_{k+1} - \mathbf{p}_k \in \mathrm{Span}(\mathbf{x}_1, \ldots, \mathbf{x}_{k+1})$.

$$\mathbf{x}_{k+1} - \mathbf{p}_k = \mathbf{x}_{k+1} - \sum_{i=1}^{k} c_i \mathbf{x}_i$$

Since $\mathbf{x}_1, \ldots, \mathbf{x}_{k+1}$ are linearly independent, it follows that $\mathbf{x}_{k+1} - \mathbf{p}_k$ is nonzero and by Theorem 5.5.7 it is orthogonal to each $\mathbf{u}_i$, $1 \leq i \leq k$. Thus $\{\mathbf{u}_1, \mathbf{u}_2, \ldots, \mathbf{u}_{k+1}\}$ is an orthonormal set of vectors in $\mathrm{Span}(\mathbf{x}_1, \ldots, \mathbf{x}_{k+1})$. Since $\mathbf{u}_1, \ldots, \mathbf{u}_{k+1}$ are linearly independent, they form a basis for $\mathrm{Span}(\mathbf{x}_1, \ldots, \mathbf{x}_{k+1})$, and consequently

$$\mathrm{Span}(\mathbf{u}_1, \ldots, \mathbf{u}_{k+1}) = \mathrm{Span}(\mathbf{x}_1, \ldots, \mathbf{x}_{k+1})$$

It follows by mathematical induction that $\{\mathbf{u}_1, \mathbf{u}_2, \ldots, \mathbf{u}_n\}$ is an orthonormal basis for V.   ◀

EXAMPLE 1. Find an orthonormal basis for P_3 if the inner product on P_3 is defined by

$$\langle p, q \rangle = \sum_{i=1}^{3} p(x_i)q(x_i)$$

where $x_1 = -1$, $x_2 = 0$, and $x_3 = 1$.

SOLUTION. Starting with the basis $\{1, x, x^2\}$, we can use the Gram–Schmidt process to generate an orthonormal basis.

$$\|1\|^2 = \langle 1, 1 \rangle = 3$$

so

$$\mathbf{u}_1 = \left(\frac{1}{\|1\|}\right) 1 = \frac{1}{\sqrt{3}}$$

Set

$$p_1 = \left\langle x, \frac{1}{\sqrt{3}}\right\rangle \frac{1}{\sqrt{3}} = \left(-1 \cdot \frac{1}{\sqrt{3}} + 0 \cdot \frac{1}{\sqrt{3}} + 1 \cdot \frac{1}{\sqrt{3}}\right)\frac{1}{\sqrt{3}} = 0$$

Therefore,

$$x - p_1 = x \qquad \text{and} \qquad \|x - p_1\|^2 = \langle x, x\rangle = 2$$

Hence

$$\mathbf{u}_2 = \frac{1}{\sqrt{2}}x$$

Finally,

$$p_2 = \left\langle x^2, \frac{1}{\sqrt{3}}\right\rangle \frac{1}{\sqrt{3}} + \left\langle x^2, \frac{1}{\sqrt{2}}x\right\rangle \frac{1}{\sqrt{2}}x = \frac{2}{3}$$

$$\|x^2 - p_2\|^2 = \left\langle x^2 - \frac{2}{3}, x^2 - \frac{2}{3}\right\rangle = \frac{2}{3}$$

and hence

$$\mathbf{u}_3 = \frac{\sqrt{6}}{2}\left(x^2 - \frac{2}{3}\right) \qquad \blacktriangleleft$$

Orthogonal polynomials will be studied in more detail in Section 7.

EXAMPLE 2. Let

$$A = \begin{bmatrix} 1 & -1 & 4 \\ 1 & 4 & -2 \\ 1 & 4 & 2 \\ 1 & -1 & 0 \end{bmatrix}$$

Find an orthonormal basis for the column space of A.

SOLUTION. The column vectors of A are linearly independent and hence form a basis for a three-dimensional subspace of R^4. The Gram–Schmidt process can be used to construct an orthonormal basis as follows. Set

$$r_{11} = \|\mathbf{a}_1\| = 2$$

$$\mathbf{q}_1 = \frac{1}{r_{11}}\mathbf{a}_1 = \left(\frac{1}{2}, \frac{1}{2}, \frac{1}{2}, \frac{1}{2}\right)^T$$

$$r_{12} = \langle\mathbf{a}_2, \mathbf{q}_1\rangle = \mathbf{q}_1^T\mathbf{a}_2 = 3$$

$$\mathbf{p}_1 = r_{12}\mathbf{q}_1 = 3\mathbf{q}_1$$

$$\mathbf{a}_2 - \mathbf{p}_1 = \left(-\frac{5}{2}, \frac{5}{2}, \frac{5}{2}, -\frac{5}{2}\right)^T$$

$$r_{22} = \|\mathbf{a}_2 - \mathbf{p}_1\| = 5$$

$$\mathbf{q}_2 = \frac{1}{r_{22}}(\mathbf{a}_2 - \mathbf{p}_1) = \left(-\frac{1}{2}, \frac{1}{2}, \frac{1}{2}, -\frac{1}{2}\right)^T$$

$$r_{13} = \langle \mathbf{a}_3, \mathbf{q}_1 \rangle = \mathbf{q}_1^T \mathbf{a}_3 = 2, \qquad r_{23} = \langle \mathbf{a}_3, \mathbf{q}_2 \rangle = \mathbf{q}_2^T \mathbf{a}_3 = -2$$

$$\mathbf{p}_2 = r_{13}\mathbf{q}_1 + r_{23}\mathbf{q}_2 = (2, 0, 0, 2)^T$$

$$\mathbf{a}_3 - \mathbf{p}_2 = (2, -2, 2, -2)^T$$

$$r_{33} = \|\mathbf{a}_3 - \mathbf{p}_2\| = 4$$

$$\mathbf{q}_3 = \frac{1}{r_{33}}(\mathbf{a}_3 - \mathbf{p}_2) = \left(\frac{1}{2}, -\frac{1}{2}, \frac{1}{2}, -\frac{1}{2}\right)^T$$

The vectors $\mathbf{q}_1, \mathbf{q}_2, \mathbf{q}_3$ form an orthonormal basis for $R(A)$. ◀

We can obtain a useful factorization of the matrix A if we keep track of all the inner products and norms computed in the Gram–Schmidt process. For the matrix in Example 2, if the r_{ij}'s are used to form a matrix

$$R = \begin{bmatrix} r_{11} & r_{12} & r_{13} \\ 0 & r_{22} & r_{23} \\ 0 & 0 & r_{33} \end{bmatrix} = \begin{bmatrix} 2 & 3 & 2 \\ 0 & 5 & -2 \\ 0 & 0 & 4 \end{bmatrix}$$

and we set

$$Q = (\mathbf{q}_1, \mathbf{q}_2, \mathbf{q}_3) = \begin{bmatrix} \frac{1}{2} & -\frac{1}{2} & \frac{1}{2} \\ \frac{1}{2} & \frac{1}{2} & -\frac{1}{2} \\ \frac{1}{2} & \frac{1}{2} & \frac{1}{2} \\ \frac{1}{2} & -\frac{1}{2} & -\frac{1}{2} \end{bmatrix}$$

then it is easily verified that $QR = A$. This result is proved in the following theorem.

▶ **THEOREM 5.6.2 (QR Factorization)** *If A is an $m \times n$ matrix of rank n, then A can be factored into a product QR, where Q is an $m \times n$ matrix with orthonormal columns and R is an $n \times n$ matrix that is upper triangular and invertible.*

▶ *Proof.* Let $\mathbf{p}_1, \ldots, \mathbf{p}_{n-1}$ be the projection vectors defined in Theorem 5.6.1, and let $\{\mathbf{q}_1, \mathbf{q}_2, \ldots, \mathbf{q}_n\}$ be the orthonormal basis of $R(A)$ derived from the Gram–Schmidt process. Define

$$r_{11} = \|\mathbf{a}_1\|$$

$$r_{kk} = \|\mathbf{a}_k - \mathbf{p}_{k-1}\| \qquad \text{for} \qquad k = 2, \ldots, n$$

and

$$r_{ik} = \mathbf{q}_i^T \mathbf{a}_k \qquad \text{for} \quad i = 1, \ldots, k-1 \qquad \text{and} \qquad k = 2, \ldots, n$$

By the Gram–Schmidt process,

(4) $r_{11}\mathbf{q}_1 = \mathbf{a}_1$

$r_{kk}\mathbf{q}_k = \mathbf{a}_k - r_{1k}\mathbf{q}_1 - r_{2k}\mathbf{q}_2 - \cdots - r_{k-1,k}\mathbf{q}_{k-1}$ for $k = 2, \ldots, n$

System (4) may be rewritten in the form

$$\mathbf{a}_1 = r_{11}\mathbf{q}_1$$

$$\mathbf{a}_2 = r_{12}\mathbf{q}_1 + r_{22}\mathbf{q}_2$$

$$\vdots$$

$$\mathbf{a}_n = r_{1n}\mathbf{q}_1 + \cdots + r_{nn}\mathbf{q}_n$$

If we set

$$Q = (\mathbf{q}_1, \mathbf{q}_2, \ldots, \mathbf{q}_n)$$

and define R to be the upper triangular matrix

$$R = \begin{bmatrix} r_{11} & r_{12} & \cdots & r_{1n} \\ 0 & r_{22} & \cdots & r_{2n} \\ \vdots & & & \\ 0 & 0 & \cdots & r_{nn} \end{bmatrix}$$

then the jth column of the product QR will be

$$Q\mathbf{r}_j = r_{1j}\mathbf{q}_1 + r_{2j}\mathbf{q}_2 + \cdots + r_{jj}\mathbf{q}_j = \mathbf{a}_j$$

for $j = 1, \ldots, n$. Therefore,

$$QR = (\mathbf{a}_1, \mathbf{a}_2, \ldots, \mathbf{a}_n) = A$$ ◀

EXAMPLE 3.   Compute the Gram–Schmidt QR factorization of the matrix

$$A = \begin{bmatrix} 1 & -2 & -1 \\ 2 & 0 & 1 \\ 2 & -4 & 2 \\ 4 & 0 & 0 \end{bmatrix}$$

SOLUTION.

Step 1. Set

$$r_{11} = \|\mathbf{a}_1\| = 5$$

$$\mathbf{q}_1 = \frac{1}{r_{11}}\mathbf{a}_1 = \left(\frac{1}{5}, \frac{2}{5}, \frac{2}{5}, \frac{4}{5}\right)^T$$

Step 2. Set

$$r_{12} = \mathbf{q}_1^T \mathbf{a}_2 = -2$$

$$\mathbf{p}_1 = r_{12}\mathbf{q}_1 = -2\mathbf{q}_1$$

$$\mathbf{a}_2 - \mathbf{p}_1 = \left(-\frac{8}{5}, \frac{4}{5}, -\frac{16}{5}, \frac{8}{5}\right)^T$$

$$r_{22} = \|\mathbf{a}_2 - \mathbf{p}_1\| = 4$$

$$\mathbf{q}_2 = \frac{1}{r_{22}}(\mathbf{a}_2 - \mathbf{p}_1) = \left(-\frac{2}{5}, \frac{1}{5}, -\frac{4}{5}, \frac{2}{5}\right)^T$$

Step 3. Set

$$r_{13} = \mathbf{q}_1^T \mathbf{a}_3 = 1, \qquad r_{23} = \mathbf{q}_2^T \mathbf{a}_3 = -1$$

$$\mathbf{p}_2 = r_{13}\mathbf{q}_1 + r_{23}\mathbf{q}_2 = \mathbf{q}_1 - \mathbf{q}_2 = \left(\frac{3}{5}, \frac{1}{5}, \frac{6}{5}, \frac{2}{5}\right)^T$$

$$\mathbf{a}_3 - \mathbf{p}_2 = \left(-\frac{8}{5}, \frac{4}{5}, \frac{4}{5}, -\frac{2}{5}\right)^T$$

$$r_{33} = \|\mathbf{a}_3 - \mathbf{p}_2\| = 2$$

$$\mathbf{q}_3 = \frac{1}{r_{33}}(\mathbf{a}_3 - \mathbf{p}_2) = \left(-\frac{4}{5}, \frac{2}{5}, \frac{2}{5}, -\frac{1}{5}\right)^T$$

At each step we have determined a column of Q and a column of R. The factorization is given by

$$A = QR = \begin{bmatrix} \frac{1}{5} & -\frac{2}{5} & -\frac{4}{5} \\ \frac{2}{5} & \frac{1}{5} & \frac{2}{5} \\ \frac{2}{5} & -\frac{4}{5} & \frac{2}{5} \\ \frac{4}{5} & \frac{2}{5} & -\frac{1}{5} \end{bmatrix} \begin{bmatrix} 5 & -2 & 1 \\ 0 & 4 & -1 \\ 0 & 0 & 2 \end{bmatrix} \qquad \blacktriangleleft$$

We saw in Section 5 that, if the columns of an $m \times n$ matrix A form an orthonormal set, then the least squares solution to $A\mathbf{x} = \mathbf{b}$ is simply $\hat{\mathbf{x}} = A^T\mathbf{b}$. If A has rank n, but its column vectors do not form an orthonormal set in R^m, then the QR factorization can be used to solve the least squares problem.

▶ **THEOREM 5.6.3** *If A is an $m \times n$ matrix of rank n, then the solution to the least squares problem $A\mathbf{x} = \mathbf{b}$ is given by $\hat{\mathbf{x}} = R^{-1}Q^T\mathbf{b}$, where Q and R are the matrices obtained from the factorization given in Theorem 5.6.2. The solution $\hat{\mathbf{x}}$ may be obtained by using back substitution to solve $R\mathbf{x} = Q^T\mathbf{b}$.*

▶*Proof.* Let $\hat{\mathbf{x}}$ be the solution to the least squares problem $A\mathbf{x} = \mathbf{b}$ guaranteed by Theorem 5.3.2. Thus $\hat{\mathbf{x}}$ is the solution to the normal equations

$$A^T A\mathbf{x} = A^T\mathbf{b}$$

If A is factored into a product QR, these equations become

$$(QR)^T QR\mathbf{x} = (QR)^T\mathbf{b}$$

or

$$R^T(Q^TQ)R\mathbf{x} = R^TQ^T\mathbf{b}$$

Since Q has orthonormal columns, it follows that $Q^TQ = I$ and hence

$$R^TR\mathbf{x} = R^TQ^T\mathbf{b}$$

Since R^T is invertible, this simplifies to

$$R\mathbf{x} = Q^T\mathbf{b} \qquad \text{or} \qquad \mathbf{x} = R^{-1}Q^T\mathbf{b} \qquad \blacktriangleleft$$

EXAMPLE 4. Find the least squares solution to

$$\begin{bmatrix} 1 & -2 & -1 \\ 2 & 0 & 1 \\ 2 & -4 & 2 \\ 4 & 0 & 0 \end{bmatrix} \begin{bmatrix} x_1 \\ x_2 \\ x_3 \end{bmatrix} = \begin{bmatrix} -1 \\ 1 \\ 1 \\ -2 \end{bmatrix}$$

SOLUTION. The coefficient matrix of this system was factored in Example 3. Using that factorization, we have

$$Q^T\mathbf{b} = \begin{bmatrix} \frac{1}{5} & \frac{2}{5} & \frac{2}{5} & \frac{4}{5} \\ -\frac{2}{5} & \frac{1}{5} & -\frac{4}{5} & \frac{2}{5} \\ -\frac{4}{5} & \frac{2}{5} & \frac{2}{5} & -\frac{1}{5} \end{bmatrix} \begin{bmatrix} -1 \\ 1 \\ 1 \\ -2 \end{bmatrix} = \begin{bmatrix} -1 \\ -1 \\ 2 \end{bmatrix}$$

The system $R\mathbf{x} = Q^T\mathbf{b}$ is easily solved by back substitution:

$$\left[\begin{array}{ccc|c} 5 & -2 & 1 & -1 \\ 0 & 4 & -1 & -1 \\ 0 & 0 & 2 & 2 \end{array} \right]$$

The solution is $\mathbf{x} = \left(-\frac{2}{5}, 0, 1\right)^T$. $\qquad \blacktriangleleft$

The Modified Gram–Schmidt Process

In Chapter 7 we will consider computer methods for solving least squares problems. The QR method of Example 4 does not in general produce accurate results when carried out with finite-precision arithmetic. In practice, there may be a loss of orthogonality due to roundoff error in computing $\mathbf{q}_1, \mathbf{q}_2, \ldots, \mathbf{q}_n$. We can achieve better numerical accuracy using a modified version of the Gram–Schmidt method. In the modified version the vector $\mathbf{q}_1$ is constructed as before:

$$\mathbf{q}_1 = \frac{1}{\|\mathbf{a}_1\|}\mathbf{a}_1$$

However, the remaining vectors $\mathbf{a}_2, \ldots, \mathbf{a}_n$ are then modified so as to be orthogonal to $\mathbf{q}_1$. This can be done by subtracting from each vector $\mathbf{a}_k$ the projection of $\mathbf{a}_k$

onto $\mathbf{q}_1$.

$$\mathbf{a}_k^{(1)} = \mathbf{a}_k - (\mathbf{q}_1^T \mathbf{a}_k)\mathbf{q}_1 \qquad k = 2, \dots, n$$

At the second step we take

$$\mathbf{q}_2 = \frac{1}{\|\mathbf{a}_2^{(1)}\|} \mathbf{a}_2^{(1)}$$

The vector $\mathbf{q}_2$ is already orthogonal to $\mathbf{q}_1$. We then modify the remaining vectors to make them orthogonal to $\mathbf{q}_2$.

$$\mathbf{a}_k^{(2)} = \mathbf{a}_k^{(1)} - (\mathbf{q}_2^T \mathbf{a}_k^{(1)})\mathbf{q}_2 \qquad k = 3, \dots, n$$

In a similar manner $\mathbf{q}_3, \mathbf{q}_4, \dots, \mathbf{q}_n$ are successively determined. At the last step we need only set

$$\mathbf{q}_n = \frac{1}{\|\mathbf{a}_n^{(n-1)}\|} \mathbf{a}_n^{(n-1)}$$

to achieve an orthonormal set $\{\mathbf{q}_1, \dots, \mathbf{q}_n\}$. The following algorithm summarizes the process.

▶ **ALGORITHM 5.6.4** *[The Modified Gram–Schmidt Process]*

> For $k = 1, 2, \dots, n$ set
>
> $\quad r_{kk} = \|\mathbf{a}_k\|$
>
> $\quad \mathbf{q}_k = \dfrac{1}{r_{kk}} \mathbf{a}_k$
>
> > For $j = k + 1, k + 2, \dots, n$, set
> >
> > $\quad r_{kj} = \mathbf{q}_k^T \mathbf{a}_j$
> >
> > $\quad \mathbf{a}_j = \mathbf{a}_j - r_{kj}\mathbf{q}_k$
> >
> > End for loop
>
> End for loop ◀

    If the modified Gram–Schmidt process is applied to the column vectors of an $m \times n$ matrix A having rank n, then, as before, we can obtain a QR factorization of A. This factorization may then be used computationally to determine the least squares solution to $A\mathbf{x} = \mathbf{b}$.

EXERCISES

1. For each of the following, use the Gram–Schmidt process to find an orthonormal basis for $R(A)$.

(a) $A = \begin{bmatrix} -1 & 3 \\ 1 & 5 \end{bmatrix}$

(b) $A = \begin{bmatrix} 2 & 5 \\ 1 & 10 \end{bmatrix}$

2. Factor each of the matrices in Exercise 1 into a product QR, where Q is an orthogonal matrix and R is upper triangular.

3. Given the basis $\{(1, 2, -2)^T, (4, 3, 2)^T, (1, 2, 1)^T\}$ for R^3, use the Gram–Schmidt process to obtain an orthonormal basis.

4. Consider the vector space $C[-1, 1]$ with inner product defined by

$$\langle f, g \rangle = \int_{-1}^{1} f(x)g(x)\, dx$$

Find an orthonormal basis for the subspace spanned by 1, x, and x^2.

5. Let

$$A = \begin{bmatrix} 2 & 1 \\ 1 & 1 \\ 2 & 1 \end{bmatrix} \qquad \text{and} \qquad \mathbf{b} = \begin{bmatrix} 12 \\ 6 \\ 18 \end{bmatrix}$$

(a) Use the Gram–Schmidt process to find an orthonormal basis for the column space of A.

(b) Factor A into a product QR, where Q has an orthonormal set of column vectors and R is upper triangular.

(c) Solve the least squares problem

$$A\mathbf{x} = \mathbf{b}$$

6. Repeat Exercise 5 using

$$A = \begin{bmatrix} 3 & -1 \\ 4 & 2 \\ 0 & 2 \end{bmatrix} \qquad \text{and} \qquad \mathbf{b} = \begin{bmatrix} 0 \\ 20 \\ 10 \end{bmatrix}$$

7. The vectors $\mathbf{x}_1 = \frac{1}{2}(1, 1, 1, -1)^T$ and $\mathbf{x}_2 = \frac{1}{6}(1, 1, 3, 5)^T$ form an orthonormal set in R^4. Extend this set to an orthonormal basis for R^4 by finding an orthonormal basis for the nullspace of

$$\begin{bmatrix} 1 & 1 & 1 & -1 \\ 1 & 1 & 3 & 5 \end{bmatrix}$$

[*Hint:* First find a basis for the nullspace and then use the Gram–Schmidt process.]

8. Use the Gram–Schmidt process to find an orthonormal basis for the subspace of R^4 spanned by $\mathbf{x}_1 = (4, 2, 2, 1)^T$, $\mathbf{x}_2 = (2, 0, 0, 2)^T$, and $\mathbf{x}_3 = (1, 1, -1, 1)^T$.

9. Repeat Exercise 8 using the modified Gram–Schmidt process and compare answers.

10. Show that, when carried out in exact arithmetic, the modified Gram–Schmidt process will produce the same orthonormal set as the classical Gram–Schmidt process.

11. What will happen if the Gram–Schmidt is applied to a set of vectors $\{\mathbf{v}_1, \mathbf{v}_2, \mathbf{v}_3\}$, where $\mathbf{v}_1$ and $\mathbf{v}_2$ are linearly independent, but $\mathbf{v}_3 \in \mathrm{Span}(\mathbf{v}_1, \mathbf{v}_2)$. Will the process fail? If so, how? Explain.

12. Let A be an $m \times n$ matrix of rank n and let $\mathbf{b} \in R^m$. If Q and R are the matrices derived from applying the Gram–Schmidt process to the column vectors of A and

$$\mathbf{p} = c_1 \mathbf{q}_1 + c_2 \mathbf{q}_2 + \cdots + c_n \mathbf{q}_n$$

is the projection of $\mathbf{b}$ onto $R(A)$, then show that:

(a) $\mathbf{c} = Q^T \mathbf{b}$ (b) $\mathbf{p} = QQ^T \mathbf{b}$ (c) $QQ^T = A(A^TA)^{-1}A^T$

7 ORTHOGONAL POLYNOMIALS

We have already seen how polynomials can be used for data fitting and for approximating continuous functions. Since both of these problems are least squares problems, they can be simplified by selecting an orthogonal basis for the class of approximating polynomials. This leads us to the concept of orthogonal polynomials.

In this section we study families of orthogonal polynomials associated with various inner products on $C[a, b]$. We will see that the polynomials in each of these classes satisfy a three-term recursion relation. This recursion relation is particularly useful in computer applications. Certain families of orthogonal polynomials have important applications in many areas of mathematics. We will refer to these polynomials as classical polynomials and examine them in more detail. In particular, the classical polynomials are solutions to certain classes of second-order linear differential equations that arise in the solution of many partial differential equations from mathematical physics.

Orthogonal Sequences

Since the proof of Theorem 5.6.1 was by induction, the Gram–Schmidt process is valid for a denumerable set. Thus, if $\mathbf{x}_1, \mathbf{x}_2, \ldots$ is a sequence of vectors in an inner product space V and $\mathbf{x}_1, \mathbf{x}_2, \ldots, \mathbf{x}_n$ are linearly independent for each n, then the Gram–Schmidt process may be used to form a sequence $\mathbf{u}_1, \mathbf{u}_2, \ldots$, where $\{\mathbf{u}_1, \mathbf{u}_2, \ldots\}$ is an orthonormal set and

$$\mathrm{Span}(\mathbf{x}_1, \mathbf{x}_2, \ldots, \mathbf{x}_n) = \mathrm{Span}(\mathbf{u}_1, \mathbf{u}_2, \ldots, \mathbf{u}_n)$$

for each n. In particular, from the sequence $1, x, x^2, \ldots$ it is possible to construct an *orthonormal sequence* $p_0(x), p_1(x), \ldots$.

Let P be the vector space of all polynomials and define the inner product $\langle, \rangle$ on P by

(1) $$\langle p, q \rangle = \int_a^b p(x)q(x)w(x)\,dx$$

where $w(x)$ is a positive continuous function. The interval can be taken as either open or closed and may be finite or infinite. If, however,

$$\int_a^b p(x)w(x)\,dx$$

is improper, we require that it converge for every $p \in P$.

▶   **DEFINITION** Let $p_0(x), p_1(x), \ldots$ be a sequence of polynomials with $\deg p_i(x) = i$ for each i. If $\langle p_i(x), p_j(x)\rangle = 0$ whenever $i \neq j$, then $\{p_n(x)\}$ is said to be a **sequence of orthogonal polynomials**. If $\langle p_i, p_j\rangle = \delta_{ij}$, then $\{p_n(x)\}$ is said to be a **sequence of orthonormal polynomials**.   ◀

▶ **THEOREM 5.7.1** *If $p_0, p_1, \ldots$ is a sequence of orthogonal polynomials, then*

(i) $p_0, \ldots, p_{n-1}$ *form a basis for P_n.*

(ii) $p_n \in P_n^\perp$ *(that is, p_n is orthogonal to every polynomial of degree less than n).*

▶*Proof.* It follows from Theorem 5.5.1 that $p_0, p_1, \ldots, p_{n-1}$ are linearly independent in P_n. Since $\dim P_n = n$, these n vectors must form a basis for P_n. Let $p(x)$ be any polynomial of degree less than n. Then

$$p(x) = \sum_{i=0}^{n-1} c_i p_i(x)$$

and hence

$$\langle p_n, p\rangle = \left\langle p_n, \sum_{i=0}^{n-1} c_i p_i\right\rangle = \sum_{i=0}^{n-1} c_i \langle p_n, p_i\rangle = 0$$

Therefore, $p_n \in P_n^\perp$.   ◀

If $\{p_0, p_1, \ldots, p_{n-1}\}$ is an orthogonal set in P_n and

$$u_i = \left(\frac{1}{\|p_i\|}\right) p_i \qquad \text{for} \quad i = 0, \ldots, n-1$$

then $\{u_0, \ldots, u_{n-1}\}$ is an orthonormal basis for P_n. Hence, if $p \in P_n$, then

$$p = \sum_{i=0}^{n-1} \langle p, u_i\rangle u_i$$

$$= \sum_{i=0}^{n-1} \left\langle p, \left(\frac{1}{\|p_i\|}\right) p_i\right\rangle \left(\frac{1}{\|p_i\|}\right) p_i$$

$$= \sum_{i=0}^{n-1} \frac{\langle p, p_i\rangle}{\langle p_i, p_i\rangle} p_i$$

Similarly, if $f \in C[a, b]$, then the best least squares approximation to f by the elements of P_n is given by

$$p = \sum_{i=0}^{n-1} \frac{\langle f, p_i \rangle}{\langle p_i, p_i \rangle} p_i$$

where $p_0, p_1, \ldots, p_{n-1}$ are orthogonal polynomials.

Another nice feature of sequences of orthogonal polynomials is that they satisfy a three-term recursion relation.

▶ **THEOREM 5.7.2** *Let $p_0, p_1, \ldots$ be a sequence of orthogonal polynomials. Let a_i denote the lead coefficient of p_i for each i, and define $p_{-1}(x)$ to be the zero polynomial. Then*

$$\alpha_{n+1} p_{n+1}(x) = (x - \beta_{n+1}) p_n(x) - \alpha_n \gamma_n p_{n-1}(x) \qquad (n \geq 0)$$

where $\alpha_0 = \gamma_0 = 1$ and

$$\alpha_n = \frac{a_{n-1}}{a_n}, \qquad \beta_n = \frac{\langle p_{n-1}, x p_{n-1} \rangle}{\langle p_{n-1}, p_{n-1} \rangle}, \qquad \gamma_n = \frac{\langle p_n, p_n \rangle}{\langle p_{n-1}, p_{n-1} \rangle} \qquad (n \geq 1)$$

▶ *Proof.* Since $p_0, p_1, \ldots, p_{n+1}$ form a basis for P_{n+2}, we can write

$$(2) \qquad\qquad x p_n(x) = \sum_{k=0}^{n+1} c_{nk} p_k(x)$$

where

$$(3) \qquad\qquad c_{nk} = \frac{\langle x p_n, p_k \rangle}{\langle p_k, p_k \rangle}$$

For any inner product defined by (1),

$$\langle xf, g \rangle = \langle f, xg \rangle$$

In particular,

$$\langle x p_n, p_k \rangle = \langle p_n, x p_k \rangle$$

It follows from Theorem 5.7.1 that if $k < n - 1$ then

$$c_{nk} = \frac{\langle x p_n, p_k \rangle}{\langle p_k, p_k \rangle} = \frac{\langle p_n, x p_k \rangle}{\langle p_k, p_k \rangle} = 0$$

Therefore, (2) simplifies to

$$x p_n(x) = c_{n,n-1} p_{n-1}(x) + c_{n,n} p_n(x) + c_{n,n+1} p_{n+1}(x)$$

This can be rewritten in the form

$$(4) \qquad c_{n,n+1} p_{n+1}(x) = (x - c_{n,n}) p_n(x) - c_{n,n-1} p_{n-1}(x)$$

Comparing the lead coefficients of the polynomials on each side of (4), we see that

$$c_{n,n+1} a_{n+1} = a_n$$

or

$$(5) \qquad\qquad c_{n,n+1} = \frac{a_n}{a_{n+1}} = \alpha_{n+1}$$

It follows from (4) that

$$c_{n,n+1}\langle p_n, p_{n+1}\rangle = \langle p_n, (x - c_{n,n})p_n\rangle - c_{n,n-1}\langle p_n, p_{n-1}\rangle$$

$$0 = \langle p_n, xp_n\rangle - c_{nn}\langle p_n, p_n\rangle$$

and hence

$$c_{nn} = \frac{\langle p_n, xp_n\rangle}{\langle p_n, p_n\rangle} = \beta_{n+1}$$

It follows from (3) that

$$\langle p_{n-1}, p_{n-1}\rangle c_{n,n-1} = \langle xp_n, p_{n-1}\rangle$$

$$= \langle p_n, xp_{n-1}\rangle$$

$$= \langle p_n, p_n\rangle c_{n-1,n}$$

and hence by (5) we have

$$c_{n,n-1} = \frac{\langle p_n, p_n\rangle}{\langle p_{n-1}, p_{n-1}\rangle}\alpha_n = \gamma_n\alpha_n \qquad \blacktriangleleft$$

In generating a sequence of orthogonal polynomials by the recursion relation in Theorem 5.7.2, we are free to choose any nonzero lead coefficient a_{n+1} that we want at each step. This is reasonable, since any nonzero multiple of a particular p_{n+1} will also be orthogonal to $p_0, \ldots, p_n$. If we were to choose our a_i's to be 1, for example, then the recursion relation would simplify to

$$p_{n+1}(x) = (x - \beta_{n+1})p_n(x) - \gamma_n p_{n-1}(x)$$

Classical Orthogonal Polynomials

Let us now look at some examples. Because of their importance we will consider the classical polynomials beginning with the simplest, the Legendre polynomials.

Legendre Polynomials

The Legendre polynomials are orthogonal with respect to the inner product

$$\langle p, q\rangle = \int_{-1}^{1} p(x)q(x)\,dx$$

Let $P_n(x)$ denote the Legendre polynomial of degree n. If we choose the lead coefficients so that $P_n(1) = 1$ for each n, then the recursion formula for the Legendre polynomials is

$$(n+1)P_{n+1}(x) = (2n+1)xP_n(x) - nP_{n-1}(x)$$

By the use of this formula, the sequence of Legendre polynomials is easily generated. The first five polynomials of the sequence are

$$P_0(x) = 1$$

$$P_1(x) = x$$

$$P_2(x) = \tfrac{1}{2}(3x^2 - 1)$$

$$P_3(x) = \tfrac{1}{2}(5x^3 - 3x)$$

$$P_4(x) = \tfrac{1}{8}(35x^4 - 30x^2 + 3)$$

Chebyshev Polynomials

The Chebyshev polynomials are orthogonal with respect to the inner product,

$$\langle p, q \rangle = \int_{-1}^{1} p(x)q(x)(1 - x^2)^{-1/2} \, dx$$

It is customary to normalize the lead coefficients so that $a_0 = 1$ and $a_k = 2^{k-1}$ for $k = 1, 2, \ldots$. The Chebyshev polynomials are denoted by $T_n(x)$ and have the interesting property that

$$T_n(\cos \theta) = \cos n\theta$$

This property, together with the trigonometric identity

$$\cos(n + 1)\theta = 2 \cos \theta \cos n\theta - \cos(n - 1)\theta$$

can be used to derive the recursion relation

$$T_{n+1}(x) = 2x T_n(x) - T_{n-1}(x)$$

for $n \geq 1$.

Jacobi Polynomials

The Legendre and Chebyshev polynomials are both special cases of the Jacobi polynomials. The Jacobi polynomials $P_n^{(\lambda, \mu)}$ are orthogonal with respect to the inner product,

$$\langle p, q \rangle = \int_{-1}^{1} p(x)q(x)(1 - x)^{\lambda}(1 + x)^{\mu} \, dx$$

where $\lambda, \mu > -1$.

Hermite Polynomials

The Hermite polynomials are defined on the interval $(-\infty, \infty)$. They are orthogonal with respect to the inner product

$$\langle p, q \rangle = \int_{-\infty}^{\infty} p(x)q(x)e^{-x^2} \, dx$$

The recursion relation for Hermite polynomials is given by

$$H_{n+1}(x) = 2x H_n(x) - 2n H_{n-1}(x)$$

Laguerre Polynomials

The Laguerre polynomials are defined on the interval $(0, \infty)$ and are orthogonal with respect to the inner product,

$$\langle p, q \rangle = \int_{0}^{\infty} p(x)q(x)x^{\lambda}e^{-x} \, dx$$

TABLE I

Chebyshev	Hermite	Laguerre ($\lambda = 0$)
$T_{n+1} = 2xT_n - T_{n-1}, n \geq 1$	$H_{n+1} = 2xH_n - 2nH_{n-1}$	$(n+1)L_{n+1}^{(0)} = (2n+1-x)L_n^{(0)} - nL_{n-1}^{(0)}$
$T_0 = H_0$	$H_0 = 1$	$L_0^{(0)} = 1$
$T_1 = x$	$H_1 = 2x$	$L_1^{(0)} = 1 - x$
$T_2 = 2x^2 - 1$	$H_2 = 4x^2 - 2$	$L_2^{(0)} = \frac{1}{2}x^2 - x + 2$
$T_3 = 4x^3 - 3x$	$H_3 = 8x^3 - 12x$	$L_3^{(0)} = \frac{1}{6}x^3 + 9x^2 - 18x + 6$

where $\lambda > -1$. The recursion relation for the Laguerre polynomials is given by

$$(n+1)L_{n+1}^{(\lambda)}(x) = (2n + \lambda + 1 - x)L_n^{(\lambda)}(x) - (n + \lambda)L_{n-1}^{(\lambda)}(x)$$

The Chebyshev, Hermite, and Laguerre polynomials are compared in Table 1.

APPLICATION I: NUMERICAL INTEGRATION

One important application of orthogonal polynomials occurs in numerical integration. In order to approximate

$$(6) \qquad \int_a^b f(x)w(x)\,dx$$

we first approximate $f(x)$ by an interpolating polynomial. We can determine a polynomial $P(x)$ that agrees with $f(x)$ at n points $x_1, \ldots, x_n$ in $[a, b]$ using *Lagrange's interpolation formula*,

$$P(x) = \sum_{i=1}^{n} f(x_i)\delta_i(x)$$

where

$$\delta_i(x) = \frac{\displaystyle\prod_{\substack{j=1 \\ j \neq i}}^{n} (x - x_j)}{\displaystyle\prod_{\substack{j=1 \\ j \neq i}}^{n} (x_i - x_j)}$$

The integral (6) is then approximated by

$$(7) \qquad \int_a^b P(x)w(x)\,dx = \sum_{i=1}^{n} A_i f(x_i)$$

where

$$A_i = \int_a^b \delta_i(x)w(x)\,dx \qquad i = 1, \ldots, n$$

It can be shown that (7) will give the exact value of the integral whenever $f(x)$ is a polynomial of degree less than n. If the points $x_1, \ldots, x_n$ are chosen properly, formula (7) will be exact for higher-degree polynomials. Indeed, it can be shown that, if $p_0, p_1, p_2, \ldots$ is a sequence of orthogonal polynomials with respect to the inner product (1) and $x_1, \ldots, x_n$ are the zeros of $p_n(x)$, then formula (7) will be exact for all polynomials of degree less than $2n$. The following theorem guarantees that the roots of p_n are all real and lie in the open interval (a, b).

▶ **THEOREM 5.7.3** *If $p_0, p_1, p_2, \ldots$ is a sequence of orthogonal polynomials with respect to the inner product* (1), *then the zeros of $p_n(x)$ are all real and distinct and lie in the interval (a, b).*

▶ *Proof.* Let $x_1, \ldots, x_m$ be the zeros of $p_n(x)$ that lie in (a, b) and for which $p_n(x)$ changes sign. Thus $p_n(x)$ must have a factor of $(x - x_i)^{k_i}$, where k_i is odd, for $i = 1, \ldots, m$. We may write

$$p_n(x) = (x - x_1)^{k_1}(x - x_2)^{k_2} \cdots (x - x_m)^{k_m} q(x)$$

where $q(x)$ does not change sign on (a, b) and $q(x_i) \neq 0$ for $i = 1, \ldots, m$. Clearly, $m \leq n$. We will show that $m = n$. Let

$$r(x) = (x - x_1)(x - x_2) \cdots (x - x_m)$$

The product

$$p_n(x)r(x) = (x - x_1)^{k_1+1}(x - x_2)^{k_2+1} \cdots (x - x_m)^{k_m+1} q(x)$$

will involve only even powers of $(x - x_i)$ for each i and hence will not change sign on (a, b). Therefore,

$$\langle p_n, r \rangle = \int_a^b p_n(x)r(x)w(x)\,dx \neq 0$$

Since p_n is orthogonal to all polynomials of degree less than n, it follows that $\deg(r(x)) = m \geq n$. ◀

Numerical integration formulas of the form (7), where the x_i's are roots of orthogonal polynomials, are called *Gaussian quadrature formulas*. The proof of exactness for polynomials of degree less than $2n$ can be found in most undergraduate numerical analysis textbooks.

EXERCISES

1. Use the recursion formulas to calculate (a) T_4, T_5 and (b) H_4, H_5.

2. Let $p_0(x)$, $p_1(x)$, and $p_2(x)$ be orthogonal with respect to the inner product

$$\langle p(x), q(x) \rangle = \int_{-1}^1 \frac{p(x)q(x)}{1 + x^2}\,dx$$

Use Theorem 5.7.2 to calculate $p_1(x)$ and $p_2(x)$ if all polynomials have lead coefficient 1.

3. Show that the Chebyshev polynomials have the following properties:

   (a) $2T_m(x)T_n(x) = T_{m+n}(x) + T_{m-n}(x)$, for $m > n$
   (b) $T_m(T_n(x)) = T_{mn}(x)$

4. Find the best quadratic least squares approximation to e^x on $[-1, 1]$ with respect to the inner product,

$$\langle f, g \rangle = \int_{-1}^{1} f(x)g(x)\,dx$$

5. Let $p_0, p_1, \ldots$ be a sequence of orthogonal polynomials and let a_n denote the lead coefficient of p_n. Prove that

$$\|p_n\|^2 = a_n \langle x^n, p_n \rangle$$

6. Let $T_n(x)$ denote the Chebyshev polynomial of degree n and define

$$U_{n-1}(x) = \frac{1}{n}T_n'(x)$$

for $n = 1, 2, \ldots$.

   (a) Compute $U_0(x)$, $U_1(x)$, and $U_2(x)$.
   (b) If $x = \cos\theta$, show that

$$U_{n-1}(x) = \frac{\sin n\theta}{\sin \theta}$$

7. Let $U_{n-1}(x)$ be defined as in Exercise 6 for $n \geq 1$ and define $U_{-1}(x) = 0$. Show that:

   (a) $T_n(x) = U_n(x) - xU_{n-1}(x)$, for $n \geq 0$
   (b) $U_n(x) = 2xU_{n-1}(x) - U_{n-2}(x)$, for $n \geq 1$

8. Show that the U_i's defined in Exercise 6 are orthogonal with respect to the inner product

$$\langle p, q \rangle = \int_{-1}^{1} p(x)q(x)(1 - x^2)^{1/2}\,dx$$

   The U_i's are called *Chebyshev polynomials of the second kind*.

9. For $n = 0, 1, 2$, show that the Legendre polynomial $P_n(x)$ satisfies the second-order equation

$$(1 - x^2)y'' - 2xy' + n(n + 1)y = 0$$

10. Prove each of the following.

    (a) $H_n'(x) = 2n\,H_{n-1}(x)$, $n = 0, 1, \ldots$
    (b) $H_n''(x) - 2x\,H_n'(x) + 2n\,H_n(x) = 0$, $n = 0, 1, \ldots$

11. Given a function $f(x)$ that passes through the points $(1, 2)$, $(2, -1)$, $(3, 4)$, use the Lagrange interpolating formula to construct a second-degree polynomial that interpolates f at the given points.

12. Show that if $f(x)$ is a polynomial of degree less than n then $f(x)$ must equal the interpolating polynomial $P(x)$ in (7), and hence the sum in (7) gives the exact value for $\int_a^b f(x)w(x)\,dx$.

13. Use the zeros of the Legendre polynomial $P_2(x)$ to obtain a two-point quadrature formula

$$\int_{-1}^1 f(x)\,dx \approx A_1 f(x_1) + A_2 f(x_2)$$

14. (a) For what degree polynomials will the quadrature formula in Exercise 13 be exact?

(b) Use the formula from Exercise 13 to approximate

$$\int_{-1}^1 (x^3 + 3x^2 + 1)\,dx \qquad \text{and} \qquad \int_{-1}^1 \frac{1}{1+x^2}\,dx$$

How do the approximations compare with the actual values?

MATLAB EXERCISES

1. Set

$$\mathbf{x} = [0:4, 4, -4, 1, 1]' \qquad \text{and} \qquad \mathbf{y} = \mathbf{ones}(9, 1)$$

(a) Use the MATLAB function **norm** to compute the values of $\|\mathbf{x}\|$, $\|\mathbf{y}\|$, $\|\mathbf{x+y}\|$ and to verify that the triangle inequality holds. Use MATLAB also to verify that the parallelogram law

$$\|\mathbf{x} + \mathbf{y}\|^2 + \|\mathbf{x} - \mathbf{y}\|^2 = 2(\|\mathbf{x}\|^2 + \|\mathbf{y}\|^2)$$

is satisfied.

(b) If

$$t = \frac{\mathbf{x}^T \mathbf{y}}{\|\mathbf{x}\|\|\mathbf{y}\|}$$

then why do we know that $|t|$ must be less than or equal to 1? Use MATLAB to compute the value of t and use the MATLAB function **acos** to compute the angle between $\mathbf{x}$ and $\mathbf{y}$. Convert the angle to degrees by multiplying by $180/\pi$. (Note that the number π is given by **pi** in MATLAB.)

(c) Use MATLAB to compute the vector projection $\mathbf{p}$ of $\mathbf{x}$ onto $\mathbf{y}$. Set $\mathbf{z} = \mathbf{x} - \mathbf{p}$ and verify that $\mathbf{z}$ is orthogonal to $\mathbf{p}$ by computing the scalar product of the two vectors. Compute $\|\mathbf{x}\|^2$ and $\|\mathbf{z}\|^2 + \|\mathbf{p}\|^2$ and verify that the Pythagorean Law is satisfied.

2. **(Least Squares Fit to a Data Set by a Linear Function)** The following table of x and y values was given in Section 4 of this chapter (see Figure 5.4.2).

x	−1.0	0.0	2.1	2.3	2.4	5.3	6.0	6.5	8.0
y	−1.02	−0.52	0.55	0.70	0.70	2.13	2.52	2.82	3.54

The nine data points are nearly linear and hence the data can be approximated by a linear function $z = c_1 x + c_2$. Enter the x and y coordinates of the data points as column vectors **x** and **y**, respectively. Set $V = [\mathbf{x}, \texttt{ones(size(x))}]$ and use the MATLAB "\" operation to compute the coefficients c_1 and c_2 as the least squares solution to the 9×2 linear system $V\mathbf{c} = \mathbf{y}$. To see the results graphically, set

$$\mathbf{w} = -1:0.1:8 \quad \text{and} \quad \mathbf{z} = c(1) * \mathbf{w} + c(2) * \texttt{ones(size(w))}$$

and plot the original data points and the least squares linear fit using the MATLAB command

$$\texttt{plot}(\mathbf{x}, \mathbf{y}, \text{'}x\text{'}, \mathbf{w}, \mathbf{z})$$

3. **(Construction of Temperature Profiles by Least Squares Polynomials)** Among the important inputs in weather forecasting models are data sets consisting of temperature values at various parts of the atmosphere. These are either measured directly using weather balloons or inferred from remote soundings taken by weather satellites. A typical set of RAOB (weather balloon) data is given next. The temperature T in kelvins may be considered as a function of p, the atmospheric pressure measured in decibars. Pressures in the range from 1 to 3 decibars correspond to the top of the atmosphere, and those in the range from 9 to 10 decibars correspond to the lower part of the atmosphere.

p	1	2	3	4	5	6	7	8	9	10
T	222	227	223	233	244	253	260	266	270	266

(a) Enter the pressure values as a column vector **p** by setting $\mathbf{p} = [1:10]'$, and enter the temperature values as a column vector **T**. To find the best least squares fit to the data by a linear function $c_1 x + c_2$, set up an overdetermined system $V\mathbf{c} = \mathbf{T}$. The coefficient matrix V can be generated in MATLAB by setting

$$V = [\mathbf{p}, \texttt{ones}(10, 1)]$$

or alternatively by setting

$$A = \texttt{vander}(\mathbf{p}); \qquad V = A(:, 9:10)$$

NOTE For any vector $\mathbf{x} = (x_1, x_2, \dots, x_{n+1})^T$, the MATLAB command `vander(x)` generates a full Vandermonde matrix of the form

$$\begin{pmatrix} x_1^n & x_1^{n-1} & \cdots & x_1 & 1 \\ x_2^n & x_2^{n-1} & \cdots & x_2 & 1 \\ \vdots & & & & \\ x_{n+1}^n & x_{n+1}^{n-1} & \cdots & x_{n+1} & 1 \end{pmatrix}$$

For a linear fit, only the last two columns of the full Vandermonde matrix are used. More information on the **vander** function can be obtained by typing **help vander**. Once V has been constructed, the least squares solution **c** to the system can then be calculated using the MATLAB "\" operation.

(b) To see how well the linear function fits the data, define a range of pressure values by setting

$$q = 1 : 0.1 : 10;$$

The corresponding function values can be determined by setting

$$z = \texttt{polyval}(\mathbf{c}, \mathbf{q});$$

We can plot the function and the data points using the command

$$\texttt{plot}(\mathbf{q}, \mathbf{z}, \mathbf{p}, \mathbf{T}, \text{'}x\text{'})$$

(c) Let us now try to obtain a better fit using a cubic polynomial approximation. Again we can calculate the coefficients of the cubic polynomial

$$c_1 x^3 + c_2 x^2 + c_3 x + c_4$$

that gives the best least squares fit to the data by finding the least squares solution to an overdetermined system $V\mathbf{c} = \mathbf{T}$. The coefficient matrix V is determined by taking the last four columns of the matrix $A = \texttt{vander}(\mathbf{p})$. To see the results graphically, again set

$$z = \texttt{polyval}(\mathbf{c}, \mathbf{q})$$

and plot the cubic function and data points using the same plot command as before. Where do you get the better fit, at the top or bottom of the atmosphere?

(d) To obtain a good fit at both the top and the bottom of the atmosphere, try using a sixth-degree polynomial. Determine the coefficients as before using the last seven columns of A. Set $z = \texttt{polyval}(\mathbf{c}, \mathbf{q})$ and plot the results.

4. (Least Squares Circles) The parametric equations for a circle with center $(3, 1)$ and radius 2 are

$$x = 3 + 2\cos t \quad y = 1 + 2\sin t$$

Set $\mathbf{t} = 0 : .5 : 6$ and use MATLAB to generate vectors of x and y coordinates for the corresponding points on the circle. Next add some noise to your points by setting

$$\mathbf{x} = \mathbf{x} + 0.1 * \texttt{rand}(1, 13) \qquad \text{and} \qquad \mathbf{y} = \mathbf{y} + 0.1 * \texttt{rand}(1, 13)$$

Use MATLAB to determine the center **c** and radius r of the circle that gives the best least squares fit to the points. Set

$$\mathbf{t1} = 0 : 0.1 : 6.3 \quad \mathbf{x1} = \mathbf{c}(1) + \mathbf{r} * \cos(\mathbf{t1}) \quad \mathbf{y1} = \mathbf{c}(2) + \mathbf{r} * \sin(\mathbf{t1})$$

and use the command

$$\texttt{plot}(\mathbf{x1}, \mathbf{y1}, \mathbf{x}, \mathbf{y}, \text{'}x\text{'})$$

to plot the circle and the data points.

5. (Fundamental Subspaces: Orthonormal Bases) The vector spaces $N(A)$, $R(A)$, $N(A^T)$, $R(A^T)$ are the four fundamental subspaces associated with a matrix A. We can use MATLAB to construct orthonormal bases for each of the fundamental subspaces associated with a given matrix. We can then construct projection matrices corresponding to each subspace.

(a) Set

$$A = \mathbf{rand}(5, 2) * \mathbf{rand}(2, 5)$$

What would you expect the rank and nullity of A to be? Explain. Use MATLAB to check your answer by computing $\mathbf{rank}(A)$ and $Z = \mathbf{null}(A)$. The columns of Z form an orthonormal basis for $N(A)$.

(b) Next set

$$Q = \mathbf{orth}(A), \qquad W = \mathbf{null}(A'), \qquad S = [Q \quad W]$$

The matrix S should be orthogonal. Why? Explain. Compute $S * S'$ and compare your result to $\mathbf{eye}(5)$. In theory, $A^T W$ and $W^T A$ should both consist entirely of zeros. Why? Explain. Use MATLAB to compute $A^T W$ and $W^T A$.

(c) Prove that if Q and W had been computed in exact arithmetic, then we would have

$$I - W W^T = Q Q^T \qquad \text{and} \qquad Q Q^T A = A$$

[*Hint:* Write $S S^T$ in terms of Q and W.] Use MATLAB to verify these identities.

(d) Prove that if Q had been calculated in exact arithmetic, then we would have $Q Q^T \mathbf{b} = \mathbf{b}$ for all $\mathbf{b} \in R(A)$. Use MATLAB to verify this by setting $\mathbf{b} = A * \mathbf{rand}(5, 1)$ and then computing $Q * Q' * \mathbf{b}$ and comparing it to $\mathbf{b}$.

(e) Since the column vectors of Q form an orthonormal basis for $R(A)$, it follows that $Q Q^T$ is the projection matrix corresponding to $R(A)$. Thus, for any $\mathbf{c} \in R^5$, the vector $\mathbf{q} = Q Q^T \mathbf{c}$ is the projection of $\mathbf{c}$ onto $R(A)$. Set $\mathbf{c} = \mathbf{rand}(5, 1)$ and compute the projection vector $\mathbf{q}$. The vector $\mathbf{r} = \mathbf{c} - \mathbf{q}$ should be in $N(A^T)$. Why? Explain. Use MATLAB to compute $A' * \mathbf{r}$.

(f) The matrix $W W^T$ is the projection matrix corresponding to $N(A^T)$. Use MATLAB to compute the projection $\mathbf{w} = W W^T \mathbf{c}$ of $\mathbf{c}$ onto $N(A^T)$ and compare the result to $\mathbf{r}$.

(g) Set $Y = \mathbf{orth}(A')$ and use it to compute the projection matrix U corresponding to $R(A^T)$. Let $\mathbf{b} = \mathbf{rand}(5, 1)$ and compute the projection vector $\mathbf{y} = U * \mathbf{b}$ of $\mathbf{b}$ onto $R(A^T)$. Compute also $U * \mathbf{y}$ and compare it to $\mathbf{y}$. The vector $\mathbf{s} = \mathbf{b} - \mathbf{y}$ should be in $N(A)$. Why? Explain. Use MATLAB to compute $A * \mathbf{s}$.

(h) Use the matrix $Z = \mathbf{null}(A)$ to compute the projection matrix V corresponding to $N(A)$. Compute $V * \mathbf{b}$ and compare it to $\mathbf{s}$.

CHAPTER TEST

In each of the following answer *true* if the statement is always true and *false* otherwise. In the case of a true statement, explain or prove your answer. In the case of a false statement, give an example to show that the statement is not always true.

1. If $\mathbf{x}$ and $\mathbf{y}$ are nonzero vectors in R^n and the vector projection of $\mathbf{x}$ onto $\mathbf{y}$ is equal to the vector projection of $\mathbf{y}$ onto $\mathbf{x}$, then $\mathbf{x}$ and $\mathbf{y}$ are linearly dependent.

2. If $\mathbf{x}$ and $\mathbf{y}$ are unit vectors in R^n and $|\mathbf{x}^T\mathbf{y}| = 1$, then $\mathbf{x}$ and $\mathbf{y}$ are linearly independent.

3. If U, V, and W are subspaces of R^3 and $U \perp V$ and $V \perp W$, then $U \perp W$.

4. It is possible to find a nonzero vector $\mathbf{y}$ in the column space of A such that $A^T\mathbf{y} = \mathbf{0}$.

5. If A is an $m \times n$ matrix, then AA^T and A^TA have the same rank.

6. If an $m \times n$ matrix A has linearly dependent columns and $\mathbf{b}$ is a vector in R^m, then $\mathbf{b}$ does not have a unique projection onto the column space of A.

7. If $N(A) = \{\mathbf{0}\}$, then the system $A\mathbf{x} = \mathbf{b}$ will have a unique least squares solution.

8. If Q_1 and Q_2 are orthogonal matrices, then Q_1Q_2 is also an orthogonal matrix.

9. If $\{\mathbf{u}_1, \mathbf{u}_2, \ldots, \mathbf{u}_k\}$ is an orthonormal set of vectors in R^n and

$$U = (\mathbf{u}_1, \mathbf{u}_2, \ldots, \mathbf{u}_k),$$

then $U^TU = I_k$ (the $k \times k$ identity matrix).

10. If $\{\mathbf{u}_1, \mathbf{u}_2, \ldots, \mathbf{u}_k\}$ is an orthonormal set of vectors in R^n and

$$U = (\mathbf{u}_1, \mathbf{u}_2, \ldots, \mathbf{u}_k),$$

then $UU^T = I_n$ (the $n \times n$ identity matrix).

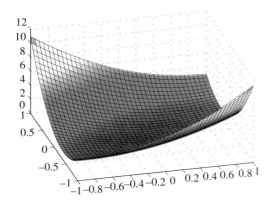

CHAPTER

6

EIGENVALUES

In Section 1 we will be concerned with the equation $A\mathbf{x} = \lambda\mathbf{x}$. This equation occurs in many applications of linear algebra. If the equation has a nonzero solution $\mathbf{x}$, then λ is said to be an *eigenvalue* of A and $\mathbf{x}$ is said to be an *eigenvector* belonging to λ.

Eigenvalues are a common part of our life whether we realize it or not. Wherever there are vibrations there are eigenvalues, the natural frequencies of the vibrations. If you have ever tuned a guitar, you have solved an eigenvalue problem. When engineers design structures, they are concerned with the frequencies of vibration of the structure. This is particularly important in earthquake prone regions such as California. The eigenvalues of a boundary value problem can be used to determine the energy states of an atom or critical loads that cause buckling in a beam. This latter application is presented in Section 1.

In Section 2 we will learn more about how to use eigenvalues and eigenvectors to solve systems of linear differential equations. We will consider a number of applications, including mixture problems, the harmonic motion of a system of springs, and the vibrations of a building. The motion of a building can be modeled by a second-order system of differential equations of the form

$$M\mathbf{Y}''(t) = K\mathbf{Y}(t)$$

where $\mathbf{Y}(t)$ is a vector whose entries are all functions of t and $\mathbf{Y}''(t)$ is the vector of functions formed by taking the second derivatives of each of the entries of $\mathbf{Y}(t)$. The solution of the equation is determined by the eigenvalues and eigenvectors of the matrix $A = M^{-1}K$.

In general, we can view eigenvalues as natural frequencies associated with linear transformations. If A is an $n \times n$ matrix, we can think of A as representing a linear transformation from R^n into itself. Eigenvalues and eigenvectors provide the key to understanding how the operator works. For example, if $\lambda > 0$, the effect of the operator on any eigenvector belonging to λ is simply a stretching or a shrinking by a constant factor. Indeed, the effect of the operator is easily determined on any linear combination of eigenvectors. In particular, if it is possible to find a basis of eigenvectors for R^n, the operator can be represented by a diagonal matrix D with respect to that basis, and the matrix A can be factored into a product XDX^{-1}. In Section 3 we see how this is done and look at a number of applications.

In Section 4 we consider matrices with complex entries. In this setting we will be concerned with matrices whose eigenvectors can be used to form an orthonormal basis for C^n (the vector space of all n-tuples of complex numbers). In Section 5 we introduce the singular value decomposition of a matrix and show four applications. Another important application of this factorization will be presented in Chapter 7.

Section 6 deals with the application of eigenvalues to quadratic equations in several variables and also with applications involving maximums and minimums of functions of several variables. In Section 7 we consider symmetric positive definite matrices. The eigenvalues of such matrices are real and positive. These matrices occur in a wide variety of applications. In particular, we will show the applications to conic sections and to optimization of functions. Finally, in Section 8 we study matrices with nonnegative entries and some applications to economics.

I   EIGENVALUES AND EIGENVECTORS

Many application problems involve applying a linear transformation repeatedly to a given vector. The key to solving these problems is to choose a coordinate system or basis that is in some sense natural for the operator and for which it will be simpler to do calculations involving the operator. With respect to these new basis vectors (*eigenvectors*) we associate scaling factors (*eigenvalues*) which represent the natural frequencies of the operator. We illustrate with a simple example.

EXAMPLE I. Let us recall Application 3 from Section 3 of Chapter 1. In a certain town, 30 percent of the married women get divorced each year and 20 percent of the single women get married each year. There are 8000 married women and 2000 single women, and the total population remains constant. Let us investigate the long-range prospects if these percentages of marriages and divorces continue indefinitely into the future.

To find the number of married and single women after 1 year, we multiply the vector $\mathbf{w}_0 = (8000, 2000)^T$ by

$$A = \begin{bmatrix} 0.7 & 0.2 \\ 0.3 & 0.8 \end{bmatrix}$$

The number of married and single women after 1 year is given by

$$\mathbf{w}_1 = A\mathbf{w}_0 = \begin{bmatrix} 0.7 & 0.2 \\ 0.3 & 0.8 \end{bmatrix} \begin{bmatrix} 8000 \\ 2000 \end{bmatrix} = \begin{bmatrix} 6000 \\ 4000 \end{bmatrix}$$

To determine the number of married and single women after 2 years, we compute

$$\mathbf{w}_2 = A\mathbf{w}_1 = A^2\mathbf{w}_0$$

and in general for n years we must compute $\mathbf{w}_n = A^n\mathbf{w}_0$.

Let us compute $\mathbf{w}_{10}$, $\mathbf{w}_{20}$, $\mathbf{w}_{30}$ in this way and round the entries of each to the nearest integer.

$$\mathbf{w}_{10} = \begin{bmatrix} 4004 \\ 5996 \end{bmatrix}, \qquad \mathbf{w}_{20} = \begin{bmatrix} 4000 \\ 6000 \end{bmatrix}, \qquad \mathbf{w}_{30} = \begin{bmatrix} 4000 \\ 6000 \end{bmatrix}$$

After a certain point we seem to always get the same answer. In fact, $\mathbf{w}_{12} = (4000, 6000)^T$ and since

$$A\mathbf{w}_{12} = \begin{bmatrix} 0.7 & 0.2 \\ 0.3 & 0.8 \end{bmatrix}\begin{bmatrix} 4000 \\ 6000 \end{bmatrix} = \begin{bmatrix} 4000 \\ 6000 \end{bmatrix}$$

it follows that all the succeeding vectors in the sequence remain unchanged. The vector $(4000, 6000)^T$ is said to be a *steady-state vector* for the process.

Suppose that initially we had different proportions of married and single women. If, for example, we had started with 10,000 married women and 0 single women, then $\mathbf{w}_0 = (10,000, 0)^T$, and we can compute $\mathbf{w}_n$ as before by multiplying $\mathbf{w}_0$ by A^n. In this case it turns out that $\mathbf{w}_{14} = (4000, 6000)^T$, and hence we still end up with the same steady-state vector.

Why does this process converge and why do we seem to get the same steady-state vector even when we change the initial vector? These questions are not difficult to answer if we choose a basis for R^2 consisting of vectors for which the effect of the linear transformation A is easily determined. In particular, if we choose a multiple of the steady-state vector, say $\mathbf{x}_1 = (2, 3)^T$, as our first basis vector, then

$$A\mathbf{x}_1 = \begin{bmatrix} 0.7 & 0.2 \\ 0.3 & 0.8 \end{bmatrix}\begin{bmatrix} 2 \\ 3 \end{bmatrix} = \begin{bmatrix} 2 \\ 3 \end{bmatrix} = \mathbf{x}_1$$

Thus $\mathbf{x}_1$ is also a steady-state vector. It is a natural basis vector to use since the effect of A on $\mathbf{x}_1$ could not be simpler. Although it would be nice to use another steady-state vector as the second basis vector, this is not possible since all the steady-state vectors turn out to be multiples of $\mathbf{x}_1$. However, if we choose $\mathbf{x}_2 = (-1, 1)^T$, then the effect of A on $\mathbf{x}_2$ is also quite simple.

$$A\mathbf{x}_2 = \begin{bmatrix} 0.7 & 0.2 \\ 0.3 & 0.8 \end{bmatrix}\begin{bmatrix} -1 \\ 1 \end{bmatrix} = \begin{bmatrix} -\frac{1}{2} \\ \frac{1}{2} \end{bmatrix} = \tfrac{1}{2}\mathbf{x}_2$$

Let us now analyze the process using $\mathbf{x}_1$ and $\mathbf{x}_2$ as our basis vectors. If we express the initial vector $\mathbf{w}_0 = (8000, 2000)^T$ as a linear combination

$$\mathbf{w}_0 = 2000\begin{bmatrix} 2 \\ 3 \end{bmatrix} - 4000\begin{bmatrix} -1 \\ 1 \end{bmatrix} = 2000\mathbf{x}_1 - 4000\mathbf{x}_2$$

then

$$\mathbf{w}_1 = A\mathbf{w}_0 = 2000A\mathbf{x}_1 - 4000A\mathbf{x}_2 = 2000\mathbf{x}_1 - 4000\left(\frac{1}{2}\right)\mathbf{x}_2$$

$$\mathbf{w}_2 = A\mathbf{w}_1 = 2000\mathbf{x}_1 - 4000\left(\frac{1}{2}\right)^2\mathbf{x}_2$$

and, in general,

$$\mathbf{w}_n = A^n\mathbf{w}_0 = 2000\mathbf{x}_1 - 4000\left(\frac{1}{2}\right)^n\mathbf{x}_2$$

The first component of this sum is the steady-state vector and the second component converges to the zero vector.

Will we always end up with the same steady-state vector for any choice of $\mathbf{w}_0$? Suppose that initially there are p married women. Since there are 10,000 women altogether, the number of single women must be $10,000 - p$. Our initial vector is then

$$\mathbf{w}_0 = \left\{ \begin{array}{c} p \\ 10,000 - p \end{array} \right\}$$

If we express $\mathbf{w}_0$ as a linear combination $c_1\mathbf{x}_1 + c_2\mathbf{x}_2$, then, as before,

$$\mathbf{w}_n = A^n\mathbf{w}_0 = c_1\mathbf{x}_1 + \left(\frac{1}{2}\right)^n c_2\mathbf{x}_2$$

The steady-state vector will be $c_1\mathbf{x}_1$. To determine c_1, we write the equation

$$c_1\mathbf{x}_1 + c_2\mathbf{x}_2 = \mathbf{w}_0$$

as a linear system

$$2c_1 - c_2 = p$$
$$3c_1 + c_2 = 10,000 - p$$

Adding the two equations, we see that $c_1 = 2000$. Thus, for any integer p in the range $0 \le p \le 10,000$, the steady-state vector turns out to be

$$2000\mathbf{x}_1 = \left[\begin{array}{c} 4000 \\ 6000 \end{array} \right] \qquad \blacktriangleleft$$

The vectors $\mathbf{x}_1$ and $\mathbf{x}_2$ were natural vectors to use in analyzing the process in Example 1 since the effect of the matrix A on each of these vectors was so simple.

$$A\mathbf{x}_1 = \mathbf{x}_1 = 1\mathbf{x}_1 \qquad \text{and} \qquad A\mathbf{x}_2 = \tfrac{1}{2}\mathbf{x}_2$$

For each of these vectors the effect of A was just to multiply the vector by a scalar. The two scalars 1 and $\frac{1}{2}$ can be thought of as the natural frequencies of the linear transformation.

In general, if a linear transformation is represented by an $n \times n$ matrix A and we can find a nonzero vector $\mathbf{x}$ so that $A\mathbf{x} = \lambda\mathbf{x}$, for some scalar λ, then, for this transformation, $\mathbf{x}$ is a natural choice to use as a basis vector for R^n, and the scalar λ defines a natural frequency corresponding to that basis vector. More precisely, we use the following terminology to refer to $\mathbf{x}$ and λ.

▶ **DEFINITION** Let A be an $n \times n$ matrix. A scalar λ is said to be an **eigenvalue** or a **characteristic value** of A if there exists a nonzero vector $\mathbf{x}$ such that $A\mathbf{x} = \lambda\mathbf{x}$. The vector $\mathbf{x}$ is said to be an **eigenvector** or a **characteristic vector** belonging to λ. ◀

EXAMPLE 2. Let

$$A = \begin{bmatrix} 4 & -2 \\ 1 & 1 \end{bmatrix} \quad \text{and} \quad \mathbf{x} = \begin{bmatrix} 2 \\ 1 \end{bmatrix}$$

Since

$$A\mathbf{x} = \begin{bmatrix} 4 & -2 \\ 1 & 1 \end{bmatrix} \begin{bmatrix} 2 \\ 1 \end{bmatrix} = \begin{bmatrix} 6 \\ 3 \end{bmatrix} = 3 \begin{bmatrix} 2 \\ 1 \end{bmatrix} = 3\mathbf{x}$$

it follows that $\lambda = 3$ is an eigenvalue of A and $\mathbf{x} = (2, 1)^T$ is an eigenvector belonging to λ. Actually, any nonzero multiple of $\mathbf{x}$ will be an eigenvector, since

$$A(\alpha\mathbf{x}) = \alpha A\mathbf{x} = \alpha\lambda\mathbf{x} = \lambda(\alpha\mathbf{x})$$

Thus, for example, $(4, 2)^T$ is also an eigenvector belonging to $\lambda = 3$.

$$\begin{bmatrix} 4 & -2 \\ 1 & 1 \end{bmatrix} \begin{bmatrix} 4 \\ 2 \end{bmatrix} = \begin{bmatrix} 12 \\ 6 \end{bmatrix} = 3 \begin{bmatrix} 4 \\ 2 \end{bmatrix} \qquad ◀$$

The equation $A\mathbf{x} = \lambda\mathbf{x}$ can be written in the form

$$(1) \qquad\qquad\qquad (A - \lambda I)\mathbf{x} = \mathbf{0}$$

Thus λ is an eigenvalue of A if and only if (1) has a nontrivial solution. The set of solutions to (1) is $N(A - \lambda I)$, which is a subspace of R^n. Thus, if λ is an eigenvalue of A, then $N(A - \lambda I) \neq \{\mathbf{0}\}$, and any nonzero vector in $N(A - \lambda I)$ is an eigenvector belonging to λ. The subspace $N(A - \lambda I)$ is called the *eigenspace* corresponding to the eigenvalue λ.

Equation (1) will have a nontrivial solution if and only if $A - \lambda I$ is singular, or, equivalently,

$$(2) \qquad\qquad\qquad \det(A - \lambda I) = 0$$

If the determinant in (2) is expanded, we obtain an nth-degree polynomial in the variable λ,

$$p(\lambda) = \det(A - \lambda I)$$

This polynomial is called the *characteristic polynomial*, and equation (2) is called the *characteristic equation* for the matrix A. The roots of the characteristic polynomial are the eigenvalues of A. If we count roots according to multiplicity, the characteristic polynomial will have exactly n roots. Thus A will have n eigenvalues, some of which may be repeated and some of which may be complex numbers. To take care of the latter case, it will be necessary to expand our field of scalars to the complex numbers and to allow complex entries for our vectors and matrices.

We have now established a number of equivalent conditions for λ to be an eigenvalue of A.

Let A be an $n \times n$ matrix and λ be a scalar. The following statements are equivalent:

(a) λ is an eigenvalue of A.

(b) $(A - \lambda I)\mathbf{x} = \mathbf{0}$ has a nontrivial solution.

(c) $N(A - \lambda I) \neq \{\mathbf{0}\}$

(d) $A - \lambda I$ is singular.

(e) $\det(A - \lambda I) = 0$

We will now use statement **(e)** to determine the eigenvalues in a number of examples.

EXAMPLE 3. Find the eigenvalues and the corresponding eigenvectors of the matrix

$$A = \begin{bmatrix} 3 & 2 \\ 3 & -2 \end{bmatrix}$$

SOLUTION. The characteristic equation is

$$\begin{vmatrix} 3 - \lambda & 2 \\ 3 & -2 - \lambda \end{vmatrix} = 0 \qquad \text{or} \qquad \lambda^2 - \lambda - 12 = 0$$

Thus, the eigenvalues of A are $\lambda_1 = 4$ and $\lambda_2 = -3$. To find the eigenvectors belonging to $\lambda_1 = 4$, we must determine the nullspace of $A - 4I$.

$$A - 4I = \begin{bmatrix} -1 & 2 \\ 3 & -6 \end{bmatrix}$$

Solving $(A - 4I)\mathbf{x} = \mathbf{0}$, we get

$$\mathbf{x} = (2x_2, x_2)^T$$

Thus, any nonzero multiple of $(2, 1)^T$ is an eigenvector belonging to λ_1, and $\{(2, 1)^T\}$ is a basis for the eigenspace corresponding to λ_1. Similarly, to find the eigenvectors for λ_2, we must solve

$$(A + 3I)\mathbf{x} = \mathbf{0}$$

In this case $\{(-1, 3)^T\}$ is a basis for $N(A + 3I)$, and any nonzero multiple of $(-1, 3)^T$ is an eigenvector belonging to λ_2. ◄

EXAMPLE 4.   Let

$$A = \begin{bmatrix} 2 & -3 & 1 \\ 1 & -2 & 1 \\ 1 & -3 & 2 \end{bmatrix}$$

Find the eigenvalues and the corresponding eigenspaces.

SOLUTION.

$$\begin{vmatrix} 2-\lambda & -3 & 1 \\ 1 & -2-\lambda & 1 \\ 1 & -3 & 2-\lambda \end{vmatrix} = -\lambda(\lambda-1)^2$$

Thus, the characteristic polynomial has roots $\lambda_1 = 0$, $\lambda_2 = \lambda_3 = 1$. The eigenspace corresponding to $\lambda_1 = 0$ is $N(A)$, which we determine in the usual manner.

$$\begin{bmatrix} 2 & -3 & 1 & \big| & 0 \\ 1 & -2 & 1 & \big| & 0 \\ 1 & -3 & 2 & \big| & 0 \end{bmatrix} \rightarrow \begin{bmatrix} 1 & 0 & -1 & \big| & 0 \\ 0 & 1 & -1 & \big| & 0 \\ 0 & 0 & 0 & \big| & 0 \end{bmatrix}$$

Setting $x_3 = \alpha$, we find that $x_1 = x_2 = x_3 = \alpha$. Thus the eigenspace corresponding to $\lambda_1 = 0$ consists of all vectors of the form $\alpha(1, 1, 1)^T$. To find the eigenspace corresponding to $\lambda = 1$, we must solve the system $(A - I)\mathbf{x} = \mathbf{0}$.

$$\begin{bmatrix} 1 & -3 & 1 & \big| & 0 \\ 1 & -3 & 1 & \big| & 0 \\ 1 & -3 & 1 & \big| & 0 \end{bmatrix} \rightarrow \begin{bmatrix} 1 & -3 & 1 & \big| & 0 \\ 0 & 0 & 0 & \big| & 0 \\ 0 & 0 & 0 & \big| & 0 \end{bmatrix}$$

Setting $x_2 = \alpha$ and $x_3 = \beta$, we get $x_1 = 3\alpha - \beta$. Thus the eigenspace corresponding to $\lambda = 1$ consists of all vectors of the form

$$\begin{bmatrix} 3\alpha - \beta \\ \alpha \\ \beta \end{bmatrix} = \alpha \begin{bmatrix} 3 \\ 1 \\ 0 \end{bmatrix} + \beta \begin{bmatrix} -1 \\ 0 \\ 1 \end{bmatrix} \qquad \blacktriangleleft$$

EXAMPLE 5.   Given

$$A = \begin{bmatrix} 1 & 2 \\ -2 & 1 \end{bmatrix}$$

Compute the eigenvalues of A and find bases for the corresponding eigenspaces.

SOLUTION.

$$\begin{vmatrix} 1-\lambda & 2 \\ -2 & 1-\lambda \end{vmatrix} = (1-\lambda)^2 + 4$$

The roots of the characteristic polynomial are $\lambda_1 = 1 + 2i$, $\lambda_2 = 1 - 2i$.

$$A - \lambda_1 I = \begin{bmatrix} -2i & 2 \\ -2 & -2i \end{bmatrix} = -2 \begin{bmatrix} i & -1 \\ 1 & i \end{bmatrix}$$

It follows that $\{(1, i)^T\}$ is a basis for the eigenspace corresponding to $\lambda_1 = 1 + 2i$. Similarly,

$$A - \lambda_2 I = \begin{bmatrix} 2i & 2 \\ -2 & 2i \end{bmatrix} = 2 \begin{bmatrix} i & 1 \\ -1 & i \end{bmatrix}$$

and $\{(1, -i)^T\}$ is a basis for $N(A - \lambda_2 I)$. ◀

APPLICATION I: STRUCTURES—BUCKLING OF A BEAM

For an example of a physical eigenvalue problem, consider the case of a beam. If a force or load is applied to one end of the beam, the beam will buckle when the load reaches a critical value. If we continue increasing the load beyond the critical value, we can expect the beam to buckle again when the load reaches a second critical value, and so on. Assume that the beam has length L and that it is positioned along the x axis in the plane with the left support of the beam at $x = 0$. Let $y(x)$ represent the vertical displacement of the beam for any point x, and assume that the beam is simply supported, that is, $y(0) = y(L) = 0$. (See Figure 6.1.1.)

The physical system for the beam is modeled by the boundary value problem

$$(3) \qquad R\frac{d^2y}{dx^2} = -Py \qquad y(0) = y(L) = 0$$

where R is the flexural rigidity of the beam and P is the load placed on the beam. A standard procedure to compute the solution $y(x)$ is to use a finite difference method to approximate the differential equation. Specifically, we partition the interval $[0, L]$ into n equal subintervals

$$0 = x_0 < x_1 < \cdots < x_n = L \qquad \left(x_j = \frac{jL}{n}, \ j = 0, \ldots, n \right)$$

FIGURE 6.1.1

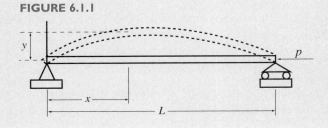

and for each j we approximate $y''(x_j)$ by a difference quotient. If we set $h = \frac{L}{n}$ and use the shorthand notation y_k for $y(x_k)$, then the standard difference approximation is given by

$$y''(x_j) \approx \frac{y_{j+1} - 2y_j + y_{j-1}}{h^2} \quad j = 1, \dots, n$$

Substituting these into equation (3), we end up with a system of n linear equations. If we multiply each equation through by $-\frac{h^2}{R}$ and set $\lambda = \frac{Ph^2}{R}$, then the system can be written as a matrix equation of the form $A\mathbf{y} = \lambda\mathbf{y}$, where

$$A = \begin{bmatrix} 2 & -1 & 0 & \cdots & 0 & 0 & 0 \\ -1 & 2 & -1 & \cdots & 0 & 0 & 0 \\ 0 & -1 & 2 & \cdots & 0 & 0 & 0 \\ \vdots & & & & & & \vdots \\ 0 & 0 & 0 & \cdots & -1 & 2 & -1 \\ 0 & 0 & 0 & \cdots & 0 & -1 & 2 \end{bmatrix}$$

The eigenvalues of this matrix will all be real and positive. (See MATLAB Exercise 24 at the end of this chapter.) For n sufficiently large, each eigenvalue λ of A can be used to approximate a critical load $P = \frac{R\lambda}{h^2}$ where buckling may occur. The most important of these critical loads is the one corresponding to the smallest eigenvalue since the beam may actually break after this load is exceeded.

Complex Eigenvalues

If A is an $n \times n$ matrix with real entries, then the characteristic polynomial of A will have real coefficients, and hence all its complex roots must occur in conjugate pairs. Thus, if $\lambda = a + bi$ ($b \neq 0$) is an eigenvalue of A, then $\bar{\lambda} = a - bi$ must also be an eigenvalue of A. Here the symbol $\bar{\lambda}$ (read *lambda bar*) is used to denote the complex conjugate of λ. A similar notation can be used for matrices. If $A = (a_{ij})$ is a matrix with complex entries, then $\overline{A} = (\overline{a_{ij}})$ is the matrix formed from A by conjugating each of its entries. We define a *real matrix* to be one with the property that $\overline{A} = A$. In general, if A and B are matrices with complex entries and the multiplication AB is possible, then $\overline{AB} = \overline{A}\,\overline{B}$ (see Exercise 17).

Not only do the complex eigenvalues of a real matrix occur in conjugate pairs, but so do the eigenvectors. Indeed, if λ is a complex eigenvalue of a real $n \times n$ matrix A and $\mathbf{z}$ is an eigenvector belonging to λ, then

$$A\bar{\mathbf{z}} = \overline{A}\,\bar{\mathbf{z}} = \overline{A\mathbf{z}} = \overline{\lambda\mathbf{z}} = \bar{\lambda}\,\bar{\mathbf{z}}$$

Thus, $\bar{\mathbf{z}}$ is an eigenvector of A belonging to $\bar{\lambda}$. In Example 5 the eigenvector computed for the eigenvalue $\lambda = 1 + 2i$ was $\mathbf{z} = (1, i)^T$, and the eigenvector computed for $\bar{\lambda} = 1 - 2i$ was $\bar{\mathbf{z}} = (1, -i)^T$.

The Product and Sum of the Eigenvalues

It is easy to determine the sum and product of the eigenvalues of an $n \times n$ matrix A. If $p(\lambda)$ is the characteristic polynomial of A, then

$$
(4) \qquad p(\lambda) = \det(A - \lambda I) = \begin{vmatrix} a_{11} - \lambda & a_{12} & \cdots & a_{1n} \\ a_{21} & a_{22} - \lambda & & a_{2n} \\ \vdots & & & \\ a_{n1} & a_{n2} & & a_{nn} - \lambda \end{vmatrix}
$$

Expanding along the first column, we get

$$
\det(A - \lambda I) = (a_{11} - \lambda)\det(M_{11}) + \sum_{i=2}^{n} a_{i1}(-1)^{i+1}\det(M_{i1})
$$

where the minors M_{i1}, $i = 2, \ldots, n$, do not contain the two diagonal elements $(a_{11} - \lambda)$ and $(a_{ii} - \lambda)$. Expanding $\det(M_{11})$ in the same manner, we conclude that

$$
(5) \qquad (a_{11} - \lambda)(a_{22} - \lambda) \cdots (a_{nn} - \lambda)
$$

is the only term in the expansion of $\det(A - \lambda I)$ involving a product of more than $n - 2$ of the diagonal elements. When (5) is expanded, the coefficient of λ^n will be $(-1)^n$. Thus the lead coefficient of $p(\lambda)$ is $(-1)^n$ and hence, if $\lambda_1, \ldots, \lambda_n$ are the eigenvalues of A, then

$$
(6) \qquad \begin{aligned} p(\lambda) &= (-1)^n(\lambda - \lambda_1)(\lambda - \lambda_2) \cdots (\lambda - \lambda_n) \\ &= (\lambda_1 - \lambda)(\lambda_2 - \lambda) \cdots (\lambda_n - \lambda) \end{aligned}
$$

It follows from (4) and (6) that

$$
\lambda_1 \cdot \lambda_2 \cdots \lambda_n = p(0) = \det(A)
$$

From (5) we also see that the coefficient of $(-\lambda)^{n-1}$ is $\displaystyle\sum_{i=1}^{n} a_{ii}$. If we determine this same coefficient using (6), we obtain $\displaystyle\sum_{i=1}^{n} \lambda_i$. It follows that

$$
\sum_{i=1}^{n} \lambda_i = \sum_{i=1}^{n} a_{ii}
$$

The sum of the diagonal elements of A is called the *trace* of A and is denoted by $\mathrm{tr}(A)$.

EXAMPLE 6. If

$$A = \begin{bmatrix} 5 & -18 \\ 1 & -1 \end{bmatrix}$$

then

$$\det(A) = -5 + 18 = 13 \qquad \text{and} \qquad \text{tr}(A) = 5 - 1 = 4$$

The characteristic polynomial of A is given by

$$\begin{vmatrix} 5 - \lambda & -18 \\ 1 & -1 - \lambda \end{vmatrix} = \lambda^2 - 4\lambda + 13$$

and hence the eigenvalues of A are $\lambda_1 = 2 + 3i$ and $\lambda_2 = 2 - 3i$. Note that

$$\lambda_1 + \lambda_2 = 4 = \text{tr}(A)$$

and

$$\lambda_1 \lambda_2 = 13 = \det(A) \qquad \blacktriangleleft$$

In the examples we have looked at so far, n has always been less than 4. For larger n it is more difficult to find the roots of the characteristic polynomial. In Chapter 7 we will learn numerical methods for computing eigenvalues. (These methods will not involve the characteristic polynomial at all.) If the eigenvalues of A have been computed using some numerical method, one way to check their accuracy is to compare their sum to the trace of A.

Similar Matrices

We close this section with an important result about the eigenvalues of similar matrices. Recall that a matrix B is said to be *similar* to a matrix A if there exists a nonsingular matrix S such that $B = S^{-1}AS$.

▶ **THEOREM 6.1.1** *Let A and B be $n \times n$ matrices. If B is similar to A, then the two matrices both have the same characteristic polynomial and consequently both have the same eigenvalues.*

▶ *Proof.* Let $p_A(x)$ and $p_B(x)$ denote the characteristic polynomials of A and B, respectively. If B is similar to A, then there exists a nonsingular matrix S such that $B = S^{-1}AS$. Thus

$$\begin{aligned} p_B(\lambda) &= \det(B - \lambda I) \\ &= \det(S^{-1}AS - \lambda I) \\ &= \det(S^{-1}(A - \lambda I)S) \\ &= \det(S^{-1}) \det(A - \lambda I) \det(S) \\ &= p_A(\lambda) \end{aligned}$$

The eigenvalues of a matrix are the roots of the characteristic polynomial. Since the two matrices have the same characteristic polynomial, they must have the same eigenvalues. ◀

EXAMPLE 7. Given

$$T = \begin{bmatrix} 2 & 1 \\ 0 & 3 \end{bmatrix} \quad \text{and} \quad S = \begin{bmatrix} 5 & 3 \\ 3 & 2 \end{bmatrix}$$

It is easily seen that the eigenvalues of T are $\lambda_1 = 2$ and $\lambda_2 = 3$. If we set $A = S^{-1}TS$, then the eigenvalues of A should be the same as those of T.

$$A = \begin{bmatrix} 2 & -3 \\ -3 & 5 \end{bmatrix} \begin{bmatrix} 2 & 1 \\ 0 & 3 \end{bmatrix} \begin{bmatrix} 5 & 3 \\ 3 & 2 \end{bmatrix} = \begin{bmatrix} -1 & -2 \\ 6 & 6 \end{bmatrix}$$

We leave it to the reader to verify that the eigenvalues of this matrix are $\lambda_1 = 2$ and $\lambda_2 = 3$. ◀

EXERCISES

1. Find the eigenvalues and the corresponding eigenspaces for each of the following matrices.

(a) $\begin{bmatrix} 3 & 2 \\ 4 & 1 \end{bmatrix}$

(b) $\begin{bmatrix} 6 & -4 \\ 3 & -1 \end{bmatrix}$

(c) $\begin{bmatrix} 3 & -1 \\ 1 & 1 \end{bmatrix}$

(d) $\begin{bmatrix} 3 & -8 \\ 2 & 3 \end{bmatrix}$

(e) $\begin{bmatrix} 1 & 1 \\ -2 & 3 \end{bmatrix}$

(f) $\begin{bmatrix} 0 & 1 & 0 \\ 0 & 0 & 1 \\ 0 & 0 & 0 \end{bmatrix}$

(g) $\begin{bmatrix} 1 & 1 & 1 \\ 0 & 2 & 1 \\ 0 & 0 & 1 \end{bmatrix}$

(h) $\begin{bmatrix} 1 & 2 & 1 \\ 0 & 3 & 1 \\ 0 & 5 & -1 \end{bmatrix}$

(i) $\begin{bmatrix} 4 & -5 & 1 \\ 1 & 0 & -1 \\ 0 & 1 & -1 \end{bmatrix}$

(j) $\begin{bmatrix} -2 & 0 & 1 \\ 1 & 0 & -1 \\ 0 & 1 & -1 \end{bmatrix}$

(k) $\begin{bmatrix} 2 & 0 & 0 & 0 \\ 0 & 2 & 0 & 0 \\ 0 & 0 & 3 & 0 \\ 0 & 0 & 0 & 4 \end{bmatrix}$

(l) $\begin{bmatrix} 3 & 0 & 0 & 0 \\ 4 & 1 & 0 & 0 \\ 0 & 0 & 2 & 1 \\ 0 & 0 & 0 & 2 \end{bmatrix}$

2. Show that the eigenvalues of a triangular matrix are the diagonal elements of the matrix.

3. Let A be an $n \times n$ matrix. Prove that A is singular if and only if $\lambda = 0$ is an eigenvalue of A.

4. Let A be a nonsingular matrix and let λ be an eigenvalue of A. Show that $1/\lambda$ is an eigenvalue of A^{-1}.

5. Let λ be an eigenvalue of A and let $\mathbf{x}$ be an eigenvector belonging to λ. Use mathematical induction to show that λ^m is an eigenvalue of A^m and $\mathbf{x}$ is an eigenvector of A^m belonging to λ^m for $m = 1, 2, \ldots$.

6. An $n \times n$ matrix A is said to be *idempotent* if $A^2 = A$. Show that if λ is an eigenvalue of an idempotent matrix, then λ must be either 0 or 1.

7. An $n \times n$ matrix is said to be *nilpotent* if $A^k = O$ for some positive integer k. Show that all eigenvalues of a nilpotent matrix are 0.

8. Let A be an $n \times n$ matrix and let $B = A - \alpha I$ for some scalar α. How do the eigenvalues of A and B compare? Explain.

9. Show that A and A^T have the same eigenvalues. Do they necessarily have the same eigenvectors? Explain.

10. Show that the matrix

$$A = \begin{bmatrix} \cos \theta & -\sin \theta \\ \sin \theta & \cos \theta \end{bmatrix}$$

will have complex eigenvalues if θ is not a multiple of π. Give a geometric interpretation of this result.

11. Let A be a 2×2 matrix. If $\operatorname{tr}(A) = 8$ and $\det(A) = 12$, what are the eigenvalues of A?

12. Let $A = (a_{ij})$ be an $n \times n$ matrix with eigenvalues $\lambda_1, \ldots, \lambda_n$. Show that

$$\lambda_j = a_{jj} + \sum_{i \neq j} (a_{ii} - \lambda_i) \qquad \text{for} \quad j = 1, \ldots, n$$

13. Let A be a 2×2 matrix and let $p(\lambda) = \lambda^2 + b\lambda + c$ be the characteristic polynomial of A. Show that $b = -\operatorname{tr}(A)$ and $c = \det(A)$.

14. Let λ be a nonzero eigenvalue of A and let $\mathbf{x}$ be an eigenvector belonging to λ. Show that $A^m \mathbf{x}$ is also an eigenvector belonging to λ for $m = 1, 2, \ldots$.

15. Let A be an $n \times n$ matrix and let λ be an eigenvalue of A. If $A - \lambda I$ has rank k, what is the dimension of the eigenspace corresponding to λ? Explain.

16. Let A be an $n \times n$ matrix. Show that a vector $\mathbf{x}$ in R^n is an eigenvector belonging to A if and only if the subspace S of R^n spanned by $\mathbf{x}$ and $A\mathbf{x}$ has dimension 1.

17. Let $\alpha = a + bi$ and $\beta = c + di$ be complex scalars and let A and B be matrices with complex entries.

    (a) Show that

$$\overline{\alpha + \beta} = \overline{\alpha} + \overline{\beta} \qquad \text{and} \qquad \overline{\alpha\beta} = \overline{\alpha}\,\overline{\beta}$$

    (b) Show that the (i, j) entries of $\overline{AB}$ and $\overline{A}\,\overline{B}$ are equal and hence that

$$\overline{AB} = \overline{A}\,\overline{B}$$

18. Let $x_1, \ldots, x_r$ be eigenvectors of an $n \times n$ matrix A and let S be the subspace of R^n spanned by $x_1, x_2, \ldots, x_r$. Show that S is invariant under A (i.e., show that $Ax \in S$ whenever $x \in S$).

19. Let $B = S^{-1}AS$ and let x be an eigenvector of B belonging to an eigenvalue λ. Show that Sx is an eigenvector of A belonging to λ.

20. Show that if two $n \times n$ matrices A and B have a common eigenvector x (but not necessarily a common eigenvalue), then x will also be an eigenvector of any matrix of the form $C = \alpha A + \beta B$.

21. Let A be an $n \times n$ matrix and let λ be a nonzero eigenvalue of A. Show that if x is an eigenvector belonging to λ, then x is in the column space of A. Hence the eigenspace corresponding to λ is a subspace of the column space of A.

22. Let $\{u_1, u_2, \ldots, u_n\}$ be an orthonormal basis for R^n and let A be a linear combination of the rank 1 matrices $u_1 u_1^T, u_2 u_2^T, \ldots, u_n u_n^T$. If

$$A = c_1 u_1 u_1^T + c_2 u_2 u_2^T + \cdots + c_n u_n u_n^T$$

show that A is a symmetric matrix with eigenvalues $c_1, c_2, \ldots, c_n$ and that u_i is an eigenvector belonging to c_i for each i.

23. Let A be a matrix whose columns all add up to a fixed constant δ. Show that δ is an eigenvalue of A.

24. Let λ_1 and λ_2 be distinct eigenvalues of A. Let x be an eigenvector of A belonging to λ_1 and let y be an eigenvector of A^T belonging to λ_2. Show that x and y are orthogonal.

25. Let A and B be $n \times n$ matrices. Show that:

(a) If λ is a nonzero eigenvalue of AB, then it is also an eigenvalue of BA.

(b) If $\lambda = 0$ is an eigenvalue of AB, then $\lambda = 0$ is also an eigenvalue of BA.

26. Prove that there do not exist $n \times n$ matrices A and B such that

$$AB - BA = I$$

[*Hint:* See Exercises 8 and 25.]

27. Let $p(\lambda) = (-1)^n (\lambda^n - a_{n-1}\lambda^{n-1} - \cdots - a_1\lambda - a_0)$ be a polynomial of degree $n \geq 1$, and let

$$C = \begin{bmatrix} a_{n-1} & a_{n-2} & \cdots & a_1 & a_0 \\ 1 & 0 & \cdots & 0 & 0 \\ 0 & 1 & \cdots & 0 & 0 \\ \vdots & & & & \\ 0 & 0 & \cdots & 1 & 0 \end{bmatrix}$$

(a) Show that if λ_i is a root of $p(\lambda) = 0$, then λ_i is an eigenvalue of C with eigenvector $x = (\lambda_i^{n-1}, \lambda_i^{n-2}, \ldots, \lambda_i, 1)^T$.

(b) Use part (a) to show that if $p(\lambda)$ has n distinct roots, then $p(\lambda)$ is the characteristic polynomial of C.

The matrix C is called the *companion matrix* of $p(\lambda)$.

28. The result given in Exercise 27(b) holds even if all the eigenvalues of $p(\lambda)$ are not distinct. Prove this as follows:

(a) Let

$$D_m(\lambda) = \begin{bmatrix} a_m & a_{m-1} & \cdots & a_1 & a_0 \\ 1 & -\lambda & \cdots & 0 & 0 \\ \vdots & & & & \\ 0 & 0 & \cdots & 1 & -\lambda \end{bmatrix}$$

and use mathematical induction to prove that

$$\det(D_m(\lambda)) = (-1)^m (a_m \lambda^m + a_{m-1} \lambda^{m-1} + \cdots + a_1 \lambda + a_0)$$

(b) Show that

$$\det(C - \lambda I) = (a_{n-1} - \lambda)(-\lambda)^{n-1} - \det(D_{n-2}) = p(\lambda)$$

2   SYSTEMS OF LINEAR DIFFERENTIAL EQUATIONS

Eigenvalues play an important role in the solution of systems of linear differential equations. In this section we see how they are used in the solution of systems of linear differential equations with constant coefficients. We begin by considering systems of first-order equations of the form

$$y_1' = a_{11} y_1 + a_{12} y_2 + \cdots + a_{1n} y_n$$
$$y_2' = a_{21} y_1 + a_{22} y_2 + \cdots + a_{2n} y_n$$
$$\vdots$$
$$y_n' = a_{n1} y_1 + a_{n2} y_2 + \cdots + a_{nn} y_n$$

where $y_i = f_i(t)$ is a function in $C^1[a, b]$ for each i. If we let

$$\mathbf{Y} = \begin{bmatrix} y_1 \\ y_2 \\ \vdots \\ y_n \end{bmatrix} \quad \text{and} \quad \mathbf{Y}' = \begin{bmatrix} y_1' \\ y_2' \\ \vdots \\ y_n' \end{bmatrix}$$

then the system can be written in the form

$$\mathbf{Y}' = A\mathbf{Y}$$

$\mathbf{Y}$ and $\mathbf{Y}'$ are both vector functions of t. Let us consider the simplest case first. When $n = 1$, the system is simply

(1) $$y' = ay$$

Clearly, any function of the form

$$y(t) = ce^{at} \qquad (c \text{ an arbitrary constant})$$

satisfies equation (1). A natural generalization of this solution for the case $n > 1$ is to take

$$\mathbf{Y} = \begin{bmatrix} x_1 e^{\lambda t} \\ x_2 e^{\lambda t} \\ \vdots \\ x_n e^{\lambda t} \end{bmatrix} = e^{\lambda t} \mathbf{x}$$

where $\mathbf{x} = (x_1, x_2, \ldots, x_n)^T$. To verify that a vector function of this form does work, we compute the derivative

$$\mathbf{Y}' = \lambda e^{\lambda t} \mathbf{x} = \lambda \mathbf{Y}$$

Now, if we choose λ to be an eigenvalue of A and $\mathbf{x}$ to be an eigenvector belonging to λ, then

$$A\mathbf{Y} = e^{\lambda t} A\mathbf{x} = \lambda e^{\lambda t} \mathbf{x} = \lambda \mathbf{Y} = \mathbf{Y}'$$

Hence $\mathbf{Y}$ is a solution to the system. Thus, if λ is an eigenvalue of A and $\mathbf{x}$ is an eigenvector belonging to λ, then $e^{\lambda t}\mathbf{x}$ is a solution of the system $\mathbf{Y}' = A\mathbf{Y}$. This will be true whether λ is real or complex. Note that if $\mathbf{Y}_1$ and $\mathbf{Y}_2$ are both solutions to $\mathbf{Y}' = A\mathbf{Y}$, then $\alpha\mathbf{Y}_1 + \beta\mathbf{Y}_2$ is also a solution, since

$$(\alpha\mathbf{Y}_1 + \beta\mathbf{Y}_2)' = \alpha\mathbf{Y}'_1 + \beta\mathbf{Y}'_2$$
$$= \alpha A\mathbf{Y}_1 + \beta A\mathbf{Y}_2$$
$$= A(\alpha\mathbf{Y}_1 + \beta\mathbf{Y}_2)$$

It follows by induction that if $\mathbf{Y}_1, \ldots, \mathbf{Y}_n$ are solutions to $\mathbf{Y}' = A\mathbf{Y}$, then any linear combination $c_1\mathbf{Y}_1 + \cdots + c_n\mathbf{Y}_n$ will also be a solution.

In general, the solutions to an $n \times n$ first-order system of the form

$$\mathbf{Y}' = A\mathbf{Y}$$

will form an n-dimensional subspace of the vector space of all continuous vector-valued functions. If, in addition, we require that $\mathbf{Y}(t)$ take on a prescribed value $\mathbf{Y}_0$ when $t = 0$, the problem will have a unique solution (see [34], p. 228). A problem of the form

$$\mathbf{Y}' = A\mathbf{Y}, \qquad \mathbf{Y}(0) = \mathbf{Y}_0$$

is called an *initial value problem*.

EXAMPLE 1. Solve the system

$$y_1' = 3y_1 + 4y_2$$
$$y_2' = 3y_1 + 2y_2$$

SOLUTION.

$$A = \begin{bmatrix} 3 & 4 \\ 3 & 2 \end{bmatrix}$$

The eigenvalues of A are $\lambda_1 = 6$ and $\lambda_2 = -1$. Solving $(A - \lambda I)\mathbf{x} = \mathbf{0}$ with $\lambda = \lambda_1$ and $\lambda = \lambda_2$, we see that $\mathbf{x}_1 = (4, 3)^T$ is an eigenvector belonging to λ_1 and $\mathbf{x}_2 = (1, -1)^T$ is an eigenvector belonging to λ_2. Thus any vector function of the form

$$\mathbf{Y} = c_1 e^{\lambda_1 t} \mathbf{x}_1 + c_2 e^{\lambda_2 t} \mathbf{x}_2$$

$$= \begin{bmatrix} 4c_1 e^{6t} + c_2 e^{-t} \\ 3c_1 e^{6t} - c_2 e^{-t} \end{bmatrix}$$

is a solution to the system. ◀

In Example 1, suppose that we require that $y_1 = 6$ and $y_2 = 1$ when $t = 0$. Thus

$$\mathbf{Y}(0) = \begin{bmatrix} 4c_1 + c_2 \\ 3c_1 - c_2 \end{bmatrix} = \begin{bmatrix} 6 \\ 1 \end{bmatrix}$$

and it follows that $c_1 = 1$ and $c_2 = 2$. Hence the solution to the initial value problem is given by

$$\mathbf{Y} = e^{6t} \mathbf{x}_1 + 2e^{-t} \mathbf{x}_2$$

$$= \begin{bmatrix} 4e^{6t} + 2e^{-t} \\ 3e^{6t} - 2e^{-t} \end{bmatrix}$$

APPLICATION 1: MIXTURES

Two tanks are connected as shown in Figure 6.2.1. Initially, tank A contains 200 liters of water in which 60 grams of salt has been dissolved, and tank B contains 200 liters of pure water. Liquid is pumped in and out of the two tanks at rates shown in the diagram. Determine the amount of salt in each tank at time t.

SOLUTION. Let $y_1(t)$ and $y_2(t)$ be the number of grams of salt in tanks A and B, respectively, at time t. Initially,

$$\mathbf{Y}(0) = \begin{bmatrix} y_1(0) \\ y_2(0) \end{bmatrix} = \begin{bmatrix} 60 \\ 0 \end{bmatrix}$$

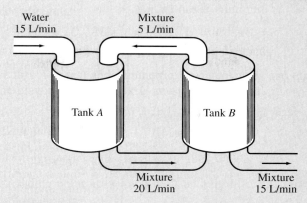

Water
15 L/min

Mixture
5 L/min

Tank *A*

Tank *B*

Mixture
20 L/min

Mixture
15 L/min

FIGURE 6.2.1

The total amount of liquid in each tank will remain at 200 liters since the amount being pumped in equals the amount being pumped out. The rate of change in the amount of salt for each tank is equal to the rate at which it is being added minus the rate at which it is being pumped out. For tank *A* the rate at which the salt is added is given by

$$(5 \text{ L/min}) \cdot \left(\frac{y_2(t)}{200} \text{g/L} \right) = \frac{y_2(t)}{40} \text{g/min}$$

and the rate at which the salt is being pumped out is

$$(20 \text{ L/min}) \cdot \left(\frac{y_1(t)}{200} \text{g/L} \right) = \frac{y_1(t)}{10} \text{g/min}$$

Thus the rate of change for tank *A* is given by

$$y_1'(t) = \frac{y_2(t)}{40} - \frac{y_1(t)}{10}$$

Similarly, for tank *B*, the rate of change is given by

$$y_2'(t) = \frac{20 y_1(t)}{200} - \frac{20 y_2(t)}{200} = \frac{y_1(t)}{10} - \frac{y_2(t)}{10}$$

To determine $y_1(t)$ and $y_2(t)$, we must solve the initial value problem

$$\mathbf{Y}' = A\mathbf{Y}, \quad \mathbf{Y}(0) = \mathbf{Y}_0$$

where

$$A = \begin{bmatrix} -\frac{1}{10} & \frac{1}{40} \\ \frac{1}{10} & -\frac{1}{10} \end{bmatrix}, \quad \mathbf{Y}_0 = \begin{bmatrix} 60 \\ 0 \end{bmatrix}$$

The eigenvalues of A are $\lambda_1 = -\frac{3}{20}$, $\lambda_2 = -\frac{1}{20}$ with corresponding eigenvectors

$$\mathbf{x}_1 = \begin{bmatrix} 1 \\ -2 \end{bmatrix} \quad \text{and} \quad \mathbf{x}_2 = \begin{bmatrix} 1 \\ 2 \end{bmatrix}$$

The solution must then be of the form

$$\mathbf{Y} = c_1 e^{-3t/20}\mathbf{x}_1 + c_2 e^{-t/20}\mathbf{x}_2$$

When $t = 0$, $\mathbf{Y} = \mathbf{Y}_0$. Thus

$$c_1\mathbf{x}_1 + c_2\mathbf{x}_2 = \mathbf{Y}_0$$

and we can find c_1 and c_2 by solving

$$\begin{bmatrix} 1 & 1 \\ -2 & 2 \end{bmatrix}\begin{bmatrix} c_1 \\ c_2 \end{bmatrix} = \begin{bmatrix} 60 \\ 0 \end{bmatrix}$$

The solution to this system is $c_1 = c_2 = 30$. Therefore, the solution to the initial value problem is

$$\mathbf{Y}(t) = \begin{bmatrix} y_1(t) \\ y_2(t) \end{bmatrix} = \begin{bmatrix} 30e^{-3t/20} + 30e^{-t/20} \\ -60e^{-3t/20} + 60e^{-t/20} \end{bmatrix} \qquad \blacktriangleleft$$

Complex Eigenvalues

Let A be a real $n \times n$ matrix with a complex eigenvalue $\lambda = a + bi$, and let $\mathbf{x}$ be an eigenvector belonging to λ. The vector $\mathbf{x}$ can be split up into its real and imaginary parts.

$$\mathbf{x} = \begin{bmatrix} \operatorname{Re} x_1 + i\operatorname{Im} x_1 \\ \operatorname{Re} x_2 + i\operatorname{Im} x_2 \\ \vdots \\ \operatorname{Re} x_n + i\operatorname{Im} x_n \end{bmatrix} = \begin{bmatrix} \operatorname{Re} x_1 \\ \operatorname{Re} x_2 \\ \vdots \\ \operatorname{Re} x_n \end{bmatrix} + i\begin{bmatrix} \operatorname{Im} x_1 \\ \operatorname{Im} x_2 \\ \vdots \\ \operatorname{Im} x_n \end{bmatrix} = \operatorname{Re}\mathbf{x} + i\operatorname{Im}\mathbf{x}$$

Since entries of A are all real, it follows that $\bar{\lambda} = a - bi$ is also an eigenvalue of A with eigenvector

$$\bar{\mathbf{x}} = \begin{bmatrix} \operatorname{Re} x_1 - i\operatorname{Im} x_1 \\ \operatorname{Re} x_2 - i\operatorname{Im} x_2 \\ \vdots \\ \operatorname{Re} x_n - i\operatorname{Im} x_n \end{bmatrix} = \operatorname{Re}\mathbf{x} - i\operatorname{Im}\mathbf{x}$$

and hence $e^{\lambda t}\mathbf{x}$ and $e^{\bar{\lambda}t}\bar{\mathbf{x}}$ are both solutions to the first-order system $\mathbf{Y}' = A\mathbf{Y}$. Any linear combination of these two solutions will also be a solution. Thus, if we set

$$\mathbf{Y}_1 = \frac{1}{2}(e^{\lambda t}\mathbf{x} + e^{\bar{\lambda}t}\bar{\mathbf{x}}) = \operatorname{Re}(e^{\lambda t}\mathbf{x})$$

and

$$\mathbf{Y}_2 = \frac{1}{2i}(e^{\lambda t}\mathbf{x} - e^{\bar{\lambda} t}\overline{\mathbf{x}}) = \text{Im}\,(e^{\lambda t}\mathbf{x})$$

then the vector functions $\mathbf{Y}_1$ and $\mathbf{Y}_2$ are real-valued solutions to $\mathbf{Y}' = A\mathbf{Y}$. Taking the real and imaginary parts of

$$e^{\lambda t}\mathbf{x} = e^{(a+ib)t}\mathbf{x}$$
$$= e^{at}(\cos bt + i\sin bt)(\text{Re}\,\mathbf{x} + i\,\text{Im}\,\mathbf{x})$$

we see that

$$\mathbf{Y}_1 = e^{at}\,[(\cos bt)\text{Re}\,\mathbf{x} - (\sin bt)\text{Im}\,\mathbf{x}]$$
$$\mathbf{Y}_2 = e^{at}\,[(\cos bt)\text{Im}\,\mathbf{x} + (\sin bt)\text{Re}\,\mathbf{x}]$$

EXAMPLE 2. Solve the system

$$y_1' = \quad y_1 + y_2$$
$$y_2' = -2y_1 + 3y_2$$

SOLUTION. Let

$$A = \begin{bmatrix} 1 & 1 \\ -2 & 3 \end{bmatrix}$$

The eigenvalues of A are $\lambda = 2+i$ and $\bar{\lambda} = 2-i$ with eigenvectors $\mathbf{x} = (1, 1+i)^T$ and $\overline{\mathbf{x}} = (1, 1-i)^T$, respectively.

$$e^{\lambda t}\mathbf{x} = \begin{bmatrix} e^{2t}(\cos t + i\sin t) \\ e^{2t}(\cos t + i\sin t)(1+i) \end{bmatrix}$$
$$= \begin{bmatrix} e^{2t}\cos t + ie^{2t}\sin t \\ e^{2t}(\cos t - \sin t) + ie^{2t}(\cos t + \sin t) \end{bmatrix}$$

Let

$$\mathbf{Y}_1 = \text{Re}\,(e^{\lambda t}\mathbf{x}) = \begin{bmatrix} e^{2t}\cos t \\ e^{2t}(\cos t - \sin t) \end{bmatrix}$$

and

$$\mathbf{Y}_2 = \text{Im}\,(e^{\lambda t}\mathbf{x}) = \begin{bmatrix} e^{2t}\sin t \\ e^{2t}(\cos t + \sin t) \end{bmatrix}$$

Any linear combination

$$\mathbf{Y} = c_1\mathbf{Y}_1 + c_2\mathbf{Y}_2$$

will be a solution to the system. ◀

If the $n \times n$ coefficient matrix A of the system $\mathbf{Y}' = A\mathbf{Y}$ has n linearly independent eigenvectors, the general solution can be obtained by the methods that have been presented. The case when A has less than n linearly independent eigenvectors is more complicated and consequently will not be covered in this book.

Higher-Order Systems

Given a second-order system of the form

$$\mathbf{Y}'' = A_1\mathbf{Y} + A_2\mathbf{Y}'$$

we may translate it into a first-order system by setting

$$y_{n+1}(t) = y_1'(t)$$
$$y_{n+2}(t) = y_2'(t)$$
$$\vdots$$
$$y_{2n}(t) = y_n'(t)$$

If we let

$$\mathbf{Y}_1 = \mathbf{Y} = (y_1, y_2, \ldots, y_n)^T$$

and

$$\mathbf{Y}_2 = \mathbf{Y}' = (y_{n+1}, \ldots, y_{2n})^T$$

then

$$\mathbf{Y}_1' = O\mathbf{Y}_1 + I\mathbf{Y}_2$$

and

$$\mathbf{Y}_2' = A_1\mathbf{Y}_1 + A_2\mathbf{Y}_2$$

The equations can be combined to give the $2n \times 2n$ first-order system

$$\begin{pmatrix} \mathbf{Y}_1' \\ \mathbf{Y}_2' \end{pmatrix} = \begin{bmatrix} O & I \\ A_1 & A_2 \end{bmatrix} \begin{bmatrix} \mathbf{Y}_1 \\ \mathbf{Y}_2 \end{bmatrix}$$

If the values of $\mathbf{Y}_1 = \mathbf{Y}$ and $\mathbf{Y}_2 = \mathbf{Y}'$ are specified when $t = 0$, then the initial value problem will have a unique solution.

EXAMPLE 3. Solve the initial value problem

$$y_1'' = 2y_1 + y_2 + y_1' + y_2'$$
$$y_2'' = -5y_1 + 2y_2 + 5y_1' - y_2'$$
$$y_1(0) = y_2(0) = y_1'(0) = 4, \qquad y_2'(0) = -4$$

SOLUTION. Set $y_3 = y_1'$ and $y_4 = y_2'$. This gives the first-order system

$$
\begin{aligned}
y_1' &= && y_3 \\
y_2' &= && y_4 \\
y_3' &= 2y_1 + y_2 + y_3 + y_4 \\
y_4' &= -5y_1 + 2y_2 + 5y_3 - y_4
\end{aligned}
$$

The coefficient matrix for this system,

$$
A = \begin{pmatrix} 0 & 0 & 1 & 0 \\ 0 & 0 & 0 & 1 \\ 2 & 1 & 1 & 1 \\ -5 & 2 & 5 & -1 \end{pmatrix}
$$

has eigenvalues

$$
\lambda_1 = 1, \qquad \lambda_2 = -1, \qquad \lambda_3 = 3, \qquad \lambda_4 = -3
$$

Corresponding to these eigenvalues are the eigenvectors

$$
\begin{aligned}
\mathbf{x}_1 &= (1, -1, 1, -1)^T, & \mathbf{x}_2 &= (1, 5, -1, -5)^T \\
\mathbf{x}_3 &= (1, 1, 3, 3)^T, & \mathbf{x}_4 &= (1, -5, -3, 15)^T
\end{aligned}
$$

Thus, the solution will be of the form

$$
c_1 \mathbf{x}_1 e^t + c_2 \mathbf{x}_2 e^{-t} + c_3 \mathbf{x}_3 e^{3t} + c_4 \mathbf{x}_4 e^{-3t}
$$

We can use the initial conditions to find c_1, c_2, c_3, and c_4. For $t = 0$, we have

$$
c_1 \mathbf{x}_1 + c_2 \mathbf{x}_2 + c_3 \mathbf{x}_3 + c_4 \mathbf{x}_4 = (4, 4, 4, -4)^T
$$

or, equivalently,

$$
\begin{pmatrix} 1 & 1 & 1 & 1 \\ -1 & 5 & 1 & -5 \\ 1 & -1 & 3 & -3 \\ -1 & -5 & 3 & 15 \end{pmatrix} \begin{pmatrix} c_1 \\ c_2 \\ c_3 \\ c_4 \end{pmatrix} = \begin{pmatrix} 4 \\ 4 \\ 4 \\ -4 \end{pmatrix}
$$

The solution to this system is $\mathbf{c} = (2, 1, 1, 0)^T$, and hence the solution to the initial value problem is

$$
\mathbf{Y} = 2\mathbf{x}_1 e^t + \mathbf{x}_2 e^{-t} + \mathbf{x}_3 e^{3t}
$$

Therefore,

$$
\begin{pmatrix} y_1 \\ y_2 \\ y_1' \\ y_2' \end{pmatrix} = \begin{pmatrix} 2e^t + e^{-t} + e^{3t} \\ -2e^t + 5e^{-t} + e^{3t} \\ 2e^t - e^{-t} + 3e^{3t} \\ -2e^t - 5e^{-t} + 3e^{3t} \end{pmatrix}
$$

◀

In general, if we have an mth-order system of the form

$$\mathbf{Y}^{(m)} = A_1\mathbf{Y} + A_2\mathbf{Y}' + \cdots + A_m\mathbf{Y}^{(m-1)}$$

where each A_i is an $n \times n$ matrix, we can transform it into a first-order system by setting

$$\mathbf{Y}_1 = \mathbf{Y}, \mathbf{Y}_2 = \mathbf{Y}'_1, \ldots, \mathbf{Y}_m = \mathbf{Y}'_{m-1}$$

We will end up with a system of the form

$$
\begin{Bmatrix}
\mathbf{Y}'_1 \\
\mathbf{Y}'_2 \\
\vdots \\
\mathbf{Y}'_{m-1} \\
\mathbf{Y}'_m
\end{Bmatrix}
=
\begin{Bmatrix}
O & I & O & \cdots & O \\
O & O & I & \cdots & O \\
\vdots & & & & \\
O & O & O & \cdots & I \\
A_1 & A_2 & A_3 & \cdots & A_m
\end{Bmatrix}
\begin{Bmatrix}
\mathbf{Y}_1 \\
\mathbf{Y}_2 \\
\vdots \\
\mathbf{Y}_{m-1} \\
\mathbf{Y}_m
\end{Bmatrix}
$$

If, in addition, we require that $\mathbf{Y}, \mathbf{Y}', \ldots, \mathbf{Y}^{(m-1)}$ take on specific values when $t = 0$, there will be exactly one solution to the problem.

If the system is simply of the form $\mathbf{Y}^{(m)} = A\mathbf{Y}$, it is usually not necessary to introduce new variables. In this case we need only calculate the mth roots of the eigenvalues of A. If λ is an eigenvalue of A, $\mathbf{x}$ is an eigenvector belonging to λ, σ is an mth root of λ, and $\mathbf{Y} = e^{\sigma t}\mathbf{x}$, then

$$\mathbf{Y}^{(m)} = \sigma^m e^{\sigma t}\mathbf{x} = \lambda\mathbf{Y}$$

and

$$A\mathbf{Y} = e^{\sigma t}A\mathbf{x} = \lambda e^{\sigma t}\mathbf{x} = \lambda\mathbf{Y}$$

Therefore, $\mathbf{Y} = e^{\sigma t}\mathbf{x}$ is a solution to the system.

APPLICATION 2: HARMONIC MOTION

In Figure 6.2.2 two masses are adjoined by springs and the ends A and B are fixed. The masses are free to move horizontally. We will assume that the three springs are uniform and that initially the system is in the equilibrium position. A force is exerted on the system to set the masses in motion. The horizontal displacements of the masses at time t will be denoted by $x_1(t)$ and $x_2(t)$, respectively. We will assume that there are no retarding forces such as friction. Then the only forces acting on mass m_1 at time t will be from the springs 1 and 2. The force from spring 1 will be

FIGURE 6.2.2

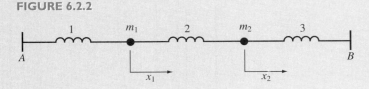

$-kx_1$ and the force from spring 2 will be $k(x_2 - x_1)$. By Newton's second law,

$$m_1 x_1''(t) = -kx_1 + k(x_2 - x_1)$$

Similarly, the only forces acting on the second mass will be from springs 2 and 3. Using Newton's second law again, we get

$$m_2 x_2''(t) = -k(x_2 - x_1) - kx_2$$

Thus we end up with the second-order system

$$x_1'' = -\frac{k}{m_1}(2x_1 - x_2)$$

$$x_2'' = -\frac{k}{m_2}(-x_1 + 2x_2)$$

Suppose now that $m_1 = m_2 = 1$, $k = 1$, and that the initial velocity of both masses is $+2$ units per second. To determine the displacements x_1 and x_2 as functions of t, we write the system in the form

(2)
$$\mathbf{X}'' = A\mathbf{X}$$

The coefficient matrix

$$A = \begin{pmatrix} -2 & 1 \\ 1 & -2 \end{pmatrix}$$

has eigenvalues $\lambda_1 = -1$ and $\lambda_2 = -3$. Corresponding to λ_1, we have the eigenvector $\mathbf{v}_1 = (1, 1)^T$ and $\sigma_1 = \pm i$. Thus $e^{it}\mathbf{v}_1$ and $e^{-it}\mathbf{v}_1$ are both solutions to (2). It follows that

$$\frac{1}{2}(e^{it} + e^{-it})\mathbf{v}_1 = (\operatorname{Re} e^{it})\mathbf{v}_1 = (\cos t)\mathbf{v}_1$$

and

$$\frac{1}{2i}(e^{it} - e^{-it})\mathbf{v}_1 = (\operatorname{Im} e^{it})\mathbf{v}_1 = (\sin t)\mathbf{v}_1$$

are both solutions to (2). Similarly, for $\lambda_2 = -3$, we have the eigenvector $\mathbf{v}_2 = (1, -1)^T$ and $\sigma_2 = \pm\sqrt{3}i$. It follows that

$$(\operatorname{Re} e^{\sqrt{3}it})\mathbf{v}_2 = (\cos \sqrt{3}t)\mathbf{v}_2$$

and

$$(\operatorname{Im} e^{\sqrt{3}it})\mathbf{v}_2 = (\sin \sqrt{3}t)\mathbf{v}_2$$

are also solutions to (2). Thus the general solution will be of the form

$$\mathbf{X}(t) = c_1(\cos t)\mathbf{v}_1 + c_2(\sin t)\mathbf{v}_1 + c_3(\cos \sqrt{3}t)\mathbf{v}_2 + c_4(\sin \sqrt{3}t)\mathbf{v}_2$$

$$= \begin{pmatrix} c_1 \cos t + c_2 \sin t + c_3 \cos \sqrt{3}t + c_4 \sin \sqrt{3}t \\ c_1 \cos t + c_2 \sin t - c_3 \cos \sqrt{3}t - c_4 \sin \sqrt{3}t \end{pmatrix}$$

At time $t = 0$, we have

$$x_1(0) = x_2(0) = 0 \qquad \text{and} \qquad x_1'(0) = x_2'(0) = 2$$

It follows that

$$
\begin{array}{ccc}
c_1 + c_3 = 0 & & c_2 + \sqrt{3}c_4 = 2 \\
& \text{and} & \\
c_1 - c_3 = 0 & & c_2 - \sqrt{3}c_4 = 2
\end{array}
$$

and hence

$$c_1 = c_3 = c_4 = 0 \qquad \text{and} \qquad c_2 = 2$$

Therefore, the solution to the initial value problem is simply

$$\mathbf{X}(t) = \begin{pmatrix} 2\sin t \\ 2\sin t \end{pmatrix}$$

The masses will oscillate with frequency 1 and amplitude 2.

APPLICATION 3: VIBRATIONS OF A BUILDING

For another example of a physical system, we consider the vibrations of a building. If the building has k stories, we can represent the horizontal deflections of the stories at time t by the vector function $\mathbf{Y}(t) = (y_1(t), y_2(t), \ldots, y_k(t))^T$. The motion of a building can be modeled by a second-order system of differential equations of the form

$$M\mathbf{Y}''(t) = K\mathbf{Y}(t)$$

The *mass matrix* M is a diagonal matrix whose entries correspond to the concentrated weights of each level. The entries of the *stiffness matrix* K are determined by the spring constants of the supporting structures. Solutions to the equation are of the form $\mathbf{Y}(t) = e^{i\sigma t}\mathbf{x}$, where $\mathbf{x}$ is an eigenvector of $A = M^{-1}K$ belonging to an eigenvalue λ, and σ is a square root of λ.

EXERCISES

1. Find the general solution to each of the following systems.

(a) $\begin{aligned} y_1' &= y_1 + y_2 \\ y_2' &= -2y_1 + 4y_2 \end{aligned}$

(b) $\begin{aligned} y_1' &= 2y_1 + 4y_2 \\ y_2' &= -y_1 - 3y_2 \end{aligned}$

(c) $\begin{aligned} y_1' &= y_1 - 2y_2 \\ y_2' &= -2y_1 + 4y_2 \end{aligned}$

(d) $\begin{aligned} y_1' &= y_1 - y_2 \\ y_2' &= y_1 + y_2 \end{aligned}$

(e) $\begin{aligned} y_1' &= 3y_1 - 2y_2 \\ y_2' &= 2y_1 + 3y_2 \end{aligned}$

(f) $\begin{aligned} y_1' &= y_1 + y_3 \\ y_2' &= 2y_2 + 6y_3 \\ y_3' &= y_2 + 3y_3 \end{aligned}$

2. Solve each of the following initial value problems.

(a) $y_1' = -y_1 + 2y_2$ $y_1(0) = 3,\ y_2(0) = 1$
   $y_2' = 2y_1 - y_2$

(b) $y_1' = y_1 - 2y_2$ $y_1(0) = 1,\ y_2(0) = -2$
   $y_2' = 2y_1 + y_2$

(c) $y_1' = 2y_1 - 6y_3$ $y_1(0) = y_2(0) = y_3(0) = 2$
   $y_2' = y_1 - 3y_3$
   $y_3' = y_2 - 2y_3$

(d) $y_1' = y_1 + 2y_3$ $y_1(0) = y_2(0) = 1,\ y_3(0) = 4$
   $y_2' = y_2 - y_3$
   $y_3' = y_1 + y_2 + y_3$

3. Given

$$\mathbf{Y} = c_1 e^{\lambda_1 t} \mathbf{x}_1 + c_2 e^{\lambda_2 t} \mathbf{x}_2 + \cdots + c_n e^{\lambda_n t} \mathbf{x}_n$$

is the solution to the initial value problem

$$\mathbf{Y}' = A\mathbf{Y}, \qquad \mathbf{Y}(0) = \mathbf{Y}_0$$

(a) Show that

$$\mathbf{Y}_0 = c_1 \mathbf{x}_1 + c_2 \mathbf{x}_2 + \cdots + c_n \mathbf{x}_n$$

(b) Let $X = (\mathbf{x}_1, \dots, \mathbf{x}_n)$ and $\mathbf{c} = (c_1, \dots, c_n)^T$. Assuming that the vectors $\mathbf{x}_1, \dots, \mathbf{x}_n$ are linearly independent, show that $\mathbf{c} = X^{-1} \mathbf{Y}_0$.

4. Two tanks each contain 100 liters of a mixture. Initially, the mixture in tank A contains 40 grams of salt while tank B contains 20 grams of salt. Liquid is pumped in and out of the tanks as shown in the figure. Determine the amount of salt in each tank at time t.

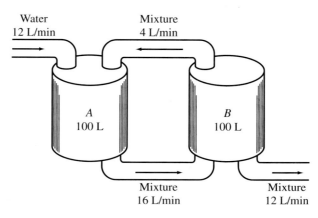

5. Find the general solution to each of the following:

(a) $y_1'' = -2y_2$ (b) $y_1'' = 2y_1 + y_2'$
   $y_2'' = y_1 + 3y_2$ $y_2'' = 2y_2 + y_1'$

6. Solve the initial value problem

$$y_1'' = -2y_2 + \ y_1' + 2y_2'$$
$$y_2'' = \quad 2y_1 + 2y_1' - \ y_2'$$
$$y_1(0) = 1, \qquad y_2(0) = 0, \qquad y_1'(0) = -3, \qquad y_2'(0) = 2$$

7. In the second application problem, assume that the solutions are of the form $x_1 = a_1 \sin \sigma t$, $x_2 = a_2 \sin \sigma t$. Substitute these expressions into the system and solve for the frequency σ and the amplitudes a_1 and a_2.

8. Solve the second application problem with the initial conditions

$$x_1(0) = x_2(0) = 1, \qquad x_1'(0) = 4, \qquad \text{and} \qquad x_2'(0) = 2$$

9. Two masses are connected by springs as shown in the diagram. Both springs have the same spring constant, and the end of the first spring is fixed. If x_1 and x_2 represent the displacements from the equilibrium position, derive a system of second-order differential equations that describes the motion of the system.

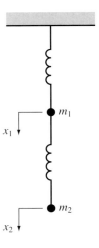

10. Three masses are connected by a series of springs between two fixed points as shown in the figure. Assume that the springs all have the same spring constant, and let $x_1(t)$, $x_2(t)$, and $x_3(t)$ represent the displacements of the respective masses at time t.

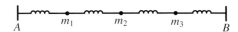

(a) Derive a system of second-order differential equations that describes the motion of this system.

(b) Solve the system if $m_1 = m_3 = \frac{1}{3}$, $m_2 = \frac{1}{4}$, $k = 1$, and

$$x_1(0) = x_2(0) = x_3(0) = 1$$
$$x_1'(0) = x_2'(0) = x_3'(0) = 0$$

11. Transform the nth-order equation

$$y^{(n)} = a_0 y + a_1 y' + \cdots + a_{n-1} y^{(n-1)}$$

into a system of first-order equations by setting $y_1 = y$ and $y_j = y'_{j-1}$ for $j = 2, \ldots, n$. Determine the characteristic polynomial of the coefficient matrix of this system.

3 DIAGONALIZATION

In this section we consider the problem of factoring an $n \times n$ matrix A into a product of the form XDX^{-1}, where D is diagonal. We will give a necessary and sufficient condition for the existence of such a factorization and look at a number of examples. We begin by showing that eigenvectors belonging to distinct eigenvalues are linearly independent.

▶ **THEOREM 6.3.1** *If $\lambda_1, \lambda_2, \ldots, \lambda_k$ are distinct eigenvalues of an $n \times n$ matrix A with corresponding eigenvectors $\mathbf{x}_1, \mathbf{x}_2, \ldots, \mathbf{x}_k$, then $\mathbf{x}_1, \ldots, \mathbf{x}_k$ are linearly independent.*

▶ **Proof.** Let r be the dimension of the subspace of R^n spanned by $\mathbf{x}_1, \ldots, \mathbf{x}_k$ and suppose that $r < k$. We may assume (reordering the $\mathbf{x}_i$'s and λ_i's if necessary) that $\mathbf{x}_1, \ldots, \mathbf{x}_r$ are linearly independent. Since $\mathbf{x}_1, \mathbf{x}_2, \ldots, \mathbf{x}_r, \mathbf{x}_{r+1}$ are linearly dependent, there exist scalars $c_1, \ldots, c_r, c_{r+1}$ not all zero such that

$$(1) \qquad c_1 \mathbf{x}_1 + \cdots + c_r \mathbf{x}_r + c_{r+1} \mathbf{x}_{r+1} = \mathbf{0}$$

Note that c_{r+1} must be nonzero; otherwise, $\mathbf{x}_1, \ldots, \mathbf{x}_r$ would be dependent. So $c_{r+1} \mathbf{x}_{r+1} \neq \mathbf{0}$ and hence $c_1, \ldots, c_r$ cannot all be zero. Multiplying (1) by A, we get

$$c_1 A\mathbf{x}_1 + \cdots + c_r A\mathbf{x}_r + c_{r+1} A\mathbf{x}_{r+1} = \mathbf{0}$$

or

$$(2) \qquad c_1 \lambda_1 \mathbf{x}_1 + \cdots + c_r \lambda_r \mathbf{x}_r + c_{r+1} \lambda_{r+1} \mathbf{x}_{r+1} = \mathbf{0}$$

Subtracting λ_{r+1} times (1) from (2) gives

$$c_1 (\lambda_1 - \lambda_{r+1}) \mathbf{x}_1 + \cdots + c_r (\lambda_r - \lambda_{r+1}) \mathbf{x}_r = \mathbf{0}$$

This contradicts the independence of $\mathbf{x}_1, \ldots, \mathbf{x}_r$. Therefore, r must equal k.   ◀

▶ **DEFINITION** An $n \times n$ matrix A is said to be **diagonalizable** if there exists a nonsingular matrix X and a diagonal matrix D such that

$$X^{-1}AX = D$$

We say that X **diagonalizes** A.   ◀

▶ **THEOREM 6.3.2** *An $n \times n$ matrix A is diagonalizable if and only if A has n linearly independent eigenvectors.*

▶*Proof.* Suppose that the matrix A has n linearly independent eigenvectors $\mathbf{x}_1, \mathbf{x}_2, \ldots, \mathbf{x}_n$. Let λ_i be the eigenvalue of A corresponding to $\mathbf{x}_i$ for each i. (Some of the λ_i's may be equal.) Let X be the matrix whose jth column vector is $\mathbf{x}_j$ for $j = 1, \ldots, n$. It follows that $A\mathbf{x}_j = \lambda_j\mathbf{x}_j$ is the jth column vector of AX. Thus

$$AX = (A\mathbf{x}_1, A\mathbf{x}_2, \ldots, A\mathbf{x}_n)$$

$$= (\lambda_1\mathbf{x}_1, \lambda_2\mathbf{x}_2, \ldots, \lambda_n\mathbf{x}_n)$$

$$= (\mathbf{x}_1, \mathbf{x}_2, \ldots, \mathbf{x}_n) \begin{bmatrix} \lambda_1 & & & \\ & \lambda_2 & & \\ & & \ddots & \\ & & & \lambda_n \end{bmatrix}$$

$$= XD$$

Since X has n linearly independent column vectors, it follows that X is nonsingular and hence

$$D = X^{-1}XD = X^{-1}AX$$

Conversely, suppose that A is diagonalizable. Then there exists a nonsingular matrix X such that $AX = XD$. If $\mathbf{x}_1, \mathbf{x}_2, \ldots, \mathbf{x}_n$ are the column vectors of X, then

$$A\mathbf{x}_j = \lambda_j\mathbf{x}_j \qquad (\lambda_j = d_{jj})$$

for each j. Thus, for each j, λ_j is an eigenvalue of A and $\mathbf{x}_j$ is an eigenvector belonging to λ_j. Since the column vectors of X are linearly independent, it follows that A has n linearly independent eigenvectors. ◀

Remarks

1. If A is diagonalizable, then the column vectors of the diagonalizing matrix X are eigenvectors of A, and the diagonal elements of D are the corresponding eigenvalues of A.

2. The diagonalizing matrix X is not unique. Reordering the columns of a given diagonalizing matrix X or multiplying them by nonzero scalars will produce a new diagonalizing matrix.

3. If A is $n \times n$ and A has n distinct eigenvalues, then A is diagonalizable. If the eigenvalues are not distinct, then A may or may not be diagonalizable depending on whether A has n linearly independent eigenvectors.

4. If A is diagonalizable, then A can be factored into a product XDX^{-1}. ◀

It follows from remark 4 that

$$A^2 = (XDX^{-1})(XDX^{-1}) = XD^2X^{-1}$$

and, in general,

$$A^k = XD^k X^{-1}$$

$$= X \begin{bmatrix} (\lambda_1)^k & & & \\ & (\lambda_2)^k & & \\ & & \ddots & \\ & & & (\lambda_n)^k \end{bmatrix} X^{-1}$$

Once we have a factorization $A = XDX^{-1}$, it is easy to compute powers of A.

EXAMPLE 1. Let

$$A = \begin{bmatrix} 2 & -3 \\ 2 & -5 \end{bmatrix}$$

The eigenvalues of A are $\lambda_1 = 1$ and $\lambda_2 = -4$. Corresponding to λ_1 and λ_2, we have the eigenvectors $\mathbf{x}_1 = (3, 1)^T$ and $\mathbf{x}_2 = (1, 2)^T$. Let

$$X = \begin{bmatrix} 3 & 1 \\ 1 & 2 \end{bmatrix}$$

It follows that

$$X^{-1} AX = \frac{1}{5} \begin{bmatrix} 2 & -1 \\ -1 & 3 \end{bmatrix} \begin{bmatrix} 2 & -3 \\ 2 & -5 \end{bmatrix} \begin{bmatrix} 3 & 1 \\ 1 & 2 \end{bmatrix}$$

$$= \begin{bmatrix} 1 & 0 \\ 0 & -4 \end{bmatrix}$$

and

$$XDX^{-1} = \begin{bmatrix} 3 & 1 \\ 1 & 2 \end{bmatrix} \begin{bmatrix} 1 & 0 \\ 0 & -4 \end{bmatrix} \begin{bmatrix} \frac{2}{5} & -\frac{1}{5} \\ -\frac{1}{5} & \frac{3}{5} \end{bmatrix} = \begin{bmatrix} 2 & -3 \\ 2 & -5 \end{bmatrix} = A \qquad \blacktriangleleft$$

EXAMPLE 2. Let

$$A = \begin{bmatrix} 3 & -1 & -2 \\ 2 & 0 & -2 \\ 2 & -1 & -1 \end{bmatrix}$$

It is easily seen that the eigenvalues of A are $\lambda_1 = 0$, $\lambda_2 = 1$, $\lambda_3 = 1$. Corresponding to $\lambda_1 = 0$, we have the eigenvector $(1, 1, 1)^T$, and corresponding to $\lambda = 1$, we have the eigenvectors $(1, 2, 0)^T$ and $(0, -2, 1)^T$. Let

$$X = \begin{bmatrix} 1 & 1 & 0 \\ 1 & 2 & -2 \\ 1 & 0 & 1 \end{bmatrix}$$

It follows that

$$XDX^{-1} = \begin{bmatrix} 1 & 1 & 0 \\ 1 & 2 & -2 \\ 1 & 0 & 1 \end{bmatrix} \begin{bmatrix} 0 & 0 & 0 \\ 0 & 1 & 0 \\ 0 & 0 & 1 \end{bmatrix} \begin{bmatrix} -2 & 1 & 2 \\ 3 & -1 & -2 \\ 2 & -1 & -1 \end{bmatrix}$$

$$= \begin{bmatrix} 3 & -1 & -2 \\ 2 & 0 & -2 \\ 2 & -1 & -1 \end{bmatrix}$$

$$= A$$

Even though $\lambda = 1$ is a multiple eigenvalue, the matrix can still be diagonalized since there are three linearly independent eigenvectors. Note also that

$$A^k = XD^k X^{-1} = XDX^{-1} = A$$

for any $k \geq 1$. ◀

If an $n \times n$ matrix A has fewer than n linearly independent eigenvectors, we say that A is *defective*. It follows from Theorem 6.3.2 that a defective matrix is not diagonalizable.

EXAMPLE 3. Let

$$A = \begin{bmatrix} 1 & 1 \\ 0 & 1 \end{bmatrix}$$

The eigenvalues of A are both equal to 1. Any eigenvector corresponding to $\lambda = 1$ must be a multiple of $\mathbf{x}_1 = (1, 0)^T$. Thus A is defective and cannot be diagonalized. ◀

EXAMPLE 4. Let

$$A = \begin{bmatrix} 2 & 0 & 0 \\ 0 & 4 & 0 \\ 1 & 0 & 2 \end{bmatrix} \quad \text{and} \quad B = \begin{bmatrix} 2 & 0 & 0 \\ -1 & 4 & 0 \\ -3 & 6 & 2 \end{bmatrix}$$

A and B both have the same eigenvalues

$$\lambda_1 = 4, \qquad \lambda_2 = \lambda_3 = 2$$

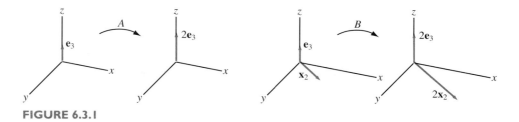

FIGURE 6.3.1

The eigenspace of A corresponding to $\lambda_1 = 4$ is spanned by $\mathbf{e}_2$, and the eigenspace corresponding to $\lambda = 2$ is spanned by $\mathbf{e}_3$. Since A has only two linearly independent eigenvectors, it is defective. On the other hand, the matrix B has eigenvector $\mathbf{x}_1 = (0, 1, 3)^T$ corresponding to $\lambda_1 = 4$ and eigenvectors $\mathbf{x}_2 = (2, 1, 0)^T$ and $\mathbf{e}_3$ corresponding to $\lambda = 2$. Thus B has three linearly independent eigenvectors and consequently is not defective. Even though $\lambda = 2$ is an eigenvalue of multiplicity 2, the matrix B is nondefective since the corresponding eigenspace has dimension 2.

Geometrically, the matrix B has the effect of stretching two linearly independent vectors by a factor of 2. We can think of the eigenvalue $\lambda = 2$ as having *geometric multiplicity* 2. On the other hand, the matrix A only stretches vectors along the z axis by a factor of 2. In this case the eigenvalue $\lambda = 2$ has algebraic multiplicity 2, but its geometric multiplicity is only 1 (see Figure 6.3.1).   ◀

APPLICATION 1: MARKOV CHAINS

In Section 1 of this chapter we studied a simple matrix model for predicting the number of married and single women in a certain town each year. Given an initial vector $\mathbf{x}_0$ whose coordinates represent the current number of married and single women, we were able to predict the number of married and single women in future years by computing

$$\mathbf{x}_1 = A\mathbf{x}_0, \ \mathbf{x}_2 = A\mathbf{x}_1, \ \mathbf{x}_3 = A\mathbf{x}_2, \ldots$$

If we scale the initial vector so that its entries indicate the proportions of the population that are married and single, then the coordinates of $\mathbf{x}_n$ will indicate the proportions of married and single women after n years. The sequence of vectors that we generate in this manner is an example of a *Markov chain*. Markov chain models occur in a wide variety of applied fields.

▶ **DEFINITION** A **stochastic process** is any sequence of experiments for which the outcome at any stage depends on chance. A **Markov process** is a stochastic process with the following properties:

   (i) The set of possible outcomes or states is finite.
   (ii) The probability of the next outcome depends only on the previous outcome.
   (iii) The probabilities are constant over time.   ◀

The following is an example of a Markov process.

TABLE I

TRANSITION PROBABILITIES FOR VEHICLE LEASING

Sedan	Sports Car	Minivan	Sport Utility	Next Lease
0.80	0.10	0.05	0.05	**Sedan**
0.10	0.80	0.05	0.05	**Sports Car**
0.05	0.05	0.80	0.10	**Minivan**
0.05	0.05	0.10	0.80	**Four-wheel**

EXAMPLE 5. [Automobile Leasing] An automobile dealer leases four types of vehicles: four-door sedans, sports cars, minivans, and sport utility vehicles. The term of the lease is 2 years. At the end of the term, customers must renegotiate the lease and choose a new vehicle.

The automobile leasing can be viewed as a process with four possible outcomes. The probability of each outcome can be estimated by reviewing records of previous leases. These records indicate that 80 percent of the customers currently leasing sedans will continue doing so in the next lease. Furthermore, 10 percent of the customers currently leasing sports cars will switch to sedans. In addition, 5 percent of the customers driving minivans or sport utility vehicles will also switch to sedans. These results are summarized in the first row of Table 1. The second row indicates the percentages of customers that will lease sports cars the next time, and the final two rows give the percentages that will lease minivans and four-wheel vehicles, respectively.

Suppose that initially there are 200 sedans leased and 100 of each of the other three types of vehicles. If we set

$$A = \begin{bmatrix} 0.80 & 0.10 & 0.05 & 0.05 \\ 0.10 & 0.80 & 0.05 & 0.05 \\ 0.05 & 0.05 & 0.80 & 0.10 \\ 0.05 & 0.05 & 0.10 & 0.80 \end{bmatrix} \qquad \mathbf{x}_0 = \begin{bmatrix} 200 \\ 100 \\ 100 \\ 100 \end{bmatrix}$$

then we can determine how many people will lease each type of vehicle 2 years later by setting

$$\mathbf{x}_1 = A\mathbf{x}_0 = \begin{bmatrix} 0.80 & 0.10 & 0.05 & 0.05 \\ 0.10 & 0.80 & 0.05 & 0.05 \\ 0.05 & 0.05 & 0.80 & 0.10 \\ 0.05 & 0.05 & 0.10 & 0.80 \end{bmatrix} \begin{bmatrix} 200 \\ 100 \\ 100 \\ 100 \end{bmatrix} = \begin{bmatrix} 180 \\ 110 \\ 105 \\ 105 \end{bmatrix}$$

We can predict the numbers for future leases by setting

$$\mathbf{x}_{n+1} = A\mathbf{x}_n \quad \text{for} \quad n = 1, 2, \ldots$$

The vectors $\mathbf{x}_i$ produced in this manner are referred to as *state vectors*, and the sequence of state vectors is called a *Markov chain*. The matrix A is referred to as a transition matrix. The entries of each column of A are nonnegative numbers that add up to 1. Each column can be viewed as a *probability vector*. For example, the first column of A corresponds to individuals currently leasing sedans. The entries in this column are the probabilities of choosing each type of vehicle when the lease is renewed.

In general, a matrix is said to be *stochastic* if its entries are nonnegative and the entries in each column add up to 1. The columns of a stochastic matrix can be viewed as probability vectors.

If we divide the entries of the initial vector by 500 (the total number of customers), then the entries of the new initial state vector $\mathbf{x}_0 = (0.40, 0.20, 0.20, 0.20)^T$ represent the proportions of the population that rent each type of vehicle. The entries of $\mathbf{x}_1$ will represent the proportions for the next lease. Thus $\mathbf{x}_0$ and $\mathbf{x}_1$ are probability vectors, and it is easily seen that the succeeding state vectors in the chain will all be probability vectors.

The long-range behavior of the process is determined by the eigenvalues and eigenvectors of the transition matrix A. The eigenvalues of A are $\lambda_1 = 1$, $\lambda_2 = 0.8$, $\lambda_3 = \lambda_4 = 0.7$ Even though A has a multiple eigenvalue, it does have four linearly independent eigenvectors and hence it can be diagonalized. If the eigenvectors are used to form a diagonalizing matrix Y, then

$$A = YDY^{-1}$$

$$= \begin{bmatrix} 1 & -1 & 0 & 1 \\ 1 & -1 & 0 & -1 \\ 1 & 1 & 1 & 0 \\ 1 & 1 & -1 & 0 \end{bmatrix} \begin{bmatrix} 1 & 0 & 0 & 0 \\ 0 & \frac{8}{10} & 0 & 0 \\ 0 & 0 & \frac{7}{10} & 0 \\ 0 & 0 & 0 & \frac{7}{10} \end{bmatrix} \begin{bmatrix} \frac{1}{4} & \frac{1}{4} & \frac{1}{4} & \frac{1}{4} \\ -\frac{1}{4} & -\frac{1}{4} & \frac{1}{4} & \frac{1}{4} \\ 0 & 0 & \frac{1}{2} & -\frac{1}{2} \\ \frac{1}{2} & -\frac{1}{2} & 0 & 0 \end{bmatrix}$$

The state vectors are computed by setting

$$\begin{aligned}
\mathbf{x}_n &= YD^nY^{-1}\mathbf{x}_0 \\
&= YD^n(0.25, -0.05, 0, 0.10)^T \\
&= Y(0.25, -0.05(0.8)^n, 0, 0.10(0.7)^n)^T
\end{aligned}$$

$$= 0.25 \begin{bmatrix} 1 \\ 1 \\ 1 \\ 1 \end{bmatrix} - 0.05(0.8)^n \begin{bmatrix} -1 \\ -1 \\ 1 \\ 1 \end{bmatrix} + 0.10(0.7)^n \begin{bmatrix} 1 \\ -1 \\ 0 \\ 0 \end{bmatrix}$$

As n increases, $\mathbf{x}_n$ approaches the steady-state vector

$$\mathbf{x} = (0.25, 0.25, 0.25, 0.25)^T$$

Thus the Markov chain model predicts that in the long run the leases will be divided equally among the four types of vehicles. ◀

Not all Markov chains converge to a steady-state vector. However, if all the entries of the transition matrix A are positive, then it can be shown that there is a unique steady-state vector $\mathbf{x}$ and that $A^n \mathbf{x}_0$ will converge to $\mathbf{x}$ for any initial probability vector $\mathbf{x}_0$. In fact, this result will hold if A^k has strictly positive entries even though A may have some 0 entries. A Markov process with transition matrix A is said to be *regular* if all the entries of some power of A are positive.

The steady-state vector $\mathbf{x}$ of a regular Markov chain is both a probability vector and also an eigenvector belonging to the eigenvalue $\lambda = 1$. The existence of such a vector is guaranteed by Perron's Theorem (see Section 8 of this chapter).

APPLICATION 2: SEX-LINKED GENES

Sex-linked genes are genes that are located on the X chromosome. For example, the gene for blue-green color blindness is a recessive sex-linked gene. To devise a mathematical model to describe color blindness in a given population, it is necessary to divide the population into two classes, males and females. Let $x_1^{(0)}$ be the proportion of genes for color blindness in the male population, and let $x_2^{(0)}$ be the proportion in the female population. [Since color blindness is recessive, the actual proportion of color-blind females will be less than $x_2^{(0)}$.] Since the male receives one X chromosome from the mother and none from the father, the proportion $x_1^{(1)}$ of color-blind males in the next generation will be the same as the proportion of recessive genes in the present generation of females. Since the female receives an X chromosome from each parent, the proportion $x_2^{(1)}$ of recessive genes in the next generation of females will be the average of $x_1^{(0)}$ and $x_2^{(0)}$. Thus

$$x_2^{(0)} = x_1^{(1)}$$
$$\tfrac{1}{2}x_1^{(0)} + \tfrac{1}{2}x_2^{(0)} = x_2^{(1)}$$

If $x_1^{(0)} = x_2^{(0)}$, the proportion will not change in future generations. Let us assume that $x_1^{(0)} \neq x_2^{(0)}$ and write the system as a matrix equation.

$$\begin{bmatrix} 0 & 1 \\ \tfrac{1}{2} & \tfrac{1}{2} \end{bmatrix} \begin{bmatrix} x_1^{(0)} \\ x_2^{(0)} \end{bmatrix} = \begin{bmatrix} x_1^{(1)} \\ x_2^{(1)} \end{bmatrix}$$

Let A denote the coefficient matrix, and let $\mathbf{x}^{(n)} = (x_1^{(n)}, x_2^{(n)})^T$ denote the proportion of color-blind genes in the male and female populations of the $(n+1)$st generation. Thus

$$\mathbf{x}^{(n)} = A^n \mathbf{x}^{(0)}$$

To compute A^n, we note that A has eigenvalues 1 and $-\tfrac{1}{2}$ and consequently can be factored into a product

$$A = \begin{bmatrix} 1 & -2 \\ 1 & 1 \end{bmatrix} \begin{bmatrix} 1 & 0 \\ 0 & -\tfrac{1}{2} \end{bmatrix} \begin{bmatrix} \tfrac{1}{3} & \tfrac{2}{3} \\ -\tfrac{1}{3} & \tfrac{1}{3} \end{bmatrix}$$

Thus

$$\mathbf{x}^{(n)} = \begin{bmatrix} 1 & -2 \\ 1 & 1 \end{bmatrix} \begin{bmatrix} 1 & 0 \\ 0 & -\frac{1}{2} \end{bmatrix}^{n} \begin{bmatrix} \frac{1}{3} & \frac{2}{3} \\ -\frac{1}{3} & \frac{1}{3} \end{bmatrix} \begin{bmatrix} x_1^{(0)} \\ x_2^{(0)} \end{bmatrix}$$

$$= \frac{1}{3} \begin{bmatrix} 1 - (-\frac{1}{2})^{n-1} & 2 + (-\frac{1}{2})^{n-1} \\ 1 - (-\frac{1}{2})^{n} & 2 + (-\frac{1}{2})^{n} \end{bmatrix} \begin{bmatrix} x_1^{(0)} \\ x_2^{(0)} \end{bmatrix}$$

and hence

$$\lim_{n \to \infty} \mathbf{x}^{(n)} = \frac{1}{3} \begin{bmatrix} 1 & 2 \\ 1 & 2 \end{bmatrix} \begin{bmatrix} x_1^{(0)} \\ x_2^{(0)} \end{bmatrix}$$

$$= \begin{bmatrix} \dfrac{x_1^{(0)} + 2x_2^{(0)}}{3} \\ \dfrac{x_1^{(0)} + 2x_2^{(0)}}{3} \end{bmatrix}$$

The proportions of genes for color blindness in the male and female populations will tend to the same value as the number of generations increases. If the proportion of color-blind men is p and over a number of generations no outsiders have entered the population, there is justification for assuming that the proportion of genes for color blindness in the female population is also p. Since color blindness is recessive, we would expect the proportion of color-blind women to be about p^2. Thus, if 1 percent of the male population is color-blind, we would expect about 0.01 percent of the female population to be color-blind.

The Exponential of a Matrix

Given a scalar a, the exponential e^a can be expressed in terms of a power series

$$e^a = 1 + a + \frac{1}{2!}a^2 + \frac{1}{3!}a^3 + \cdots$$

Similarly, for any $n \times n$ matrix A, we can define the *matrix exponential* e^A in terms of the convergent power series

(3) $$e^A = I + A + \frac{1}{2!}A^2 + \frac{1}{3!}A^3 + \cdots$$

The matrix exponential (3) occurs in a wide variety of applications. In the case of a diagonal matrix

$$D = \begin{bmatrix} \lambda_1 & & & \\ & \lambda_2 & & \\ & & \ddots & \\ & & & \lambda_n \end{bmatrix}$$

the matrix exponential is easy to compute:

$$e^D = \lim_{m \to \infty} \left(I + D + \frac{1}{2!}D^2 + \cdots + \frac{1}{m!}D^m \right)$$

$$= \lim_{m \to \infty} \begin{bmatrix} \sum_{k=1}^{m} \frac{1}{k!}\lambda_1^k & & \\ & \ddots & \\ & & \sum_{k=1}^{m} \frac{1}{k!}\lambda_n^k \end{bmatrix}$$

$$= \begin{bmatrix} e^{\lambda_1} & & & \\ & e^{\lambda_2} & & \\ & & \ddots & \\ & & & e^{\lambda_n} \end{bmatrix}$$

It is more difficult to compute the matrix exponential for a general $n \times n$ matrix A. If, however, A is diagonalizable, then

$$A^k = XD^k X^{-1} \quad \text{for} \quad k = 1, 2, \ldots$$
$$e^A = X \left(I + D + \frac{1}{2!}D^2 + \frac{1}{3!}D^3 + \cdots \right) X^{-1}$$
$$= X e^D X^{-1}$$

EXAMPLE 6.   Compute e^A for

$$A = \begin{bmatrix} -2 & -6 \\ 1 & 3 \end{bmatrix}$$

SOLUTION.   The eigenvalues of A are $\lambda_1 = 1$ and $\lambda_2 = 0$ with eigenvectors $\mathbf{x}_1 = (-2, 1)^T$ and $\mathbf{x}_2 = (-3, 1)^T$. Thus

$$A = XDX^{-1} = \begin{bmatrix} -2 & -3 \\ 1 & 1 \end{bmatrix} \begin{bmatrix} 1 & 0 \\ 0 & 0 \end{bmatrix} \begin{bmatrix} 1 & 3 \\ -1 & -2 \end{bmatrix}$$

and

$$e^A = X e^D X^{-1} = \begin{bmatrix} -2 & -3 \\ 1 & 1 \end{bmatrix} \begin{bmatrix} e^1 & 0 \\ 0 & e^0 \end{bmatrix} \begin{bmatrix} 1 & 3 \\ -1 & -2 \end{bmatrix}$$

$$= \begin{bmatrix} 3 - 2e & 6 - 6e \\ e - 1 & 3e - 2 \end{bmatrix}$$

◀

The matrix exponential can be applied to the initial value problem

(4) $\mathbf{Y}' = A\mathbf{Y}, \qquad \mathbf{Y}(0) = \mathbf{Y}_0$

studied in Section 2. In the case of one equation in one unknown,

$$y' = ay, \quad y(0) = y_0$$

the solution is

(5)
$$y = e^{at} y_0$$

We can generalize this and express the solution to (4) in terms of the matrix exponential e^{At}, where $At = tA$ (i.e., t times the matrix A). In general, a power series can be differentiated term by term within its radius of convergence. Since the expansion of e^{At} has infinite radius of convergence, we have

$$
\begin{aligned}
\frac{d}{dt} e^{At} &= \frac{d}{dt} \left(I + tA + \frac{1}{2!} t^2 A^2 + \frac{1}{3!} t^3 A^3 + \cdots \right) \\
&= \left(A + tA^2 + \frac{1}{2!} t^2 A^3 + \cdots \right) \\
&= A \left(I + tA + \frac{1}{2!} t^2 A^2 + \cdots \right) \\
&= A e^{At}
\end{aligned}
$$

If, as in (5), we set

$$\mathbf{Y}(t) = e^{At} \mathbf{Y}_0$$

then

$$\mathbf{Y}' = A e^{At} \mathbf{Y}_0 = A\mathbf{Y}$$

and

$$\mathbf{Y}(0) = \mathbf{Y}_0$$

Thus, the solution to

$$\mathbf{Y}' = A\mathbf{Y}, \qquad \mathbf{Y}(0) = \mathbf{Y}_0$$

is simply

(6)
$$\mathbf{Y} = e^{At} \mathbf{Y}_0$$

Although the form of this solution looks different from the solutions in Section 2, there is really no difference. In Section 2 the solution was expressed in the form

$$c_1 e^{\lambda_1 t} \mathbf{x}_1 + c_2 e^{\lambda_2 t} \mathbf{x}_2 + \cdots + c_n e^{\lambda_n t} \mathbf{x}_n$$

where $\mathbf{x}_i$ was an eigenvector belonging to λ_i for $i = 1, \ldots, n$. The c_i's that satisfied the initial conditions were determined by solving a system

$$X\mathbf{c} = \mathbf{Y}_0$$

with coefficient matrix $X = (\mathbf{x}_1, \ldots, \mathbf{x}_n)$.

If A is diagonalizable, we can write (6) in the form

$$\mathbf{Y} = X e^{Dt} X^{-1} \mathbf{Y}_0$$

Thus,

$$\mathbf{Y} = Xe^{Dt}\mathbf{c}$$

$$= (\mathbf{x}_1, \mathbf{x}_2, \ldots, \mathbf{x}_n) \begin{bmatrix} c_1 e^{\lambda_1 t} \\ c_2 e^{\lambda_2 t} \\ \vdots \\ c_n e^{\lambda_n t} \end{bmatrix}$$

$$= c_1 e^{\lambda_1 t} \mathbf{x}_1 + \cdots + c_n e^{\lambda_n t} \mathbf{x}_n$$

To summarize, the solution to the initial value problem (4) is given by

$$\mathbf{Y} = e^{At}\mathbf{Y}_0$$

If A is diagonalizable, this solution can be written in the form

$$\mathbf{Y} = Xe^{Dt} X^{-1}\mathbf{Y}_0$$
$$= c_1 e^{\lambda_1 t} \mathbf{x}_1 + c_2 e^{\lambda_2 t} \mathbf{x}_2 + \cdots + c_n e^{\lambda_n t} \mathbf{x}_n \quad (\mathbf{c} = X^{-1}\mathbf{Y}_0)$$

EXAMPLE 7.   Use the matrix exponential to solve the initial value problem

$$\mathbf{Y}' = A\mathbf{Y}, \qquad \mathbf{Y}(0) = \mathbf{Y}_0$$

where

$$A = \begin{bmatrix} 3 & 4 \\ 3 & 2 \end{bmatrix}, \qquad \mathbf{Y}_0 = \begin{bmatrix} 6 \\ 1 \end{bmatrix}$$

(This problem was solved in Example 1 of Section 2.)

SOLUTION.   The eigenvalues of A are $\lambda_1 = 6$ and $\lambda_2 = -1$ with eigenvectors $\mathbf{x}_1 = (4, 3)^T$ and $\mathbf{x}_2 = (1, -1)^T$. Thus

$$A = XDX^{-1} = \begin{bmatrix} 4 & 1 \\ 3 & -1 \end{bmatrix} \begin{bmatrix} 6 & 0 \\ 0 & -1 \end{bmatrix} \begin{bmatrix} \frac{1}{7} & \frac{1}{7} \\ \frac{3}{7} & -\frac{4}{7} \end{bmatrix}$$

and the solution is given by

$$\mathbf{Y} = e^{At}\mathbf{Y}_0$$
$$= Xe^{Dt} X^{-1}\mathbf{Y}_0$$

$$= \begin{bmatrix} 4 & 1 \\ 3 & -1 \end{bmatrix} \begin{bmatrix} e^{6t} & 0 \\ 0 & e^{-t} \end{bmatrix} \begin{bmatrix} \frac{1}{7} & \frac{1}{7} \\ \frac{3}{7} & -\frac{4}{7} \end{bmatrix} \begin{bmatrix} 6 \\ 1 \end{bmatrix}$$

$$= \begin{bmatrix} 4e^{6t} + 2e^{-t} \\ 3e^{6t} - 2e^{-t} \end{bmatrix}$$

Compare this to Example 1 in Section 2.   ◀

EXAMPLE 8. Use the matrix exponential to solve the initial value problem

$$\mathbf{Y}' = A\mathbf{Y}, \qquad \mathbf{Y}(0) = \mathbf{Y}_0$$

where

$$A = \begin{bmatrix} 0 & 1 & 0 \\ 0 & 0 & 1 \\ 0 & 0 & 0 \end{bmatrix}, \qquad \mathbf{Y}_0 = \begin{bmatrix} 2 \\ 1 \\ 4 \end{bmatrix}$$

SOLUTION. Since the matrix A is defective, we will use the definition of the matrix exponential to compute e^{At}. Note that $A^3 = O$, so

$$e^{At} = I + tA + \frac{1}{2!}t^2 A^2$$

$$= \begin{bmatrix} 1 & t & t^2/2 \\ 0 & 1 & t \\ 0 & 0 & 1 \end{bmatrix}$$

The solution to the initial value problem is given by

$$\mathbf{Y} = e^{At}\mathbf{Y}_0$$

$$= \begin{bmatrix} 1 & t & t^2/2 \\ 0 & 1 & t \\ 0 & 0 & 1 \end{bmatrix} \begin{bmatrix} 2 \\ 1 \\ 4 \end{bmatrix}$$

$$= \begin{bmatrix} 2 + t + 2t^2 \\ 1 + 4t \\ 4 \end{bmatrix} \qquad \blacktriangleleft$$

EXERCISES

1. In each of the following, factor the matrix A into a product XDX^{-1}, where D is diagonal.

(a) $A = \begin{bmatrix} 0 & 1 \\ 1 & 0 \end{bmatrix}$

(b) $A = \begin{bmatrix} 5 & 6 \\ -2 & -2 \end{bmatrix}$

(c) $A = \begin{bmatrix} 2 & -8 \\ 1 & -4 \end{bmatrix}$

(d) $A = \begin{bmatrix} 2 & 2 & 1 \\ 0 & 1 & 2 \\ 0 & 0 & -1 \end{bmatrix}$

(e) $A = \begin{bmatrix} 1 & 0 & 0 \\ -2 & 1 & 3 \\ 1 & 1 & -1 \end{bmatrix}$

(f) $A = \begin{bmatrix} 1 & 2 & -1 \\ 2 & 4 & -2 \\ 3 & 6 & -3 \end{bmatrix}$

2. For each of the matrices in Exercise 1, use the XDX^{-1} factorization to compute A^6.

3. For each of the nonsingular matrices in Exercise 1, use the XDX^{-1} factorization to compute A^{-1}.

4. For each of the following, find a matrix B such that $B^2 = A$.

(a) $A = \begin{bmatrix} 2 & 1 \\ -2 & -1 \end{bmatrix}$

(b) $A = \begin{bmatrix} 9 & -5 & 3 \\ 0 & 4 & 3 \\ 0 & 0 & 1 \end{bmatrix}$

5. Let A be a nondefective $n \times n$ matrix with diagonalizing matrix X. Show that the matrix $Y = (X^{-1})^T$ diagonalizes A^T.

6. Let A be a diagonalizable matrix whose eigenvalues are all either 1 or -1. Show that $A^{-1} = A$.

7. Show that any 3×3 matrix of the form

$$\begin{bmatrix} a & 1 & 0 \\ 0 & a & 1 \\ 0 & 0 & b \end{bmatrix}$$

is defective.

8. For each of the following, find all possible values of the scalar α that make the matrix defective or show that no such values exist.

(a) $\begin{bmatrix} 1 & 1 & 0 \\ 1 & 1 & 0 \\ 0 & 0 & \alpha \end{bmatrix}$

(b) $\begin{bmatrix} 1 & 1 & 1 \\ 1 & 1 & 1 \\ 0 & 0 & \alpha \end{bmatrix}$

(c) $\begin{bmatrix} 1 & 2 & 0 \\ 2 & 1 & 0 \\ 2 & -1 & \alpha \end{bmatrix}$

(d) $\begin{bmatrix} 4 & 6 & -2 \\ -1 & -1 & 1 \\ 0 & 0 & \alpha \end{bmatrix}$

9. Let A be a 4×4 matrix and let λ be an eigenvalue of multiplicity 3. If $A - \lambda I$ has rank 1, is A defective? Explain.

10. Let A be an $n \times n$ matrix with positive real eigenvalues $\lambda_1 > \lambda_2 > \cdots > \lambda_n$. Let $\mathbf{x}_i$ be an eigenvector belonging to λ_i for each i, and let $\mathbf{x} = \alpha_1 \mathbf{x}_1 + \cdots + \alpha_n \mathbf{x}_n$.

(a) Show that $A^m \mathbf{x} = \displaystyle\sum_{i=1}^{n} \alpha_i \lambda_i^m \mathbf{x}_i$.

(b) If $\lambda_1 = 1$, show that $\displaystyle\lim_{m \to \infty} A^m \mathbf{x} = \alpha_1 \mathbf{x}_1$.

11. Let A be an $n \times n$ matrix with an eigenvalue λ of multiplicity n. Show that A is diagonalizable if and only if $A = \lambda I$.

12. Show that a nonzero nilpotent matrix is defective.

13. Let A be a diagonalizable matrix and let X be the diagonalizing matrix. Show that the column vectors of X that correspond to nonzero eigenvalues of A form a basis for $R(A)$.

14. It follows from Exercise 13 that for a diagonalizable matrix the number of nonzero eigenvalues (counted according to multiplicity) equals the rank of the matrix. Give an example of a defective matrix whose rank is not equal to the number of nonzero eigenvalues.

15. Let A be an $n \times n$ matrix and let λ be an eigenvalue of A whose eigenspace has dimension k, where $1 < k < n$. Any basis $\{\mathbf{x}_1, \ldots, \mathbf{x}_k\}$ for the eigenspace can be extended to a basis $\{\mathbf{x}_1, \ldots, \mathbf{x}_n\}$ for R^n. Let $X = (\mathbf{x}_1, \ldots, \mathbf{x}_n)$ and $B = X^{-1}AX$.

   (a) Show that B is of the form

$$
\begin{bmatrix}
\lambda I & B_{12} \\
O & B_{22}
\end{bmatrix}
$$

   where I is the $k \times k$ identity matrix.

   (b) Use Theorem 6.1.1 to show that λ is an eigenvalue of A with multiplicity at least k.

16. Let $\mathbf{x}, \mathbf{y}$ be nonzero vectors in R^n, $n \geq 2$, and let $A = \mathbf{x}\mathbf{y}^T$. Show that:

   (a) Zero is an eigenvalue of A with $n - 1$ linearly independent eigenvectors and consequently has multiplicity at least $n - 1$ (see Exercise 15).

   (b) The remaining eigenvalue of A is

$$
\lambda_n = \operatorname{tr} A = \mathbf{x}^T \mathbf{y}
$$

   and $\mathbf{x}$ is an eigenvector belonging to λ_n.

   (c) If $\lambda_n = \mathbf{x}^T \mathbf{y} \neq 0$, then A is diagonalizable.

17. Let A be a diagonalizable $n \times n$ matrix. Prove that if B is any matrix that is similar to A, then B is diagonalizable.

18. Show that if A and B are two $n \times n$ matrices that both have the same diagonalizing matrix X, then $AB = BA$.

19. Each year employees at a company are given the option of donating to a local charity as part of a payroll deduction plan. In general, 80 percent of the employees enrolled in the plan in any one year will choose to sign up again the following year, and 30 percent of the unenrolled will choose to enroll the

following year. Determine the transition matrix for the Markov process and find the steady-state vector. What percentage of employees would you expect to find enrolled in the program in the long run?

20. The city of Mawtookit maintains a constant population of 300,000 people from year to year. A political science study estimated that there were 150,000 Independents, 90,000 Democrats, and 60,000 Republicans in the town. It was also estimated that each year 20 percent of the Independents become Democrats and 10 percent become Republicans. Similarly, 20 percent of the Democrats become Independents and 10 percent become Republicans, while 10 percent of the Republicans defect to the Democrats and 10 percent become Independents each year. Let

$$\mathbf{x} = \begin{bmatrix} 150{,}000 \\ 90{,}000 \\ 60{,}000 \end{bmatrix}$$

and let $\mathbf{x}^{(1)}$ be a vector representing the number of people in each group after 1 year.

(a) Find a matrix A such that $A\mathbf{x} = \mathbf{x}^{(1)}$.
(b) Show that $\lambda_1 = 1.0$, $\lambda_2 = 0.5$, and $\lambda_3 = 0.7$ are the eigenvalues of A, and factor A into a product XDX^{-1}, where D is diagonal.
(c) Which group will dominate in the long run? Justify your answer by computing $\lim_{n \to \infty} A^n \mathbf{x}$.

21. Show that if A is a stochastic matrix then $\lambda = 1$ is an eigenvalue of A.

22. A matrix A is said to be *doubly stochastic* if both A and A^T are stochastic. Show that if A is an $n \times n$ doubly stochastic matrix then $\mathbf{x} = \left(\frac{1}{n}, \frac{1}{n}, \ldots, \frac{1}{n} \right)^T$ is an eigenvector belonging to $\lambda = 1$.

23. Use the definition of the matrix exponential to compute e^A for each of the following matrices:

(a) $A = \begin{bmatrix} 1 & 1 \\ -1 & -1 \end{bmatrix}$ (b) $A = \begin{bmatrix} 1 & 1 \\ 0 & 1 \end{bmatrix}$ (c) $A = \begin{bmatrix} 1 & 0 & -1 \\ 0 & 1 & 0 \\ 0 & 0 & 1 \end{bmatrix}$

24. Compute e^A for each of the following matrices:

(a) $A = \begin{bmatrix} -2 & -1 \\ 6 & 3 \end{bmatrix}$ (b) $A = \begin{bmatrix} 3 & 4 \\ -2 & -3 \end{bmatrix}$ (c) $A = \begin{bmatrix} 1 & 1 & 1 \\ -1 & -1 & -1 \\ 1 & 1 & 1 \end{bmatrix}$

25. In each of the following, solve the initial value problem $\mathbf{Y}' = A\mathbf{Y}$, $\mathbf{Y}(0) = \mathbf{Y}_0$ by computing $e^{At}\mathbf{Y}_0$.

(a) $A = \begin{bmatrix} 1 & -2 \\ 0 & -1 \end{bmatrix}$, $\quad \mathbf{Y}_0 = \begin{bmatrix} 1 \\ 1 \end{bmatrix}$

(b) $A = \begin{bmatrix} 2 & 3 \\ -1 & -2 \end{bmatrix}$, $\quad \mathbf{Y}_0 = \begin{bmatrix} -4 \\ 2 \end{bmatrix}$

(c) $A = \begin{bmatrix} 1 & 1 & 1 \\ 0 & 0 & 1 \\ 0 & 0 & -1 \end{bmatrix}$, $\quad \mathbf{Y}_0 = \begin{bmatrix} 1 \\ 1 \\ 1 \end{bmatrix}$

(d) $A = \begin{bmatrix} 1 & 1 & 1 \\ 1 & 0 & 1 \\ -1 & -1 & -1 \end{bmatrix}$, $\quad \mathbf{Y}_0 = \begin{bmatrix} 1 \\ 1 \\ -1 \end{bmatrix}$

26. Let λ be an eigenvalue of an $n \times n$ matrix A, and let $\mathbf{x}$ be an eigenvector belonging to λ. Show that e^λ is an eigenvalue of e^A and $\mathbf{x}$ is an eigenvector of e^A belonging to e^λ.

27. Show that e^A is nonsingular for any diagonalizable matrix A.

28. Let A be a diagonalizable matrix with characteristic polynomial

$$p(\lambda) = a_1 \lambda^n + a_2 \lambda^{n-1} + \cdots + a_{n+1}$$

(a) If D is a diagonal matrix whose diagonal entries are the eigenvalues of A, show that

$$p(D) = a_1 D^n + a_2 D^{n-1} + \cdots + a_{n+1} I = O$$

(b) Show that $p(A) = O$.

(c) Show that if $a_{n+1} \neq 0$, then A is nonsingular and $A^{-1} = q(A)$ for some polynomial q of degree less than n.

4 HERMITIAN MATRICES

Let C^n denote the vector space of all n-tuples of complex numbers. The set C of all complex numbers will be taken as our field of scalars. We have already seen that a matrix A with real entries may have complex eigenvalues and eigenvectors. In this section we study matrices with complex entries and look at the complex analogues of symmetric and orthogonal matrices.

Complex Inner Products

If $\alpha = a + bi$ is a complex scalar, the length of α is given by

$$|\alpha| = \sqrt{\overline{\alpha}\alpha} = \sqrt{a^2 + b^2}$$

The length of a vector $\mathbf{z} = (z_1, z_2, \ldots, z_n)^T$ in C^n is given by

$$\|\mathbf{z}\| = \left(|z_1|^2 + |z_2|^2 + \cdots + |z_n|^2\right)^{1/2}$$
$$= (\bar{z}_1 z_1 + \bar{z}_2 z_2 + \cdots + \bar{z}_n z_n)^{1/2}$$
$$= \left(\bar{\mathbf{z}}^T \mathbf{z}\right)^{1/2}$$

As a notational convenience, we write $\mathbf{z}^H$ for the transpose of $\bar{\mathbf{z}}$. Thus

$$\bar{\mathbf{z}}^T = \mathbf{z}^H \qquad \text{and} \qquad \|\mathbf{z}\| = (\mathbf{z}^H \mathbf{z})^{1/2}$$

▶ **DEFINITION** Let V be a vector space over the complex numbers. An **inner product** on V is an operation that assigns to each pair of vectors $\mathbf{z}$ and $\mathbf{w}$ in V a complex number $\langle \mathbf{z}, \mathbf{w} \rangle$ satisfying the following conditions.

(i) $\langle \mathbf{z}, \mathbf{z} \rangle \geq 0$ with equality if and only if $\mathbf{z} = \mathbf{0}$.
(ii) $\langle \mathbf{z}, \mathbf{w} \rangle = \overline{\langle \mathbf{w}, \mathbf{z} \rangle}$ for all $\mathbf{z}$ and $\mathbf{w}$ in V.
(iii) $\langle \alpha \mathbf{z} + \beta \mathbf{w}, \mathbf{u} \rangle = \alpha \langle \mathbf{z}, \mathbf{u} \rangle + \beta \langle \mathbf{w}, \mathbf{u} \rangle$. ◀

Note that for a complex inner product space $\langle \mathbf{z}, \mathbf{w} \rangle = \overline{\langle \mathbf{w}, \mathbf{z} \rangle}$, rather than $\langle \mathbf{w}, \mathbf{z} \rangle$. If we make the proper modifications to allow for this, the theorems on real inner product spaces in Chapter 5, Section 5, will all be valid for complex inner product spaces. In particular, let us recall Theorem 5.5.2; if $\{\mathbf{u}_1, \ldots, \mathbf{u}_n\}$ is an orthonormal basis for a real inner product space V and

$$\mathbf{x} = \sum_{i=1}^{n} c_i \mathbf{u}_i$$

then

$$c_i = \langle \mathbf{u}_i, \mathbf{x} \rangle = \langle \mathbf{x}, \mathbf{u}_i \rangle \qquad \text{and} \qquad \|\mathbf{x}\|^2 = \sum_{i=1}^{n} c_i^2$$

In the case of a complex inner product space, if $\{\mathbf{w}_1, \ldots, \mathbf{w}_n\}$ is an orthonormal basis and

$$\mathbf{z} = \sum_{i=1}^{n} c_i \mathbf{w}_i$$

then

$$c_i = \langle \mathbf{z}, \mathbf{w}_i \rangle, \bar{c}_i = \langle \mathbf{w}_i, \mathbf{z} \rangle \qquad \text{and} \qquad \|\mathbf{z}\|^2 = \sum_{i=1}^{n} c_i \bar{c}_i$$

We can define an inner product on C^n by

(1) $$\langle \mathbf{z}, \mathbf{w} \rangle = \mathbf{w}^H \mathbf{z}$$

for all $\mathbf{z}$ and $\mathbf{w}$ in C^n. We leave it to the reader to verify that (1) actually does define an inner product on C^n. The complex inner product space C^n is quite similar to the real inner product space R^n. The main difference is that in the complex case it is necessary to conjugate before transposing when taking an inner product.

	R^n	C^n
	$\langle \mathbf{x}, \mathbf{y} \rangle = \mathbf{y}^T \mathbf{x}$	$\langle \mathbf{z}, \mathbf{w} \rangle = \mathbf{w}^H \mathbf{z}$
	$\mathbf{x}^T \mathbf{y} = \overline{\mathbf{y}^T \mathbf{x}}$	$\mathbf{z}^H \mathbf{w} = \overline{\mathbf{w}^H \mathbf{z}}$
	$\|\mathbf{x}\|^2 = \mathbf{x}^T \mathbf{x}$	$\|\mathbf{z}\|^2 = \mathbf{z}^H \mathbf{z}$

EXAMPLE I. If

$$\mathbf{z} = \begin{bmatrix} 5 + i \\ 1 - 3i \end{bmatrix} \quad \text{and} \quad \mathbf{w} = \begin{bmatrix} 2 + i \\ -2 + 3i \end{bmatrix}$$

then

$$\mathbf{w}^H \mathbf{z} = (2 - i, \ -2 - 3i) \begin{bmatrix} 5 + i \\ 1 - 3i \end{bmatrix} = (11 - 3i) + (-11 + 3i) = 0$$

$$\mathbf{z}^H \mathbf{z} = |5 + i|^2 + |1 - 3i|^2 = 36$$

$$\mathbf{w}^H \mathbf{w} = |2 + i|^2 + |-2 + 3i|^2 = 18$$

It follows that $\mathbf{z}$ and $\mathbf{w}$ are orthogonal and

$$\|\mathbf{z}\| = 6, \qquad \|\mathbf{w}\| = 3\sqrt{2} \qquad \blacktriangleleft$$

Hermitian Matrices

Let $M = (m_{ij})$ be an $m \times n$ matrix with $m_{ij} = a_{ij} + ib_{ij}$ for each i and j. We may write M in the form

$$M = A + iB$$

where $A = (a_{ij})$ and $B = (b_{ij})$ have real entries. We define the conjugate of M by

$$\overline{M} = A - iB$$

Thus, $\overline{M}$ is the matrix formed by conjugating each of the entries of M. The transpose of $\overline{M}$ will be denoted by M^H. The vector space of all $m \times n$ matrices with complex entries is denoted by $C^{m \times n}$. If A and B are elements of $C^{m \times n}$ and $C \in C^{n \times r}$, then the following rules are easily verified (see Exercise 7).

I. $(A^H)^H = A$
II. $(\alpha A + \beta B)^H = \overline{\alpha} A^H + \overline{\beta} B^H$
III. $(AC)^H = C^H A^H$

▶ **DEFINITION** A matrix M is said to be **Hermitian** if $M = M^H$.   ◀

EXAMPLE 2. The matrix

$$M = \begin{bmatrix} 3 & 2-i \\ 2+i & 4 \end{bmatrix}$$

is Hermitian, since

$$M^H = \begin{bmatrix} \overline{3} & \overline{2-i} \\ \overline{2+i} & \overline{4} \end{bmatrix}^T = \begin{bmatrix} 3 & 2-i \\ 2+i & 4 \end{bmatrix} = M \qquad ◀$$

If M is a matrix with real entries, then $M^H = M^T$. In particular, if M is a real symmetric matrix, then M is Hermitian. Thus we may view Hermitian matrices as the complex analogue of real symmetric matrices. Hermitian matrices have many nice properties, as we shall see in the following theorem.

▶ **THEOREM 6.4.1** *The eigenvalues of a Hermitian matrix are all real. Furthermore, eigenvectors belonging to distinct eigenvalues are orthogonal.*

▶ *Proof.* Let A be a Hermitian matrix. Let λ be an eigenvalue of A and let $\mathbf{x}$ be an eigenvector belonging to λ. If $\alpha = \mathbf{x}^H A \mathbf{x}$, then

$$\overline{\alpha} = \alpha^H = (\mathbf{x}^H A \mathbf{x})^H = \mathbf{x}^H A \mathbf{x} = \alpha$$

Thus, α is real. It follows that

$$\alpha = \mathbf{x}^H A \mathbf{x} = \mathbf{x}^H \lambda \mathbf{x} = \lambda \|\mathbf{x}\|^2$$

and hence

$$\lambda = \frac{\alpha}{\|\mathbf{x}\|^2}$$

is real. If $\mathbf{x}_1$ and $\mathbf{x}_2$ are eigenvectors belonging to distinct eigenvalues λ_1 and λ_2, respectively, then

$$(A\mathbf{x}_1)^H \mathbf{x}_2 = \mathbf{x}_1^H A^H \mathbf{x}_2 = \mathbf{x}_1^H A \mathbf{x}_2 = \lambda_2 \mathbf{x}_1^H \mathbf{x}_2$$

and

$$(A\mathbf{x}_1)^H \mathbf{x}_2 = (\mathbf{x}_2^H A \mathbf{x}_1)^H = (\lambda_1 \mathbf{x}_2^H \mathbf{x}_1)^H = \lambda_1 \mathbf{x}_1^H \mathbf{x}_2$$

Thus,

$$\lambda_1 \mathbf{x}_1^H \mathbf{x}_2 = \lambda_2 \mathbf{x}_1^H \mathbf{x}_2$$

and, since $\lambda_1 \neq \lambda_2$, it follows that

$$\langle \mathbf{x}_2, \mathbf{x}_1 \rangle = \mathbf{x}_1^H \mathbf{x}_2 = 0 \qquad ◀$$

▶ **DEFINITION** An $n \times n$ matrix U is said to be **unitary** if its column vectors form an orthonormal set in C^n.   ◀

Thus, U is unitary if and only if $U^H U = I$. If U is unitary, then, since the column vectors are orthonormal, U must have rank n. It follows that

$$U^{-1} = IU^{-1} = U^H U U^{-1} = U^H$$

A real unitary matrix is an orthogonal matrix.

▶ **COROLLARY 6.4.2** *If the eigenvalues of a Hermitian matrix A are distinct, then there exists a unitary matrix U that diagonalizes A.*

▶*Proof.* Let $\mathbf{x}_i$ be an eigenvector belonging to λ_i for each eigenvalue λ_i of A. Let $\mathbf{u}_i = (1/\|\mathbf{x}_i\|)\mathbf{x}_i$. Thus $\mathbf{u}_i$ is a unit eigenvector belonging to λ_i for each i. It follows from Theorem 6.4.1 that $\{\mathbf{u}_1, \ldots, \mathbf{u}_n\}$ is an orthonormal set in C^n. Let U be the matrix whose ith column vector is $\mathbf{u}_i$ for each i; U is unitary and U diagonalizes A. ◀

EXAMPLE 3. Let

$$A = \begin{bmatrix} 2 & 1-i \\ 1+i & 1 \end{bmatrix}$$

Find a unitary matrix U that diagonalizes A.

SOLUTION. The eigenvalues of A are $\lambda_1 = 3$ and $\lambda_2 = 0$ with corresponding eigenvectors $\mathbf{x}_1 = (1 - i, 1)^T$ and $\mathbf{x}_2 = (-1, 1 + i)^T$. Let

$$\mathbf{u}_1 = \frac{1}{\|\mathbf{x}_1\|}\mathbf{x}_1 = \frac{1}{\sqrt{3}}(1 - i, 1)^T$$

and

$$\mathbf{u}_2 = \frac{1}{\|\mathbf{x}_2\|}\mathbf{x}_2 = \frac{1}{\sqrt{3}}(-1, 1 + i)^T$$

Thus,

$$U = \frac{1}{\sqrt{3}} \begin{bmatrix} 1-i & -1 \\ 1 & 1+i \end{bmatrix}$$

and

$$U^H A U = \frac{1}{3} \begin{bmatrix} 1+i & 1 \\ -1 & 1-i \end{bmatrix} \begin{bmatrix} 2 & 1-i \\ 1+i & 1 \end{bmatrix} \begin{bmatrix} 1-i & -1 \\ 1 & 1+i \end{bmatrix}$$

$$= \begin{bmatrix} 3 & 0 \\ 0 & 0 \end{bmatrix}$$ ◀

Actually, Corollary 6.4.2 is valid even if the eigenvalues of A are not distinct. To show this, we will first prove the following theorem.

▶ **THEOREM 6.4.3 (Schur's Theorem)** *For each $n \times n$ matrix A, there exists a unitary matrix U such that $U^H A U$ is upper triangular.*

▶*Proof.* The proof is by induction on n. The result is obvious if $n = 1$. Assume that the hypothesis holds for $k \times k$ matrices, and let A be a $(k + 1) \times (k + 1)$ matrix. Let λ_1 be an eigenvalue of A, and let $\mathbf{w}_1$ be a unit eigenvector belonging to λ_1. Using the Gram–Schmidt process, construct $\mathbf{w}_2, \ldots, \mathbf{w}_{k+1}$ such that $\{\mathbf{w}_1, \ldots, \mathbf{w}_{k+1}\}$ is an orthonormal basis for C^{k+1}. Let W be the matrix whose ith column vector is $\mathbf{w}_i$ for $i = 1, \ldots, k+1$. Thus, by construction, W is unitary. The first column of $W^H A W$ will be $W^H A \mathbf{w}_1$.

$$W^H A \mathbf{w}_1 = \lambda_1 W^H \mathbf{w}_1 = \lambda_1 \mathbf{e}_1$$

Thus, $W^H A W$ is a matrix of the form

$$\begin{pmatrix} \lambda_1 & \times & \times & \cdots & \times \\ \hline 0 & & & & \\ \vdots & & & M & \\ 0 & & & & \end{pmatrix}$$

where M is a $k \times k$ matrix. By the induction hypothesis, there exists a $k \times k$ unitary matrix V_1 such that $V_1^H M V_1 = T_1$, where T_1 is triangular. Let

$$V = \begin{pmatrix} 1 & 0 & \cdots & 0 \\ \hline 0 & & & \\ \vdots & & V_1 & \\ 0 & & & \end{pmatrix}$$

Here V is unitary and

$$V^H W^H A W V = \begin{pmatrix} \lambda_1 & \times & \cdots & \times \\ \hline 0 & & & \\ \vdots & & V_1^H M V_1 & \\ 0 & & & \end{pmatrix}$$

$$= \begin{pmatrix} \lambda_1 & \times & \cdots & \times \\ \hline 0 & & & \\ \vdots & & T_1 & \\ 0 & & & \end{pmatrix}$$

$$= T$$

Let $U = WV$. The matrix U is unitary, since

$$U^H U = (WV)^H WV = V^H W^H WV = I$$

and $U^H A U = T$.   ◀

The factorization $A = UTU^H$ is often referred to as the *Schur decomposition* of A. In the case that A is Hermitian, the matrix T will be diagonal.

▶ **THEOREM 6.4.4 (Spectral Theorem)** *If A is Hermitian, then there exists a unitary matrix U that diagonalizes A.*

▶*Proof.* By Theorem 6.4.3, there is a unitary matrix U such that $U^H A U = T$, where T is upper triangular.

$$T^H = (U^H A U)^H = U^H A^H U = U^H A U = T$$

Therefore, T is Hermitian and consequently must be diagonal.   ◀

In the case that A is real and symmetric, its eigenvalues and eigenvectors must be real. Thus the diagonalizing matrix U must be orthogonal.

▶ **COROLLARY 6.4.5** *If A is a real symmetric matrix, then there is an orthogonal matrix U that diagonalizes A, that is, $U^T A U = D$, where D is diagonal.*   ◀

EXAMPLE 4.   Given

$$A = \begin{bmatrix} 0 & 2 & -1 \\ 2 & 3 & -2 \\ -1 & -2 & 0 \end{bmatrix}$$

find an orthogonal matrix U that diagonalizes A.

SOLUTION.   The characteristic polynomial

$$p(\lambda) = -\lambda^3 + 3\lambda^2 + 9\lambda + 5 = (1 + \lambda)^2 (5 - \lambda)$$

has roots $\lambda_1 = \lambda_2 = -1$, $\lambda_3 = 5$. Computing eigenvectors in the usual way, we see that $\mathbf{x}_1 = (1, 0, 1)^T$ and $\mathbf{x}_2 = (-2, 1, 0)^T$ form a basis for the eigenspace $N(A + I)$. We can apply the Gram–Schmidt process to obtain an orthonormal basis for the eigenspace corresponding to $\lambda_1 = \lambda_2 = -1$.

$$\mathbf{u}_1 = \frac{1}{\|\mathbf{x}_1\|} \mathbf{x}_1 = \frac{1}{\sqrt{2}} (1, 0, 1)^T$$

$$\mathbf{p} = \left(\mathbf{x}_2^T \mathbf{u}_1 \right) \mathbf{u}_1 = -\sqrt{2} \mathbf{u}_1 = (-1, 0, 1)^T$$

$$\mathbf{x}_2 - \mathbf{p} = (-1, 1, 1)^T$$

$$\mathbf{u}_2 = \frac{1}{\|\mathbf{x}_2 - \mathbf{p}\|} (\mathbf{x}_2 - \mathbf{p}) = \frac{1}{\sqrt{3}} (-1, 1, 1)^T$$

The eigenspace corresponding to $\lambda_3 = 5$ is spanned by $\mathbf{x}_3 = (-1, -2, 1)^T$. Since $\mathbf{x}_3$ must be orthogonal to $\mathbf{u}_1$ and $\mathbf{u}_2$ (Theorem 6.4.1), we need only normalize

$$\mathbf{u}_3 = \frac{1}{\|\mathbf{x}_3\|} \mathbf{x}_3 = \frac{1}{\sqrt{6}} (-1, -2, 1)^T$$

Thus $\{\mathbf{u}_1, \mathbf{u}_2, \mathbf{u}_3\}$ is an orthonormal set and

$$U = \begin{bmatrix} \dfrac{1}{\sqrt{2}} & -\dfrac{1}{\sqrt{3}} & -\dfrac{1}{\sqrt{6}} \\[2ex] 0 & \dfrac{1}{\sqrt{3}} & -\dfrac{2}{\sqrt{6}} \\[2ex] \dfrac{1}{\sqrt{2}} & \dfrac{1}{\sqrt{3}} & \dfrac{1}{\sqrt{6}} \end{bmatrix}$$

diagonalizes A.   ◀

It follows from Theorem 6.4.4 that each Hermitian matrix A can be factored into a product UDU^H, where U is unitary and D is diagonal. Since U diagonalizes A, it follows that the diagonal elements of D are the eigenvalues of A and the column vectors of U are eigenvectors of A. Thus A cannot be defective. It has a complete set of eigenvectors that form an orthonormal basis for C^n. This is in a sense the ideal situation. We have seen how to express a vector as a linear combination of orthonormal basis elements (Theorem 5.5.2), and the action of A on any linear combination of eigenvectors can easily be determined. Thus, if A has an orthonormal set of eigenvectors $\{\mathbf{u}_1, \ldots, \mathbf{u}_n\}$ and $\mathbf{x} = c_1\mathbf{u}_1 + \cdots + c_n\mathbf{u}_n$, then

$$A\mathbf{x} = c_1\lambda_1\mathbf{u}_1 + \cdots + c_n\lambda_n\mathbf{u}_n$$

Furthermore,

$$c_i = \langle \mathbf{x}, \mathbf{u}_i \rangle = \mathbf{u}_i^H \mathbf{x}$$

or, equivalently, $\mathbf{c} = U^H\mathbf{x}$. Thus,

$$A\mathbf{x} = \lambda_1(\mathbf{u}_1^H\mathbf{x})\mathbf{u}_1 + \cdots + \lambda_n(\mathbf{u}_n^H\mathbf{x})\mathbf{u}_n$$

Normal Matrices

There are non-Hermitian matrices that possess complete orthonormal sets of eigenvectors. For example, skew symmetric and skew Hermitian matrices have this property. (A is *skew Hermitian* if $A^H = -A$.) In general, if A is any matrix with a complete orthonormal set of eigenvectors, then $A = UDU^H$, where U is unitary and D is a diagonal matrix (whose diagonal elements may be complex). In general, $D^H \neq D$ and, consequently,

$$A^H = UD^HU^H \neq A$$

However,

$$AA^H = UDU^HUD^HU^H = UDD^HU^H$$

and

$$A^HA = UD^HU^HUDU^H = UD^HDU^H$$

Since

$$D^HD = DD^H = \begin{bmatrix} |\lambda_1|^2 & & & \\ & |\lambda_2|^2 & & \\ & & \ddots & \\ & & & |\lambda_n|^2 \end{bmatrix}$$

it follows that

$$AA^H = A^HA$$

▶ **DEFINITION** A matrix A is said to be **normal** if $AA^H = A^HA$.   ◀

We have shown that if a matrix has a complete orthonormal set of eigenvectors, then it is normal. The converse is also true.

▶ **THEOREM 6.4.6** *A matrix A is normal if and only if A possesses a complete orthonormal set of eigenvectors.*

▶*Proof.* In view of the preceding remarks, we need only show that a normal matrix A has a complete orthonormal set of eigenvectors. By Theorem 6.4.3, there exists a unitary matrix U and a triangular matrix T such that $T = U^H A U$. We claim that T is also normal.

$$T^H T = U^H A^H U U^H A U = U^H A^H A U$$

and

$$T T^H = U^H A U U^H A^H U = U^H A A^H U$$

Since $A^H A = A A^H$, it follows that $T^H T = T T^H$. Comparing the diagonal elements of $T T^H$ and $T^H T$, we see that

$$|t_{11}|^2 + |t_{12}|^2 + |t_{13}|^2 + \cdots + |t_{1n}|^2 = |t_{11}|^2$$
$$|t_{22}|^2 + |t_{23}|^2 + \cdots + |t_{2n}|^2 = |t_{12}|^2 + |t_{22}|^2$$
$$\vdots$$
$$|t_{nn}|^2 = |t_{1n}|^2 + |t_{2n}|^2 + |t_{3n}|^2 + \cdots + |t_{nn}|^2$$

It follows that $t_{ij} = 0$ whenever $i \neq j$. Thus U diagonalizes A and the column vectors of U are eigenvectors of A.   ◀

EXERCISES

1. For each of the following pairs of vectors $\mathbf{z}$ and $\mathbf{w}$ in C^2, compute (i) $\|\mathbf{z}\|$, (ii) $\|\mathbf{w}\|$, (iii) $\langle \mathbf{z}, \mathbf{w} \rangle$, and (iv) $\langle \mathbf{w}, \mathbf{z} \rangle$.

(a) $\mathbf{z} = \begin{bmatrix} 4 + 2i \\ 4i \end{bmatrix}$, $\quad \mathbf{w} = \begin{bmatrix} -2 \\ 2 + i \end{bmatrix}$

(b) $\mathbf{z} = \begin{bmatrix} 1 + i \\ 2i \\ 3 - i \end{bmatrix}$, $\quad \mathbf{w} = \begin{bmatrix} 2 - 4i \\ 5 \\ 2i \end{bmatrix}$

2. Let

$$\mathbf{z}_1 = \begin{bmatrix} \dfrac{1+i}{2} \\ \dfrac{1-i}{2} \end{bmatrix} \quad \text{and} \quad \mathbf{z}_2 = \begin{bmatrix} \dfrac{i}{\sqrt{2}} \\ -\dfrac{1}{\sqrt{2}} \end{bmatrix}$$

(a) Show that $\{\mathbf{z}_1, \mathbf{z}_2\}$ is an orthonormal set in C^2.

(b) Write the vector $\mathbf{z} = \begin{bmatrix} 2 + 4i \\ -2i \end{bmatrix}$ as a linear combination of $\mathbf{z}_1$ and $\mathbf{z}_2$.

3. Let $\{\mathbf{u}_1, \mathbf{u}_2\}$ be an orthonormal basis for C^2, and let $\mathbf{z} = (4+2i)\mathbf{u}_1 + (6-5i)\mathbf{u}_2$.

(a) What are the values of $\mathbf{u}_1^H \mathbf{z}$, $\mathbf{z}^H \mathbf{u}_1$, $\mathbf{u}_2^H \mathbf{z}$, and $\mathbf{z}^H \mathbf{u}_2$?

(b) Determine the value of $\|\mathbf{z}\|$.

4. Which of the following matrices are Hermitian? Normal?

(a) $\begin{bmatrix} 1-i & 2 \\ 2 & 3 \end{bmatrix}$
(b) $\begin{bmatrix} 1 & 2-i \\ 2+i & -1 \end{bmatrix}$

(c) $\begin{bmatrix} \dfrac{1}{\sqrt{2}} & -\dfrac{1}{\sqrt{2}} \\ \dfrac{1}{\sqrt{2}} & \dfrac{1}{\sqrt{2}} \end{bmatrix}$
(d) $\begin{bmatrix} \dfrac{1}{\sqrt{2}}i & \dfrac{1}{\sqrt{2}} \\ \dfrac{1}{\sqrt{2}} & -\dfrac{1}{\sqrt{2}}i \end{bmatrix}$

(e) $\begin{bmatrix} 0 & i & 1 \\ i & 0 & -2+i \\ -1 & 2+i & 0 \end{bmatrix}$
(f) $\begin{bmatrix} 3 & 1+i & i \\ 1-i & 1 & 3 \\ -i & 3 & 1 \end{bmatrix}$

5. Find an orthogonal or unitary diagonalizing matrix for each of the following:

(a) $\begin{bmatrix} 2 & 1 \\ 1 & 2 \end{bmatrix}$
(b) $\begin{bmatrix} 1 & 3+i \\ 3-i & 4 \end{bmatrix}$
(c) $\begin{bmatrix} 2 & i & 0 \\ -i & 2 & 0 \\ 0 & 0 & 2 \end{bmatrix}$

(d) $\begin{bmatrix} 2 & 1 & 1 \\ 1 & 3 & -2 \\ 1 & -2 & 3 \end{bmatrix}$
(e) $\begin{bmatrix} 0 & 0 & 1 \\ 0 & 1 & 0 \\ 1 & 0 & 0 \end{bmatrix}$
(f) $\begin{bmatrix} 1 & 1 & 1 \\ 1 & 1 & 1 \\ 1 & 1 & 1 \end{bmatrix}$

(g) $\begin{bmatrix} 4 & 2 & -2 \\ 2 & 1 & -1 \\ -2 & -1 & 1 \end{bmatrix}$

6. Show that the diagonal entries of a Hermitian matrix must be real.

7. Let A and C be matrices in $C^{m \times n}$ and let $B \in C^{n \times r}$. Prove each of the following rules:

(a) $(A^H)^H = A$
(b) $(\alpha A + \beta C)^H = \bar{\alpha} A^H + \bar{\beta} C^H$

(c) $(AB)^H = B^H A^H$

8. Show that

$$\langle \mathbf{z}, \mathbf{w} \rangle = \mathbf{w}^H \mathbf{z}$$

defines an inner product on C^n.

9. Let $\{\mathbf{u}_1, \ldots, \mathbf{u}_n\}$ be an orthonormal basis for a complex inner product space V, and let $\mathbf{z}$ and $\mathbf{w}$ be elements of V. Show that

$$\langle \mathbf{z}, \mathbf{w} \rangle = \sum_{i=1}^{n} (\langle \mathbf{z}, \mathbf{u}_i \rangle \cdot \langle \mathbf{u}_i, \mathbf{w} \rangle)$$

10. Given

$$A = \begin{bmatrix} 4 & 0 & 0 \\ 0 & 1 & i \\ 0 & -i & 1 \end{bmatrix}$$

find a matrix B such that $B^H B = A$.

11. Let U be a unitary matrix. Prove:

   (a) U is normal.
   (b) $\|U\mathbf{x}\| = \|\mathbf{x}\|$ for all $\mathbf{x} \in C^n$.
   (c) If λ is an eigenvalue of U, then $|\lambda| = 1$.

12. Let $\mathbf{u}$ be a unit vector in C^n and define $U = I - 2\mathbf{u}\mathbf{u}^H$. Show that U is both unitary and Hermitian and, consequently, is its own inverse.

13. Show that if a matrix U is both unitary and Hermitian, then any eigenvalue of U must equal either 1 or -1.

14. Let A be a 2×2 matrix with Schur decomposition UTU^H and suppose that $t_{12} \neq 0$. Show that:

   (a) The eigenvalues of A are $\lambda_1 = t_{11}$ and $\lambda_2 = t_{22}$.
   (b) $\mathbf{u}_1$ is an eigenvector of A belonging to $\lambda_1 = t_{11}$.
   (c) $\mathbf{u}_2$ is not an eigenvector of A belonging to $\lambda_2 = t_{22}$.

15. Show that $M = A + iB$ (A and B real matrices) is skew Hermitian if and only if A is skew symmetric and B is symmetric.

16. Show that if A is skew Hermitian and λ is an eigenvalue of A, then λ is purely imaginary (i.e., $\lambda = bi$, where b is real).

17. Let A be a real 2×2 matrix with the property that $a_{21}a_{12} > 0$, and let

$$r = \sqrt{a_{21}/a_{12}} \quad \text{and} \quad S = \begin{bmatrix} r & 0 \\ 0 & 1 \end{bmatrix}$$

   Compute $B = SAS^{-1}$. What can you conclude about the eigenvalues and eigenvectors of B? What can you conclude about the eigenvalues and eigenvectors of A? Explain.

18. Let $p(x) = -x^3 + cx^2 + (c+3)x + 1$, where c is a real number. Let C denote the companion matrix of $p(x)$,

$$C = \begin{bmatrix} c & c+3 & 1 \\ 1 & 0 & 0 \\ 0 & 1 & 0 \end{bmatrix}$$

and let

$$A = \begin{bmatrix} -1 & 2 & -c-3 \\ 1 & -1 & c+2 \\ -1 & 1 & -c-1 \end{bmatrix}$$

(a) Compute $A^{-1}CA$.

(b) Use the result from part (a) to prove that $p(x)$ will have only real roots regardless of the value of c.

19. Let A be a Hermitian matrix with eigenvalues $\lambda_1, \ldots, \lambda_n$ and orthonormal eigenvectors $\mathbf{u}_1, \ldots, \mathbf{u}_n$. Show that

$$A = \lambda_1 \mathbf{u}_1 \mathbf{u}_1^H + \lambda_2 \mathbf{u}_2 \mathbf{u}_2^H + \cdots + \lambda_n \mathbf{u}_n \mathbf{u}_n^H$$

20. Let

$$A = \begin{bmatrix} 0 & 1 \\ 1 & 0 \end{bmatrix}$$

Write A as a sum $\lambda_1 \mathbf{u}_1 \mathbf{u}_1^T + \lambda_2 \mathbf{u}_2 \mathbf{u}_2^T$, where λ_1 and λ_2 are eigenvalues and $\mathbf{u}_1$ and $\mathbf{u}_2$ are orthonormal eigenvectors.

21. Let A be a Hermitian matrix with eigenvalues $\lambda_1 \geq \lambda_2 \geq \cdots \geq \lambda_n$ and orthonormal eigenvectors $\mathbf{u}_1, \ldots, \mathbf{u}_n$. For any nonzero vector $\mathbf{x}$ in R^n the *Rayleigh quotient* $\rho(\mathbf{x})$ is defined by

$$\rho(\mathbf{x}) = \frac{\langle A\mathbf{x}, \mathbf{x} \rangle}{\langle \mathbf{x}, \mathbf{x} \rangle} = \frac{\mathbf{x}^H A \mathbf{x}}{\mathbf{x}^H \mathbf{x}}$$

(a) If $\mathbf{x} = c_1 \mathbf{u}_1 + \cdots + c_n \mathbf{u}_n$, show that

$$\rho(\mathbf{x}) = \frac{|c_1|^2 \lambda_1 + |c_2|^2 \lambda_2 + \cdots + |c_n|^2 \lambda_n}{\|\mathbf{c}\|^2}$$

(b) Show that

$$\lambda_n \leq \rho(\mathbf{x}) \leq \lambda_1$$

(c) Show that

$$\max_{\mathbf{x} \neq \mathbf{0}} \rho(\mathbf{x}) = \lambda_1 \qquad \text{and} \qquad \min_{\mathbf{x} \neq \mathbf{0}} \rho(\mathbf{x}) = \lambda_n$$

5 THE SINGULAR VALUE DECOMPOSITION

In many applications it is necessary to either determine the rank of a matrix or to determine whether the matrix is deficient in rank. Theoretically, we can use Gaussian elimination to reduce the matrix to row echelon form and then count the number of nonzero rows. However, this approach is not practical when working in finite precision arithmetic. If A is rank deficient and U is the computed echelon form, then, because of roundoff errors in the elimination process, it is unlikely that U will have the proper number of nonzero rows. In practice, the coefficient matrix A usually involves some error. This may be due to errors in the data or to the finite number system. Thus, it is generally more practical to ask whether A is "close" to a rank deficient matrix. However, it may well turn out that A is close to being rank deficient and the computed row echelon form U is not.

In this section we assume throughout that A is an $m \times n$ matrix with $m \geq n$. (This assumption is made for convenience only; all the results will also hold if $m < n$.) We will present a method for determining how close A is to a matrix of smaller rank. The method involves factoring A into a product $U \Sigma V^T$, where U is an $m \times m$ orthogonal matrix, V is an $n \times n$ orthogonal matrix, and Σ is an $m \times n$ matrix whose off-diagonal entries are all 0's and whose diagonal elements satisfy

$$\sigma_1 \geq \sigma_2 \geq \cdots \geq \sigma_n \geq 0$$

$$\Sigma = \begin{pmatrix} \sigma_1 & & & \\ & \sigma_2 & & \\ & & \ddots & \\ & & & \sigma_n \\ & & & \\ & & & \end{pmatrix}$$

The σ_i's determined by this factorization are unique and are called the *singular values* of A. The factorization $U \Sigma V^T$ is called the *singular value decomposition* of A. We will show that the rank of A equals the number of nonzero singular values, and that the magnitudes of the nonzero singular values provide a measure of how close A is to a matrix of lower rank.

We begin by showing that such a decomposition is always possible.

▶ **THEOREM 6.5.1 (The SVD Theorem)** *If A is an $m \times n$ matrix, then A has a singular value decomposition.*

▶*Proof.* $A^T A$ is a symmetric $n \times n$ matrix. Therefore, its eigenvalues are all real and it has an orthogonal diagonalizing matrix V. Furthermore, its eigenvalues must all be nonnegative. To see this, let λ be an eigenvalue of $A^T A$ and $\mathbf{x}$ be an eigenvector belonging to λ. It follows that

$$\|A\mathbf{x}\|^2 = \mathbf{x}^T A^T A \mathbf{x} = \lambda \mathbf{x}^T \mathbf{x} = \lambda \|\mathbf{x}\|^2$$

Hence,

$$\lambda = \frac{\|A\mathbf{x}\|^2}{\|\mathbf{x}\|^2} \geq 0$$

We may assume that the columns of V have been ordered so that the corresponding eigenvalues satisfy

$$\lambda_1 \geq \lambda_2 \geq \cdots \geq \lambda_n \geq 0$$

The singular values of A are given by

$$\sigma_j = \sqrt{\lambda_j} \qquad j = 1, \ldots, n$$

Let r denote the rank of A. The matrix $A^T A$ will also have rank r. Since $A^T A$ is symmetric, its rank equals the number of nonzero eigenvalues. Thus

$$\lambda_1 \geq \lambda_2 \geq \cdots \geq \lambda_r > 0 \qquad \text{and} \qquad \lambda_{r+1} = \lambda_{r+2} = \cdots = \lambda_n = 0$$

The same relation holds for the singular values

$$\sigma_1 \geq \sigma_2 \geq \cdots \geq \sigma_r > 0 \qquad \text{and} \qquad \sigma_{r+1} = \sigma_{r+2} = \cdots = \sigma_n = 0$$

Now, let

$$V_1 = (\mathbf{v}_1, \ldots, \mathbf{v}_r), \qquad V_2 = (\mathbf{v}_{r+1}, \ldots, \mathbf{v}_n)$$

and

(1)
$$\Sigma_1 = \begin{pmatrix} \sigma_1 & & & \\ & \sigma_2 & & \\ & & \ddots & \\ & & & \sigma_r \end{pmatrix}$$

Thus Σ_1 is an $r \times r$ diagonal matrix whose diagonal entries are the nonzero singular values $\sigma_1, \ldots, \sigma_r$. The $m \times n$ matrix Σ is then given by

$$\Sigma = \begin{bmatrix} \Sigma_1 & O \\ O & O \end{bmatrix}$$

The column vectors of V_2 are eigenvectors of $A^T A$ belonging to $\lambda = 0$. Thus

$$A^T A \mathbf{v}_j = \mathbf{0} \qquad j = r+1, \ldots, n$$

and, consequently, the column vectors of V_2 form an orthonormal basis for $N(A^T A) = N(A)$. Therefore,

$$A V_2 = O$$

and, since V is an orthogonal matrix, it follows that

$$I = V V^T = V_1 V_1^T + V_2 V_2^T$$

(2)
$$A = AI = A V_1 V_1^T + A V_2 V_2^T = A V_1 V_1^T$$

So far we have shown how to construct the matrices V and Σ of the singular value decomposition. To complete the proof, we must show how to construct an $m \times m$ orthogonal matrix U such that

$$A = U \Sigma V^T$$

or, equivalently,

(3)
$$AV = U\Sigma$$

Comparing the first r columns of each side of (3), we see that

$$A \mathbf{v}_j = \sigma_j \mathbf{u}_j \qquad j = 1, \ldots, r$$

Thus, if we define

(4)
$$\mathbf{u}_j = \frac{1}{\sigma_j} A\mathbf{v}_j \qquad j = 1, \ldots, r$$

and

$$U_1 = (\mathbf{u}_1, \ldots, \mathbf{u}_r)$$

then it follows that

(5)
$$AV_1 = U_1 \Sigma_1$$

The column vectors of U_1 form an orthonormal set since

$$\mathbf{u}_i^T \mathbf{u}_j = \left(\frac{1}{\sigma_i} \mathbf{v}_i^T A^T \right) \left(\frac{1}{\sigma_j} A\mathbf{v}_j \right) \qquad 1 \leq i \leq r, \quad 1 \leq j \leq r$$

$$= \frac{1}{\sigma_i \sigma_j} \mathbf{v}_i^T \left(A^T A \mathbf{v}_j \right)$$

$$= \frac{\sigma_j}{\sigma_i} \mathbf{v}_i^T \mathbf{v}_j$$

$$= \delta_{ij}$$

It follows from (4) that each $\mathbf{u}_j$, $1 \leq j \leq r$, is in the column space of A. The dimension of the column space is r, so $\mathbf{u}_1, \ldots, \mathbf{u}_r$ form an orthonormal basis for $R(A)$. The vector space $R(A)^{\perp} = N(A^T)$ has dimension $m - r$. Let $\{\mathbf{u}_{r+1}, \mathbf{u}_{r+2}, \ldots, \mathbf{u}_m\}$ be an orthonormal basis for $N(A^T)$ and set

$$U_2 = (\mathbf{u}_{r+1}, \mathbf{u}_{r+2}, \ldots, \mathbf{u}_m)$$

$$U = \begin{bmatrix} U_1 & U_2 \end{bmatrix}$$

It follows from Theorem 5.2.2 that $\mathbf{u}_1, \ldots, \mathbf{u}_m$ form an orthonormal basis for R^m. Hence U is an orthogonal matrix. We still must show that $U\Sigma V^T$ actually equals A. This follows from (5) and (2) since

$$U\Sigma V^T = \begin{bmatrix} U_1 & U_2 \end{bmatrix} \begin{bmatrix} \Sigma_1 & O \\ O & O \end{bmatrix} \begin{bmatrix} V_1^T \\ V_2^T \end{bmatrix}$$

$$= U_1 \Sigma_1 V_1^T$$

$$= AV_1 V_1^T$$

$$= A \qquad \blacktriangleleft$$

Observations Let A be an $m \times n$ matrix with a singular value decomposition $U\Sigma V^T$.

1. The singular values $\sigma_1, \ldots, \sigma_n$ of A are unique; however, the matrices U and V are not unique.
2. Since V diagonalizes $A^T A$, it follows that the $\mathbf{v}_j$'s are eigenvectors of $A^T A$.
3. Since $AA^T = U\Sigma\Sigma^T U^T$, it follows that U diagonalizes AA^T and that the $\mathbf{u}_j$'s are eigenvectors of AA^T.
4. Comparing the jth columns of each side of the equation

$$AV = U\Sigma$$

we get

$$A\mathbf{v}_j = \sigma_j \mathbf{u}_j \qquad j = 1, \ldots, n$$

Similarly,

$$A^T U = V \Sigma^T$$

and hence

$$A^T \mathbf{u}_j = \sigma_j \mathbf{v}_j \qquad \text{for } j = 1, \ldots, n$$
$$A^T \mathbf{u}_j = \mathbf{0} \qquad \text{for } j = n+1, \ldots, m$$

The $\mathbf{v}_j$'s are called the *right singular vectors* of A, and the $\mathbf{u}_j$'s are called the *left singular vectors* of A.

5. If A has rank r, then

   (i) $\mathbf{v}_1, \ldots, \mathbf{v}_r$ form an orthonormal basis for $R(A^T)$.

   (ii) $\mathbf{v}_{r+1}, \ldots, \mathbf{v}_n$ form an orthonormal basis for $N(A)$.

   (iii) $\mathbf{u}_1, \ldots, \mathbf{u}_r$ form an orthonormal basis for $R(A)$.

   (iv) $\mathbf{u}_{r+1}, \ldots, \mathbf{u}_m$ form an orthonormal basis for $N(A^T)$.

6. The rank of the matrix A is equal to the number of its nonzero singular values (where singular values are counted according to multiplicity). The reader should be careful not to make a similar assumption about eigenvalues. The matrix

$$M = \begin{bmatrix} 0 & 1 & 0 & 0 \\ 0 & 0 & 1 & 0 \\ 0 & 0 & 0 & 1 \\ 0 & 0 & 0 & 0 \end{bmatrix}$$

for example, has rank 3 even though all of its eigenvalues are 0.

7. In the case that A has rank $r < n$, if we set

$$U_1 = (\mathbf{u}_1, \mathbf{u}_2, \ldots, \mathbf{u}_r) \qquad V_1 = (\mathbf{v}_1, \mathbf{v}_2, \ldots, \mathbf{v}_r)$$

and define Σ_1 as in equation (1), then

(6) $$A = U_1 \Sigma_1 V_1^T$$

The factorization (6) is called the *compact form of the singular value decomposition* of A. This form is useful in many applications.

EXAMPLE I.   Let

$$A = \begin{bmatrix} 1 & 1 \\ 1 & 1 \\ 0 & 0 \end{bmatrix}$$

Compute the singular values and the singular value decomposition of A.

SOLUTION.   The matrix

$$A^T A = \begin{bmatrix} 2 & 2 \\ 2 & 2 \end{bmatrix}$$

has eigenvalues $\lambda_1 = 4$ and $\lambda_2 = 0$. Consequently, the singular values of A are $\sigma_1 = \sqrt{4} = 2$ and $\sigma_2 = 0$. The eigenvalue λ_1 has eigenvectors of the form $\alpha(1, 1)^T$, and λ_2 has eigenvectors $\beta(1, -1)^T$. Therefore, the orthogonal matrix

$$V = \frac{1}{\sqrt{2}} \begin{bmatrix} 1 & 1 \\ 1 & -1 \end{bmatrix}$$

diagonalizes $A^T A$. From observation 4 it follows that

$$\mathbf{u}_1 = \frac{1}{\sigma_1} A \mathbf{v}_1 = \frac{1}{2} \begin{bmatrix} 1 & 1 \\ 1 & 1 \\ 0 & 0 \end{bmatrix} \begin{bmatrix} \dfrac{1}{\sqrt{2}} \\ \dfrac{1}{\sqrt{2}} \end{bmatrix} = \begin{bmatrix} \dfrac{1}{\sqrt{2}} \\ \dfrac{1}{\sqrt{2}} \\ 0 \end{bmatrix}$$

The remaining column vectors of U must form an orthonormal basis for $N(A^T)$. We can compute a basis $\{\mathbf{x}_2, \mathbf{x}_3\}$ for $N(A^T)$ in the usual way.

$$\mathbf{x}_2 = (1, -1, 0)^T \qquad \text{and} \qquad \mathbf{x}_3 = (0, 0, 1)^T$$

Since these vectors are already orthogonal, it is not necessary to use the Gram–Schmidt process to obtain an orthonormal basis. We need only set

$$\mathbf{u}_2 = \frac{1}{\|\mathbf{x}_2\|} \mathbf{x}_2 = \left(\frac{1}{\sqrt{2}}, -\frac{1}{\sqrt{2}}, 0 \right)^T$$

$$\mathbf{u}_3 = \mathbf{x}_3 = (0, 0, 1)^T$$

It follows then that

$$A = U \Sigma V^T = \begin{bmatrix} \dfrac{1}{\sqrt{2}} & \dfrac{1}{\sqrt{2}} & 0 \\ \dfrac{1}{\sqrt{2}} & -\dfrac{1}{\sqrt{2}} & 0 \\ 0 & 0 & 1 \end{bmatrix} \begin{bmatrix} 2 & 0 \\ 0 & 0 \\ 0 & 0 \end{bmatrix} \begin{bmatrix} \dfrac{1}{\sqrt{2}} & \dfrac{1}{\sqrt{2}} \\ \dfrac{1}{\sqrt{2}} & -\dfrac{1}{\sqrt{2}} \end{bmatrix} \qquad \blacktriangleleft$$

If A is an $m \times n$ matrix of rank r and $0 < k < r$, we can use the singular value decomposition to find a matrix in $R^{m \times n}$ of rank k that is closest to A with respect to the Frobenius norm. Let $\mathcal{M}$ be the set of all $m \times n$ matrices of rank k or less. It can be shown that there is a matrix X in $\mathcal{M}$ such that

(7) $$\|A - X\|_F = \min_{S \in \mathcal{M}} \|A - S\|_F$$

We will not prove this, since the proof is beyond the scope of this book. Assuming that the minimum is achieved, we will show how such a matrix X can be derived from the singular value decomposition of A. The following lemma will be useful.

▶ **LEMMA 6.5.2** *If A is an $m \times n$ matrix and Q is an $m \times m$ orthogonal matrix, then*

$$\|QA\|_F = \|A\|_F$$

▶ *Proof.*

$$\|QA\|_F^2 = \|(Q\mathbf{a}_1, Q\mathbf{a}_2, \ldots, Q\mathbf{a}_n)\|_F^2$$
$$= \sum_{i=1}^{n} \|Q\mathbf{a}_i\|_2^2$$
$$= \sum_{i=1}^{n} \|\mathbf{a}_i\|_2^2$$
$$= \|A\|_F^2 \qquad \blacktriangleleft$$

It follows from the lemma that, if A has singular value decomposition $U\Sigma V^T$, then

$$\|A\|_F = \|\Sigma V^T\|_F$$

Since

$$\|\Sigma V^T\|_F = \|(\Sigma V^T)^T\|_F = \|V\Sigma^T\|_F = \|\Sigma^T\|_F$$

it follows that

$$\|A\|_F = \left(\sigma_1^2 + \sigma_2^2 + \cdots + \sigma_n^2\right)^{1/2}$$

▶ **THEOREM 6.5.3** *Let $A = U\Sigma V^T$ be an $m \times n$ matrix, and let $\mathcal{M}$ denote the set of all $m \times n$ matrices of rank k or less, where $0 < k < \text{rank}(A)$. If X is a matrix in $\mathcal{M}$ satisfying (7), then*

$$\|A - X\|_F = \left(\sigma_{k+1}^2 + \sigma_{k+2}^2 + \cdots + \sigma_n^2\right)^{1/2}$$

In particular, if $A' = U\Sigma' V^T$, where

$$\Sigma' = \begin{bmatrix} \sigma_1 & & & \\ & \ddots & & O \\ & & \sigma_k & \\ \hline & O & & O \end{bmatrix} = \begin{bmatrix} \Sigma_k & O \\ O & O \end{bmatrix}$$

then

$$\|A - A'\|_F = \left(\sigma_{k+1}^2 + \cdots + \sigma_n^2\right)^{1/2} = \min_{S \in \mathcal{M}} \|A - S\|_F$$

▶**Proof.** Let X be a matrix in $\mathcal{M}$ satisfying (7). Since $A' \in \mathcal{M}$, it follows that

(8)
$$\|A - X\|_F \leq \|A - A'\|_F = \left(\sigma_{k+1}^2 + \cdots + \sigma_n^2\right)^{1/2}$$

We will show that

$$\|A - X\|_F \geq \left(\sigma_{k+1}^2 + \cdots + \sigma_n^2\right)^{1/2}$$

and hence that equality holds in (8). Let $Q\Omega P^T$ be the singular value decomposition of X.

$$\Omega = \left[\begin{array}{ccccc|c} \omega_1 & & & & & \\ & \omega_2 & & & & \\ & & \ddots & & & O \\ & & & \omega_k & & \\ \hline & & O & & & O \end{array}\right] = \left[\begin{array}{cc} \Omega_k & O \\ O & O \end{array}\right]$$

If we set $B = Q^T A P$, then $A = QBP^T$, and it follows that

$$\|A - X\|_F = \|Q(B - \Omega)P^T\|_F = \|B - \Omega\|_F$$

Let us partition B in the same manner as Ω.

$$B = \left[\begin{array}{c|c} \overbrace{B_{11}}^{k \times k} & \overbrace{B_{12}}^{k \times (n-k)} \\ \hline \underbrace{B_{21}}_{(m-k) \times k} & \underbrace{B_{22}}_{(m-k) \times (n-k)} \end{array}\right]$$

It follows that

$$\|A - X\|_F^2 = \|B_{11} - \Omega_k\|_F^2 + \|B_{12}\|_F^2 + \|B_{21}\|_F^2 + \|B_{22}\|_F^2$$

We claim $B_{12} = O$. If not, then define

$$Y = Q \left[\begin{array}{cc} B_{11} & B_{12} \\ O & O \end{array}\right] P^T$$

The matrix Y is in $\mathcal{M}$ and

$$\|A - Y\|_F^2 = \|B_{21}\|_F^2 + \|B_{22}\|_F^2 < \|A - X\|_F^2$$

This contradicts the definition of X. Therefore, $B_{12} = O$. In a similar manner it can be shown that B_{21} must equal O. If we set

$$Z = Q \left[\begin{array}{cc} B_{11} & O \\ O & O \end{array}\right] P^T$$

then $Z \in \mathcal{M}$ and

$$\|A - Z\|_F^2 = \|B_{22}\|_F^2 \leq \|B_{11} - \Omega_k\|_F^2 + \|B_{22}\|_F^2 = \|A - X\|_F^2$$

It follows from the definition of X that B_{11} must equal Ω_k. If B_{22} has singular value decomposition $U_1 \Lambda V_1^T$, then

$$\|A - X\|_F = \|B_{22}\|_F = \|\Lambda\|_F$$

Let

$$U_2 = \begin{bmatrix} I_k & O \\ O & U_1 \end{bmatrix} \qquad \text{and} \qquad V_2 = \begin{bmatrix} I_k & O \\ O & V_1 \end{bmatrix}$$

Now

$$U_2^T Q^T A P V_2 = \begin{bmatrix} \Omega_k & O \\ O & \Lambda \end{bmatrix}$$

$$A = (QU_2) \begin{bmatrix} \Omega_k & O \\ O & \Lambda \end{bmatrix} (PV_2)^T$$

and hence it follows that the diagonal elements of Λ are singular values of A. Thus

$$\|A - X\|_F = \|\Lambda\|_F \geq \left(\sigma_{k+1}^2 + \cdots + \sigma_n^2 \right)^{1/2}$$

It follows from (8) that

$$\|A - X\|_F = \left(\sigma_{k+1}^2 + \cdots + \sigma_n^2 \right)^{1/2} = \|A - A'\|_F \qquad \blacktriangleleft$$

If A has singular value decomposition $U\Sigma V^T$, then we can think of A as a product of $U\Sigma$ times V^T. If we partition $U\Sigma$ into columns and V^T into rows, then

$$U\Sigma = (\sigma_1 \mathbf{u}_1, \sigma_2 \mathbf{u}_2, \ldots, \sigma \mathbf{u}_n)$$

and we can represent A by an outer product expansion

(9) $$A = \sigma_1 \mathbf{u}_1 \mathbf{v}_1^T + \sigma_2 \mathbf{u}_2 \mathbf{v}_2^T + \cdots + \sigma_n \mathbf{u}_n \mathbf{v}_n^T$$

If A is of rank n, then

$$A' = U \begin{bmatrix} \sigma_1 & & & & \\ & \sigma_2 & & & \\ & & \ddots & & \\ & & & \sigma_{n-1} & \\ & & & & 0 \end{bmatrix} V^T$$

$$= \sigma_1 \mathbf{u}_1 \mathbf{v}_1^T + \sigma_2 \mathbf{u}_2 \mathbf{v}_2^T + \cdots + \sigma_{n-1} \mathbf{u}_{n-1} \mathbf{v}_{n-1}^T$$

will be the matrix of rank $n - 1$ that is closest to A with respect to the Frobenius norm. Similarly,

$$A'' = \sigma_1 \mathbf{u}_1 \mathbf{v}_1^T + \sigma_2 \mathbf{u}_2 \mathbf{v}_2^T \cdots + \sigma_{n-2} \mathbf{u}_{n-2} \mathbf{v}_{n-2}^T$$

will be the nearest matrix of rank $n - 2$, and so on. In particular, if A is a nonsingular $n \times n$ matrix, then A' is singular and $\|A - A'\|_F = \sigma_n$. Thus σ_n may be taken as a measure of how close a matrix is to being singular.

The reader should be careful not to use the value of $\det(A)$ as a measure of how close A is to being singular. If, for example, A is the 100×100 diagonal matrix whose diagonal entries are all $\frac{1}{2}$, then $\det(A) = 2^{-100}$; however, $\sigma_{100} = \frac{1}{2}$. On the other hand, the matrix in the following example is very close to being singular even though its determinant is 1 and all its eigenvalues are equal to 1.

EXAMPLE 2. Let A be an $n \times n$ upper triangular matrix whose diagonal elements are all 1 and whose entries above the main diagonal are all -1.

$$A = \begin{pmatrix} 1 & -1 & -1 & \cdots & -1 & -1 \\ 0 & 1 & -1 & \cdots & -1 & -1 \\ 0 & 0 & 1 & \cdots & -1 & -1 \\ \vdots & & & & & \\ 0 & 0 & 0 & \cdots & 1 & -1 \\ 0 & 0 & 0 & \cdots & 0 & 1 \end{pmatrix}$$

Notice that $\det(A) = \det(A^{-1}) = 1$ and all the eigenvalues of A are 1. However, if n is large, then A is close to being singular. To see this, let

$$B = \begin{pmatrix} 1 & -1 & -1 & \cdots & -1 & -1 \\ 0 & 1 & -1 & \cdots & -1 & -1 \\ 0 & 0 & 1 & \cdots & -1 & -1 \\ \vdots & & & & & \\ 0 & 0 & 0 & \cdots & 1 & -1 \\ \frac{-1}{2^{n-2}} & 0 & 0 & \cdots & 0 & 1 \end{pmatrix}$$

The matrix B must be singular, since the system $B\mathbf{x} = \mathbf{0}$ has a nontrivial solution $\mathbf{x} = (2^{n-2}, 2^{n-3}, \dots, 2^0, 1)^T$. The matrices A and B only differ in the $(n, 1)$ position.

$$\|A - B\|_F = \frac{1}{2^{n-2}}$$

It follows from Theorem 6.5.3 that

$$\sigma_n = \min_{X \text{ singular}} \|A - X\|_F \leq \|A - B\|_F = \frac{1}{2^{n-2}}$$

Thus, if $n = 100$, then $\sigma_n \leq 1/2^{98}$ and, consequently, A is very close to singular. ◀

APPLICATION 1: NUMERICAL RANK

In most practical applications, matrix computations are carried out by computers using finite precision arithmetic. If the computations involve a nonsingular matrix that is *very close* to be being singular, then the matrix will behave computationally exactly like a singular matrix. In this case, computed solutions to linear systems may have no digits of accuracy whatsoever. More generally, if an $m \times n$ matrix A is *close enough* to a matrix of rank r, where $r < \min(m, n)$, then A will behave like a rank r matrix in finite precision arithmetic. The singular values provide a way of measuring how close a matrix is to matrices of lower rank; however, we must clarify what we mean by very close. We must decide how close is close enough. The answer depends on the machine precision of the computer that is being used.

Machine precision can be measured in terms of the unit roundoff error for the machine. Another name for the unit roundoff is the *machine epsilon*. To understand this concept, we need to know how computers represent numbers. If the computer uses number base beta and keeps track of n digits, then it will represent a real number x by a *floating-point number*, denoted $fl(x)$, of the form $\pm 0.d_1 d_2 \ldots d_n \times \beta^k$, where the digits d_i are integers with $0 \le d_1 < \beta$. For example, $-0.54321469 \times 10^{25}$ is an eight-digit, base 10, floating-point number, and $0.110100111001 \times 2^{-9}$ is a twelve-digit, base 2 floating-point number. In Section 1 of Chapter 7 we will discuss floating-point numbers in more detail and give a precise definition of the *machine epsilon*. It turns out that the machine epsilon, ϵ, is the smallest floating-point number that will serve as a bound for the relative error whenever we approximate a real number by a floating-point number; that is, for any real number x,

$$(10) \qquad \left| \frac{fl(x) - x}{x} \right| < \epsilon$$

For eight-digit, base 10, floating-point arithmetic, the machine epsilon is 5×10^{-8}. For twelve-digit, base 2, floating-point arithmetic, the machine epsilon is $(\frac{1}{2})^{-12}$, and in general for n-digit base β arithmetic, the machine epsilon is $\frac{1}{2} \times \beta^{-n+1}$.

In light of (10) the machine epsilon is the natural choice as a basic unit for measuring roundoff errors. Suppose that A is a matrix of rank n, but k of its singular values are less than a "small" multiple of the machine epsilon. Then A is close enough to matrices of rank $n - k$ so that it is impossible to tell the difference when doing floating-point arithmetic. In this case we would say that A has *numerical rank* $n - k$. The multiple of the machine epsilon that we use to determine numerical rank depends on the dimensions of the matrix and on its largest singular value. The following definition of numerical rank is the one used by the software package MATLAB. In fact, the MATLAB command **rank(A)** will compute the numerical rank of A, rather than the exact rank.

▶ **DEFINITION** The **numerical rank** of an $m \times n$ matrix is the number of singular values of the matrix that are greater than $\sigma_1 \max(m, n)\epsilon$, where σ_1 is the largest singular value of A and ϵ is the machine epsilon. ◀

EXAMPLE 3. Suppose that A is 5×5 matrix with singular values

$$\sigma_1 = 4, \quad \sigma_2 = 1, \quad \sigma_3 = 10^{-12}, \quad \sigma_4 = 3.1 \times 10^{-14}, \quad \sigma_5 = 2.6 \times 10^{-15}$$

and suppose that the machine epsilon is 5×10^{-15}. To determine the numerical rank, we compare the singular values to

$$\sigma_1 \max(m, n)\epsilon = 4 \cdot 5 \cdot 5 \times 10^{-15} = 10^{-13}$$

Since three of the singular values are greater than 10^{-13}, the matrix has numerical rank 3. ◀

APPLICATION 2: DIGITAL IMAGE PROCESSING

A video image or photograph can be digitized by breaking it up into a rectangular array of cells (or pixels) and measuring the gray level of each cell. This information can be stored and transmitted using an $n \times n$ matrix A. The entries of A are nonnegative numbers corresponding to the measures of the gray levels. Because the gray levels of any one cell generally turn out to be close to the gray levels of its neighboring cells, it is possible to reduce the amount of storage necessary from n^2 to a multiple of n. Generally, the matrix A will have many small singular values. Consequently, A can be approximated by a matrix of much lower rank.

FIGURE 6.5.1   Courtesy Oakridge National Laboratory

Original 176 by 260 Image

Rank 5 Approximation to Image

Rank 15 Approximation to Image

Rank 30 Approximation to Image

If A has singular value decomposition $U \Sigma V^T$, then A can be represented by the outer product expansion

$$A = \sigma_1 \mathbf{u}_1 \mathbf{v}_1^T + \sigma_2 \mathbf{u}_2 \mathbf{v}_2^T + \cdots + \sigma_n \mathbf{u}_n \mathbf{v}_n^T$$

The closest matrix of rank k is obtained by truncating this sum after the first k terms:

$$A_k = \sigma_1 \mathbf{u}_1 \mathbf{v}_1^T + \sigma_2 \mathbf{u}_2 \mathbf{v}_2^T + \cdots + \sigma_k \mathbf{u}_k \mathbf{v}_k^T$$

The total storage for A_k is $k(2n + 1)$. We can choose k to be considerably less than n and still have the digital image corresponding to A_k very close to the original. For typical choices of k, the storage required for A_k will be less than 20 percent of the amount of storage necessary for the entire matrix A.

Figure 6.5.1 shows an image corresponding to a 176×260 matrix A and three images corresponding to lower-rank approximations of A. The gentlemen in the picture are (left to right) James H. Wilkinson, Wallace Givens, and George Forsythe (three pioneers in the field of numerical linear algebra).

APPLICATION 3: INFORMATION RETRIEVAL—LATENT SEMANTIC INDEXING

We return again to the information retrieval application discussed Chapter 1, Section 3 and Chapter 5, Section 1. In this application a database of documents is represented by a database matrix Q. To search the database, we form a unit search vector $\mathbf{x}$ and set $\mathbf{y} = Q^T \mathbf{x}$. The documents that best match the search criteria are those corresponding to the entries of $\mathbf{y}$ that are closest to 1.

Because of the problems of polysemy and synonymy, we can think of our database as an approximation. Some of the entries of the database matrix may contain extraneous components due to multiple meanings of words, and some may miss including components because of synonymy. Suppose that it were possible to correct for these problems and come up with a perfect database matrix P. If we set $E = Q - P$, then, since $Q = P + E$, we can think of E as a matrix representing the errors in our database matrix Q. Unfortunately, E is unknown, so we cannot determine P exactly. However, if we can find a simpler approximation Q_1 for Q, then Q_1 will also be an approximation for P. Thus $Q_1 = P + E_1$ for some error matrix E_1. In the method of *latent semantic indexing* (LSI), the database matrix Q is approximated by a matrix Q_1 with lower rank. The idea behind the method is that the lower-rank matrix may still provide a good approximation to P and, because of its simpler structure, may actually involve less error; that is, $\|E_1\| < \|E\|$.

The lower-rank approximation can be obtained by truncating the outer product expansion of the singular value decomposition of Q. This is equivalent to setting

$$\sigma_{r+1} = \sigma_{r+2} = \cdots = \sigma_n = 0$$

and then setting $Q_1 = U_1 \Sigma_1 V_1^T$, the compact form of the singular value decomposition of the rank r matrix. Furthermore, if $r < \min(m, n)/2$, then this factorization is computationally more efficient to use and the searches will be speeded up. The speed of computation is proportional to the amount of arithmetic involved. The matrix vector multiplication $Q^T \mathbf{x}$ requires a total of mn scalar multiplications (m multiplications for each of the n entries of the product). On the other hand, $Q_1^T = V_1 \Sigma_1 U_1^T$, and the multiplication $Q_1^T \mathbf{x} = V_1(\Sigma_1(U_1 \mathbf{x}^T))$ requires a total of $r(m+n+1)$ scalar multiplications. For example, if $m = n = 1000$ and $r = 200$, then

$$mn = 10^6 \qquad \text{and} \qquad r(m+n+1) = 200 \cdot 2001 = 400{,}200$$

The search with the lower-rank matrix should be more than twice as fast.

APPLICATION 4: PSYCHOLOGY — PRINCIPAL COMPONENT ANALYSIS

In Section 1 of Chapter 5 we saw how psychologist Charles Spearman used a correlation matrix to compare scores on a series of aptitude tests. Based on the observed correlations, Spearman concluded that the test results provided evidence of common basic underlying functions. Further work by psychologists to identify the common factors that make up intelligence has led to development of an area of study known as *factor analysis*.

Predating Spearman's work by a few years is a 1901 paper by Karl Pearson analyzing a correlation matrix derived from measuring seven physical variables for each of 3000 criminals. This study contains the roots of a method popularized by H. Hotelling in a well-known paper published in 1933. The method is known as principal component analysis.

To see the basic idea of this method, assume that a series of n aptitude tests are administered to a group of m individuals and that the deviations from the mean for the tests form the columns of an $m \times n$ matrix X. Although, in practice, column vectors of X are positively correlated, the hypothetical factors that account for the scores should be uncorrelated. Thus, we wish to introduce mutually orthogonal vectors $\mathbf{y}_1, \mathbf{y}_2, \ldots, \mathbf{y}_r$ corresponding to the hypothetical factors. We require that the vectors span $R(X)$, and hence the number of vectors, r, should be equal to the rank of X. Furthermore, we wish to number these vectors in decreasing order of variance.

The first principal component vector $\mathbf{y}_1$ should account for the most variance. Since $\mathbf{y}_1$ is in the column space of X, we can represent it as a product $X\mathbf{v}_1$ for some $\mathbf{v}_1 \in R^n$, The covariance matrix is

$$S = \frac{1}{n-1} X^T X$$

and the variance of $\mathbf{y}_1$ is given by

$$\text{var}(\mathbf{y}_1) = \frac{(X\mathbf{v}_1)^T X\mathbf{v}_1}{n-1} = \mathbf{v}_1^T S \mathbf{v}_1$$

The vector $\mathbf{v}_1$ is chosen to maximize $\mathbf{v}^T S \mathbf{v}$ over all unit vectors $\mathbf{v}$. This can be accomplished by choosing $\mathbf{v}_1$ to a unit eigenvector of $X^T X$ belonging to its maximum eigenvalue λ_1. (See Exercise 21 of Chapter 6, Section 4.) The eigenvectors of $X^T X$ are the right singular vectors of X. Thus, $\mathbf{v}_1$ is the right singular vector of X corresponding to the largest singular value $\sigma_1 = \sqrt{\lambda_1}$. If $\mathbf{u}_1$ is the corresponding left singular vector, then

$$\mathbf{y}_1 = X\mathbf{v}_1 = \sigma_1 \mathbf{u}_1$$

The second principal component vector must be of the form $\mathbf{y}_2 = X\mathbf{v}_2$. It can be shown that the vector that maximizes $\mathbf{v}^T S \mathbf{v}$ over all unit vectors that are orthogonal to $\mathbf{v}_1$ is just the second right singular vector $\mathbf{v}_2$ of X. If we choose $\mathbf{v}_2$ in this way and $\mathbf{u}_2$ is the corresponding left singular vector, then

$$\mathbf{y}_2 = X\mathbf{v}_2 = \sigma_2 \mathbf{u}_2$$

and since

$$\mathbf{y}_1^T \mathbf{y}_2 = \sigma_1 \sigma_2 \mathbf{u}_1^T \mathbf{u}_2 = 0$$

it follows that $\mathbf{y}_1$ and $\mathbf{y}_2$ are orthogonal. The remaining $\mathbf{y}_i$'s are determined in a similar manner.

In general, the singular value decomposition solves the principal component problem. If X has rank r and singular value decomposition $X = U_1 \Sigma_1 V_1^T$ (in compact form), then the principal component vectors are given by

$$\mathbf{y}_1 = \sigma_1 \mathbf{u}_1, \ \ \mathbf{y}_2 = \sigma_2 \mathbf{u}_2, \ \ \ldots, \ \ \mathbf{y}_r = \sigma_r \mathbf{u}_r$$

The left singular vectors $\mathbf{u}_1, \ldots, \mathbf{u}_n$ are the normalized principal component vectors. If we set $W = \Sigma_1 V_1^T$, then

$$X = U_1 \Sigma_1 V_1^T = U_1 W$$

The columns of the matrix U_1 correspond to the hypothetical intelligence factors. The entries in each column measure how well the individual students exhibit that particular intellectual ability. The matrix W measures to what extent each test depends on the hypothetical factors.

EXERCISES

1. Show that A and A^T have the same nonzero singular values. How are their singular value decompositions related?

2. Use the method of Example 1 to find the singular value decomposition of each of the following matrices.

(a) $\begin{bmatrix} 1 & 1 \\ 2 & 2 \end{bmatrix}$ (b) $\begin{bmatrix} 2 & -2 \\ 1 & 2 \end{bmatrix}$ (c) $\begin{bmatrix} 1 & 3 \\ 3 & 1 \\ 0 & 0 \\ 0 & 0 \end{bmatrix}$ (d) $\begin{bmatrix} 2 & 0 & 0 \\ 0 & 2 & 1 \\ 0 & 1 & 2 \\ 0 & 0 & 0 \end{bmatrix}$

3. For each of the matrices in Exercise 2:

  (a) Determine the rank.

  (b) Find the closest (with respect to the Frobenius norm) matrix of rank 1.

4. Given

$$
A = \begin{pmatrix} -2 & 8 & 20 \\ 14 & 19 & 10 \\ 2 & -2 & 1 \end{pmatrix} = \begin{pmatrix} \frac{3}{5} & -\frac{4}{5} & 0 \\ \frac{4}{5} & \frac{3}{5} & 0 \\ 0 & 0 & 1 \end{pmatrix} \begin{pmatrix} 30 & 0 & 0 \\ 0 & 15 & 0 \\ 0 & 0 & 3 \end{pmatrix} \begin{pmatrix} \frac{1}{3} & \frac{2}{3} & \frac{2}{3} \\ \frac{2}{3} & \frac{1}{3} & -\frac{2}{3} \\ \frac{2}{3} & -\frac{2}{3} & \frac{1}{3} \end{pmatrix}
$$

Find the closest (with respect to the Frobenius norm) matrices of rank 1 and rank 2 to A.

5. The matrix

$$
A = \begin{pmatrix} 2 & 5 & 4 \\ 6 & 3 & 0 \\ 6 & 3 & 0 \\ 2 & 5 & 4 \end{pmatrix}
$$

has singular value decomposition

$$
\begin{pmatrix} \frac{1}{2} & \frac{1}{2} & \frac{1}{2} & \frac{1}{2} \\ \frac{1}{2} & -\frac{1}{2} & -\frac{1}{2} & \frac{1}{2} \\ \frac{1}{2} & -\frac{1}{2} & \frac{1}{2} & -\frac{1}{2} \\ \frac{1}{2} & \frac{1}{2} & -\frac{1}{2} & -\frac{1}{2} \end{pmatrix} \begin{pmatrix} 12 & 0 & 0 \\ 0 & 6 & 0 \\ 0 & 0 & 0 \\ 0 & 0 & 0 \end{pmatrix} \begin{pmatrix} \frac{2}{3} & \frac{2}{3} & \frac{1}{3} \\ -\frac{2}{3} & \frac{1}{3} & \frac{2}{3} \\ \frac{1}{3} & -\frac{2}{3} & \frac{2}{3} \end{pmatrix}
$$

  (a) Use the singular value decomposition to find orthonormal bases for $R(A^T)$ and $N(A)$.

  (b) Use the singular value decomposition to find orthonormal bases for $R(A)$ and $N(A^T)$.

6. Prove that, if A is a symmetric matrix with eigenvalues $\lambda_1, \lambda_2, \ldots, \lambda_n$, then the singular values of A are $|\lambda_1|, |\lambda_2|, \ldots, |\lambda_n|$.

7. Let A be an $m \times n$ matrix with singular value decomposition $U \Sigma V^T$, and suppose that A has rank r, where $r < n$. Show that $\{\mathbf{v}_1, \ldots, \mathbf{v}_r\}$ is an orthonormal basis for $R(A^T)$.

8. Show that if σ is a singular value of A, then there exists a nonzero vector $\mathbf{x}$ such that

$$
\sigma = \frac{\|A\mathbf{x}\|_2}{\|\mathbf{x}\|_2}
$$

9. Let A be an $m \times n$ matrix of rank n with singular value decomposition $U \Sigma V^T$. Let Σ^+ denote the $n \times m$ matrix

$$
\begin{pmatrix}
\frac{1}{\sigma_1} & & & & \\
& \frac{1}{\sigma_2} & & & \\
& & \ddots & & O \\
& & & \frac{1}{\sigma_n} &
\end{pmatrix}
$$

and define $A^+ = V \Sigma^+ U^T$. Show that $\hat{\mathbf{x}} = A^+ \mathbf{b}$ satisfies the normal equations $A^T A \mathbf{x} = A^T \mathbf{b}$.

10. Let A^+ be defined as in Exercise 9 and let $P = A A^+$. Show that $P^2 = P$ and $P^T = P$.

6 QUADRATIC FORMS

By this time the reader should be well aware of the important role that matrices play in the study of linear equations. In this section we will see that matrices also play an important role in the study of quadratic equations. With each quadratic equation we can associate a vector function $f(\mathbf{x}) = \mathbf{x}^T A \mathbf{x}$. Such a vector function is called a *quadratic form*. Quadratic forms arise in a wide variety of applied problems. They are particularly important in the study of optimization theory.

▶ **DEFINITION** A **quadratic equation** in two variables x and y is an equation of the form

(1) $$ ax^2 + 2bxy + cy^2 + dx + ey + f = 0 $$

Equation (1) may be rewritten in the form

(2) $$ \begin{pmatrix} x & y \end{pmatrix} \begin{pmatrix} a & b \\ b & c \end{pmatrix} \begin{pmatrix} x \\ y \end{pmatrix} + \begin{pmatrix} d & e \end{pmatrix} \begin{pmatrix} x \\ y \end{pmatrix} + f = 0 $$

Let

$$ \mathbf{x} = \begin{pmatrix} x \\ y \end{pmatrix} \quad \text{and} \quad A = \begin{pmatrix} a & b \\ b & c \end{pmatrix} $$

The term

$$ \mathbf{x}^T A \mathbf{x} = ax^2 + 2bxy + cy^2 $$

is called the **quadratic form** associated with (1).   ◀

Conic Sections

The graph of an equation of the form (1) is called a *conic section*. [If there are no ordered pairs (x, y) that satisfy (1), we say that the equation represents an imaginary conic.] If the graph of (1) consists of a single point, a line, or a pair of lines, we say that (1) represents a degenerate conic. Of more interest are the nondegenerate conics. Graphs of nondegenerate conics turn out to be circles, ellipses, parabolas, or hyperbolas (see Figure 6.6.1). The graph of a conic is particularly easy to sketch when its equation can be put into one of the following standard forms:

(a) $x^2 + y^2 = r^2$ (circle)

(b) $\dfrac{x^2}{\alpha^2} + \dfrac{y^2}{\beta^2} = 1$ (ellipse)

(c) $\dfrac{x^2}{\alpha^2} - \dfrac{y^2}{\beta^2} = 1$ or $\dfrac{y^2}{\alpha^2} - \dfrac{x^2}{\beta^2} = 1$ (hyperbola)

(d) $x^2 = \alpha y$ or $y^2 = \alpha x$ (parabola)

where α, β, and r are nonzero real numbers. Note that the circle is a special case of the ellipse ($\alpha = \beta = r$). A conic section is said to be in *standard position* if its equation can be put into one of these four standard forms. The graphs of (i), (ii), and (iii) in Figure 6.6.1 will all be symmetric to both coordinate axes and the origin. We say that these curves are centered at the origin. A parabola in standard position will have its vertex at the origin and will be symmetric to one of the axes.

What about the conics that are not in standard position? Let us consider the following cases.

CASE 1

The conic section has been translated horizontally from the standard position. This occurs when the x^2 and x terms in (1) both have nonzero coefficients.

CASE 2

The conic section has been translated vertically from the standard position. This occurs when the y^2 and y terms in (1) have nonzero coefficients (i.e., $c \neq 0$ and $e \neq 0$).

FIGURE 6.6.1

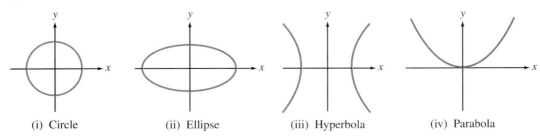

(i) Circle   (ii) Ellipse   (iii) Hyperbola   (iv) Parabola

CASE 3

The conic section has been rotated from the standard position by an angle θ that is not a multiple of $90°$. This occurs when the coefficient of the xy term is nonzero (i.e., $b \neq 0$).

In general, we may have any one or combination of these three cases. In order to graph a conic section that is not in standard position, we usually find a new set of axes x' and y' such that the conic section is in standard position with respect to the new axes. This is not difficult if the conic has only been translated horizontally or vertically, in which case the new axes can be found by completing the squares. The following example illustrates how this is done.

EXAMPLE I.   Sketch the graph of the equation

$$9x^2 - 18x + 4y^2 + 16y - 11 = 0$$

SOLUTION.   To see how to choose our new axis system, we complete the squares.

$$9(x^2 - 2x + 1) + 4(y^2 + 4y + 4) - 11 = 9 + 16$$

This equation can be simplified to the form

$$\frac{(x-1)^2}{2^2} + \frac{(y+2)^2}{3^2} = 1$$

If we let

$$x' = x - 1 \qquad \text{and} \qquad y' = y + 2$$

the equation becomes

$$\frac{(x')^2}{2^2} + \frac{(y')^2}{3^2} = 1$$

which is in standard form with respect to the variables x' and y'. Thus the graph, as shown in Figure 6.6.2, will be an ellipse that is in standard position in the $x'y'$ axis system. The center of the ellipse will be at the origin of the $x'y'$ plane [i.e., at the point $(x, y) = (1, -2)$]. The equation of the x' axis is simply $y' = 0$, which is the

FIGURE 6.6.2

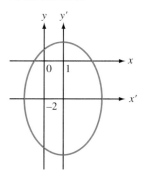

equation of the line $y = -2$ in the xy plane. Similarly, the y' axis coincides with the line $x = 1$.   ◄

There is little problem if the center or vertex of the conic section has been translated. If, however, the conic section has also been rotated from the standard position, it is necessary to change coordinates so that the equation in terms of the new coordinates x' and y' involves no $x'y'$ term. Let $\mathbf{x} = (x, y)^T$ and $\mathbf{x}' = (x', y')^T$. Since the new coordinates differ from the old coordinates by a rotation, we have

$$\mathbf{x} = Q\mathbf{x}' \qquad \text{or} \qquad \mathbf{x}' = Q^T\mathbf{x}$$

where

$$Q = \begin{pmatrix} \cos\theta & \sin\theta \\ -\sin\theta & \cos\theta \end{pmatrix} \qquad \text{or} \qquad Q^T = \begin{pmatrix} \cos\theta & -\sin\theta \\ \sin\theta & \cos\theta \end{pmatrix}$$

If $0 < \theta < \pi$, the matrix Q corresponds to a rotation of θ radians in the clockwise direction, and Q^T corresponds to a rotation of θ radians in the counterclockwise direction (see Example 2 in Section 2 of Chapter 4). With this change of variables, (2) becomes

$$(3) \qquad (\mathbf{x}')^T(Q^T A Q)\mathbf{x}' + \begin{bmatrix} d' & e' \end{bmatrix}\mathbf{x}' + f = 0$$

where $\begin{bmatrix} d' & e' \end{bmatrix} = \begin{bmatrix} d & e \end{bmatrix} Q$. This equation will involve no $x'y'$ term if and only if $Q^T A Q$ is diagonal. Since A is symmetric, it is possible to find a pair of orthonormal eigenvectors $\mathbf{q}_1 = (x_1, -y_1)^T$ and $\mathbf{q}_2 = (y_1, x_1)^T$. Thus, if we set $\cos\theta = x_1$ and $\sin\theta = y_1$, then

$$Q = \begin{pmatrix} \mathbf{q}_1 & \mathbf{q}_2 \end{pmatrix} = \begin{pmatrix} x_1 & y_1 \\ -y_1 & x_1 \end{pmatrix}$$

diagonalizes A and (3) simplifies to

$$\lambda_1(x')^2 + \lambda_2(y')^2 + d'x' + e'y' + f = 0$$

EXAMPLE 2. Consider the conic section

$$3x^2 + 2xy + 3y^2 - 8 = 0$$

This equation can be written in the form

$$\begin{pmatrix} x & y \end{pmatrix} \begin{bmatrix} 3 & 1 \\ 1 & 3 \end{bmatrix} \begin{bmatrix} x \\ y \end{bmatrix} = 8$$

The matrix

$$\begin{bmatrix} 3 & 1 \\ 1 & 3 \end{bmatrix}$$

has eigenvalues $\lambda = 2$ and $\lambda = 4$ with corresponding unit eigenvectors

$$\left(\frac{1}{\sqrt{2}}, -\frac{1}{\sqrt{2}}\right)^T \qquad \text{and} \qquad \left(\frac{1}{\sqrt{2}}, \frac{1}{\sqrt{2}}\right)^T$$

Let

$$Q = \begin{bmatrix} \dfrac{1}{\sqrt{2}} & \dfrac{1}{\sqrt{2}} \\ -\dfrac{1}{\sqrt{2}} & \dfrac{1}{\sqrt{2}} \end{bmatrix} = \begin{bmatrix} \cos 45° & \sin 45° \\ -\sin 45° & \cos 45° \end{bmatrix}$$

and set

$$\begin{bmatrix} x \\ y \end{bmatrix} = \begin{bmatrix} \dfrac{1}{\sqrt{2}} & \dfrac{1}{\sqrt{2}} \\ -\dfrac{1}{\sqrt{2}} & \dfrac{1}{\sqrt{2}} \end{bmatrix} \begin{bmatrix} x' \\ y' \end{bmatrix}$$

Thus,

$$Q^T A Q = \begin{bmatrix} 2 & 0 \\ 0 & 4 \end{bmatrix}$$

and the equation of the conic becomes

$$2(x')^2 + 4(y')^2 = 8$$

or

$$\frac{(x')^2}{4} + \frac{(y')^2}{2} = 1$$

In the new coordinate system the direction of the x' axis is determined by the point $x' = 1$, $y' = 0$. To translate this to the xy coordinate system, we multiply

$$\begin{bmatrix} \dfrac{1}{\sqrt{2}} & \dfrac{1}{\sqrt{2}} \\ -\dfrac{1}{\sqrt{2}} & \dfrac{1}{\sqrt{2}} \end{bmatrix} \begin{bmatrix} 1 \\ 0 \end{bmatrix} = \begin{bmatrix} \dfrac{1}{\sqrt{2}} \\ -\dfrac{1}{\sqrt{2}} \end{bmatrix} = \mathbf{q}_1$$

The x' axis will be in the direction of $\mathbf{q}_1$. Similarly, to find the direction of the y' axis, we multiply

$$Q\mathbf{e}_2 = \mathbf{q}_2$$

The eigenvectors that form the columns of Q tell us the directions of the new coordinate axes (see Figure 6.6.3). ◀

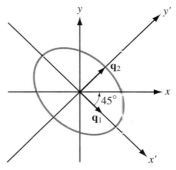

FIGURE 6.6.3

EXAMPLE 3. Given the quadratic equation

$$3x^2 + 2xy + 3y^2 + 8\sqrt{2}y - 4 = 0$$

find a change of coordinates so that the resulting equation represents a conic in standard position.

SOLUTION. The xy term is eliminated in the same manner as in Example 2. In this case, using the rotation matrix

$$Q = \begin{bmatrix} \dfrac{1}{\sqrt{2}} & \dfrac{1}{\sqrt{2}} \\[2mm] -\dfrac{1}{\sqrt{2}} & \dfrac{1}{\sqrt{2}} \end{bmatrix}$$

the equation becomes

$$2(x')^2 + 4(y')^2 + \begin{bmatrix} 0 & 8\sqrt{2} \end{bmatrix} Q \begin{bmatrix} x' \\ y' \end{bmatrix} = 4$$

or

$$(x')^2 - 4x' + 2(y')^2 + 4y' = 2$$

If we complete the square, we get

$$(x' - 2)^2 + 2(y' + 1)^2 = 8$$

If we set $x'' = x' - 2$ and $y'' = y' + 1$ (see Figure 6.6.4), the equation simplifies to

$$\frac{(x'')^2}{8} + \frac{(y'')^2}{4} = 1 \qquad \blacktriangleleft$$

To summarize, a quadratic equation in the variables x and y can be written in the form

$$\mathbf{x}^T A \mathbf{x} + B \mathbf{x} + f = 0$$

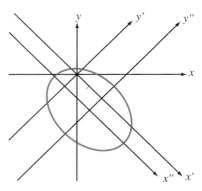

FIGURE 6.6.4

where $\mathbf{x} = (x, y)^T$, A is a 2×2 symmetric matrix, B is a 1×2 matrix, and f is a scalar. If A is nonsingular, then by rotating and translating the axes it is possible to rewrite the equation in the form

$$(4) \qquad \qquad \lambda_1(x')^2 + \lambda_2(y')^2 + f' = 0$$

where λ_1 and λ_2 are the eigenvalues of A. If (4) represents a real nondegenerate conic, it will be either an ellipse or a hyperbola, depending on whether λ_1 and λ_2 agree in sign or differ in sign. If A is singular and exactly one of its eigenvalues is zero, the quadratic equation can be reduced to either

$$\lambda_1(x')^2 + e'y' + f' = 0 \qquad \text{or} \qquad \lambda_2(y')^2 + d'x' + f' = 0$$

These equations will represent parabolas, provided that e' and d' are nonzero.

There is no reason to limit ourselves to two variables. We could just as well have quadratic equations and quadratic forms in any number of variables. Indeed, a *quadratic equation in n variables* $x_1, \ldots , x_n$ is one of the form

$$(5) \qquad \qquad \mathbf{x}^T A \mathbf{x} + B \mathbf{x} + \alpha = 0$$

where $\mathbf{x} = (x_1, \ldots , x_n)^T$, A is an $n \times n$ symmetric matrix, B is a $1 \times n$ matrix, and α is a scalar. The vector function

$$f(\mathbf{x}) = \mathbf{x}^T A \mathbf{x} = \sum_{i=1}^{n} \left(\sum_{j=1}^{n} a_{ij} x_j \right) x_i$$

is the *quadratic form in n variables* associated with the quadratic equation.

In the case of three unknowns, if

$$\mathbf{x} = \begin{bmatrix} x \\ y \\ z \end{bmatrix}, \qquad A = \begin{bmatrix} a & d & e \\ d & b & f \\ e & f & c \end{bmatrix}, \qquad B = \begin{bmatrix} g \\ h \\ i \end{bmatrix}$$

then (5) becomes

$$ax^2 + by^2 + cz^2 + 2dxy + 2exz + 2fyz + gx + hy + iz + \alpha = 0$$

The graph of a quadratic equation in three variables is called a *quadric surface*.

There are four basic types of nondegenerate quadric surfaces:

1. Ellipsoids
2. Hyperboloids (of one or two sheets)
3. Cones
4. Paraboloids (either elliptic or hyperbolic)

As in the two-dimensional case, we can use translations and rotations to transform the equation into standard form,

$$\lambda_1 (x')^2 + \lambda_2 (y')^2 + \lambda_3 (z')^2 + \alpha = 0$$

where λ_1, λ_2, λ_3 are the eigenvalues of A. For the general n-dimensional case, the quadratic form can always be translated to a simpler diagonal form. More precisely, we have the following theorem.

▶ **THEOREM 6.6.1 (Principal Axes Theorem)** *If A is a real symmetric $n \times n$ matrix, then there is a change of variables $\mathbf{u} = Q^T \mathbf{x}$ such that $\mathbf{x}^T A \mathbf{x} = \mathbf{u}^T D \mathbf{u}$, where D is a diagonal matrix.*

▶*Proof.* If A is a real symmetric matrix, then by Corollary 6.4.5 there is an orthogonal matrix Q that diagonalizes A; that is, $Q^T A Q = D$ (diagonal). If we set $\mathbf{u} = Q^T \mathbf{x}$, then $\mathbf{x} = Q \mathbf{u}$ and

$$\mathbf{x}^T A \mathbf{x} = \mathbf{u}^T Q^T A Q \mathbf{u} = \mathbf{u}^T D \mathbf{u} \qquad ◀$$

Optimization: An Application to the Calculus

Let us consider the problem of maximizing and minimizing functions of several variables. In particular, we would like to determine the nature of the critical points of a real-valued vector function $w = F(\mathbf{x})$. If the function is a quadratic form, $w = \mathbf{x}^T A \mathbf{x}$, then $\mathbf{0}$ is a critical point. Whether it is a maximum, minimum, or saddle point depends on the eigenvalues of A. More generally, if the function to be extremized is sufficiently differentiable, it behaves locally like a quadratic form. Thus each critical point can be tested by determining the signs of the eigenvalues of the matrix of an associated quadratic form.

▶ **DEFINITION** Let $F(\mathbf{x})$ be a real-valued vector function on R^n. A point $\mathbf{x}_0$ in R^n is said to be a **stationary point** of F if all the first partial derivatives of F at $\mathbf{x}_0$ exist and are zero. ◀

If $F(\mathbf{x})$ has either a local maximum or a local minimum at a point $\mathbf{x}_0$ and the first partials of F exist at $\mathbf{x}_0$, they will all be zero. Thus, if $F(\mathbf{x})$ has first partials everywhere, its local maxima and minima will occur at stationary points.

Consider the quadratic form

$$f(x, y) = ax^2 + 2bxy + cy^2$$

The first partials of f are

$$f_x = 2ax + 2by$$
$$f_y = 2bx + 2cy$$

Setting these equal to zero, we see that $(0, 0)$ is a stationary point. Moreover, if the matrix

$$A = \begin{bmatrix} a & b \\ b & c \end{bmatrix}$$

is nonsingular, this will be the only critical point. Thus, if A is nonsingular, f will have either a global minimum, a global maximum, or a saddle point at $(0, 0)$.

Let us write f in the form

$$f(\mathbf{x}) = \mathbf{x}^T A \mathbf{x} \qquad \text{where} \qquad \mathbf{x} = \begin{bmatrix} x \\ y \end{bmatrix}$$

Since $f(\mathbf{0}) = 0$, it follows that f will have a global minimum at $\mathbf{0}$ if and only if

$$\mathbf{x}^T A \mathbf{x} > 0 \qquad \text{for all} \qquad \mathbf{x} \neq \mathbf{0}$$

and f will have a global maximum at $\mathbf{0}$ if and only if

$$\mathbf{x}^T A \mathbf{x} < 0 \qquad \text{for all} \qquad \mathbf{x} \neq \mathbf{0}$$

If $\mathbf{x}^T A \mathbf{x}$ changes sign, then $\mathbf{0}$ is a saddle point.

In general, if f is a quadratic form in n variables, then for each $\mathbf{x} \in R^n$

$$f(\mathbf{x}) = \mathbf{x}^T A \mathbf{x}$$

where A is a symmetric $n \times n$ matrix.

▶ **DEFINITION** A quadratic form $f(\mathbf{x}) = \mathbf{x}^T A \mathbf{x}$ is said to be **definite** if it takes on only one sign as $\mathbf{x}$ varies over all nonzero vectors in R^n. The form is **positive definite** if $\mathbf{x}^T A \mathbf{x} > 0$ for all nonzero $\mathbf{x}$ in R^n and **negative definite** if $\mathbf{x}^T A \mathbf{x} < 0$ for all nonzero $\mathbf{x}$ in R^n. A quadratic form is said to be **indefinite** if it takes on values that differ in sign. If $f(\mathbf{x}) = \mathbf{x}^T A \mathbf{x} \geq 0$ and assumes the value 0 for some $\mathbf{x} \neq \mathbf{0}$, then $f(\mathbf{x})$ is said to be **positive semidefinite**. If $f(\mathbf{x}) \leq 0$ and assumes the value 0 for some $\mathbf{x} \neq \mathbf{0}$, then $f(\mathbf{x})$ is said to be **negative semidefinite**. ◀

Whether the quadratic form is positive definite or negative definite depends on the matrix A. If the quadratic form is positive definite, we say simply that A is positive definite. The preceding definition can then be restated as follows.

▶ **DEFINITION** A real symmetric matrix A is said to be

(i) **Positive definite** if $\mathbf{x}^T A \mathbf{x} > 0$ for all nonzero $\mathbf{x}$ in R^n
(ii) **Negative definite** if $\mathbf{x}^T A \mathbf{x} < 0$ for all nonzero $\mathbf{x}$ in R^n
(iii) **Positive semidefinite** if $\mathbf{x}^T A \mathbf{x} \geq 0$ for all nonzero $\mathbf{x}$ in R^n
(iv) **Negative semidefinite** if $\mathbf{x}^T A \mathbf{x} \leq 0$ for all nonzero $\mathbf{x}$ in R^n ◀

If A is nonsingular, then $\mathbf{0}$ will be the only stationary point of $f(\mathbf{x}) = \mathbf{x}^T A \mathbf{x}$. It will be a global minimum if A is positive definite and a global maximum if A is negative definite. If A is indefinite, then $\mathbf{0}$ is a saddle point. To classify the stationary point, we must then classify the matrix A. There are a number of ways of determining whether a matrix is positive definite. We will study some of these methods in the next section. The following theorem gives perhaps the most important characterization of positive definite matrices.

▶ **THEOREM 6.6.2** *Let A be a real symmetric $n \times n$ matrix. Then A is positive definite if and only if all its eigenvalues are positive.*

▶*Proof.* If A is positive definite and λ is an eigenvalue of A, then for any eigenvector $\mathbf{x}$ belonging to λ

$$\mathbf{x}^T A\mathbf{x} = \lambda \mathbf{x}^T \mathbf{x} = \lambda \|\mathbf{x}\|^2$$

Hence

$$\lambda = \frac{\mathbf{x}^T A\mathbf{x}}{\|\mathbf{x}\|^2} > 0$$

Conversely, suppose that all the eigenvalues of A are positive. Let $\{\mathbf{x}_1, \ldots, \mathbf{x}_n\}$ be an orthonormal set of eigenvectors of A. If $\mathbf{x}$ is any nonzero vector in R^n, then $\mathbf{x}$ can be written in the form

$$\mathbf{x} = \alpha_1 \mathbf{x}_1 + \alpha_2 \mathbf{x}_2 + \cdots + \alpha_n \mathbf{x}_n$$

where

$$\alpha_i = \mathbf{x}^T \mathbf{x}_i \quad \text{for} \quad i = 1, \ldots, n \quad \text{and} \quad \sum_{i=1}^{n} (\alpha_i)^2 = \|\mathbf{x}\|^2 > 0$$

It follows that

$$\mathbf{x}^T A\mathbf{x} = (\alpha_1 \mathbf{x}_1 + \cdots + \alpha_n \mathbf{x}_n)^T (\alpha_1 \lambda_1 \mathbf{x}_1 + \cdots + \alpha_n \lambda_n \mathbf{x}_n)$$
$$= \sum_{i=1}^{n} (\alpha_i)^2 \lambda_i$$
$$\geq (\min \lambda_i) \|\mathbf{x}\|^2 > 0$$

and hence A is positive definite.   ◀

If the eigenvalues of A are all negative, then $-A$ must be positive definite and, consequently, A must be negative definite. If A has eigenvalues that differ in sign, then A is indefinite. Indeed, if λ_1 is a positive eigenvalue of A and $\mathbf{x}_1$ is an eigenvector belonging to λ_1, then

$$\mathbf{x}_1^T A\mathbf{x}_1 = \lambda_1 \mathbf{x}_1^T \mathbf{x}_1 = \lambda_1 \|\mathbf{x}_1\|^2 > 0$$

and if λ_2 is a negative eigenvalue with eigenvector $\mathbf{x}_2$, then

$$\mathbf{x}_2^T A\mathbf{x}_2 = \lambda_2 \mathbf{x}_2^T \mathbf{x}_2 = \lambda_2 \|\mathbf{x}_2\|^2 < 0$$

EXAMPLE 4.   The graph of the quadratic form $f(x, y) = 2x^2 - 4xy + 5y^2$ is pictured in Figure 6.6.5. It is not entirely clear from the graph if the stationary point $(0, 0)$ is a global minimum or a saddle point. We can use the matrix A of the

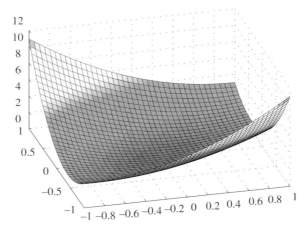

FIGURE 6.6.5

quadratic to decide the issue.

$$A = \begin{bmatrix} 2 & -2 \\ -2 & 5 \end{bmatrix}$$

The eigenvalues of A are $\lambda_1 = 6$ and $\lambda_2 = 1$. Since both eigenvalues are positive, it follows that A is positive definite and hence the stationary point $(0, 0)$ is a global minimum. ◀

Suppose now that we have a function $F(x, y)$ with a stationary point (x_0, y_0). If F has continuous third partials in a neighborhood of (x_0, y_0), it can be expanded in a Taylor series about that point.

$$
\begin{aligned}
F(x_0 + h, y_0 + k) &= F(x_0, y_0) + \big[h F_x(x_0, y_0) + k F_y(x_0, y_0)\big] \\
&\quad + \tfrac{1}{2}\big[h^2 F_{xx}(x_0, y_0) + 2hk F_{xy}(x_0, y_0) + k^2 F_{yy}(x_0, y_0)\big] + R \\
&= F(x_0, y_0) + \tfrac{1}{2}(ah^2 + 2bhk + ck^2) + R
\end{aligned}
$$

where

$$ a = F_{xx}(x_0, y_0), \qquad b = F_{xy}(x_0, y_0), \qquad c = F_{yy}(x_0, y_0) $$

and the remainder R is given by

$$
\begin{aligned}
R &= \tfrac{1}{6}\big[h^3 F_{xxx}(\mathbf{z}) + 3h^2 k F_{xyy}(\mathbf{z}) + 3hk^2 F_{xyy}(\mathbf{z}) + k^3 F_{yyy}(\mathbf{z})\big] \\
\mathbf{z} &= (x_0 + \theta h, y_0 + \theta k), \qquad 0 < \theta < 1
\end{aligned}
$$

If h and k are sufficiently small, $|R|$ will be less than the modulus of $\tfrac{1}{2}\{ah^2 + 2bhk + ck^2\}$, and hence $[F(x_0 + h, y_0 + k) - F(x_0, y_0)]$ will have the same sign as $(ah^2 + 2bhk + ck^2)$. The expression

$$ f(h, k) = ah^2 + 2bhk + ck^2 $$

is a quadratic form in the variables h and k. Thus $F(x, y)$ will have a local minimum (maximum) at (x_0, y_0) if and only if $f(h, k)$ has a minimum (maximum) at $(0, 0)$. Let

$$H = \begin{bmatrix} a & b \\ b & c \end{bmatrix} = \begin{bmatrix} F_{xx}(x_0, y_0) & F_{xy}(x_0, y_0) \\ F_{xy}(x_0, y_0) & F_{yy}(x_0, y_0) \end{bmatrix}$$

and let λ_1 and λ_2 be the eigenvalues of H. If H is nonsingular, then λ_1 and λ_2 are nonzero and we can classify the stationary points as follows:

(i) F has a minimum at (x_0, y_0) if $\lambda_1 > 0$, $\lambda_2 > 0$.
(ii) F has a maximum at (x_0, y_0) if $\lambda_1 < 0$, $\lambda_2 < 0$.
(iii) F has a saddle point at (x_0, y_0) if λ_1 and λ_2 differ in sign.

EXAMPLE 5. The graph of the function

$$F(x, y) = \tfrac{1}{3}x^3 + xy^2 - 4xy + 1$$

is pictured in Figure 6.6.6. Although all the stationary points lie in the region shown, it is difficult to distinguish them from the graph. However, we can solve for the stationary points analytically and then classify each stationary point by examining the corresponding matrix of second partial derivatives.

SOLUTION. The first partials of F are

$$F_x = x^2 + y^2 - 4y$$
$$F_y = 2xy - 4x = 2x(y - 2)$$

Setting $F_y = 0$, we get $x = 0$ or $y = 2$. Setting $F_x = 0$, we see that if $x = 0$, then y must either be 0 or 4, and if $y = 2$, then $x = \pm 2$. Thus $(0, 0)$, $(0, 4)$, $(2, 2)$, $(-2, 2)$ are the stationary points of F. To classify the stationary points, we compute

FIGURE 6.6.6

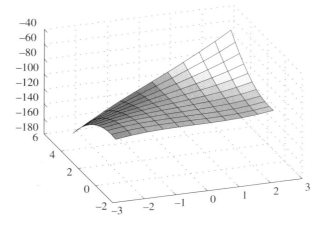

TABLE I			
Stationary Point (x_0, y_0)	λ_1	λ_2	**Description**
$(0, 0)$	4	-4	Saddle point
$(0, 4)$	4	-4	Saddle point
$(2, 2)$	4	4	Local minimum
$(-2, 2)$	-4	-4	Local maximum

the second partials.

$$F_{xx} = 2x, \qquad F_{xy} = 2y - 4, \qquad F_{yy} = 2x$$

For each stationary point (x_0, y_0) we determine the eigenvalues of

$$\begin{bmatrix} 2x_0 & 2y_0 - 4 \\ 2y_0 - 4 & 2x_0 \end{bmatrix}$$

These values are summarized in Table 1. ◀

We can now generalize our method of classifying stationary points to functions of more than two variables. Let $F(\mathbf{x}) = F(x_1, \ldots, x_n)$ be a real-valued function whose third partial derivatives are all continuous. Let $\mathbf{x}_0$ be a stationary point of F and define the matrix $H = H(\mathbf{x}_0)$ by

$$h_{ij} = F_{x_i x_j}(\mathbf{x}_0)$$

$H(\mathbf{x}_0)$ is called the *Hessian* of F at $\mathbf{x}_0$.

The stationary point can be classified as follows:

   (i) $\mathbf{x}_0$ is a local minimum of F if $H(\mathbf{x}_0)$ is positive definite.

   (ii) $\mathbf{x}_0$ is a local maximum of F if $H(\mathbf{x}_0)$ is negative definite.

   (iii) $\mathbf{x}_0$ is a saddle point of F if $H(\mathbf{x}_0)$ is indefinite.

EXAMPLE 6.   Find the local minima of the function

$$F(x, y, z) = x^2 + xz - 3\cos y + z^2$$

SOLUTION.   The first partials of F are

$$F_x = 2x + z$$
$$F_y = 3\sin y$$
$$F_z = x + 2z$$

It follows that (x, y, z) is a stationary point of F if and only if $x = z = 0$ and $y = n\pi$, where n is an integer. Let $\mathbf{x}_0 = (0, 2k\pi, 0)^T$. The Hessian of F at $\mathbf{x}_0$ is given by

$$H(\mathbf{x}_0) = \begin{bmatrix} 2 & 0 & 1 \\ 0 & 3 & 0 \\ 1 & 0 & 2 \end{bmatrix}$$

The eigenvalues of $H(\mathbf{x}_0)$ are 3, 3, and 1. Since the eigenvalues are all positive, it follows that $H(\mathbf{x}_0)$ is positive definite and hence F has a local minimum at $\mathbf{x}_0$. On the other hand, at a stationary point of the form $\mathbf{x}_1 = (0, (2k-1)\pi, 0)^T$ the Hessian will be

$$H(\mathbf{x}_1) = \begin{bmatrix} 2 & 0 & 1 \\ 0 & -3 & 0 \\ 1 & 0 & 2 \end{bmatrix}$$

The eigenvalues of $H(\mathbf{x}_1)$ are -3, 3, and 1. It follows that $H(\mathbf{x}_1)$ is indefinite and hence $\mathbf{x}_1$ is a saddle point of F. ◀

EXERCISES

1. Find the matrix associated with each of the following quadratic forms:

   (a) $3x^2 - 5xy + y^2$
   (b) $2x^2 + 3y^2 + z^2 + xy - 2xz + 3yz$
   (c) $x^2 + 2y^2 + z^2 + xy - 2xz + 3yz$

2. Reorder the eigenvalues in Example 2 so that $\lambda_1 = 4$ and $\lambda_2 = 2$ and rework the example. In what quadrants will the positive x' and y' axes lie? Sketch the graph and compare it to Figure 6.6.3.

3. In each of the following, find a suitable change of coordinates (i.e., a rotation and/or a translation) so that the resulting conic section is in standard form; identify the curve and sketch the graph.

   (a) $x^2 + xy + y^2 - 6 = 0$
   (b) $3x^2 + 8xy + 3y^2 + 28 = 0$
   (c) $-3x^2 + 6xy + 5y^2 - 24 = 0$
   (d) $x^2 + 2xy + y^2 + 3x + y - 1 = 0$

4. Let λ_1 and λ_2 be the eigenvalues of

   $$A = \begin{bmatrix} a & b \\ b & c \end{bmatrix}$$

   What kind of conic section will the equation

   $$ax^2 + 2bxy + cy^2 = 1$$

   represent if $\lambda_1 \lambda_2 < 0$? Explain.

5. Let A be a symmetric 2×2 matrix and let α be a nonzero scalar for which the equation $\mathbf{x}^T A \mathbf{x} = \alpha$ is consistent. Show that the corresponding conic section will be nondegenerate if and only if A is nonsingular.

6. Which of the following matrices are positive definite? Negative definite? Indefinite?

(a) $\begin{bmatrix} 3 & 2 \\ 2 & 2 \end{bmatrix}$
(b) $\begin{bmatrix} 3 & 4 \\ 4 & 1 \end{bmatrix}$
(c) $\begin{bmatrix} 3 & \sqrt{2} \\ \sqrt{2} & 4 \end{bmatrix}$

(d) $\begin{bmatrix} -2 & 0 & 1 \\ 0 & -1 & 0 \\ 1 & 0 & -2 \end{bmatrix}$
(e) $\begin{bmatrix} 1 & 2 & 1 \\ 2 & 1 & 1 \\ 1 & 1 & 2 \end{bmatrix}$
(f) $\begin{bmatrix} 2 & 0 & 0 \\ 0 & 5 & 3 \\ 0 & 3 & 5 \end{bmatrix}$

7. For each of the following functions, determine whether the given stationary point corresponds to a local minimum, local maximum, or saddle point.

(a) $f(x, y) = 3x^2 - xy + y^2$ $(0, 0)$
(b) $f(x, y) = \sin x + y^3 + 3xy + 2x - 3y$ $(0, -1)$
(c) $f(x, y) = \frac{1}{3}x^3 - \frac{1}{3}y^3 + 3xy + 2x - 2y$ $(1, -1)$
(d) $f(x, y) = \dfrac{y}{x^2} + \dfrac{x}{y^2} + xy$ $(1, 1)$
(e) $f(x, y, z) = x^3 + xyz + y^2 - 3x$ $(1, 0, 0)$
(f) $f(x, y, z) = -\frac{1}{4}(x^{-4} + y^{-4} + z^{-4}) + yz - x - 2y - 2z$ $(1, 1, 1)$

8. Show that if A is symmetric positive definite, then $\det(A) > 0$. Give an example of a 2×2 matrix with positive determinant that is not positive definite.

9. Show that if A is a symmetric positive definite matrix, then A is nonsingular and A^{-1} is also positive definite.

10. Let A be a singular $n \times n$ matrix. Show that $A^T A$ is positive semidefinite, but not positive definite.

11. Let A be a symmetric $n \times n$ matrix with eigenvalues $\lambda_1, \dots, \lambda_n$. Show that there exists an orthonormal set of vectors $\{\mathbf{x}_1, \dots, \mathbf{x}_n\}$ such that

$$\mathbf{x}^T A \mathbf{x} = \sum_{i=1}^{n} \lambda_i \left(\mathbf{x}^T \mathbf{x}_i \right)^2$$

for each $\mathbf{x} \in R^n$.

12. Let A be a symmetric positive definite matrix. Show that the diagonal elements of A must all be positive.

13. Let A be a symmetric positive definite $n \times n$ matrix and let S be a nonsingular $n \times n$ matrix. Show that $S^T A S$ is positive definite.

14. Let A be a symmetric positive definite $n \times n$ matrix. Show that A can be factored into a product $Q Q^T$, where Q is an $n \times n$ matrix whose columns are mutually orthogonal. [*Hint:* See Corollary 6.4.5.]

7 POSITIVE DEFINITE MATRICES

In Section 6 we saw that a symmetric matrix is positive definite if and only if its eigenvalues are all positive. These types of matrices occur in a wide variety of applications. They frequently arise in the numerical solution of boundary value problems by finite difference methods or by finite element methods. Because of their importance in applied mathematics, we devote this section to studying their properties.

Recall that a symmetric $n \times n$ matrix A is positive definite if $\mathbf{x}^T A \mathbf{x} > 0$ for all nonzero vectors $\mathbf{x}$ in R^n. In Theorem 6.6.2, symmetric positive definite matrices were characterized by the condition that all their eigenvalues are positive. This characterization can be used to establish the following properties.

Property I. If A is a symmetric positive definite matrix, then A is nonsingular.

Property II. If A is a symmetric positive definite matrix, then $\det(A) > 0$.

If A were singular, $\lambda = 0$ would be an eigenvalue. Since all the eigenvalues of A are positive, A must be nonsingular. The second property also follows from Theorem 6.6.2, since

$$\det(A) = \lambda_1 \cdots \lambda_n > 0$$

Given an $n \times n$ matrix A, let A_r denote the matrix formed by deleting the last $n - r$ rows and columns of A. A_r is called the *leading principal submatrix* of A of order r. We can now state a third property of positive definite matrices.

Property III. If A is a symmetric positive definite matrix, then the leading principal submatrices $A_1, A_2, \ldots, A_n$ of A are all positive definite.

▶ *Proof.* To show that A_r is positive definite, $1 \leq r \leq n$, let $\mathbf{x}_r = (x_1, \ldots, x_r)^T$ be any nonzero vector in R^r and set

$$\mathbf{x} = (x_1, \ldots, x_r, 0, \ldots, 0)^T$$

Since

$$\mathbf{x}_r^T A_r \mathbf{x}_r = \mathbf{x}^T A \mathbf{x} > 0$$

it follows that A_r is positive definite. ◀

An immediate consequence of properties **I**, **II**, and **III** is that, if A_r is a leading principal submatrix of a symmetric positive definite matrix A, then A_r is nonsingular and $\det(A_r) > 0$. This has significance when related to the Gaussian elimination process. In general, if A is an $n \times n$ matrix whose leading principal submatrices are all nonsingular, then A can be reduced to upper triangular form using only row operation III; that is, the diagonal elements will never be 0 in the elimination process, so the reduction can be completed without interchanging rows.

Property IV. If A is a symmetric positive definite matrix, then A can be reduced to upper triangular form using only row operation III and the pivot elements will all be positive.

Let us illustrate property **IV** in the case of a 4×4 symmetric positive definite matrix A. Note first that

$$a_{11} = \det(A_1) > 0$$

$$\begin{pmatrix} a_{11} & x & x & x \\ x & a_{22} & x & x \\ x & x & a_{33} & x \\ x & x & x & a_{44} \end{pmatrix} \xrightarrow{1} \begin{pmatrix} a_{11} & x & x & x \\ 0 & a_{22}^{(1)} & x & x \\ 0 & x & a_{33}^{(1)} & x \\ 0 & x & x & a_{44}^{(1)} \end{pmatrix} \xrightarrow{2} \begin{pmatrix} a_{11} & x & x & x \\ 0 & a_{22}^{(1)} & x & x \\ 0 & 0 & a_{33}^{(2)} & x \\ 0 & 0 & x & a_{44}^{(2)} \end{pmatrix} \xrightarrow{3} \begin{pmatrix} a_{11} & x & x & x \\ 0 & a_{22}^{(1)} & x & x \\ 0 & 0 & a_{33}^{(2)} & x \\ 0 & 0 & 0 & a_{44}^{(3)} \end{pmatrix}$$

$$A \qquad\qquad A^{(1)} \qquad\qquad A^{(2)} \qquad\qquad A^{(3)} = U$$

FIGURE 6.7.1

so a_{11} can be used as a pivot element and row 1 is the first pivot row. Let $a_{22}^{(1)}$ denote the entry in the $(2, 2)$ position after the last three elements of column 1 have been eliminated (see Figure 6.7.1). At this step, the submatrix A_2 has been transformed into a matrix

$$\begin{bmatrix} a_{11} & a_{12} \\ 0 & a_{22}^{(1)} \end{bmatrix}$$

Since this was accomplished using only row operation III, the value of the determinant remains unchanged. Thus

$$\det(A_2) = a_{11}a_{22}^{(1)}$$

and hence

$$a_{22}^{(1)} = \frac{\det(A_2)}{a_{11}} = \frac{\det(A_2)}{\det(A_1)} > 0$$

Since $a_{22}^{(1)} \neq 0$, it can be used as a pivot in the second step of the elimination process. After step 2, the matrix A_3 has been transformed into

$$\begin{bmatrix} a_{11} & a_{12} & a_{13} \\ 0 & a_{22}^{(1)} & a_{23}^{(1)} \\ 0 & 0 & a_{33}^{(2)} \end{bmatrix}$$

Since only row operation III was used,

$$\det(A_3) = a_{11}a_{22}^{(1)}a_{33}^{(2)}$$

and hence

$$a_{33}^{(2)} = \frac{\det(A_3)}{a_{11}a_{22}^{(1)}} = \frac{\det(A_3)}{\det(A_2)} > 0$$

Thus, $a_{33}^{(2)}$ can be used as a pivot in the last step. After step 3 the remaining diagonal entry will be

$$a_{44}^{(3)} = \frac{\det(A_4)}{\det(A_3)} > 0$$

In general, if an $n \times n$ matrix A can be reduced to an upper triangular form U without row interchanges, then A can be factored into a product LU, where L is lower triangular with 1's on the diagonal. The (i, j) entry of L below the diagonal will be the multiple of the ith row that was subtracted from the jth row during the elimination process. We illustrate with a 3×3 example.

EXAMPLE 1. Let

$$A = \begin{pmatrix} 4 & 2 & -2 \\ 2 & 10 & 2 \\ -2 & 2 & 5 \end{pmatrix}$$

The matrix L is determined as follows. At the first step of the elimination process $\frac{1}{2}$ times the first row is subtracted from the second row and $-\frac{1}{2}$ times the first row is subtracted from the third. Corresponding to these operations, we set $l_{21} = \frac{1}{2}$ and $l_{31} = -\frac{1}{2}$. After step 1 we obtain the matrix

$$A^{(1)} = \begin{pmatrix} 4 & 2 & -2 \\ 0 & 9 & 3 \\ 0 & 3 & 4 \end{pmatrix}$$

The final elimination is carried out by subtracting $\frac{1}{3}$ times the second row from the third row. Corresponding to this step, we set $l_{32} = \frac{1}{3}$. After step 2 we end up with the upper triangular matrix,

$$U = A^{(2)} = \begin{pmatrix} 4 & 2 & -2 \\ 0 & 9 & 3 \\ 0 & 0 & 3 \end{pmatrix}$$

The matrix L is given by

$$L = \begin{pmatrix} 1 & 0 & 0 \\ \frac{1}{2} & 1 & 0 \\ -\frac{1}{2} & \frac{1}{3} & 1 \end{pmatrix}$$

and we can verify that the product $LU = A$.

$$\begin{pmatrix} 1 & 0 & 0 \\ \frac{1}{2} & 1 & 0 \\ -\frac{1}{2} & \frac{1}{3} & 1 \end{pmatrix} \begin{pmatrix} 4 & 2 & -2 \\ 0 & 9 & 3 \\ 0 & 0 & 3 \end{pmatrix} = \begin{pmatrix} 4 & 2 & -2 \\ 2 & 10 & 2 \\ -2 & 2 & 5 \end{pmatrix}$$

To see why this factorization works, let us view that process in terms of elementary matrices. Row operation III was applied three times during the process. This is equivalent to multiplying A on the left by three elementary matrices E_1, E_2, E_3.

Thus $E_3 E_2 E_1 A = U$.

$$\begin{bmatrix} 1 & 0 & 0 \\ 0 & 1 & 0 \\ 0 & \frac{1}{3} & 1 \end{bmatrix} \begin{bmatrix} 1 & 0 & 0 \\ 0 & 1 & 0 \\ \frac{1}{2} & 0 & 1 \end{bmatrix} \begin{bmatrix} 1 & 0 & 0 \\ -\frac{1}{2} & 1 & 0 \\ 0 & 0 & 1 \end{bmatrix} \begin{bmatrix} 4 & 2 & -2 \\ 2 & 10 & 2 \\ -2 & 2 & 5 \end{bmatrix}$$

$$= \begin{bmatrix} 4 & 2 & -2 \\ 0 & 9 & 3 \\ 0 & 0 & 3 \end{bmatrix}$$

Since the elementary matrices are nonsingular, it follows that

$$A = (E_1^{-1} E_2^{-1} E_3^{-1}) U$$

When the inverse elementary matrices are multiplied in this order, the result is a lower triangular matrix L with 1's on the diagonal. The entries below the diagonal of L will just be the multiples that were subtracted during the elimination process.

$$E_1^{-1} E_2^{-1} E_3^{-1} = \begin{bmatrix} 1 & 0 & 0 \\ \frac{1}{2} & 1 & 0 \\ 0 & 0 & 1 \end{bmatrix} \begin{bmatrix} 1 & 0 & 0 \\ 0 & 1 & 0 \\ -\frac{1}{2} & 0 & 1 \end{bmatrix} \begin{bmatrix} 1 & 0 & 0 \\ 0 & 1 & 0 \\ 0 & \frac{1}{3} & 1 \end{bmatrix}$$

$$= \begin{bmatrix} 1 & 0 & 0 \\ \frac{1}{2} & 1 & 0 \\ -\frac{1}{2} & \frac{1}{3} & 1 \end{bmatrix} \qquad \blacktriangleleft$$

Given an LU factorization of a matrix A, it is possible to go one step further and factor U into a product DU_1, where D is diagonal and U_1 is upper triangular with 1's on the diagonal.

$$DU_1 = \begin{bmatrix} u_{11} & & & \\ & u_{22} & & \\ & & \ddots & \\ & & & u_{nn} \end{bmatrix} \begin{bmatrix} 1 & \frac{u_{12}}{u_{11}} & \frac{u_{13}}{u_{11}} & \cdots & \frac{u_{1n}}{u_{11}} \\ & 1 & \frac{u_{23}}{u_{22}} & \cdots & \frac{u_{2n}}{u_{22}} \\ & & & & \vdots \\ & & & & 1 \end{bmatrix}$$

It follows, then, that $A = LDU_1$. In general, if A can be factored into a product of the form LDU, where L is lower triangular, D is diagonal, U is upper triangular, and L and U both have 1's along the diagonal, then such a factorization will be unique (see Exercise 7).

If A is a symmetric positive definite matrix, then A can be factored into a product $LU = LDU_1$. The diagonal elements of D are the entries $u_{11}, \ldots, u_{nn}$, which were the pivot elements in the elimination process. By property **IV**, these elements are all positive. Furthermore, since A is symmetric,

$$LDU_1 = A = A^T = (LDU_1)^T = U_1^T D^T L^T$$

It follows from the uniqueness of the LDU factorization that $L^T = U_1$. Thus

$$A = LDL^T$$

This important factorization is often used in numerical computations. There are efficient algorithms that make use of this factorization in solving symmetric positive definite linear systems.

Property V. If A is a symmetric positive definite matrix, then A can be factored into a product LDL^T, where L is lower triangular with 1's along the diagonal and D is a diagonal matrix whose diagonal entries are all positive.

EXAMPLE 2. We saw in Example 1 that

$$A = \begin{bmatrix} 4 & 2 & -2 \\ 2 & 10 & 2 \\ -2 & 2 & 5 \end{bmatrix}$$

$$= \begin{bmatrix} 1 & 0 & 0 \\ \frac{1}{2} & 1 & 0 \\ -\frac{1}{2} & \frac{1}{3} & 1 \end{bmatrix} \begin{bmatrix} 4 & 2 & -2 \\ 0 & 9 & 3 \\ 0 & 0 & 3 \end{bmatrix} = LU$$

Factoring out the diagonal entries of U, we get

$$A = \begin{bmatrix} 1 & 0 & 0 \\ \frac{1}{2} & 1 & 0 \\ -\frac{1}{2} & \frac{1}{3} & 1 \end{bmatrix} \begin{bmatrix} 4 & 0 & 0 \\ 0 & 9 & 0 \\ 0 & 0 & 3 \end{bmatrix} \begin{bmatrix} 1 & \frac{1}{2} & -\frac{1}{2} \\ 0 & 1 & \frac{1}{3} \\ 0 & 0 & 1 \end{bmatrix} = LDL^T \qquad \blacktriangleleft$$

Since the diagonal elements $u_{11}, \ldots, u_{nn}$ are positive, it is possible to go one step further with the factorization. Let

$$D^{1/2} = \begin{bmatrix} \sqrt{u_{11}} & & & \\ & \sqrt{u_{22}} & & \\ & & \ddots & \\ & & & \sqrt{u_{nn}} \end{bmatrix}$$

and set $L_1 = LD^{1/2}$. Then

$$A = LDL^T = LD^{1/2}(D^{1/2})^T L^T = L_1 L_1^T$$

This factorization is known as the *Cholesky decomposition* of A.

Property VI. (Cholesky decomposition)   If A is a symmetric positive definite matrix, then A can be factored into a product LL^T, where L is lower triangular with positive diagonal elements.

EXAMPLE 3.   Let A be the matrix from Examples 1 and 2. If we set

$$L_1 = LD^{1/2} = \begin{bmatrix} 1 & 0 & 0 \\ \frac{1}{2} & 1 & 0 \\ -\frac{1}{2} & \frac{1}{3} & 1 \end{bmatrix} \begin{bmatrix} 2 & 0 & 0 \\ 0 & 3 & 0 \\ 0 & 0 & \sqrt{3} \end{bmatrix} = \begin{bmatrix} 2 & 0 & 0 \\ 1 & 3 & 0 \\ -1 & 1 & \sqrt{3} \end{bmatrix}$$

then

$$L_1 L_1^T = \begin{bmatrix} 2 & 0 & 0 \\ 1 & 3 & 0 \\ -1 & 1 & \sqrt{3} \end{bmatrix} \begin{bmatrix} 2 & 1 & -1 \\ 0 & 3 & 1 \\ 0 & 0 & \sqrt{3} \end{bmatrix}$$

$$= \begin{bmatrix} 4 & 2 & -2 \\ 2 & 10 & 2 \\ -2 & 2 & 5 \end{bmatrix} = A \qquad \blacktriangleleft$$

The matrix $A = LL^T$ could also be written in terms of the upper triangular matrix $R = L^T$. Indeed, if $R = L^T$, then $A = LL^T = R^T R$. On the other hand, it is not difficult to show that any product $B^T B$ will be positive definite provided that B is nonsingular. Putting all these results together, we have the following theorem.

▶ **THEOREM 6.7.1**   *Let A be a symmetric $n \times n$ matrix. The following are equivalent.*

(a) *A is positive definite.*

(b) *The leading principal submatrices $A_1, \ldots, A_n$ all have positive determinants.*

(c) *A can be reduced to upper triangular form using only row operation III, and the pivot elements will all be positive.*

(d) *A has a Cholesky factorization LL^T (where L is lower triangular with positive diagonal entries).*

(e) *A can be factored into a product $B^T B$ for some nonsingular matrix B.*

▶ *Proof.*   We have already shown that (a) implies (b), (b) implies (c), and (c) implies (d). To see that (d) implies (e), assume that $A = LL^T$. If we set $B = L^T$, then B is nonsingular and

$$A = LL^T = B^T B$$

Finally, to show that (e) $\Rightarrow$ (a), assume that $A = B^T B$, where B is nonsingular. Let $\mathbf{x}$ be any nonzero vector in R^n and set $\mathbf{y} = B\mathbf{x}$. Since B is nonsingular, $\mathbf{y} \neq \mathbf{0}$ and it follows that

$$\mathbf{x}^T A\mathbf{x} = \mathbf{x}^T B^T B\mathbf{x} = \mathbf{y}^T \mathbf{y} = \|\mathbf{y}\|^2 > 0$$

Thus A is positive definite. $\blacktriangleleft$

Analogous results to Theorem 6.7.1 are not valid for positive semidefiniteness. For example, consider the matrix

$$A = \begin{pmatrix} 1 & 1 & -3 \\ 1 & 1 & -3 \\ -3 & -3 & 5 \end{pmatrix}$$

The leading principal submatrices all have nonnegative determinants

$$\det(A_1) = 1, \qquad \det(A_2) = 0, \qquad \det(A_3) = 0$$

but A is not positive semidefinite since it has a negative eigenvalue $\lambda = -1$. Indeed, $\mathbf{x} = (1, 1, 1)^T$ is an eigenvector belonging to $\lambda = -1$ and

$$\mathbf{x}^T A \mathbf{x} = -3$$

EXERCISES

1. For each of the following, compute the determinants of all the leading principal submatrices and use them to determine whether the matrix is positive definite.

   (a) $\begin{bmatrix} 2 & -1 \\ -1 & 2 \end{bmatrix}$

   (b) $\begin{bmatrix} 3 & 4 \\ 4 & 2 \end{bmatrix}$

   (c) $\begin{bmatrix} 6 & 4 & -2 \\ 4 & 5 & 3 \\ -2 & 3 & 6 \end{bmatrix}$

   (d) $\begin{bmatrix} 4 & 2 & 1 \\ 2 & 3 & -2 \\ 1 & -2 & 5 \end{bmatrix}$

2. Let A be a 3×3 symmetric positive definite matrix and suppose that $\det(A_1) = 3$, $\det(A_2) = 6$, and $\det(A_3) = 8$. What would the pivot elements be in the reduction of A to triangular form assuming that only row operation III is used in the reduction process?

3. Let

$$A = \begin{bmatrix} 2 & -1 & 0 & 0 \\ -1 & 2 & -1 & 0 \\ 0 & -1 & 2 & -1 \\ 0 & 0 & -1 & 2 \end{bmatrix}$$

   (a) Compute the LU factorization of A.
   (b) Explain why A must be positive definite.

4. For each of the following, factor the given matrix into a product LDL^T, where L is lower triangular with 1's on the diagonal and D is a diagonal matrix.

(a) $\begin{bmatrix} 4 & 2 \\ 2 & 10 \end{bmatrix}$

(b) $\begin{bmatrix} 9 & -3 \\ -3 & 2 \end{bmatrix}$

(c) $\begin{bmatrix} 16 & 8 & 4 \\ 8 & 6 & 0 \\ 4 & 0 & 7 \end{bmatrix}$

(d) $\begin{bmatrix} 9 & 3 & -6 \\ 3 & 4 & 1 \\ -6 & 1 & 9 \end{bmatrix}$

5. Find the Cholesky decomposition LL^T for each of the matrices in Exercise 4.

6. Let A be an $n \times n$ symmetric positive definite matrix. For each $\mathbf{x}, \mathbf{y} \in R^n$, define

$$\langle \mathbf{x}, \mathbf{y} \rangle = \mathbf{x}^T A \mathbf{y}$$

Show that $\langle\ ,\ \rangle$ defines an inner product on R^n.

7. Let A be a nonsingular $n \times n$ matrix, and suppose that $A = L_1 D_1 U_1 = L_2 D_2 U_2$, where L_1 and L_2 are lower triangular, D_1 and D_2 are diagonal, U_1 and U_2 are upper triangular, and L_1, L_2, U_1, U_2 all have 1's along the diagonal. Show that $L_1 = L_2$, $D_1 = D_2$, and $U_1 = U_2$. [*Hint:* L_2^{-1} is lower triangular and U_1^{-1} is upper triangular. Compare both sides of the equation $D_2^{-1} L_2^{-1} L_1 D_1 = U_2 U_1^{-1}$.]

8. Let A be a symmetric positive definite matrix and let Q be an orthogonal diagonalizing matrix. Use the factorization $A = QDQ^T$ to find a nonsingular matrix B such that $B^T B = A$.

9. Let B be an $m \times n$ matrix of rank n. Show that $B^T B$ is positive definite.

10. Let A be a symmetric $n \times n$ matrix. Show that e^A is symmetric and positive definite.

11. Show that if B is a symmetric nonsingular matrix, then B^2 is positive definite.

12. Let

$$A = \begin{bmatrix} 1 & -\frac{1}{2} \\ -\frac{1}{2} & 1 \end{bmatrix} \quad \text{and} \quad B = \begin{bmatrix} 1 & -1 \\ 0 & 1 \end{bmatrix}$$

(a) Show that A is positive definite and that $\mathbf{x}^T A \mathbf{x} = \mathbf{x}^T B \mathbf{x}$ for all $\mathbf{x} \in R^2$.

(b) Show that B is positive definite, but B^2 is not positive definite.

13. Let A be an $n \times n$ symmetric negative definite matrix.

(a) What will the sign of $\det(A)$ be if n is even? If n is odd?

(b) Show that the leading principal submatrices of A are negative definite.

(c) Show that the determinants of the leading principal submatrices of A alternate in sign.

14. Let A be a symmetric positive definite $n \times n$ matrix.

(a) If $k < n$, the leading principal submatrices A_k and A_{k+1} are both positive definite and, consequently, have Cholesky factorizations $L_k L_k^T$ and $L_{k+1} L_{k+1}^T$. If A_{k+1} is expressed in the form

$$A_{k+1} = \begin{pmatrix} A_k & \mathbf{y}_k \\ \mathbf{y}_k^T & \beta_k \end{pmatrix}$$

where $\mathbf{y}_k \in R^k$ and β_k is a scalar, show that L_{k+1} is of the form

$$L_{k+1} = \begin{pmatrix} L_k & \mathbf{0} \\ \mathbf{x}_k^T & \alpha_k \end{pmatrix}$$

and determine $\mathbf{x}_k$ and α_k in terms of L_k, $\mathbf{y}_k$, and β_k.

(b) The leading principal submatrix A_1 has Cholesky decomposition $L_1 L_1^T$, where $L_1 = (\sqrt{a_{11}})$. Explain how part (a) can be used to compute successively the Cholesky factorizations of $A_2, \ldots, A_n$. Devise an algorithm that computes $L_2, L_3, \ldots, L_n$ in a single loop. Since $A = A_n$, the Cholesky decomposition of A will be $L_n L_n^T$. (This algorithm is efficient in that it uses approximately half the amount of arithmetic that would generally be necessary to compute an LU factorization.)

8 NONNEGATIVE MATRICES

In many of the types of linear systems that occur in applications, the entries of the coefficient matrix represent nonnegative quantities. This section deals with the study of such matrices and some of their properties.

▶ **DEFINITION** An $n \times n$ matrix A with real entries is said to be **nonnegative** if $a_{ij} \geq 0$ for each i and j and **positive** if $a_{ij} > 0$ for each i and j.

Similarly, a vector $\mathbf{x} = (x_1, \ldots, x_n)^T$ is said to be **nonnegative** if each $x_i \geq 0$ and **positive** if each $x_i > 0$. ◀

For an example of one of the applications of nonnegative matrices, we consider the Leontief input–output models.

APPLICATION 1: THE OPEN MODEL

Suppose that there are n industries producing n different products. Each industry requires input of the products from the other industries and possibly even of its own product. In the open model it is assumed that there is an additional demand for each of the products from an outside sector. The problem is to determine the output of each of the industries that is necessary to meet the total demand.

We will show that this problem can be represented by a linear system of equations and that the system has a unique nonnegative solution. Let a_{ij} denote the amount of input from the ith industry necessary to produce one unit of output in the jth industry. By a unit of input or output, we mean one dollar's worth of the product. Thus the total cost of producing one dollar's worth of the jth product will be

$$a_{1j} + a_{2j} + \cdots + a_{nj}$$

Since the entries of A are all nonnegative, this sum is equal to $\|\mathbf{a}_j\|_1$. Clearly, production of the jth product will not be profitable unless $\|\mathbf{a}_j\|_1 < 1$. Let d_i denote the demand of the open sector for the ith product. Finally, let x_i represent the amount of output of the ith product necessary to meet the total demand. If the jth industry is to have an output of x_j, it will need an input of $a_{ij}x_j$ units from the ith industry. Thus the total demand for the ith product will be

$$a_{i1}x_1 + a_{i2}x_2 + \cdots + a_{in}x_n + d_i$$

and hence we require that

$$x_i = a_{i1}x_1 + a_{i2}x_2 + \cdots + a_{in}x_n + d_i$$

for $i = 1, \ldots, n$. This leads to the system

$$
\begin{aligned}
(1 - a_{11})x_1 + & (-a_{12})x_2 + \cdots + & (-a_{1n})x_n = d_1 \\
(-a_{21})x_1 + & (1 - a_{22})x_2 + \cdots + & (-a_{2n})x_n = d_2 \\
& \vdots & \\
(-a_{n1})x_1 + & (-a_{n2})x_2 + \cdots + & (1 - a_{nn})x_n = d_n
\end{aligned}
$$

which may be written in the form

$$(1) \qquad\qquad (I - A)\mathbf{x} = \mathbf{d}$$

The entries of A have two important properties:

(i) $a_{ij} \geq 0$ for each i and j.

(ii) $\|\mathbf{a}_j\|_1 = \displaystyle\sum_{i=1}^{n} a_{ij} < 1$ for each j.

The vector $\mathbf{x}$ must not only be a solution to (1); it must also be nonnegative. (It would not make any sense to have a negative output.)

To show that the system has a unique nonnegative solution, we need to make use of a matrix norm that is related to the one-norm for vectors that was introduced in Section 4 of Chapter 5. The matrix norm is also referred to as the *one-norm* and is denoted by $\| \cdot \|_1$. The definition and properties of the one-norm for matrices are studied in Section 4 of Chapter 7. In that section we will show that for any $m \times n$ matrix B

$$(2) \qquad \|B\|_1 = \max_{1 \leq j \leq n} \left(\sum_{i=1}^{m} |b_{ij}| \right) = \max(\|\mathbf{b}_1\|_1, \|\mathbf{b}_2\|_1, \ldots, \|\mathbf{b}_n\|_1)$$

It will also be shown that the one-norm satisfies the following multiplicative properties:

(3)
$$\|BC\|_1 \leq \|B\|_1\|C\|_1 \quad \text{for any matrix} \quad C \in R^{n \times r}$$

$$\|B\mathbf{x}\|_1 \leq \|B\|_1\|\mathbf{x}\|_1 \quad \text{for any} \quad \mathbf{x} \in R^n$$

In particular, if A is an $n \times n$ matrix satisfying conditions (i) and (ii), then it follows from (2) that $\|A\|_1 < 1$. Furthermore, if λ is any eigenvalue of A and $\mathbf{x}$ is an eigenvector belonging to λ, then

$$|\lambda|\|\mathbf{x}\|_1 = \|\lambda\mathbf{x}\|_1 = \|A\mathbf{x}\|_1 \leq \|A\|_1\|\mathbf{x}\|_1$$

and hence

$$|\lambda| \leq \|A\|_1 < 1$$

Thus 1 is not an eigenvalue of A. It follows that $I - A$ is nonsingular and hence the system (1) has a unique solution

$$\mathbf{x} = (I - A)^{-1}\mathbf{d}$$

We would like to show that this solution must be nonnegative. To do this, we will show that $(I-A)^{-1}$ is nonnegative. First note that, as a consequence of multiplicative property (3), we have

$$\|A^m\|_1 \leq \|A\|_1^m$$

Since $\|A\|_1 < 1$, it follows that

$$\|A^m\|_1 \to 0 \quad \text{as} \quad m \to \infty$$

and hence A^m approaches the zero matrix as $m \to \infty$.
  Since

$$(I - A)(I + A + \cdots + A^m) = I - A^{m+1}$$

it follows that

$$I + A + \cdots + A^m = (I - A)^{-1} - (I - A)^{-1}A^{m+1}$$

As $m \to \infty$,

$$(I - A)^{-1} - (I - A)^{-1}A^{m+1} \to (I - A)^{-1}$$

and hence the series $I + A + \cdots + A^m$ converges to $(I - A)^{-1}$ as $m \to \infty$. By condition (i), $I + A + \cdots + A^m$ is nonnegative for each m, and therefore $(I - A)^{-1}$ must be nonnegative. Since $\mathbf{d}$ is nonnegative, it follows that the solution $\mathbf{x}$ must be nonnegative. We see, then, that conditions (i) and (ii) guarantee that the system (1) will have a unique nonnegative solution $\mathbf{x}$.

As you have probably guessed, there is also a closed version of the Leontief input–output model. In the closed version it is assumed that each industry must produce enough output to meet the input needs of only the other industries and itself. The open sector is ignored. Thus, in place of the system (1), we have

$$(I - A)\mathbf{x} = \mathbf{0}$$

and we require that $\mathbf{x}$ be a positive solution. The existence of such an $\mathbf{x}$ in this case is a much deeper result than in the open version and requires some more advanced theorems.

▶ **THEOREM 6.8.1 (Perron)** *If A is a positive $n \times n$ matrix, A has a positive real eigenvalue r with the following properties:*

(i) *r is a simple root of the characteristic equation.*
(ii) *r has a positive eigenvector $\mathbf{x}$.*
(iii) *If λ is any other eigenvalue of A, then $|\lambda| < r$.* ◀

The Perron theorem may be thought of as a special case of a more general theorem due to Frobenius. The Frobenius theorem applies to *irreducible* nonnegative matrices.

▶ **DEFINITION** A nonnegative matrix A is said to be **reducible** if there exists a partition of the index set $\{1, 2, \ldots, n\}$ into nonempty disjoint sets I_1 and I_2 such that $a_{ij} = 0$ whenever $i \in I_1$ and $j \in I_2$. Otherwise, A is said to be **irreducible**. ◀

EXAMPLE 1. Let A be a matrix of the form

$$\begin{bmatrix} \times & \times & 0 & 0 & \times \\ \times & \times & 0 & 0 & \times \\ \times & \times & \times & \times & \times \\ \times & \times & \times & \times & \times \\ \times & \times & 0 & 0 & \times \end{bmatrix}$$

Let $I_1 = \{1, 2, 5\}$ and $I_2 = \{3, 4\}$. Then $I_1 \cup I_2 = \{1, 2, 3, 4, 5\}$ and $a_{ij} = 0$ whenever $i \in I_1$ and $j \in I_2$. Therefore, A is reducible. If P is the permutation matrix formed by interchanging the third and fifth rows of the identity matrix I, then

$$PA = \begin{bmatrix} \times & \times & 0 & 0 & \times \\ \times & \times & 0 & 0 & \times \\ \times & \times & 0 & 0 & \times \\ \times & \times & \times & \times & \times \\ \times & \times & \times & \times & \times \end{bmatrix}$$

and

$$PAP^T = \begin{bmatrix} \times & \times & \times & 0 & 0 \\ \times & \times & \times & 0 & 0 \\ \times & \times & \times & 0 & 0 \\ \hline \times & \times & \times & \times & \times \\ \times & \times & \times & \times & \times \end{bmatrix}$$

In general, it can be shown that an $n \times n$ matrix A is reducible if and only if there exists a permutation matrix P such that PAP^T is a matrix of the form

$$\left[\begin{array}{c|c} B & O \\ \hline X & C \end{array}\right]$$

where B and C are square matrices. ◄

▶ **THEOREM 6.8.2 (Frobenius)** *If A is an irreducible nonnegative matrix, then A has a positive real eigenvalue r with the following properties:*

(i) *r has a positive eigenvector $\mathbf{x}$.*

(ii) *If λ is any other eigenvalue of A, then $|\lambda| \leq r$. The eigenvalues of modulus r are all simple roots of the characteristic equation. Indeed, if there are m eigenvalues of modulus r, they must be of the form*

$$\lambda_k = re^{2k\pi i/m} \qquad k = 0, 1, \ldots, m-1$$ ◄

The proof of this theorem is beyond the scope of the text. We refer the reader to Gantmacher [3, Vol. 2]. Perron's theorem follows as a special case of the Frobenius theorem.

APPLICATION 2: THE CLOSED MODEL

In the closed Leontief input–output model, we assume that there is no demand from the open sector, and we wish to find outputs to satisfy the demands of all n industries. Thus, defining the x_i's and the a_{ij}'s as in the open model, we have

$$x_i = a_{i1}x_1 + a_{i2}x_2 + \cdots + a_{in}x_n$$

for $i = 1, \ldots, n$. The resulting system may be written in the form

(4) $$(A - I)\mathbf{x} = \mathbf{0}$$

As before, we have the condition

(i) $$a_{ij} \geq 0$$

Since there is no open sector, the amount of output from the jth industry should be the same as the total input for that industry. Thus

$$x_j = \sum_{i=1}^{n} a_{ij} x_j$$

and hence we have as our second condition

(ii)
$$\sum_{i=1}^{n} a_{ij} = 1 \qquad j = 1, \ldots, n$$

Condition (ii) implies that $A - I$ is singular since the sum of its row vectors is **0**. Therefore, 1 is an eigenvalue of A, and since $\|A\|_1 = 1$, it follows that all the eigenvalues of A have moduli less than or equal to 1. Let us assume that enough of the coefficients of A are nonzero so that A is irreducible. Then, by Theorem 6.8.2, $\lambda = 1$ has a positive eigenvector **x**. Thus any positive multiple of **x** will be a positive solution to (4).

EXERCISES

1. Find the eigenvalues of each of the following matrices and verify that conditions (i), (ii), and (iii) of Theorem 6.8.1 hold.

(a) $\begin{bmatrix} 2 & 3 \\ 2 & 1 \end{bmatrix}$ (b) $\begin{bmatrix} 4 & 2 \\ 2 & 7 \end{bmatrix}$ (c) $\begin{bmatrix} 1 & 2 & 4 \\ 2 & 4 & 1 \\ 1 & 2 & 4 \end{bmatrix}$

2. Find the eigenvalues of each of the following matrices and verify that conditions (i) and (ii) of Theorem 6.8.2 hold.

(a) $\begin{bmatrix} 2 & 3 \\ 1 & 0 \end{bmatrix}$ (b) $\begin{bmatrix} 0 & 2 \\ 2 & 0 \end{bmatrix}$ (c) $\begin{bmatrix} 0 & 0 & 8 \\ 1 & 0 & 0 \\ 0 & 1 & 0 \end{bmatrix}$

3. Find the output vector **x** in the open version of the Leontief input–output model if

$$A = \begin{bmatrix} 0.2 & 0.4 & 0.4 \\ 0.4 & 0.2 & 0.2 \\ 0.0 & 0.2 & 0.2 \end{bmatrix} \qquad \text{and} \qquad \mathbf{d} = \begin{bmatrix} 16,000 \\ 8,000 \\ 24,000 \end{bmatrix}$$

4. Consider the closed version of the Leontief input–output model with input matrix

$$A = \begin{bmatrix} 0.5 & 0.4 & 0.1 \\ 0.5 & 0.0 & 0.5 \\ 0.0 & 0.6 & 0.4 \end{bmatrix}$$

If $\mathbf{x} = (x_1, x_2, x_3)^T$ is any output vector for this model, how are the coordinates x_1, x_2, and x_3 related?

5. Prove: If $A^m = O$ for some positive integer m, then $I - A$ is nonsingular.

6. Let

$$A = \begin{bmatrix} 0 & 1 & 1 \\ 0 & -1 & 1 \\ 0 & -1 & 1 \end{bmatrix}$$

(a) Compute $(I - A)^{-1}$.

(b) Compute A^2 and A^3. Verify that $(I - A)^{-1} = I + A + A^2$.

7. Which of the following matrices are reducible? For each reducible matrix, find a permutation matrix P such that PAP^T is of the form

$$\begin{bmatrix} B & O \\ \hline X & C \end{bmatrix}$$

where B and C are square matrices.

(a) $\begin{bmatrix} 1 & 1 & 1 & 0 \\ 1 & 1 & 1 & 0 \\ 1 & 1 & 1 & 1 \\ 1 & 1 & 1 & 1 \end{bmatrix}$

(b) $\begin{bmatrix} 1 & 0 & 1 & 1 \\ 1 & 1 & 1 & 1 \\ 1 & 0 & 1 & 1 \\ 1 & 0 & 1 & 1 \end{bmatrix}$

(c) $\begin{bmatrix} 1 & 0 & 1 & 0 & 0 \\ 0 & 1 & 1 & 1 & 1 \\ 1 & 0 & 1 & 0 & 0 \\ 1 & 1 & 0 & 1 & 1 \\ 1 & 1 & 1 & 1 & 1 \end{bmatrix}$

(d) $\begin{bmatrix} 1 & 1 & 1 & 1 & 1 \\ 1 & 1 & 0 & 0 & 1 \\ 1 & 1 & 1 & 1 & 1 \\ 1 & 1 & 0 & 0 & 1 \\ 1 & 1 & 0 & 0 & 1 \end{bmatrix}$

8. Let A be a nonnegative irreducible 3×3 matrix whose eigenvalues satisfy $\lambda_1 = 2 = |\lambda_2| = |\lambda_3|$. Determine λ_2 and λ_3.

9. Let

$$A = \begin{bmatrix} B & O \\ \hline O & C \end{bmatrix}$$

where B and C are square matrices.

(a) If λ is an eigenvalue of B with eigenvector $\mathbf{x} = (x_1, \ldots, x_k)^T$, show that λ is also an eigenvalue of A with eigenvector $\tilde{\mathbf{x}} = (x_1, \ldots, x_k, 0, \ldots, 0)^T$.

(b) If B and C are positive matrices, show that A has a positive real eigenvalue r with the property that $|\lambda| < r$ for any eigenvalue $\lambda \neq r$. Show also that the multiplicity of r is at most 2 and that r has a nonnegative eigenvector.

(c) If $B = C$, show that the eigenvalue r in part (b) has multiplicity 2 and possesses a positive eigenvector.

10. Prove that a 2×2 matrix A is reducible if and only if $a_{12}a_{21} = 0$.

11. Prove the Frobenius theorem in the case where A is a 2×2 matrix.

12. We can show that, for an $n \times n$ stochastic matrix $\lambda_1 = 1$ is the dominant eigenvalue. (See Exercise 12 of Chapter 7 Section 4.) The remaining eigenvalues must satisfy

$$|\lambda_j| \leq 1 \quad j = 2, \dots, n$$

Show that if A is an $n \times n$ stochastic matrix with the property that A^k is a positive matrix some positive integer k then

$$|\lambda_j| < 1 \quad j = 2, \dots, n$$

13. Let A be an $n \times n$ positive stochastic matrix with dominant eigenvalue $\lambda_1 = 1$ and linearly independent eigenvectors $\mathbf{x}_1, \mathbf{x}_2, \dots, \mathbf{x}_n$, and let $\mathbf{y}_0$ be an initial probability vector for a Markov chain

$$\mathbf{y}_0, \quad \mathbf{y}_1 = A\mathbf{y}_0, \quad \mathbf{y}_2 = A\mathbf{y}_1, \quad \dots$$

(a) Show that $\lambda_1 = 1$ has a positive eigenvector $\mathbf{x}_1$.
(b) Show that $\|\mathbf{y}_j\|_1 = 1, \quad j = 0, 1, \dots$.
(c) Show that if $\mathbf{y}_0$ is expressed as a linear combination of the eigenvectors of A

$$\mathbf{y}_0 = c_1\mathbf{x}_1 + c_2\mathbf{x}_2 + \dots + c_n\mathbf{x}_n$$

then the component c_1 in the direction of the positive eigenvector $\mathbf{x}_1$ must be nonzero.
(d) Show that the state vectors $\mathbf{y}_j$ of the Markov chain converge to a steady-state vector.
(e) Show that

$$c_1 = \frac{1}{\|\mathbf{x}_1\|_1}$$

and hence the steady-state vector is independent of the initial probability vector $\mathbf{y}_0$.

14. Would the results of parts (c) and (d) in Exercise 13 be valid if the stochastic matrix A was not a positive matrix? Answer this same question in the case when A is a nonnegative stochastic matrix and, for some positive integer k, the matrix A^k is positive. Explain your answers.

MATLAB EXERCISES

Visualizing Eigenvalues

MATLAB has a utility for visualizing the actions of linear operators that map the plane into itself. The utility is invoked using the command **eigshow**. This command opens a figure window that shows a unit vector **x** and also A**x**, the image of **x** under A. The matrix A can be specified as an input argument of the **eigshow** command or selected from the menu at the top of the figure window. To see the effect of the operator A on other unit vectors, point your mouse to the tip of the vector **x** and use it to drag the vector **x** around the unit circle in a counterclockwise direction. As **x** moves you will see how its image A**x** changes. In this exercise we will use the **eigshow** utility to investigate the eigenvalues and eigenvectors of the matrices in the **eigshow** menu.

1. The top matrix on the menu is the diagonal matrix

$$A = \begin{bmatrix} \frac{5}{4} & 0 \\ 0 & \frac{3}{4} \end{bmatrix}$$

   Initially, when you select this matrix, the vectors **x** and A**x** should both be aligned along positive the x axis. What information about an eigenvalue–eigenvector pair is apparent from the initial figure positions? Explain. Rotate **x** counterclockwise until **x** and A**x** are parallel, that is, until they both lie along the same line through the origin. What can you conclude about the second eigenvalue–eigenvector pair? Repeat this experiment with the second matrix. How can you determine the eigenvalues and eigenvalues of a 2×2 diagonal matrix by inspection without doing any computations? Does this also work for 3×3 diagonal matrices? Explain.

2. The third matrix on the menu is just the identity matrix I. How do **x** and I**x** compare geometrically as you rotate **x** around the unit circle? What can you conclude about the eigenvalues and eigenvectors in this case?

3. The fourth matrix has 0's on the diagonal and 1's in the off-diagonal positions. Rotate the vector **x** around the unit circle and note when **x** and A**x** are parallel. Based on these observations, determine the eigenvalues and the corresponding unit eigenvectors. Check your answers by multiplying the matrix times the eigenvector to verify that A**x** $= \lambda$**x**.

4. The next matrix in the **eigshow** menu looks the same as the previous except the $(2, 1)$ entry has been changed to -1. Rotate the vector **x** completing around the unit circle. Are **x** and A**x** ever parallel? Does A have any real eigenvectors? What can you conclude about the nature of the eigenvalues and eigenvectors of this matrix?

5. Investigate the next three matrices on the menu (the sixth, seventh and eighth). In each case, try to estimate geometrically the eigenvalues and eigenvectors and make your guesses for the eigenvalues consistent with the trace of the matrix. Use MATLAB to compute the eigenvalues and eigenvectors of the sixth matrix by setting

$$[X, D] = \textbf{eig}([0.25, \ 0.75\,; 1, \ 0.50\,])$$

   The column vectors of X are the eigenvectors of the matrix and the diagonal entries of D are the eigenvalues. Check the eigenvalues and eigenvectors of the other two matrices in the same way.

6. Investigate the ninth matrix on the menu. What can you conclude about the nature of its eigenvalues and eigenvectors? Check your conclusions by computing the eigenvalues and eigenvectors using the **eig** command.

7. Investigate the next three matrices on the menu. You should note that for the last two of these matrices the two eigenvalues are equal. For each matrix, how are the eigenvectors related? Use MATLAB to compute the eigenvalues and eigenvectors of these matrices.

8. The last item on the **eigshow** menu will generate a random 2×2 matrix each time that it is invoked. Try using the random matrix 10 times and in each case determine whether the eigenvalues are real. What percentage of the 10 random matrices had real eigenvalues? What is the likelihood that two real eigenvalues of a random matrix will turn out to be exactly equal? Explain.

Critical Loads for a Beam

9. Consider the application relating to critical loads for a beam from Section 1. For simplicity, we will assume that the beam has length 1 and that its flexural rigidity is also 1. Following the method described in the application, if the interval $[0, 1]$ is partitioned into n subintervals; then the problem can be translated into a matrix equation $A\mathbf{y} = \lambda \mathbf{y}$. The critical load for the beam can be approximated by setting $P = sn^2$, where s is the smallest eigenvalue of A. For $n = 100, 200, 400$, form the coefficient matrix by setting

$$A = \mathbf{eye}(n) - \mathbf{diag}(\mathbf{ones}(n - 1, 1), 1) - \mathbf{diag}(\mathbf{ones}(n - 1, 1), -1)$$

In each case, determine the smallest eigenvalue of A by setting

$$s = \mathbf{min}(\mathbf{eig}(A))$$

and then compute the corresponding approximation to the critical load.

Diagonalizable and Defective Matrices

10. Construct a symmetric matrix A by setting

$$A = \mathbf{round}(5 * \mathbf{rand}(6)); \qquad A = A + A'$$

Compute the eigenvalues of A by setting $\mathbf{e} = \mathbf{eig}(A)$.

(a) The trace of A can be computed using the MATLAB command **trace**(A), and the sum of the eigenvalues of A can be computed using the command **sum**$(\mathbf{e})$. Compute both of these quantities and compare the results. Use the command **prod**$(\mathbf{e})$ to compute the product of the eigenvalues of A and compare the result to **det**(A).

(b) Compute the eigenvectors of A by setting $[X, D] = \mathbf{eig}(A)$. Use MATLAB to compute $X^{-1}AX$ and compare the result to D. Compute also A^{-1} and $XD^{-1}X^{-1}$ and compare the results.

11. Set

$$A = \mathbf{ones}(10) + \mathbf{eye}(10)$$

(a) What is the rank of $A - I$? Why must $\lambda = 1$ be an eigenvalue of multiplicity 9? Compute the trace of A using the MATLAB function **trace**. The remaining eigenvalue λ_{10} must equal 11. Why? Explain. Compute the eigenvalues of A by setting $\mathbf{e} = \mathbf{eig}(A)$. Examine the eigenvalues using **format long**. How many digits of accuracy are there in the computed eigenvalues?

(b) The MATLAB routine for computing eigenvalues is based on the QR algorithm described in Section 6 of Chapter 7. We can also compute the eigenvalues of A by computing the roots of its characteristic polynomial. To determine the coefficients of the characteristic polynomial of A, set $\mathbf{p} = \mathbf{poly}(A)$. The characteristic polynomial of A should have integer coefficients. Why? Explain. If we set $\mathbf{p} = \mathbf{round}(\mathbf{p})$, we should end up with the exact coefficients of the characteristic polynomial of A. Compute the roots of $\mathbf{p}$ by setting

$$\mathbf{r} = \mathbf{roots}(\mathbf{p})$$

and display the results using **format long**. How many digits of accuracy are there in the computed results? Which method of computing eigenvalues is more accurate, using the **eig** function or computing the roots of the characteristic polynomial?

12. Consider the matrices

$$A = \begin{bmatrix} 5 & -3 \\ 3 & -5 \end{bmatrix} \quad \text{and} \quad B = \begin{bmatrix} 5 & -3 \\ 3 & 5 \end{bmatrix}$$

Note that the two matrices are the same except for their (2, 2) entries.

(a) Use MATLAB to compute the eigenvalues of A and B. Do they have the same type of eigenvalues? The eigenvalues of the matrices are the roots of their characteristic polynomials. Use the following MATLAB commands to form the polynomials and plot their graphs on the same axis system

```
p = poly(A);          q = poly(B);
x = -8 : 0.1 : 8;     z = zeros(size(x));
y = polyval(p, x);    w = polyval(q, x);
plot(x, y, x, w, x, z)   hold on
```

The **hold on** command is used so that subsequent plots in part (b) will be added to the current figure. How can you use the graph to estimate the eigenvalues of A? What does the graph tell you about the eigenvalues of B? Explain.

(b) To see how the eigenvalues change as the (2, 2) entry changes, let us construct a matrix C with a variable (2, 2) entry. Set

$$t = \mathbf{sym}('t') \qquad C = [5, -3; 3, t - 5]$$

As t goes from 0 to 10, the (2, 2) entries of these matrices go from -5 to 5. Use the following MATLAB commands to plot the graphs of the characteristic polynomials for the intermediate matrices corresponding to $t = 1, 2, \ldots, 9$.

```
p = poly(C)
for j = 1 : 9
    s = subs(p, t, j);
    ezplot(s, [−10, 10])
    axis([−10, 10, −20, 220])
    pause(2)
end
```

Which of these intermediate matrices have real eigenvalues and which have complex eigenvalues? The characteristic polynomial of the symbolic matrix C is a quadratic polynomial whose coefficients are functions of t. To find exactly where the eigenvalues change from real to complex, write the discriminant of the quadratic as a function of t and then find its roots. One root should be in the interval $(0, 10)$. Plug that value of t back into the matrix C and determine the eigenvalues of the matrix. Explain how these results correspond to your graph. Solve for the eigenvectors by hand. Is the matrix diagonalizable?

13. Set

$$B = \texttt{toeplitz}(0 : -1 : -3, 0 : 3)$$

The matrix B is not symmetric and hence it is not guaranteed to be diagonalizable. Use MATLAB to verify that the rank of B equals 2. Explain why 0 must be an eigenvalue of B and the corresponding eigenspace must have dimension 2. Set $[X, D] = \texttt{eig}(B)$. Compute $X^{-1}BX$ and compare the result to D. Compute also XD^5X^{-1} and compare the result to B^5.

14. Set

$$C = \texttt{triu}(\texttt{ones}(4), 1) + \texttt{diag}([1, -1], -2)$$

$$[X, D] = \texttt{eig}(C)$$

Compute $X^{-1}CX$ and compare the result to D. Is C diagonalizable? Compute the rank of X and the condition number of X. If the condition number of X is large, the computed values for the eigenvalues may not be accurate. Compute the reduced row echelon form of C. Explain why 0 must be an eigenvalue of C and the corresponding eigenspace must have dimension 1. Use MATLAB to compute C^4. It should equal the zero matrix. Given that $C^4 = O$, what can you conclude about the actual values of the other three eigenvalues of C? Explain. Is C defective? Explain.

15. Construct a defective matrix by setting

$$A = \texttt{ones}(6); \qquad A = A - \texttt{tril}(A) - \texttt{triu}(A, 2)$$

It is easily seen that $\lambda = 0$ is the only eigenvalue of A and that its eigenspace is spanned by $\mathbf{e}_1$. Verify this by using MATLAB to compute the eigenvalues and eigenvectors of A. Examine the eigenvectors using **format long**. Are the computed eigenvectors multiples of $\mathbf{e}_1$? Now perform a similarity transformation on A. Set

$$Q = \texttt{orth}(\texttt{rand}(6)); \qquad \text{and} \qquad B = Q' * A * Q$$

If the computations had been done in exact arithmetic, the matrix B would be similar to A and hence defective. Use MATLAB to compute the eigenvalues of B and a matrix X consisting of the eigenvectors of B. Determine the rank of X. Is the computed matrix B defective? Because of roundoff error, a more reasonable question to ask is whether the computed matrix B is close to being defective (i.e., are the column vectors of X close to being linearly dependent)? To answer this, use MATLAB to compute **rcond**(X), the reciprocal of the condition number of X. A value of **rcond** close to zero indicates that X is nearly rank deficient.

16. Generate a matrix A by setting

$$B = [-1, \ -1; \ 1, \ 1], \qquad A = [\mathbf{zeros}(2), \ \mathbf{eye}(2); \ \mathbf{eye}(2), \ B]$$

(a) The matrix A should have eigenvalues $\lambda_1 = 1$ and $\lambda_2 = -1$. Use MATLAB to verify this by computing the reduced row echelon forms of $A - I$ and $A + I$. What are the dimensions of the eigenspaces of λ_1 and λ_2?

(b) It is easily seen that **trace**$(A) = 0$ and **det**$(A) = 1$. Verify these results using MATLAB. Use the values of the trace and determinant to prove that 1 and -1 are actually both double eigenvalues. Is A defective? Explain.

(c) Set $\mathbf{e} = \mathbf{eig}(A)$ and examine the eigenvalues using **format long**. How many digits of accuracy are there in the computed eigenvalues? Set $[X, \ D] = \mathbf{eig}(A)$ and compute the condition number of X. The log of the condition number gives an estimate of how many digits of accuracy are lost in the computation of the eigenvalues of A.

(d) Compute the rank of X. Are the computed eigenvectors linearly independent? Use MATLAB to compute $X^{-1}AX$. Does the computed matrix X diagonalize A?

Application: Sex-Linked Genes

17. Suppose that 10,000 men and 10,000 women settle on an island in the Pacific that has been opened to development. Suppose also that a medical study of the settlers finds that 200 of the men are color-blind and only 9 of the women are color-blind. Let $x(1)$ denote the proportion of genes for color blindness in the male population and let $x(2)$ be the proportion for the female population. Assume that $x(1)$ is equal to the proportion of color-blind males and that $x(2)^2$ is equal to the proportion of color-blind females. Determine $x(1)$ and $x(2)$ and enter them in MATLAB as a column vector $\mathbf{x}$. Enter also the matrix A from Application 2 of Section 3. Set MATLAB to **format long** and use the matrix A to compute the proportions of genes for color blindness for each sex in generations 5, 10, 20, and 40. What are the limiting percentages of genes for color blindness for this population? In the long run, what percent of males and what percent of females will be color-blind?

Similarity

18. Set

$$S = \mathbf{round}(10 * \mathbf{rand}(5)); \quad S = \mathbf{triu}(S, 1) + \mathbf{eye}(5)$$
$$S = S' * S \qquad\qquad\qquad T = \mathbf{inv}(S)$$

(a) The exact inverse of S should have integer entries. Why? Explain. Check the entries of T using **format long**. Round the entries of T to the nearest integer by setting $T = \textbf{round}(T)$. Compute $T * S$ and compare to **eye**(5).

(b) Set

$$A = \textbf{triu}(\textbf{ones}(5), 1) + \textbf{diag}(1:5), \qquad B = S * A * T$$

The matrices A and B both have the eigenvalues 1, 2, 3, 4, 5. Use MATLAB to compute the eigenvalues of B. How many digits of accuracy are there in the computed eigenvalues? Use MATLAB to compute and compare each of the following:

(i) $\det(A)$ and $\det(B)$
(ii) **trace**(A) and **trace**(B)
(iii) SA^2T and B^2
(iv) $SA^{-1}T$ and B^{-1}

Hermitian Matrices

19. Construct a complex Hermitian matrix by setting

$$j = \textbf{sqrt}(-1); \quad A = \textbf{rand}(5) + j * \textbf{rand}(5); \quad A = (A + A')/2$$

(a) The eigenvalues of A should be real. Why? Compute the eigenvalues and examine your results using **format long**. Are the computed eigenvalues real? Compute also the eigenvectors by setting

$$[X, D] = \textbf{eig}(A)$$

What type of matrix would you expect X to be? Use the MATLAB command $X' * X$ to compute $X^H X$. Do the results agree with your expectations?

(b) Set

$$E = D + j * \textbf{eye}(5) \qquad \text{and} \qquad B = X * E / X$$

What type of matrix would you expect B to be? Use MATLAB to compute $B^H B$ and BB^H. How do these two matrices compare?

Visualizing the Singular Value Decomposition

In some of the earlier exercises we used MATLAB's **eigshow** command to look at geometric interpretations of the eigenvalues and eigenvectors of 2×2 matrices. The eigshow facility also has an **svdshow** mode that we can use to visualize the singular values and singular vectors of a nonsingular 2×2 matrix. Before using the **svdshow** facility, we first establish some basic relations between the right and left singular vectors.

20. Let A be a nonsingular 2×2 matrix with singular value decomposition $A = USV^T$ and singular values $s_1 = s_{11}, s_2 = s_{22}$. Explain why each of the following are true.

(a) $AV = US$
(b) $A\mathbf{v}_1 = s_1\mathbf{u}_1$ and $A\mathbf{v}_2 = s_2\mathbf{u}_2$.

   (c) $\mathbf{v}_1$ and $\mathbf{v}_2$ are orthogonal unit vectors and the images $A\mathbf{v}_1$ and $A\mathbf{v}_2$ are also orthogonal.

   (d) $\|A\mathbf{v}_1\| = s_1$ and $\|A\mathbf{v}_2\| = s_2$.

Set

$$A = [1, 1; 0.5, -0.5]$$

and use MATLAB to verify each of the above statements.

21. Use the command **eigshow**(A) to apply the eigshow utility to the matrix A. Click on the **eig/(svd)** button to switch into the **svdshow** mode. The display in the figure window should show a pair of orthogonal vectors $\mathbf{x}$, $\mathbf{y}$ and their images $A\mathbf{x}$ and $A\mathbf{y}$. Initially, the images of $\mathbf{x}$ and $\mathbf{y}$ should not be orthogonal. Use the mouse to rotate the $\mathbf{x}$ and $\mathbf{y}$ vectors counterclockwise until their images $A\mathbf{x}$ and $A\mathbf{y}$ become orthogonal. When the images are orthogonal, $\mathbf{x}$ and $\mathbf{y}$ are right singular vectors of A. When $\mathbf{x}$ and $\mathbf{y}$ are right singular vectors, how are the singular values and left singular vectors related to the images $A\mathbf{x}$ and $A\mathbf{y}$? Explain. Note that when you rotate a full $360°$ the image of the unit circle traces out as an ellipse. How do the singular values and singular vectors relate to the axes of the ellipse?

Optimization

22. Use the following MATLAB commands to construct a symbolic function

syms x y
$$f = (y + 1)\^3 + x * y\^2 + y\^2 - 4 * x * y - 4 * y + 1$$

Compute the first partials of f and the Hessian of f by setting

$$fx = \textbf{diff}(f, x), \; fy = \textbf{diff}(f, y)$$
$$H = [\textbf{diff}(fx, x), \textbf{diff}(fx, y); \textbf{diff}(fy, x), \textbf{diff}(fy, y)]$$

We can evaluate the Hessian for a given (x, y) pair using the **subs** command. For example, to evaluate the Hessian when $x = 3$ and $y = 5$, set

H1 = **subs**(H,[x,y],[3,5])

Use the MATLAB command **solve**(fx, fy) to determine vectors $\mathbf{x}$ and $\mathbf{y}$ containing the x and y coordinates of the stationary points. Evaluate the Hessian at each stationary point and then determine whether the stationary point is a local maximum, local minimum, or saddle point.

Positive Definite Matrices

23. Set

$$C = \textbf{ones}(6) + 7 * \textbf{eye}(6) \qquad \text{and} \qquad [X, D] = \textbf{eig}(C)$$

   (a) Even though $\lambda = 7$ is an eigenvalue of multiplicity 5, the matrix C cannot be defective. Why? Explain. Check that C is not defective by computing the rank of X. Compute also $X^T X$. What type of matrix is X? Explain. Compute also the rank of $C - 7I$. What can you conclude about the dimension of the eigenspace corresponding to $\lambda = 7$? Explain.

(b) The matrix C should be symmetric positive definite. Why? Explain. Thus C should have a Cholesky factorization LL^T. Use the MATLAB command **flops(0)** to set the flop count to 0. The MATLAB command $R = $ **chol**(C) will generate an upper triangular matrix R that is equal to L^T. Compute R in this manner and check the value of **flops** to see how many floating-point operations were necessary. Set $L = R'$ and use MATLAB to verify that

$$C = LL^T = R^TR$$

(c) Reset the flop count to 0 and compute the LU factorization of C by setting

$$[\,L \quad U\,] = \mathbf{lu}(C)$$

Check the value of **flops** to see how many floating-point operations were required. Which computation is more efficient, the Cholesky factorization or the LU factorization? Set

$$D = \mathbf{diag}(\mathbf{sqrt}(\mathbf{diag}(U))) \qquad \text{and} \qquad W = (L * D)'$$

How do R and W compare?

24. For various values of k, form an $k \times k$ matrix A by setting

$$A = 2 * \mathbf{eye}(k) - \mathbf{diag}(\mathbf{ones}(k - 1, 1), 1) - \mathbf{diag}(\mathbf{ones}(k - 1, 1), -1)$$

In each case, compute the LU factorization of A and the determinant of A. If A is an $n \times n$ matrix of this form, what will its LU factorization be? What will its determinant be? Why must the matrix be positive definite?

25. For any positive integer n, the MATLAB command $P = \mathbf{pascal}(n)$ will generate an $n \times n$ matrix P whose entries are given by

$$p_{ij} = \begin{cases} 1 & \text{if } i = 1 \quad \text{or} \quad j = 1 \\ p_{i-1,j} + p_{i,j-1} & \text{if } i > 1 \quad \text{and} \quad j > 1 \end{cases}$$

The name **pascal** refers to Pascal's triangle, a triangular array of numbers that is used to generate binomial coefficients. The entries of the matrix P form a section of Pascal's triangle.

(a) Set

$$P = \mathbf{pascal}(6)$$

and compute the value of its determinant. Now subtract 1 from the $(6, 6)$ entry of P by setting

$$P(6, 6) = P(6, 6) - 1$$

and compute the determinant of the new matrix P. What is the overall effect of subtracting 1 from the $(6, 6)$ entry of the 6×6 Pascal matrix?

(b) In part (a) we saw that the determinant of the 6×6 Pascal matrix is 1, but if we subtract 1 from the $(6, 6)$ entry, the matrix becomes singular. Will this happen in general for $n \times n$ Pascal matrices? To answer this question, consider the cases $n = 4, 8, 12$. In each case, set $P = \texttt{pascal}(n)$ and compute its determinant. Next subtract 1 from the (n, n) entry and compute the determinant of the resulting matrix. Does the property that we discovered in part (a) appear to hold for Pascal matrices in general?

(c) Set

$$P = \texttt{pascal}(8)$$

and examine its leading principal submatrices. Assuming that all Pascal matrices have determinants equal to 1, why must P be positive definite? Compute the upper triangular Cholesky factor R of P. How can the nonzero entries of R be generated as a Pascal triangle? In general, how is the determinant of a positive definite matrix related to the determinant of one of its Cholesky factors? Why must $\det(P) = 1$?

(d) Set

$$R(8, 8) = 0 \qquad \text{and} \qquad Q = R' * R$$

The matrix Q should be singular. Why? Explain. Why must the matrices P and Q be the same except for the $(8, 8)$ entry? Why must $q_{88} = p_{88} - 1$? Explain. Verify the relation between P and Q by computing the difference $P - Q$.

CHAPTER TEST

In each of the following answer *true* if the statement is always true and *false* otherwise. In the case of a true statement, explain or prove your answer. In the case of a false statement, give an example to show that the statement is not always true.

1. If A is an $n \times n$ matrix, then A and A^2 have the same eigenvectors.

2. If A is an $n \times n$ matrix, then A and A^T have the same eigenvectors.

3. If A has eigenvalues of multiplicity greater than 1, then A must be defective.

4. If A is a 4×4 matrix of rank 3 and $\lambda = 0$ is an eigenvalue of multiplicity 3, then A is diagonalizable.

5. If A is a 4×4 matrix of rank 1 and $\lambda = 0$ is an eigenvalue of multiplicity 3, then A is defective.

6. The rank of an $n \times n$ matrix A is equal to the number of nonzero eigenvalues of A, where eigenvalues are counted according to multiplicity.

7. The rank of an $m \times n$ matrix A is equal to the number of nonzero singular values of A, where singular values are counted according to multiplicity.

8. If an $n \times n$ matrix A has Schur decomposition $A = UTU^H$, then the eigenvalues of A are $t_{11}, t_{22}, \ldots, t_{nn}$.

9. If A is symmetric positive definite, then A is nonsingular and A^{-1} is also symmetric positive definite.

10. If A is symmetric and $\det(A) > 0$, then A is positive definite.

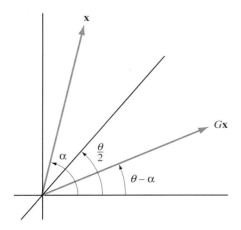

CHAPTER
7

NUMERICAL LINEAR ALGEBRA

In this chapter we consider computer methods for solving linear algebra problems. To understand these methods, you should be familiar with the type of number system used by the computer. When data are read into the computer, they are translated into the finite number system of the computer. This translation will usually involve some roundoff error. Further roundoff errors will occur when the algebraic operations of the algorithm are carried out. Because of this we cannot expect to get the exact solution to the original problem. The best we can hope for is a good approximation to a slightly perturbed problem. Suppose, for example, that we wanted to solve $A\mathbf{x} = \mathbf{b}$. When the entries of A and $\mathbf{b}$ are read into the computer roundoff errors will generally occur. Thus, the program will actually be attempting to compute a good approximation to the solution of a system of the form $(A + E)\mathbf{x} = \hat{\mathbf{b}}$. An algorithm is said to be *stable* if it will produce a good approximation to a slightly perturbed problem. Algorithms that ordinarily would converge to the solution using exact arithmetic could very well fail to be stable, owing to the growth of error in the algebraic processes.

Even using a stable algorithm, we may encounter problems that are very sensitive to perturbations. For example, if A is "nearly singular," the exact solutions of $A\mathbf{x} = \mathbf{b}$ and $(A + E)\mathbf{x} = \mathbf{b}$ may vary greatly even though all the entries of E are small. The major part of this chapter is devoted to numerical methods for solving linear systems. We will pay particular attention to the growth of error and to the sensitivity of systems to small changes.

Another problem that is very important in numerical applications is the problem of finding the eigenvalues of a matrix. Two iterative methods for computing eigenvalues are presented in Section 6. The second of these methods is the powerful QR algorithm, which makes use of the special types of orthogonal transformations presented in Section 5.

In Section 7 we will look at numerical methods for solving least squares problems. In the case where the coefficient matrix is rank deficient, we will make use of the singular value decomposition to find the particular least squares solution that has the smallest 2-norm. The Golub-Reinsch algorithm for computing the singular value decomposition will also be presented in this section.

I   FLOATING-POINT NUMBERS

In solving a numerical problem on a computer, we do not usually expect to get the exact answer. Some amount of error is inevitable. Roundoff error may occur initially when the data are represented in the finite number system of the computer. Further roundoff error may occur whenever arithmetic operations are used. Like a cancer, these errors may grow to such an extent that the computed solution may be completely unreliable. To avoid this, we must understand how computational errors occur. To do this, we must be familiar with the type of numbers used by the computer.

▶ **DEFINITION**   A **floating-point number** in base b is a number of the form

$$\pm \left(\frac{d_1}{b} + \frac{d_2}{b^2} + \cdots + \frac{d_t}{b^t} \right) \times b^e$$

where $t, d_1, d_2, \ldots, d_t, b, e$ are all integers and

$$0 \le d_i \le b - 1 \qquad i = 1, \ldots, t \qquad \blacktriangleleft$$

The integer t refers to the number of digits and this depends on the word length of the computer. The exponent e is restricted to be within certain bounds, $L \le e \le U$, which also depend on the particular computer. Most computers use base 2, although some use other bases, such as 8 or 16. Hand calculators generally use base 10.

EXAMPLE I.   The following are five-digit decimal (base 10) floating-point numbers:

$$0.53216 \times 10^{-4}$$

$$-0.81724 \times 10^{21}$$

$$0.00112 \times 10^{8}$$

$$0.11200 \times 10^{6}$$

Note that the numbers 0.00112×10^{8} and 0.11200×10^{6} are equal. Thus the floating-point representation of a number need not be unique. Floating-point numbers that are written with no leading zeros are said to be normalized.   ◀

EXAMPLE 2. $(0.236)_8 \times 8^2$ and $(0.132)_8 \times 8^4$ are normalized three-digit base 8 floating-point numbers. Here $(0.236)_8$ represents

$$\frac{2}{8} + \frac{3}{8^2} + \frac{6}{8^3}$$

Thus $(0.236)_8 \times 8^2$ is the base 8 floating-point representation of the decimal number 19.75. Similarly,

$$(0.132)_8 \times 8^4 = \left(\frac{1}{8} + \frac{3}{8^2} + \frac{2}{8^3}\right) \times 8^4 = 720 \qquad \blacktriangleleft$$

To better understand the type of number system that we are working with, it may help to look at a very simple example.

EXAMPLE 3. Suppose that $t = 1$, $L = -1$, $U = 1$, and $b = 10$. There are altogether 55 one-digit floating-point numbers in this system. These are

$$0, \pm 0.1 \times 10^{-1}, \pm 0.2 \times 10^{-1}, \ldots, \pm 0.9 \times 10^{-1}$$

$$\pm 0.1 \times 10^0, \pm 0.2 \times 10^0, \ldots, \pm 0.9 \times 10^0$$

$$\pm 0.1 \times 10^1, \pm 0.2 \times 10^1, \ldots, \pm 0.9 \times 10^1$$

Although all these numbers lie in the interval $[-9, 9]$, over one-third of the numbers have absolute value less than 0.1 and over two-thirds of the numbers have absolute value less than 1. Figure 7.1.1 illustrates how the floating-point numbers in the interval $[0, 2]$ are distributed. $\qquad \blacktriangleleft$

Most real numbers have to be rounded off in order to be represented as t-digit floating-point numbers. The difference between the floating-point number x' and the original number x is called the *roundoff error*. The size of the roundoff error is perhaps more meaningful when compared to the size of the original number.

▶ **DEFINITION** If x is a real number and x' is its floating-point approximation, then the difference $x' - x$ is called the **absolute error** and the quotient $(x' - x)/x$ is called the **relative error**. $\qquad \blacktriangleleft$

Real number x	Four-digit decimal floating-point number x	Absolute error, $x' - x$	Relative error, $(x' - x)/x$
62,133	0.6213×10^5	-3	$\dfrac{-3}{62,133} \approx -4.8 \times 10^{-5}$
0.12658	0.1266×10^0	2×10^{-5}	$\dfrac{1}{6329} \approx 1.6 \times 10^{-4}$
47.213	0.4721×10^2	-3.0×10^{-3}	$\dfrac{-0.003}{47.213} \approx -6.4 \times 10^{-5}$
π	0.3142×10^1	$3.142 - \pi \approx 4 \times 10^{-4}$	$\dfrac{3.142 - \pi}{\pi} \approx 1.3 \times 10^{-4}$

FIGURE 7.1.1

When arithmetic operations are applied to floating-point numbers, additional roundoff errors may occur.

EXAMPLE 4. Let $a' = 0.263 \times 10^4$ and $b' = 0.466 \times 10^1$ be three-digit decimal floating-point numbers. If these numbers are added, the exact sum will be

$$a' + b' = 0.263446 \times 10^4$$

However, the floating-point representation of this sum is 0.263×10^4. This then should be the computed sum. We will denote the floating-point sum by $fl(a' + b')$. The absolute error in the sum is

$$fl(a' + b') - (a' + b') = -4.46$$

and the relative error is

$$\frac{-4.46}{0.26344 \times 10^4} \approx -0.17 \times 10^{-2}$$

The actual value of $a'b'$ is 11,729.8; however, $fl(a'b')$ is 0.117×10^5. The absolute error in the product is -29.8 and the relative error is approximately -0.25×10^{-2}. Floating-point subtraction and division can be done in a similar manner. ◀

The relative error in approximating a number x by its floating-point representation x' is usually denoted by the symbol δ. Thus

(1) $$\delta = \frac{x' - x}{x} \qquad \text{or} \qquad x' = x(1 + \delta)$$

$|\delta|$ can be bounded by a positive constant ϵ, called the *machine precision* or the *machine epsilon*. The machine epsilon is defined to be the smallest floating-point number ϵ for which

$$fl(1 + \epsilon) > 1$$

For example, if the computer uses three-digit decimal floating-point numbers, then

$$fl(1 + 0.499 \times 10^{-2}) = 1$$

while

$$fl(1 + 0.500 \times 10^{-2}) = 1.01$$

In this case the machine epsilon would be 0.500×10^{-2}.

It follows from (1) that if a' and b' are two floating-point numbers, then

$$fl(a' + b') = (a' + b')(1 + \delta_1)$$
$$fl(a'b') = (a'b')(1 + \delta_2)$$
$$fl(a' - b') = (a' - b')(1 + \delta_3)$$
$$fl(a' \div b') = (a' \div b')(1 + \delta_4)$$

The δ_i's are relative errors and will all have absolute values less than ϵ. Note in Example 4 that $\delta_1 \approx -0.17 \times 10^{-2}$, $\delta_2 \approx -0.25 \times 10^{-2}$, and $\epsilon = 0.5 \times 10^{-2}$.

If the numbers you are working with involve some slight errors, arithmetic operations may compound these errors. If two numbers agree to k decimal places and one number is subtracted from the other, there will be a loss of significant digits in your answer. In this case the relative error in the difference will be many times as great as the relative error in either of the numbers.

EXAMPLE 5. Let $c = 3.4215298$ and $d = 3.4213851$. Calculate $c - d$ and $1/(c - d)$ using six-digit decimal floating-point arithmetic.

SOLUTION.

I. The first step is to represent c and d by six-digit decimal floating-point numbers.

$$c' = 0.342153 \times 10^1$$

$$d' = 0.342139 \times 10^1$$

The relative errors in c and d are

$$\frac{c' - c}{c} \approx 0.6 \times 10^{-7} \qquad \text{and} \qquad \frac{d' - d}{d} \approx 1.4 \times 10^{-6}$$

II. $fl(c' - d') = c' - d' = 0.140000 \times 10^{-3}$. The actual value of $c - d$ is 0.1447×10^{-3}. The absolute and relative errors in approximating $c - d$ by $fl(c' - d')$ are

$$fl(c' - d') - (c - d) = -0.47 \times 10^{-5}$$

and

$$\frac{fl(c' - d') - (c - d)}{c - d} \approx -3.2 \times 10^{-2}$$

Note that the magnitude of the relative error in the difference is more than 10^4 times the relative error in either c or d.

III. $fl[1/(c' - d')] = 0.714286 \times 10^4$, and the correct answer to six significant figures is

$$\frac{1}{c - d} \approx 0.691085 \times 10^4$$

The absolute and relative errors are approximately 232 and 0.03.   ◀

EXERCISES

1. Find the three-digit decimal floating-point representation of each of the following numbers.
   (a) 2312   (b) 32.56   (c) 0.01277   (d) 82,431

2. Find the absolute error and the relative error when each of the real numbers in Exercise 1 is approximated by a three-digit decimal floating-point number.

3. Represent each of the following as five-digit base 2 floating-point numbers.

   (a) 21   (b) $\frac{3}{8}$ (c) 9.872   (d) −0.1

4. Do each of the following using four-digit decimal floating-point arithmetic and calculate the absolute and relative errors in your answers.

   (a) $10{,}420 + 0.0018$ (b) $10{,}424 - 10{,}416$
   (c) $0.12347 - 0.12342$ (d) $(3626.6) \cdot (22.656)$

5. Let $x_1 = 94{,}210$, $x_2 = 8631$, $x_3 = 1440$, $x_4 = 133$, and $x_5 = 34$. Calculate each of the following using four-digit decimal floating-point arithmetic.

   (a) $(((x_1 + x_2) + x_3) + x_4) + x_5$
   (b) $x_1 + ((x_2 + x_3) + (x_4 + x_5))$
   (c) $(((x_5 + x_4) + x_3) + x_2) + x_1$

6. What would the machine epsilon be for a computer that uses five-digit base 2 floating-point arithmetic?

7. How many floating-point numbers are there in the system if $t = 2$, $L = -2$, $U = 2$, and $b = 2$?

2   GAUSSIAN ELIMINATION

In this section we discuss the problem of solving a system of n linear equations in n unknowns using Gaussian elimination. Gaussian elimination is generally considered to be the most efficient computational method since it involves the least amount of arithmetic operations.

Gaussian Elimination Without Interchanges

Let $A = A^{(1)} = (a_{ij}^{(1)})$ be a nonsingular matrix. Then A can be reduced to triangular form using row operations I and III. For simplicity, let us assume that the reduction can be done using only row operation III.

$$
A^{(1)} = \begin{pmatrix}
a_{11}^{(1)} & a_{12}^{(1)} & \cdots & a_{1n}^{(1)} \\
a_{21}^{(1)} & a_{22}^{(1)} & \cdots & a_{2n}^{(1)} \\
\vdots & & & \\
a_{n1}^{(1)} & a_{n2}^{(1)} & \cdots & a_{nn}^{(1)}
\end{pmatrix}
$$

Step 1. Let $m_{k1} = a_{k1}^{(1)}/a_{11}^{(1)}$ for $k = 2, \ldots, n$ [by our assumption, $a_{11}^{(1)} \neq 0$]. The first step of the elimination process is to apply row operation III $n - 1$ times to eliminate the entries below the diagonal in the first column of A. Note that m_{k1} is the multiple of the first row that is to be subtracted from the kth row. The new matrix obtained will be

$$A^{(2)} = \begin{bmatrix} a_{11}^{(1)} & a_{12}^{(1)} & \cdots & a_{1n}^{(1)} \\ 0 & a_{22}^{(2)} & \cdots & a_{2n}^{(2)} \\ \vdots & & & \\ 0 & a_{n2}^{(2)} & \cdots & a_{nn}^{(2)} \end{bmatrix}$$

where

$$a_{kj}^{(2)} = a_{kj}^{(1)} - m_{k1} a_{1j}^{(1)} \qquad (2 \le k \le n, \ 2 \le j \le n)$$

The first step of the elimination process requires $n - 1$ divisions, $(n - 1)^2$ multiplications, and $(n - 1)^2$ additions/subtractions.

Step 2. If $a_{22}^{(2)} \ne 0$, then it can be used as a pivot element to eliminate $a_{32}^{(2)}, \ldots,$ $a_{n2}^{(2)}$. For $k = 3, \ldots, n$, set

$$m_{k2} = \frac{a_{k2}^{(2)}}{a_{22}^{(2)}}$$

and subtract m_{k2} times the second row of $A^{(2)}$ from the kth row. The new matrix obtained will be

$$A^{(3)} = \begin{bmatrix} a_{11}^{(1)} & a_{12}^{(1)} & a_{13}^{(1)} & \cdots & a_{1n}^{(1)} \\ 0 & a_{22}^{(2)} & a_{23}^{(2)} & \cdots & a_{2n}^{(2)} \\ 0 & 0 & a_{33}^{(3)} & \cdots & a_{3n}^{(3)} \\ \vdots & \vdots & \vdots & & \vdots \\ 0 & 0 & a_{n3}^{(3)} & \cdots & a_{nn}^{(3)} \end{bmatrix}$$

The second step requires $n - 2$ divisions, $(n - 2)^2$ multiplications, and $(n - 2)^2$ additions/subtractions.

After $n - 1$ steps we will end up with a triangular matrix $U = A^{(n)}$. The operation count for the entire process can be determined as follows:

Divisions: $(n - 1) + (n - 2) + \cdots + 1 = \dfrac{n(n - 1)}{2}$

Multiplications: $(n - 1)^2 + (n - 2)^2 + \cdots + 1^2 = \dfrac{n(2n - 1)(n - 1)}{6}$

Additions and/or subtractions: $(n - 1)^2 + \cdots + 1^2 = \dfrac{n(2n - 1)(n - 1)}{6}$

The elimination process is summarized in the following algorithm.

▶ **ALGORITHM** 7.2.1 *[Gaussian elimination without interchanges]*

For $i = 1, 2, \ldots, n-1$

    For $k = i+1, \ldots, n$

$$\text{Set } m_{ki} = \frac{a_{ki}^{(i)}}{a_{ii}^{(i)}} \quad \text{[provided that } a_{ii}^{(i)} \neq 0]$$

        For $j = i+1, \ldots, n$

$$\text{Set } a_{kj}^{(i+1)} = a_{kj}^{(i)} - m_{ki} a_{ij}^{(i)}$$

        End for loop

    End for loop

End for loop   ◀

To solve the system $A\mathbf{x} = \mathbf{b}$, we could augment A by $\mathbf{b}$. Thus $\mathbf{b}$ would be stored in an extra column of A. The reduction process could then be done using Algorithm 7.2.1 and letting j run from $i+1$ to $n+1$ instead of from $i+1$ to n. The triangular system could then be solved using back substitution.

Most of the work involved in solving a system $A\mathbf{x} = \mathbf{b}$ occurs in the reduction of A to triangular form. Suppose that after having solved $A\mathbf{x} = \mathbf{b}$ we want to solve another system, $A\mathbf{x} = \mathbf{b}_1$. We know the triangular form U from the first system, and consequently we would like to be able to solve the new system without having to go through the entire reduction process again. We can do this if we make use of the LU factorization discussed in Section 4 of Chapter 1 (see Example 6 of that section). To compute the LU factorization, we must keep track of the numbers m_{ki} used in Algorithm 7.2.1. These numbers are called the *multipliers*. The multiplier m_{ki} is the multiple of the ith row that is subtracted from the kth row during the ith step of the reduction process. To see how the multipliers can be used to solve $A\mathbf{x} = \mathbf{b}_1$, it is helpful to view the reduction process in terms of matrix multiplications.

Triangular Factorization

The first step of the reduction process involves multiplying A by $n-1$ elementary matrices,

$$A^{(2)} = E_{n1} \cdots E_{31} E_{21} A^{(1)}$$

where

$$E_{k1} = \begin{bmatrix} 1 & & & & & & \\ 0 & 1 & & & & & \\ \vdots & & \ddots & & & & \\ -m_{k1} & 0 & \cdots & 1 & & & \\ \vdots & & & & \ddots & & \\ 0 & 0 & \cdots & 0 & \cdots & 1 \end{bmatrix}$$

Each E_{k1} is nonsingular, with

$$
E_{k1}^{-1} = \begin{bmatrix}
1 & & & & & \\
0 & 1 & & & & \\
\vdots & & \ddots & & & \\
m_{k1} & 0 & \cdots & 1 & & \\
\vdots & & & & \ddots & \\
0 & 0 & \cdots & 0 & \cdots & 1
\end{bmatrix}
$$

Let

$$
M_1 = E_{n1} \cdots E_{31} E_{21} = \begin{bmatrix}
1 & & & & \\
-m_{21} & 1 & & & \\
-m_{31} & 0 & 1 & & \\
\vdots & & & \ddots & \\
-m_{n1} & 0 & 0 & \cdots & 1
\end{bmatrix}
$$

Thus, $A^{(2)} = M_1 A$. The matrix M_1 is nonsingular and

$$
M_1^{-1} = E_{21}^{-1} E_{31}^{-1} \cdots E_{n1}^{-1} = \begin{bmatrix}
1 & & & & \\
m_{21} & 1 & & & \\
m_{31} & 0 & 1 & & \\
\vdots & & & \ddots & \\
m_{n1} & 0 & 0 & \cdots & 1
\end{bmatrix}
$$

Similarly,

$$
\begin{aligned}
A^{(3)} &= E_{n2} \cdots E_{42} E_{32} A^{(2)} \\
&= M_2 A^{(2)} \\
&= M_2 M_1 A
\end{aligned}
$$

where

$$
M_2 = E_{n2} \cdots E_{32} = \begin{bmatrix}
1 & & & & \\
0 & 1 & & & \\
0 & -m_{32} & 1 & & \\
\vdots & \vdots & \vdots & \ddots & \\
0 & -m_{n2} & 0 & \cdots & 1
\end{bmatrix}
$$

and

$$M_2^{-1} = \begin{bmatrix} 1 & & & & \\ 0 & 1 & & & \\ 0 & m_{32} & 1 & & \\ \vdots & \vdots & \vdots & \ddots & \\ 0 & m_{n2} & 0 & \cdots & 1 \end{bmatrix}$$

At the end of the reduction process we have

$$U = A^{(n)} = M_{n-1} \cdots M_2 M_1 A$$

It follows that

$$A = M_1^{-1} M_2^{-1} \cdots M_{n-1}^{-1} U$$

The M_j^{-1}'s multiply out to give the following lower triangular matrix when they are taken in this order:

$$L = M_1^{-1} M_2^{-1} \cdots M_{n-1}^{-1} = \begin{bmatrix} 1 & 0 & 0 & \cdots & 0 \\ m_{21} & 1 & 0 & \cdots & 0 \\ m_{31} & m_{32} & 1 & \cdots & 0 \\ \vdots & & & & \\ m_{n1} & m_{n2} & m_{n3} & \cdots & 1 \end{bmatrix}$$

Thus, $A = LU$, where L is lower triangular and U is upper triangular.

EXAMPLE I. Let

$$A = \begin{bmatrix} 2 & 3 & 1 \\ 4 & 1 & 4 \\ 3 & 4 & 6 \end{bmatrix}$$

The elimination can be carried out in two steps:

$$\begin{bmatrix} 2 & 3 & 1 \\ 4 & 1 & 4 \\ 3 & 4 & 6 \end{bmatrix} \xrightarrow{1} \begin{bmatrix} 2 & 3 & 1 \\ 0 & -5 & 2 \\ 0 & -\frac{1}{2} & \frac{9}{2} \end{bmatrix} \xrightarrow{2} \begin{bmatrix} 2 & 3 & 1 \\ 0 & -5 & 2 \\ 0 & 0 & 4.3 \end{bmatrix}$$

The multipliers for step 1 were $m_{21} = 2$, $m_{31} = \frac{3}{2}$ and the multiplier for step 2 was $m_{32} = \frac{1}{10}$. Let

$$L = \begin{bmatrix} 1 & 0 & 0 \\ m_{21} & 1 & 0 \\ m_{31} & m_{32} & 1 \end{bmatrix} = \begin{bmatrix} 1 & 0 & 0 \\ 2 & 1 & 0 \\ \frac{3}{2} & \frac{1}{10} & 1 \end{bmatrix}$$

and

$$U = \begin{bmatrix} 2 & 3 & 1 \\ 0 & -5 & 2 \\ 0 & 0 & 4.3 \end{bmatrix}$$

The reader may verify that $LU = A$.   ◀

Once A has been reduced to triangular form and the factorization LU has been determined, the system $A\mathbf{x} = \mathbf{b}$ can be solved in two steps.

Step 1. *Forward Substitution.*   The system $A\mathbf{x} = \mathbf{b}$ can be written in the form

$$LU\mathbf{x} = \mathbf{b}$$

Let $\mathbf{y} = U\mathbf{x}$. It follows that

$$L\mathbf{y} = LU\mathbf{x} = \mathbf{b}$$

Thus, we can find $\mathbf{y}$ by solving the lower triangular system

$$
\begin{aligned}
y_1 &= b_1 \\
m_{21}y_1 + y_2 &= b_2 \\
m_{31}y_1 + m_{32}y_2 + y_3 &= b_3 \\
&\;\;\vdots \\
m_{n1}y_1 + m_{n2}y_2 + m_{n3}y_3 + \cdots + y_n &= b_n
\end{aligned}
$$

It follows from the first equation that $y_1 = b_1$. This value can be used in the second equation to solve for y_2. The values of y_1 and y_2 can be used in the third equation to solve for y_3, and so on. This method of solving a lower triangular system is called *forward substitution*.

Step 2. *Back Substitution.*   Once $\mathbf{y}$ has been determined, we need only solve the upper triangular system $U\mathbf{x} = \mathbf{y}$ to find the solution $\mathbf{x}$ to the system. The upper triangular system is solved by back substitution.

EXAMPLE 2.   Solve the system

$$
\begin{aligned}
2x_1 + 3x_2 + x_3 &= -4 \\
4x_1 + x_2 + 4x_3 &= 9 \\
3x_1 + 4x_2 + 6x_3 &= 0
\end{aligned}
$$

SOLUTION.   The coefficient matrix for this system is the matrix A in Example 1. Since L and U have been determined, the system can be solved using forward and backward substitution.

$$\begin{bmatrix} 1 & 0 & 0 & | & -4 \\ 2 & 1 & 0 & | & 9 \\ \frac{3}{2} & \frac{1}{10} & 1 & | & 0 \end{bmatrix}$$

$y_1 = -4$

$y_2 = 9 - 2y_1 = 17$

$y_3 = 0 - \frac{3}{2}y_1 - \frac{1}{10}y_2 = 4.3$

$$\begin{bmatrix} 2 & 3 & 1 & | & -4 \\ 0 & -5 & 2 & | & 17 \\ 0 & 0 & 4.3 & | & 4.3 \end{bmatrix}$$

$2x_1 + 3x_2 + x_3 = -4 \qquad x_1 = 2$

$-5x_2 + 2x_3 = 17 \qquad x_2 = -3$

$4.3x_3 = 4.3 \qquad x_3 = 1$

The solution to the system is $\mathbf{x} = (2, -3, 1)^T$. ◀

▶ **ALGORITHM 7.2.2** *[Forward and back substitution]*

For $k = 1, \ldots, n$

$\qquad$ *Set* $y_k = b_k - \displaystyle\sum_{i=1}^{k-1} m_{ki} y_i$

End for loop

For $k = n, n - 1, \ldots, 1$

$\qquad$ *Set* $x_k = \dfrac{y_k - \displaystyle\sum_{j=k+1}^{n} u_{kj} x_j}{u_{kk}}$

End for loop ◀

Operation Count. Algorithm 7.2.2 requires n divisions, $n(n-1)$ multiplications, and $n(n-1)$ additions/subtractions. The total operation count for solving a system $A\mathbf{x} = \mathbf{b}$ using Algorithms 7.2.1 and 7.2.2 is then

$\qquad\qquad$ Multiplications/divisions: $\qquad \frac{1}{3}n^3 + n^2 - \frac{1}{3}n$

$\qquad\qquad$ Additions/subtractions: $\qquad \frac{1}{3}n^3 + \frac{1}{2}n^2 - \frac{5}{6}n$

In both cases the $\frac{1}{3}n^3$ is the dominant term. We will say that solving a system by Gaussian elimination involves roughly $\frac{1}{3}n^3$ multiplications/divisions and $\frac{1}{3}n^3$ additions/subtractions.

Storage. It is not necessary to store the multipliers in a separate matrix L. Each multiplier m_{ki} can be stored in the matrix A in place of the entry $a_{ki}^{(i)}$ eliminated. At the end of the reduction process, A is being used to store the m_{ki}'s and the u_{ij}'s.

$$\begin{bmatrix} u_{11} & u_{12} & \cdots & u_{1,n-1} & u_{1n} \\ m_{21} & u_{22} & \cdots & u_{2,n-1} & u_{2n} \\ \vdots & & & & \\ m_{n1} & m_{n2} & \cdots & m_{n,n-1} & u_{nn} \end{bmatrix}$$

Algorithm 7.2.1 breaks down if at any step $a_{kk}^{(k)}$ is 0. If this happens, it is necessary to perform row interchanges. In the next section we will see how to incorporate interchanges into our elimination algorithm.

EXERCISES

1. Let

$$A = \begin{bmatrix} 1 & 1 & 1 \\ 2 & 4 & 1 \\ -3 & 1 & -2 \end{bmatrix}$$

Factor A into a product LU, where L is lower triangular with 1's along the diagonal and U is upper triangular.

2. Let A be the matrix in Exercise 1. Use the LU factorization of A to solve $A\mathbf{x} = \mathbf{b}$ for each of the following choices of $\mathbf{b}$:

   (a) $(4, 3, -13)^T$ (b) $(3, 1, -10)^T$ (c) $(7, 23, 0)^T$

3. Let A and B be $n \times n$ matrices and let $\mathbf{x} \in R^n$.

   (a) How many scalar additions and multiplications are necessary to compute the product $A\mathbf{x}$?
   (b) How many scalar additions and multiplications are necessary to compute the product AB?
   (c) How many scalar additions and multiplications are necessary to compute $(AB)\mathbf{x}$? To compute $A(B\mathbf{x})$?

4. Let $A \in R^{m \times n}$, $B \in R^{n \times r}$, and let $\mathbf{x}, \mathbf{y} \in R^n$. Suppose that the product $A\mathbf{x}\mathbf{y}^T B$ is computed in the following ways:

   (i) $(A(\mathbf{x}\mathbf{y}^T))B$ (ii) $(A\mathbf{x})(\mathbf{y}^T B)$ (iii) $((A\mathbf{x})\mathbf{y}^T)B$

   (a) How many scalar additions and multiplications are necessary for each of these computations?
   (b) Compare the number of scalar additions and multiplications for each of the three methods when $m = 5$, $n = 4$, $r = 3$. Which method is most efficient in this case?

5. Let E_{ki} be the elementary matrix formed by subtracting α times the ith row of the identity matrix from the kth row.

(a) Show that $E_{ki} = I - \alpha e_k e_i^T$.

(b) Let $E_{ji} = I - \beta e_j e_i^T$. Show that $E_{ji} E_{ki} = I - (\alpha e_k + \beta e_j) e_i^T$.

(c) Show that $E_{ki}^{-1} = I + \alpha e_k e_i^T$.

6. Let A be an $n \times n$ matrix with triangular factorization LU. Show that

$$\det(A) = u_{11} u_{22} \cdots u_{nn}$$

7. If A is a symmetric $n \times n$ matrix with triangular factorization LU, then A can be factored further into a product LDL^T (where D is diagonal). Devise an algorithm, similar to Algorithm 7.2.2, for solving $LDL^T \mathbf{x} = \mathbf{b}$.

8. Write an algorithm for solving the tridiagonal system

$$\begin{bmatrix} a_1 & b_1 & & & & \\ c_1 & a_2 & \ddots & & & \\ & \ddots & \ddots & & \\ & & \ddots & a_{n-1} & b_{n-1} \\ & & & c_{n-1} & a_n \end{bmatrix} \begin{bmatrix} x_1 \\ x_2 \\ \vdots \\ x_{n-1} \\ x_n \end{bmatrix} = \begin{bmatrix} d_1 \\ d_2 \\ \vdots \\ d_{n-1} \\ d_n \end{bmatrix}$$

using Gaussian elimination with the diagonal elements as pivots. How many additions/subtractions and multiplications/divisions are necessary?

9. Let $A = LU$, where L is lower triangular with 1's on the diagonal and U is upper triangular.

(a) How many scalar additions and multiplications are necessary to solve $L\mathbf{y} = \mathbf{e}_j$ using forward substitution?

(b) How many additions/subtractions and multiplications/divisions are necessary to solve $A\mathbf{x} = \mathbf{e}_j$? The solution $\mathbf{x}_j$ to $A\mathbf{x} = \mathbf{e}_j$ will be the jth column of A^{-1}.

(c) Given the factorization $A = LU$, how many additional multiplications/divisions and additions/subtractions are needed to compute A^{-1}?

10. Suppose that A^{-1} and the LU factorization of A have already been determined. How many scalar additions and multiplications are necessary to compute $A^{-1}\mathbf{b}$? Compare this with the number of operations required to solve $LU\mathbf{x} = \mathbf{b}$ using Algorithm 7.2.2. Suppose that we have a number of systems to solve with the same coefficient matrix A. Is it worthwhile to compute A^{-1}? Explain.

11. Let A be a 3×3 matrix and assume that A can be transformed into a lower triangular matrix L using only column operations of type III, that is,

$$AE_1 E_2 E_3 = L$$

where E_1, E_2, E_3 are elementary matrices of type III. Let

$$U = (E_1 E_2 \bar{E}_3)^{-1}$$

Show that U is upper triangular with 1's on the diagonal and $A = LU$. (This illustrates a column version of Gaussian elimination.)

3   PIVOTING STRATEGIES

In this section we present an algorithm for Gaussian elimination with row interchanges. At each step of the algorithm it will be necessary to choose a pivotal row. We can often avoid unnecessarily large error accumulations by choosing the pivotal rows in a reasonable manner.

Gaussian Elimination with Interchanges

Consider the following example.

EXAMPLE I.   Let

$$A = \begin{bmatrix} 6 & -4 & 2 \\ 4 & 2 & 1 \\ 2 & -1 & 1 \end{bmatrix}$$

We wish to reduce A to triangular form using row operations I and III. To keep track of the interchanges, we will use a row vector **p**. The coordinates of **p** will be denoted by $p(1)$, $p(2)$, $p(3)$. Initially, we set $\mathbf{p} = (1, 2, 3)$. Suppose that at the first step of the reduction process the third row is chosen as the pivotal row. Instead of interchanging the first and third rows, we will interchange the first and third entries of **p**. Setting $p(1) = 3$ and $p(3) = 1$, the vector **p** becomes $(3, 2, 1)$. The vector **p** is used to keep track of the reordering of the rows. We can think of **p** as a renumbering of the rows. The actual physical reordering of the rows can be deferred until the end of the reduction process.

$$\begin{array}{c} \text{row} \\ \begin{array}{c} p(3) \\ p(2) \\ p(1) \end{array} \begin{bmatrix} 6 & -4 & 2 \\ 4 & 2 & 1 \\ \mathbf{2} & \mathbf{-1} & \mathbf{1} \end{bmatrix} \rightarrow \begin{bmatrix} 0 & -1 & -1 \\ 0 & 4 & -1 \\ 2 & -1 & 1 \end{bmatrix} \end{array}$$

If at the second step row $p(3)$ is chosen as the pivotal row, the entries of $p(3)$ and $p(2)$ are switched. The final step of the elimination process is then carried out.

$$\begin{array}{c} p(2) \\ p(3) \\ p(1) \end{array} \begin{bmatrix} 0 & \mathbf{-1} & \mathbf{-1} \\ 0 & 4 & -1 \\ 2 & -1 & 1 \end{bmatrix} \rightarrow \begin{bmatrix} 0 & -1 & -1 \\ 0 & 0 & -5 \\ 2 & -1 & 1 \end{bmatrix}$$

If the rows are reordered in the order $(p(1), p(2), p(3)) = (3, 1, 2)$, the resulting matrix will be in triangular form.

$$\begin{array}{c} p(1) = 3 \\ p(2) = 1 \\ p(3) = 2 \end{array} \begin{bmatrix} 2 & -1 & 1 \\ 0 & -1 & -1 \\ 0 & 0 & -5 \end{bmatrix}$$

Had the rows been written in the order $(3, 1, 2)$ to begin with, the reduction would have been exactly the same except that there would have been no need for interchanges. Reordering the rows of A in the order $(3, 1, 2)$ is the same as premultiplying A by the permutation matrix:

$$P = \begin{bmatrix} 0 & 0 & 1 \\ 1 & 0 & 0 \\ 0 & 1 & 0 \end{bmatrix}$$

Let us perform the reduction on A and PA simultaneously and compare the results. The multipliers used in the reduction process were 3, 2, and -4. These will be stored in the place of the terms eliminated and enclosed in boxes to distinguish them from the other entries of the matrix.

$$A = \begin{bmatrix} 6 & -4 & 2 \\ 4 & 2 & 1 \\ 2 & -1 & 1 \end{bmatrix} \rightarrow \begin{bmatrix} \boxed{3} & -1 & -1 \\ \boxed{2} & 4 & -1 \\ 2 & -1 & 1 \end{bmatrix} \rightarrow \begin{bmatrix} \boxed{3} & \boxed{-1} & -1 \\ 2 & -4 & -5 \\ 2 & -1 & 1 \end{bmatrix}$$

$$PA = \begin{bmatrix} 2 & -1 & 1 \\ 6 & -4 & 2 \\ 4 & 2 & 1 \end{bmatrix} \rightarrow \begin{bmatrix} 2 & -1 & 1 \\ \boxed{3} & -1 & -1 \\ \boxed{2} & 4 & -1 \end{bmatrix} \rightarrow \begin{bmatrix} 2 & -1 & 1 \\ \boxed{3} & \boxed{-1} & -1 \\ 2 & -4 & -5 \end{bmatrix}$$

If the rows of the reduced form of A are reordered, the resulting reduced matrices will be the same. The reduced form of PA now contains the information necessary to determine its triangular factorization. Indeed,

$$PA = LU$$

where

$$L = \begin{bmatrix} 1 & 0 & 0 \\ 3 & 1 & 0 \\ 2 & -4 & 1 \end{bmatrix} \quad \text{and} \quad U = \begin{bmatrix} 2 & -1 & 1 \\ 0 & -1 & -1 \\ 0 & 0 & -5 \end{bmatrix} \quad \blacktriangleleft$$

On the computer it is not necessary to actually interchange the rows of A. We simply treat row $p(k)$ as the kth row and use $a_{p(k),j}$ in place of $a_{k,j}$.

▶ **ALGORITHM 7.3.1** *[Gaussian elimination with interchanges]*

> For $i = 1, \ldots, n$
>
> > Set $p(i) = i$
>
> → End for loop

For $i = 1, \ldots, n$

> (1) *Choose a pivot element $a_{p(j),i}$ from the elements*
> $$a_{p(i),i}, a_{p(i+1),i}, \ldots, a_{p(n),i}$$
> *(Strategies for doing this will be discussed later in this section.)*
> (2) *Switch the ith and jth entries of p.*
> (3) For $k = i+1, \ldots, n$
> > Set $m_{p(k),i} = a_{p(k),i}/a_{p(i),i}$
> > For $j = i+1, \ldots, n$
> > > Set $a_{p(k),j} = a_{p(k),j} - m_{p(k),i} a_{p(i),j}$
> > → End for loop
> → End for loop

→ End for loop   ◀

Remarks

1. The multiplier $m_{p(k),i}$ is stored in the position of the element $a_{p(k),i}$ being eliminated.
2. The vector **p** can be used to form a permutation matrix P whose ith row is the $p(i)$th row of the identity matrix.
3. The matrix PA can be factored into a product LU, where

$$l_{ki} = \begin{cases} m_{p(k),i} & \text{if} \quad k > i \\ 1 & \text{if} \quad k = i \\ 0 & \text{if} \quad k < i \end{cases} \quad \text{and} \quad u_{ki} = \begin{cases} a_{p(k),i} & \text{if} \quad k \le i \\ 0 & \text{if} \quad k > i \end{cases}$$

4. Since P is nonsingular, the system $A\mathbf{x} = \mathbf{b}$ is equivalent to the system $PA\mathbf{x} = P\mathbf{b}$. Let $\mathbf{c} = P\mathbf{b}$. Since $PA = LU$, it follows that the system is equivalent to

$$LU\mathbf{x} = \mathbf{c}$$

5. If $PA = LU$, then $A = P^{-1}LU = P^{T}LU$.   ◀

It follows from Remarks 4 and 5 that if $A = P^{T}LU$, then the system $A\mathbf{x} = \mathbf{b}$ can be solved in three steps:

Step 1. *Reordering.*   Reorder the entries of **b** to form $\mathbf{c} = P\mathbf{b}$.
Step 2. *Forward Substitution.*   Solve the system $L\mathbf{y} = \mathbf{c}$ for **y**.
Step 3. *Back Substitution.*   Solve $U\mathbf{x} = \mathbf{y}$.

EXAMPLE 2. Solve the system

$$6x_1 - 4x_2 + 2x_3 = -2$$
$$4x_1 + 2x_2 + x_3 = 4$$
$$2x_1 - x_2 + x_3 = -1$$

SOLUTION. The coefficient matrix of this system is the matrix A from Example 1. P, L, and U have already been determined and they can be used to solve the system as follows:

Step 1. $\mathbf{c} = P\mathbf{b} = (-1, -2, 4)^T$

Step 2.

$$y_1 = -1 \qquad y_1 = -1$$
$$3y_1 + y_2 = -2 \qquad y_2 = -2 + 3 = 1$$
$$2y_1 - 4y_2 + y_3 = 4 \qquad y_3 = 4 + 2 + 4 = 10$$

Step 3.

$$2x_1 - x_2 + x_3 = -1 \qquad x_1 = 1$$
$$ - x_2 - x_3 = 1 \qquad x_2 = 1$$
$$ - 5x_3 = 10 \qquad x_3 = -2$$

The solution to the system is $\mathbf{x} = (1, 1, -2)^T$. ◀

It is possible to do Gaussian elimination without row interchanges if the diagonal entries $a_{ii}^{(i)}$ are nonzero at each step. However, in finite-precision arithmetic, pivots $a_{ii}^{(i)}$ that are near 0 can cause problems.

EXAMPLE 3. Consider the system

$$0.0001x_1 + 2x_2 = 4$$
$$x_1 + x_2 = 3$$

The exact solution to the system is

$$\mathbf{x} = \left(\frac{2}{1.9999}, \frac{3.9997}{1.9999} \right)^T$$

When rounded off to four decimal places, the solution is $(1.0001, 1.9999)^T$. Let us solve the system using three-digit decimal floating-point arithmetic.

$$\begin{bmatrix} 0.0001 & 2 & 4 \\ 1 & 1 & 3 \end{bmatrix} \rightarrow \begin{bmatrix} 0.0001 & 2 & 4 \\ 0 & -0.200 \times 10^5 & -0.400 \times 10^5 \end{bmatrix}$$

The computed solution is $\mathbf{x}' = (0, 2)^T$. There is a 100 percent error in the x_1 coordinate. On the other hand, if we interchange rows to avoid the small pivot, then three-digit decimal arithmetic gives

$$\begin{bmatrix} 1 & 1 & 3 \\ 0.0001 & 2 & 4 \end{bmatrix} \rightarrow \begin{bmatrix} 1 & 1 & 3 \\ 0 & 2.00 & 4.00 \end{bmatrix}$$

In this case the computed solution is $\mathbf{x}' = (1, 2)^T$.   ◀

If the pivot $a_{ii}^{(i)}$ is small in absolute value, the multipliers $m_{ki} = a_{ki}^{(i)}/a_{ii}^{(i)}$ will be large in absolute value. If there is an error in the computed value of $a_{ii}^{(i)}$, it will be multiplied by m_{ki}. In general, large multipliers contribute to the propagation of error. On the other hand, multipliers that are less than 1 in absolute value generally retard the growth of error. By careful selection of the pivot elements, we can try to avoid small pivots and at the same time keep the multipliers less than 1 in modulus. The most commonly used strategy for doing this is called *partial pivoting*.

Partial Pivoting

At the ith step of the reduction process, there are $n - i + 1$ candidates for the pivot element:

$$a_{p(i),i}, a_{p(i+1),i}, \ldots, a_{p(n),i}$$

Choose the candidate $a_{p(j),i}$ with maximum modulus

$$|a_{p(j),i}| = \max_{i \leq k \leq n} |a_{p(k),i}|$$

and interchange the ith and jth entries of $\mathbf{p}$. The pivot element $a_{p(i),i}$ has the property

$$|a_{p(i),i}| \geq |a_{p(k),i}|$$

for $k = i + 1, \ldots, n$. Thus the multipliers will all satisfy

$$|m_{p(k),i}| = \left| \frac{a_{p(k),i}}{a_{p(i),i}} \right| \leq 1$$

We could always carry things one step further and do *complete pivoting*. In complete pivoting the pivot element is chosen to be the element of maximum modulus among all the elements in the remaining rows and columns. In this case we must keep track of both the rows and the columns. At the ith step the element $a_{p(j)q(k)}$ is chosen so that

$$|a_{p(j)q(k)}| = \max_{\substack{i \leq s \leq n \\ i \leq t \leq n}} |a_{p(s)q(t)}|$$

The ith and jth entries of $\mathbf{p}$ are interchanged and the ith and kth entries of $\mathbf{q}$ are interchanged. The new pivot element is $a_{p(i)q(i)}$. The major drawback to complete pivoting is that at each step we must search for a pivot element among $(n - i + 1)^2$ elements of A. This may be too costly in terms of computer time.

EXERCISES

1. Let

$$A = \begin{bmatrix} 0 & 3 & 1 \\ 1 & 2 & -2 \\ 2 & 5 & 4 \end{bmatrix} \quad \text{and} \quad \mathbf{b} = \begin{bmatrix} 1 \\ 7 \\ -1 \end{bmatrix}$$

(a) Reorder the rows of $(A|\mathbf{b})$ in the order $(2, 3, 1)$ and then solve this reordered system.

(b) Factor A into a product $P^T LU$, where P is the permutation matrix corresponding to the reordering in part (a).

2. Let A be the matrix in Exercise 1. Use the factorization $P^T LU$ to solve $A\mathbf{x} = \mathbf{c}$ for each of the following choices of $\mathbf{c}$:

(a) $(8, 1, 20)^T$ (b) $(-9, -2, -7)^T$ (c) $(4, 1, 11)^T$

3. Let

$$A = \begin{bmatrix} 1 & 8 & 6 \\ -1 & -4 & 5 \\ 2 & 4 & -6 \end{bmatrix} \quad \text{and} \quad \mathbf{b} = \begin{bmatrix} 8 \\ 1 \\ 4 \end{bmatrix}$$

Solve the system $A\mathbf{x} = \mathbf{b}$ using partial pivoting. If P is the permutation matrix corresponding to the pivoting strategy, factor PA into a product LU.

4. Let

$$A = \begin{bmatrix} 3 & 2 \\ 2 & 4 \end{bmatrix} \quad \text{and} \quad \mathbf{b} = \begin{bmatrix} 5 \\ -2 \end{bmatrix}$$

Solve the system $A\mathbf{x} = \mathbf{b}$ using complete pivoting. Let P be the permutation matrix determined by the pivot rows and Q the permutation matrix determined by the pivot columns. Factor PAQ into a product LU.

5. Let A be the matrix in Exercise 4 and let $\mathbf{c} = (6, -4)^T$. Solve the system $A\mathbf{x} = \mathbf{c}$ in two steps:

(a) Set $\mathbf{z} = Q^T \mathbf{x}$ and solve $LU\mathbf{z} = P\mathbf{c}$ for $\mathbf{z}$.

(b) Calculate $\mathbf{x} = Q\mathbf{z}$.

6. Given

$$A = \begin{bmatrix} 5 & 4 & 7 \\ 2 & -4 & 3 \\ 2 & 8 & 6 \end{bmatrix}, \quad \mathbf{b} = \begin{bmatrix} 2 \\ -5 \\ 4 \end{bmatrix}, \quad \mathbf{c} = \begin{bmatrix} 5 \\ -4 \\ 2 \end{bmatrix}$$

(a) Solve the system $A\mathbf{x} = \mathbf{b}$ using complete pivoting.

(b) Let P be the permutation matrix determined by the pivot rows and let Q be the permutation matrix determined by the pivot columns. Factor PAQ into a product LU.

(c) Use the LU factorization from part (b) to solve the system $A\mathbf{x} = \mathbf{c}$.

7. The exact solution to the system

$$0.6000x_1 + 2000x_2 = 2003$$
$$0.3076x_1 - 0.4010x_2 = 1.137$$

is $\mathbf{x} = (5, 1)^T$. Suppose that the calculated value of $\mathbf{x}_2$ is $x_2' = 1 + \epsilon$. Use this value in the first equation and solve for x_1. What will the error be? Calculate the relative error in x_1 if $\epsilon = 0.001$.

8. Solve the system in Exercise 7 using four-digit decimal floating-point arithmetic and Gaussian elimination with partial pivoting.

9. Solve the system in Exercise 7 using four-digit decimal floating-point arithmetic and Gaussian elimination with complete pivoting.

10. Use four-digit decimal floating-point arithmetic and scale the system in Exercise 7 by multiplying the first equation through by $1/2000$ and the second equation through by $1/0.4010$. Solve the scaled system using partial pivoting.

4 MATRIX NORMS AND CONDITION NUMBERS

In this section we are concerned with the accuracy of computed solutions to linear systems. How accurate can we expect the computed solutions to be, and how can we test their accuracy? The answer to these questions depends largely on how sensitive the coefficient matrix of the system is to small changes. The sensitivity of the matrix can be measured in terms of its *condition number*. The condition number of a nonsingular matrix is defined in terms of its norm and the norm of its inverse. Before discussing condition numbers, it is necessary to establish some important results regarding the standard types of matrix norms.

Matrix Norms

Just as vector norms are used to measure the size of vectors, matrix norms can be used to measure the size of matrices. In Section 4 of Chapter 5 we introduced a norm on $R^{m \times n}$ that was induced by an inner product on $R^{m \times n}$. This norm was referred to as the Frobenius norm and was denoted by $\| \cdot \|_F$. We showed that the Frobenius norm of a matrix A could be computed by taking the square root of the sum of the squares of all its entries.

$$(1) \qquad \qquad \|A\|_F = \left(\sum_{j=1}^{n} \sum_{i=1}^{m} a_{ij}^2 \right)^{1/2}$$

Actually, equation (1) defines a family of matrix norms since it defines a norm on $R^{m \times n}$ for any choice of m and n. The Frobenius norm has a number of important properties.

I. If $\mathbf{a}_j$ represents the jth column vector of A, then

$$\|A\|_F = \left(\sum_{j=1}^{n} \sum_{i=1}^{m} a_{ij}^2 \right)^{1/2} = \left(\sum_{j=1}^{n} \|\mathbf{a}_j\|_2^2 \right)^{1/2}$$

II. If $\mathbf{a}(i, :)$ represents the ith row vector of A, then

$$\|A\|_F = \left(\sum_{i=1}^{m} \sum_{j=1}^{n} a_{ij}^2 \right)^{1/2} = \left(\sum_{i=1}^{m} \|\mathbf{a}(i, :)^T\|_2^2 \right)^{1/2}$$

III. If $\mathbf{x} \in R^n$, then

$$\|A\mathbf{x}\|_2 = \left[\sum_{i=1}^{m} \left(\sum_{j=1}^{n} a_{ij} x_j \right)^2 \right]^{1/2} = \left[\sum_{i=1}^{m} (\mathbf{a}(i, :)\mathbf{x})^2 \right]^{1/2}$$

$$\leq \left[\sum_{i=1}^{m} \|\mathbf{x}\|_2^2 \|\mathbf{a}(i, :)^T\|_2^2 \right]^{1/2} \qquad \text{(Cauchy–Schwarz)}$$

$$= \|A\|_F \|\mathbf{x}\|_2$$

IV. If $B = (\mathbf{b}_1, \ldots, \mathbf{b}_r)$ is an $n \times r$ matrix, it follows from properties I and III that

$$\|AB\|_F = \|(A\mathbf{b}_1, A\mathbf{b}_2, \ldots, A\mathbf{b}_r)\|_F$$

$$= \left(\sum_{i=1}^{r} \|A\mathbf{b}_i\|_2^2 \right)^{1/2}$$

$$\leq \|A\|_F \left(\sum_{i=1}^{r} \|\mathbf{b}_i\|_2^2 \right)^{1/2}$$

$$= \|A\|_F \|B\|_F$$

There are many other norms that we could use for $R^{m \times n}$ in addition to the Frobenius norm. Any norm used must satisfy the three conditions that define norms in general.

(i) $\|A\| \geq 0$ and $\|A\| = 0$ if and only if $A = O$
(ii) $\|\alpha A\| = |\alpha| \|A\|$
(iii) $\|A + B\| \leq \|A\| + \|B\|$

The families of matrix norms that turn out to be most useful also satisfy the additional property

(iv) $\|AB\| \leq \|A\| \|B\|$

Consequently, we will only consider families of norms that have this additional property. One important consequence of property (iv) is that

$$\|A^n\| \leq \|A\|^n$$

In particular, if $\|A\| < 1$, then $\|A^n\| \to 0$ as $n \to \infty$.

In general, a matrix norm $\| \cdot \|_M$ on $R^{m \times n}$ and a vector norm $\| \cdot \|_V$ on R^n are said to be *compatible* if

$$\|A\mathbf{x}\|_V \leq \|A\|_M \|\mathbf{x}\|_V$$

for every $\mathbf{x} \in R^n$. In particular, it follows from property III that the matrix norm $\| \cdot \|_F$ and the vector norm $\| \cdot \|_2$ are compatible. For each of the standard vector norms we can define a compatible matrix norm by using the vector norm to compute an operator norm for the matrix. The matrix norm defined in this way is said to be *subordinate* to the vector norm.

Subordinate Matrix Norms

We can think of each $m \times n$ matrix as a linear operator from R^n to R^m. For any family of vector norms we can define an *operator norm* by comparing $\|A\mathbf{x}\|$ and $\|\mathbf{x}\|$ for each nonzero $\mathbf{x}$ and taking

$$(2) \qquad \|A\| = \max_{\mathbf{x} \neq \mathbf{0}} \frac{\|A\mathbf{x}\|}{\|\mathbf{x}\|}$$

It can be shown that there is a particular $\mathbf{x}_0$ in R^n that maximizes $\|A\mathbf{x}\|/\|\mathbf{x}\|$, but the proof is beyond the scope of this book. Assuming that $\|A\mathbf{x}\|/\|\mathbf{x}\|$ can always be maximized, we will show that (2) actually does define a norm on $R^{m \times n}$. To do this, we must verify that each of the three conditions of the definition is satisfied.

(i) For each $\mathbf{x} \neq \mathbf{0}$,

$$\frac{\|A\mathbf{x}\|}{\|\mathbf{x}\|} \geq 0$$

and, consequently,

$$\|A\| = \max_{\mathbf{x} \neq \mathbf{0}} \frac{\|A\mathbf{x}\|}{\|\mathbf{x}\|} \geq 0$$

If $\|A\| = 0$, then $A\mathbf{x} = \mathbf{0}$ for every $\mathbf{x} \in R^n$. This implies that

$$\mathbf{a}_j = A\mathbf{e}_j = \mathbf{0} \qquad \text{for} \qquad j = 1, \ldots, n$$

and hence A must be the zero matrix.

(ii) $\|\alpha A\| = \displaystyle\max_{\mathbf{x} \neq \mathbf{0}} \frac{\|\alpha A\mathbf{x}\|}{\|\mathbf{x}\|} = |\alpha| \max_{\mathbf{x} \neq \mathbf{0}} \frac{\|A\mathbf{x}\|}{\|\mathbf{x}\|} = |\alpha| \, \|A\|$

(iii) If $\mathbf{x} \neq \mathbf{0}$, then

$$\begin{aligned}
\|A + B\| &= \max_{\mathbf{x} \neq \mathbf{0}} \frac{\|(A + B)\mathbf{x}\|}{\|\mathbf{x}\|} \\
&\leq \max_{\mathbf{x} \neq \mathbf{0}} \frac{\|A\mathbf{x}\| + \|B\mathbf{x}\|}{\|\mathbf{x}\|} \\
&\leq \max_{\mathbf{x} \neq \mathbf{0}} \frac{\|A\mathbf{x}\|}{\|\mathbf{x}\|} + \max_{\mathbf{x} \neq \mathbf{0}} \frac{\|B\mathbf{x}\|}{\|\mathbf{x}\|} \\
&= \|A\| + \|B\|
\end{aligned}$$

Thus, (2) defines a norm on $R^{m \times n}$. For each family of vector norms $\| \cdot \|$, we can then define a family of matrix norms by (2). The matrix norms defined by (2) are said to be *subordinate* to the vector norms $\| \cdot \|$.

▶ **THEOREM 7.4.1** *If the family of matrix norms $\| \cdot \|_M$ is subordinate to the family of vector norms $\| \cdot \|_V$, then $\| \cdot \|_M$ and $\| \cdot \|_V$ are compatible and the matrix norms $\| \cdot \|_M$ satisfy property* (iv).

▶*Proof.* If $\mathbf{x}$ is any nonzero vector in R^n, then

$$\frac{\|A\mathbf{x}\|_V}{\|\mathbf{x}\|_V} \leq \max_{\mathbf{y} \neq \mathbf{0}} \frac{\|A\mathbf{y}\|_V}{\|\mathbf{y}\|_V} = \|A\|_M$$

and hence

$$\|A\mathbf{x}\|_V \leq \|A\|_M \|\mathbf{x}\|_V$$

Since this last inequality is also valid if $\mathbf{x} = \mathbf{0}$, it follows that $\| \cdot \|_M$ and $\| \cdot \|_V$ are compatible. If B is an $n \times r$ matrix, then, since $\| \cdot \|_M$ and $\| \cdot \|_V$ are compatible, we have

$$\|AB\mathbf{x}\|_V \leq \|A\|_M \|B\mathbf{x}\|_V \leq \|A\|_M \|B\|_M \|\mathbf{x}\|_V$$

Thus, for all $\mathbf{x} \neq \mathbf{0}$,

$$\frac{\|AB\mathbf{x}\|_V}{\|\mathbf{x}\|_V} \leq \|A\|_M \|B\|_M$$

and hence

$$\|AB\|_M = \max_{\mathbf{x} \neq \mathbf{0}} \frac{\|AB\mathbf{x}\|_V}{\|\mathbf{x}\|_V} \leq \|A\|_M \|B\|_M \qquad ◀$$

It is a simple matter to compute the Frobenius norm of a matrix. For example, if

$$A = \begin{bmatrix} 4 & 2 \\ 0 & 4 \end{bmatrix}$$

then

$$\|A\|_F = (4^2 + 0^2 + 2^2 + 4^2)^{1/2} = 6$$

On the other hand, it is not so obvious how to compute $\|A\|$ if $\| \cdot \|$ is a subordinate matrix norm. It turns out that the matrix norm

$$\|A\|_2 = \max_{\mathbf{x} \neq \mathbf{0}} \frac{\|A\mathbf{x}\|_2}{\|\mathbf{x}\|_2}$$

is difficult to compute; however,

$$\|A\|_1 = \max_{\mathbf{x} \neq \mathbf{0}} \frac{\|A\mathbf{x}\|_1}{\|\mathbf{x}\|_1}$$

and

$$\|A\|_\infty = \max_{\mathbf{x} \neq \mathbf{0}} \frac{\|A\mathbf{x}\|_\infty}{\|\mathbf{x}\|_\infty}$$

can be easily calculated.

▶ **THEOREM 7.4.2** *If A is an $m \times n$ matrix, then*

$$\|A\|_1 = \max_{1 \le j \le n} \left(\sum_{i=1}^{m} |a_{ij}| \right)$$

and

$$\|A\|_\infty = \max_{1 \le i \le m} \left(\sum_{j=1}^{n} |a_{ij}| \right)$$

▶ *Proof.* We will prove that

$$\|A\|_1 = \max_{1 \le j \le n} \left(\sum_{i=1}^{m} |a_{ij}| \right)$$

and leave the proof of the second statement as an exercise. Let

$$\alpha = \max_{1 \le j \le n} \sum_{i=1}^{m} |a_{ij}| = \sum_{i=1}^{m} |a_{ik}|$$

That is, k is the index of the column where the maximum occurs. Let $\mathbf{x}$ be an arbitrary vector in R^n; then

$$A\mathbf{x} = \left(\sum_{j=1}^{n} a_{1j}x_j, \sum_{j=1}^{n} a_{2j}x_j, \dots, \sum_{j=1}^{n} a_{mj}x_j \right)^T$$

and it follows that

$$\|A\mathbf{x}\|_1 = \sum_{i=1}^{m} \left| \sum_{j=1}^{n} a_{ij}x_j \right|$$

$$\le \sum_{i=1}^{m} \sum_{j=1}^{n} |a_{ij}x_j|$$

$$= \sum_{j=1}^{n} \left(|x_j| \sum_{i=1}^{m} |a_{ij}| \right)$$

$$\le \alpha \sum_{j=1}^{n} |x_j|$$

$$= \alpha \|\mathbf{x}\|_1$$

Thus, for any nonzero $\mathbf{x}$ in R^n,

$$\frac{\|A\mathbf{x}\|_1}{\|\mathbf{x}\|_1} \le \alpha$$

and hence

(3) $$\|A\|_1 = \max_{\mathbf{x} \ne \mathbf{0}} \frac{\|A\mathbf{x}\|_1}{\|\mathbf{x}\|_1} \le \alpha$$

On the other hand,

$$\|A\mathbf{e}_k\|_1 = \|\mathbf{a}_k\|_1 = \alpha$$

Since $\|\mathbf{e}_k\|_1 = 1$, it follows that

(4) $$\|A\|_1 = \max_{\mathbf{x} \ne \mathbf{0}} \frac{\|A\mathbf{x}\|_1}{\|\mathbf{x}\|_1} \ge \frac{\|A\mathbf{e}_k\|_1}{\|\mathbf{e}_k\|_1} = \alpha$$

Together (3) and (4) imply that $\|A\|_1 = \alpha$.   ◀

EXAMPLE 1. Let

$$A = \begin{pmatrix} -3 & 2 & 4 & -3 \\ 5 & -2 & -3 & 5 \\ 2 & 1 & -6 & 4 \\ 1 & 1 & 1 & 1 \end{pmatrix}$$

Then

$$\|A\|_1 = |4| + |-3| + |-6| + |1| = 14$$

and

$$\|A\|_\infty = |5| + |-2| + |-3| + |5| = 15 \qquad \blacktriangleleft$$

The 2-norm of a matrix is more difficult to compute since it depends on the singular values of the matrix. In fact, the 2-norm of a matrix is its largest singular value.

▶ **THEOREM 7.4.3** *If A is an $m \times n$ matrix with singular value decomposition $U \Sigma V^T$, then*

$$\|A\|_2 = \sigma_1 \qquad (\text{the largest singular value})$$

▶*Proof.* Since U and V are orthogonal,

$$\|A\|_2 = \|U \Sigma V^T\|_2 = \|\Sigma\|_2$$

(See Exercise 31.) Now

$$\|\Sigma\|_2 = \max_{\mathbf{x} \neq \mathbf{0}} \frac{\|\Sigma \mathbf{x}\|_2}{\|\mathbf{x}\|_2}$$

$$= \max_{\mathbf{x} \neq \mathbf{0}} \frac{\left(\sum_{i=1}^{n} (\sigma_i x_i)^2 \right)^{1/2}}{\left(\sum_{i=1}^{n} x_i^2 \right)^{1/2}}$$

$$\leq \sigma_1$$

However, if we choose $\mathbf{x} = \mathbf{e}_1$, then

$$\frac{\|\Sigma \mathbf{x}\|_2}{\|\mathbf{x}\|_2} = \sigma_1$$

and hence it follows that

$$\|A\|_2 = \|\Sigma\|_2 = \sigma_1 \qquad \blacktriangleleft$$

▶ **COROLLARY 7.4.4** *If $A = U \Sigma V^T$ is nonsingular, then*

$$\|A^{-1}\|_2 = \frac{1}{\sigma_n}$$

▶**Proof.** The singular values of $A^{-1} = V\Sigma^{-1}U^T$ arranged in decreasing order are

$$\frac{1}{\sigma_n} \geq \frac{1}{\sigma_{n-1}} \geq \cdots \geq \frac{1}{\sigma_1}$$

Therefore,

$$\|A^{-1}\|_2 = \frac{1}{\sigma_n} \qquad ◀$$

Condition Numbers

Matrix norms can be used to estimate the sensitivity of linear systems to small changes in the coefficient matrix. Consider the following example.

EXAMPLE 2.   Solve the following system:

(5)
$$\begin{aligned} 2.0000x_1 + 2.0000x_2 &= 6.0000 \\ 2.0000x_1 + 2.0005x_2 &= 6.0010 \end{aligned}$$

If we use five-digit decimal floating-point arithmetic, the computed solution will be the exact solution $\mathbf{x} = (1, 2)^T$. Suppose, however, that we are forced to use four-digit decimal floating-point numbers. Thus, in place of (5), we have

(6)
$$\begin{aligned} 2.000x_1 + 2.000x_2 &= 6.000 \\ 2.000x_1 + 2.001x_2 &= 6.001 \end{aligned}$$

The computed solution to system (6) is the exact solution $\mathbf{x}' = (2, 1)^T$.

The systems (5) and (6) agree except for the coefficient a_{22}. The relative error in this coefficient is

$$\frac{a'_{22} - a_{22}}{a_{22}} \approx 0.00025$$

However, the relative errors in the coordinates of the solutions $\mathbf{x}$ and $\mathbf{x}'$ are

$$\frac{x'_1 - x_1}{x_1} = 1.0 \qquad \text{and} \qquad \frac{x'_2 - x_2}{x_2} = -0.5 \qquad ◀$$

▶ **DEFINITION** A matrix A is said to be **ill-conditioned** if relatively small changes in the entries of A can cause relatively large changes in the solutions to $A\mathbf{x} = \mathbf{b}$. A is said to be **well-conditioned** if relatively small changes in the entries of A result in relatively small changes in the solutions to $A\mathbf{x} = \mathbf{b}$. ◀

If the matrix A is ill-conditioned, the computed solution to $A\mathbf{x} = \mathbf{b}$ generally will not be very accurate. Even if the entries of A can be represented exactly as floating-point numbers, small roundoff errors occurring in the reduction process may have a drastic effect on the computed solution. On the other hand, if the matrix is well-conditioned and the proper pivoting strategy is used, we should be able to compute solutions quite accurately. In general, the accuracy of the solution depends on the conditioning of the matrix. If we could measure the conditioning of A, this measure could be used to derive a bound for the relative error in the computed solution.

Let A be an $n \times n$ nonsingular matrix and consider the system $A\mathbf{x} = \mathbf{b}$. If $\mathbf{x}$ is the exact solution to the system and $\mathbf{x}'$ is the calculated solution, then the error can be represented by the vector $\mathbf{e} = \mathbf{x} - \mathbf{x}'$. If $\|\cdot\|$ is a norm on R^n, then $\|\mathbf{e}\|$ is a measure of the absolute error and $\|\mathbf{e}\|/\|\mathbf{x}\|$ is a measure of the relative error. In general, we have no way of determining the exact values of $\|\mathbf{e}\|$ and $\|\mathbf{e}\|/\|\mathbf{x}\|$. One possible way of testing the accuracy of $\mathbf{x}'$ is to put it back into the original system and to see how close $\mathbf{b}' = A\mathbf{x}'$ comes to $\mathbf{b}$. The vector

$$\mathbf{r} = \mathbf{b} - \mathbf{b}' = \mathbf{b} - A\mathbf{x}'$$

is called the *residual* and can be easily calculated. The quantity

$$\frac{\|\mathbf{b} - A\mathbf{x}'\|}{\|\mathbf{b}\|} = \frac{\|\mathbf{r}\|}{\|\mathbf{b}\|}$$

is called the *relative residual*. Is the relative residual a good estimate of the relative error? The answer to this depends on the conditioning of A. In Example 2 the residual for the computed solution $\mathbf{x}' = (2, 1)^T$ is

$$\mathbf{r} = \mathbf{b} - A\mathbf{x}' = (0, 0.0005)^T$$

The relative residual in terms of the ∞-norm is

$$\frac{\|\mathbf{r}\|_\infty}{\|\mathbf{b}\|_\infty} = \frac{0.0005}{6.0010} \approx 0.000083$$

and the relative error is given by

$$\frac{\|\mathbf{e}\|_\infty}{\|\mathbf{x}\|_\infty} = 0.5$$

The relative error is more than 6000 times the relative residual. In general, we will show that if A is ill-conditioned the relative residual may be much smaller than the relative error. On the other hand, for well-conditioned matrices, the relative residual and the relative error are quite close. To show this, we need to make use of matrix norms. Recall that if $\|\cdot\|$ is a compatible matrix norm on $R^{n \times n}$ then, for any $n \times n$ matrix C and any vector $\mathbf{y} \in R^n$, we have

(7) $$\|C\mathbf{y}\| \leq \|C\| \, \|\mathbf{y}\|$$

Now

$$\mathbf{r} = \mathbf{b} - A\mathbf{x}' = A\mathbf{x} - A\mathbf{x}' = A\mathbf{e}$$

and, consequently,

$$\mathbf{e} = A^{-1}\mathbf{r}$$

It follows from property (7) that

$$\|\mathbf{e}\| \leq \|A^{-1}\| \, \|\mathbf{r}\|$$

and

$$\|\mathbf{r}\| = \|A\mathbf{e}\| \leq \|A\| \, \|\mathbf{e}\|$$

Therefore,

(8)
$$\frac{\|\mathbf{r}\|}{\|A\|} \leq \|\mathbf{e}\| \leq \|A^{-1}\| \, \|\mathbf{r}\|$$

Now $\mathbf{x}$ is the exact solution to $A\mathbf{x} = \mathbf{b}$, and hence $\mathbf{x} = A^{-1}\mathbf{b}$. By the same reasoning used to derive (8), we have

(9)
$$\frac{\|\mathbf{b}\|}{\|A\|} \leq \|\mathbf{x}\| \leq \|A^{-1}\| \, \|\mathbf{b}\|$$

It follows from (8) and (9) that

$$\frac{1}{\|A\| \, \|A^{-1}\|} \frac{\|\mathbf{r}\|}{\|\mathbf{b}\|} \leq \frac{\|\mathbf{e}\|}{\|\mathbf{x}\|} \leq \|A\| \, \|A^{-1}\| \frac{\|\mathbf{r}\|}{\|\mathbf{b}\|}$$

The number $\|A\| \, \|A^{-1}\|$ is called the *condition number* of A and will be denoted by $\mathrm{cond}(A)$. Thus

(10)
$$\frac{1}{\mathrm{cond}(A)} \frac{\|\mathbf{r}\|}{\|\mathbf{b}\|} \leq \frac{\|\mathbf{e}\|}{\|\mathbf{x}\|} \leq \mathrm{cond}(A) \frac{\|\mathbf{r}\|}{\|\mathbf{b}\|}$$

Inequality (10) relates the size of the relative error $\|\mathbf{e}\|/\|\mathbf{x}\|$ to the relative residual $\|\mathbf{r}\|/\|\mathbf{b}\|$. If the condition number is close to 1, the relative error and the relative residual will be close. If the condition number is large, the relative error could be many times as large as the relative residual.

EXAMPLE 3. Let

$$A = \begin{pmatrix} 3 & 3 \\ 4 & 5 \end{pmatrix}$$

Then

$$A^{-1} = \frac{1}{3} \begin{pmatrix} 5 & -3 \\ -4 & 3 \end{pmatrix}$$

$\|A\|_\infty = 9$ and $\|A^{-1}\|_\infty = \frac{8}{3}$. (We use $\|\cdot\|_\infty$ because it is easy to calculate.) Thus

$$\mathrm{cond}_\infty(A) = 9 \cdot \tfrac{8}{3} = 24$$

Theoretically, the relative error in the calculated solution to a system $A\mathbf{x} = \mathbf{b}$ could be as much as 24 times the relative residual. ◀

EXAMPLE 4. Suppose that $\mathbf{x}' = (2.0, 0.1)^T$ is the calculated solution to

$$3x_1 + 3x_2 = 6$$
$$4x_1 + 5x_2 = 9$$

Determine the residual $\mathbf{r}$ and the relative residual $\|\mathbf{r}\|_\infty/\|\mathbf{b}\|_\infty$.

SOLUTION.

$$\mathbf{r} = \begin{bmatrix} 6 \\ 9 \end{bmatrix} - \begin{bmatrix} 3 & 3 \\ 4 & 5 \end{bmatrix} \begin{bmatrix} 2.0 \\ 0.1 \end{bmatrix} = \begin{bmatrix} -0.3 \\ 0.5 \end{bmatrix}$$

$$\frac{\|\mathbf{r}\|_\infty}{\|\mathbf{b}\|_\infty} = \frac{0.5}{9} = \frac{1}{18}$$ ◀

We can see by inspection that the actual solution to the system in the example above is $\mathbf{x} = \begin{bmatrix} 1 \\ 1 \end{bmatrix}$. The error $\mathbf{e}$ is given by

$$\mathbf{e} = \mathbf{x} - \mathbf{x}' = \begin{bmatrix} -1.0 \\ 0.9 \end{bmatrix}$$

The relative error is given by

$$\frac{\|\mathbf{e}\|_\infty}{\|\mathbf{x}\|_\infty} = \frac{1.0}{1} = 1$$

The relative error is 18 times the relative residual. This is not surprising, since $\text{cond}(A) = 24$. The results are similar using $\| \cdot \|_1$. In this case

$$\frac{\|\mathbf{r}\|_1}{\|\mathbf{b}\|_1} = \frac{0.8}{15} = \frac{4}{75} \quad \text{and} \quad \frac{\|\mathbf{e}\|_1}{\|\mathbf{x}\|_1} = \frac{1.9}{2} = \frac{19}{20}$$

The condition number of a nonsingular matrix actually gives us valuable information about the conditioning of A. Let A' be a new matrix formed by altering the entries of A slightly. Let $E = A' - A$. Thus $A' = A + E$, where the entries of E are small relative to the entries of A. A will be ill-conditioned if for some such E the solutions to $A'\mathbf{x} = \mathbf{b}$ and $A\mathbf{x} = \mathbf{b}$ vary greatly. Let $\mathbf{x}'$ be the solution to $A'\mathbf{x} = \mathbf{b}$ and $\mathbf{x}$ be the solution to $A\mathbf{x} = \mathbf{b}$. The condition number allows us to compare the change in solution relative to $\mathbf{x}'$, to the relative change in the matrix A.

$$\mathbf{x} = A^{-1}\mathbf{b} = A^{-1}A'\mathbf{x}' = A^{-1}(A + E)\mathbf{x}' = \mathbf{x}' + A^{-1}E\mathbf{x}'$$

Hence

$$\mathbf{x} - \mathbf{x}' = A^{-1}E\mathbf{x}'$$

Using inequality (7), we see that

$$\|\mathbf{x} - \mathbf{x}'\| \le \|A^{-1}\| \, \|E\| \, \|\mathbf{x}'\|$$

or

(11) $$\frac{\|\mathbf{x} - \mathbf{x}'\|}{\|\mathbf{x}'\|} \le \|A^{-1}\| \, \|E\| = \text{cond}(A)\frac{\|E\|}{\|A\|}$$

Let us return to Example 2 and see how the inequality (11) applies. Let A and A' be the two coefficient matrices in Example 2:

$$E = A' - A = \begin{bmatrix} 0 & 0 \\ 0 & 0.0005 \end{bmatrix}$$

and

$$A^{-1} = \begin{bmatrix} 2000.5 & -2000 \\ -2000 & 2000 \end{bmatrix}$$

In terms of the ∞-norm, the relative error in A is

$$\frac{\|E\|_\infty}{\|A\|_\infty} = \frac{0.0005}{4.0005} \approx 0.0001$$

and the condition number is

$$\text{cond}(A) = \|A\|_\infty \|A^{-1}\|_\infty = (4.0005)(4000.5) \approx 16{,}004$$

The bound on the relative error given in (11) is then

$$\text{cond}(A)\frac{\|E\|}{\|A\|} = \|A^{-1}\|\,\|E\| = (4000.5)(0.0005) \approx 2$$

The actual relative error for the systems in Example 2 is

$$\frac{\|\mathbf{x} - \mathbf{x}'\|_\infty}{\|\mathbf{x}'\|_\infty} = \frac{1}{2}$$

If A is a nonsingular $n \times n$ matrix and we compute its condition number using the 2-norm, then

$$\text{cond}_2(A) = \|A\|_2\|A^{-1}\|_2 = \frac{\sigma_1}{\sigma_n}$$

If σ_n is small, then $\text{cond}_2(A)$ will be large. The smallest singular value, σ_n, is a measure of how close the matrix is to being singular. Thus, the closer the matrix is to being singular the more ill-conditioned it is. If the coefficient matrix of a linear system is close to being singular, then small changes in the matrix due to roundoff errors could result in drastic changes to the solution of the system. To illustrate the relation between conditioning and nearness to singularity, let us look again at an example from Chapter 6

EXAMPLE 5. In Section 5 of Chapter 6, we saw that the nonsingular 100×100 matrix

$$A = \begin{bmatrix} 1 & -1 & -1 & \cdots & -1 & -1 \\ 0 & 1 & -1 & \cdots & -1 & -1 \\ 0 & 0 & 1 & \cdots & -1 & -1 \\ \vdots & & & & & \\ 0 & 0 & 0 & \cdots & 1 & -1 \\ 0 & 0 & 0 & \cdots & 0 & 1 \end{bmatrix}$$

is actually very close to being singular, and to make it singular, we need only change the value of the (100, 1) entry of A from 0 to $-\frac{1}{2^{98}}$. It follows from Theorem 6.5.3 that

$$\sigma_n = \min_{X \text{ singular}} \|A - X\|_F \le \frac{1}{2^{98}}$$

so $\text{cond}_2(A)$ must be very large. Using the infinity norm, it is even easier to see that A is extremely ill-conditioned. The inverse of A is given by

$$A^{-1} = \begin{bmatrix} 1 & 1 & 2 & 4 & \cdots & 2^{98} \\ 0 & 1 & 1 & 2 & \cdots & 2^{97} \\ & \vdots & & & & \\ 0 & 0 & 0 & 0 & \cdots & 2^1 \\ 0 & 0 & 0 & 0 & \cdots & 2^0 \\ 0 & 0 & 0 & 0 & \cdots & 1 \end{bmatrix}$$

The infinity norms of A and A^{-1} are both determined using the first rows of the matrices.

$$\text{cond}_\infty A = \|A\|_\infty \|A^{-1}\|_\infty = 100 \times 2^{99} \approx 6.34 \times 10^{31} \qquad \blacktriangleleft$$

EXERCISES

1. Determine $\| \cdot \|_F$, $\| \cdot \|_\infty$, and $\| \cdot \|_1$ for each of the following matrices.

(a) $\begin{bmatrix} 1 & 0 \\ 0 & 1 \end{bmatrix}$ (b) $\begin{bmatrix} 1 & 4 \\ -2 & 2 \end{bmatrix}$ (c) $\begin{bmatrix} \frac{1}{2} & \frac{1}{2} \\ \frac{1}{2} & \frac{1}{2} \end{bmatrix}$

(d) $\begin{bmatrix} 0 & 5 & 1 \\ 2 & 3 & 1 \\ 1 & 2 & 2 \end{bmatrix}$ (e) $\begin{bmatrix} 5 & 0 & 5 \\ 4 & 1 & 0 \\ 3 & 2 & 1 \end{bmatrix}$

2. Let

$$A = \begin{bmatrix} 2 & 0 \\ 0 & -2 \end{bmatrix} \quad \text{and} \quad \mathbf{x} = \begin{bmatrix} x_1 \\ x_2 \end{bmatrix}$$

Write $\|A\mathbf{x}\|_2 / \|\mathbf{x}\|_2$ in terms of x_1 and x_2. Determine the value of $\|A\|_2$.

3. Let

$$A = \begin{bmatrix} 1 & 0 \\ 0 & 0 \end{bmatrix}$$

Show that $\|A\|_2 = 1$.

4. Let I denote the $n \times n$ identity matrix. Determine the values of $\|I\|_1$, $\|I\|_\infty$, and $\|I\|_F$.

5. Let $\| \cdot \|_M$ denote a matrix norm on $R^{n \times n}$, $\| \cdot \|_V$ denote a vector norm on R^n, and I be the $n \times n$ identity matrix. Show that:

   (a) If $\| \cdot \|_M$ and $\| \cdot \|_V$ are compatible, then $\|I\|_M \geq 1$.
   (b) If $\| \cdot \|_M$ is subordinate to $\| \cdot \|_V$, then $\|I\|_M = 1$.

6. Given

$$
A = \begin{pmatrix} 3 & -1 & -2 \\ -1 & 2 & -7 \\ 4 & 1 & 4 \end{pmatrix}
$$

   (a) Determine $\|A\|_\infty$.
   (b) Find a vector $\mathbf{x}$ whose coordinates are each ± 1 such that $\|A\mathbf{x}\|_\infty = \|A\|_\infty$. (Note that $\|\mathbf{x}\|_\infty = 1$, so $\|A\|_\infty = \|A\mathbf{x}\|_\infty / \|\mathbf{x}\|_\infty$.)

7. Theorem 7.4.2 states that

$$
\|A\|_\infty = \max_{1 \leq i \leq m} \left(\sum_{j=1}^n |a_{ij}| \right)
$$

   Prove this in two steps.

   (a) Show first that

$$
\|A\|_\infty \leq \max_{1 \leq i \leq m} \left(\sum_{j=1}^n |a_{ij}| \right)
$$

   (b) Construct a vector $\mathbf{x}$ whose coordinates are each ± 1 such that

$$
\frac{\|A\mathbf{x}\|_\infty}{\|\mathbf{x}\|_\infty} = \|A\mathbf{x}\|_\infty = \max_{1 \leq i \leq m} \left(\sum_{j=1}^n |a_{ij}| \right)
$$

8. Show that $\|A\|_F = \|A^T\|_F$.

9. Let A be a symmetric $n \times n$ matrix. Show that $\|A\|_\infty = \|A\|_1$.

10. Let A be a 5×4 matrix with singular values $\sigma_1 = 5$, $\sigma_2 = 3$, and $\sigma_3 = \sigma_4 = 1$. Determine the values of $\|A\|_2$ and $\|A\|_F$.

11. Let A be an $m \times n$ matrix.

    (a) Show that $\|A\|_2 \leq \|A\|_F$.
    (b) Under what circumstances will $\|A\|_2 = \|A\|_F$?

12. Let $\| \cdot \|$ denote the family of vector norms and let $\| \cdot \|_M$ be a subordinate matrix norm. Show that

$$
\|A\|_M = \max_{\|\mathbf{x}\|=1} \|A\mathbf{x}\|
$$

13. Let A be an $n \times n$ matrix and let $\| \cdot \|_M$ be a matrix norm that is compatible with some vector norm on R^n. If λ is an eigenvalue of A, show that $|\lambda| \leq \|A\|_M$.

14. Use the result from Exercise 13 to show that if λ is an eigenvalue of a stochastic matrix, then $|\lambda| \leq 1$.

15. Let A be an $n \times n$ matrix and $\mathbf{x} \in R^n$. Prove:

(a) $\|A\mathbf{x}\|_\infty \leq n^{1/2}\|A\|_2\|\mathbf{x}\|_\infty$

(b) $\|A\mathbf{x}\|_2 \leq n^{1/2}\|A\|_\infty\|\mathbf{x}\|_2$

(c) $n^{-1/2}\|A\|_2 \leq \|A\|_\infty \leq n^{1/2}\|A\|_2$

16. Let A be a symmetric $n \times n$ matrix with orthonormal eigenvectors $\mathbf{u}_1, \ldots, \mathbf{u}_n$. Let $\mathbf{x} \in R^n$ and let $c_i = \mathbf{u}_i^T\mathbf{x}$ for $i = 1, 2, \ldots, n$. Show that

(a) $\|A\mathbf{x}\|_2^2 = \displaystyle\sum_{i=1}^{n}(\lambda_i c_i)^2$

(b) If $\mathbf{x} \neq \mathbf{0}$, then

$$\min_{1\leq i\leq n} |\lambda_i| \leq \frac{\|A\mathbf{x}\|_2}{\|\mathbf{x}\|_2} \leq \max_{1\leq i\leq n} |\lambda_i|$$

(c) $\|A\|_2 = \displaystyle\max_{1\leq i\leq n} |\lambda_i|$

17. Let

$$A = \begin{bmatrix} 1 & -0.99 \\ -1 & 1 \end{bmatrix}$$

Find A^{-1} and $\text{cond}_\infty(A)$.

18. Solve the two systems below and compare the two solutions. Are the coefficient matrices well-conditioned? Ill-conditioned? Explain.

$$1.0x_1 + 2.0x_2 = 1.12 \qquad 1.000x_1 + 2.011x_2 = 1.120$$

$$2.0x_1 + 3.9x_2 = 2.16 \qquad 2.000x_1 + 3.982x_2 = 2.160$$

19. Let

$$A = \begin{bmatrix} 1 & 0 & 1 \\ 2 & 2 & 3 \\ 1 & 1 & 2 \end{bmatrix}$$

Calculate $\text{cond}_\infty(A) = \|A\|_\infty\|A^{-1}\|_\infty$.

20. Let A be a nonsingular $n \times n$ matrix and let $\|\cdot\|_M$ denote a matrix norm that is compatible with some vector norm on R^n. Show that

$$\text{cond}_M(A) \geq 1$$

21. Let

$$A_n = \begin{bmatrix} 1 & 1 \\ 1 & 1 - \dfrac{1}{n} \end{bmatrix}$$

for each positive integer n. Calculate:

(a) A_n^{-1} (b) $\text{cond}_\infty(A_n)$ (c) $\lim\limits_{n\to\infty} \text{cond}_\infty(A_n)$

22. If A is a 5×3 matrix with $\|A\|_2 = 8$, $\text{cond}_2(A) = 2$, and $\|A\|_F = 12$, determine the singular values of A.

23. Let

$$A = \begin{pmatrix} 3 & 2 \\ 1 & 1 \end{pmatrix} \qquad \text{and} \qquad b = \begin{pmatrix} 5 \\ 2 \end{pmatrix}$$

The computed solution using two-digit decimal floating-point arithmetic is $x = (1.1, 0.88)^T$.

(a) Determine the residual vector r and the value of the relative residual $\|r\|_\infty / \|b\|_\infty$.
(b) Find the value of $\text{cond}_\infty(A)$.
(c) Without computing the exact solution, use the results from parts (a) and (b) to obtain bounds for the relative error in the computed solution.
(d) Compute the exact solution x and determine the actual relative error. Compare this to the bounds derived in part (c).

24. Let

$$A = \begin{pmatrix} -0.50 & 0.75 & -0.25 \\ -0.50 & 0.25 & 0.25 \\ 1.00 & -0.50 & 0.50 \end{pmatrix}$$

Calculate $\text{cond}_1(A) = \|A\|_1 \|A^{-1}\|_1$.

25. Let A be the matrix in Exercise 24 and let

$$A' = \begin{pmatrix} -0.5 & 0.8 & -0.3 \\ -0.5 & 0.3 & 0.3 \\ 1.0 & -0.5 & 0.5 \end{pmatrix}$$

Let x and x' be the solutions to $Ax = b$ and $A'x = b$, respectively, for some $b \in R^3$. Find a bound for the relative error $(\|x - x'\|_1)/\|x'\|_1$.

26. Given

$$A = \begin{pmatrix} 1 & -1 & -1 & -1 \\ 0 & 1 & -1 & -1 \\ 0 & 0 & 1 & -1 \\ 0 & 0 & 0 & 1 \end{pmatrix}, \qquad b = \begin{pmatrix} 5.00 \\ 1.02 \\ 1.04 \\ 1.10 \end{pmatrix}$$

An approximate solution to $Ax = b$ is calculated by rounding the entries of b to the nearest integer and then solving the rounded system using integer arithmetic. The calculated solution is $x' = (12, 4, 2, 1)^T$. Let r denote the residual vector.

(a) Determine the values of $\|\mathbf{r}\|_\infty$ and $\text{cond}_\infty(A)$.

(b) Use your answer to part (a) to find an upper bound for the relative error in the solution.

(c) Compute the exact solution $\mathbf{x}$ and determine the relative error $\dfrac{\|\mathbf{x} - \mathbf{x}'\|_\infty}{\|\mathbf{x}\|_\infty}$.

27. Let A and B be nonsingular $n \times n$ matrices. Show that
$$\text{cond}(AB) \leq \text{cond}(A)\,\text{cond}(B)$$

28. Let D be a nonsingular $n \times n$ diagonal matrix. Show that
$$\text{cond}_1(D) = \text{cond}_\infty(D) = \frac{d_{\max}}{d_{\min}}$$
where
$$d_{\max} = \max_{1 \leq i \leq n} |d_{ii}| \qquad \text{and} \qquad d_{\min} = \min_{1 \leq i \leq n} |d_{ii}|$$

29. Let D, $d_{\max}$, and $d_{\min}$ be defined as in Exercise 28. Show that:

(a) $\|D\|_2 = d_{\max}$ (b) $\text{cond}_2(D) = \dfrac{d_{\max}}{d_{\min}}$

30. Let Q be an $n \times n$ orthogonal matrix. Show that:

(a) $\|Q\|_2 = 1$

(b) $\text{cond}_2(Q) = 1$

(c) For any $\mathbf{b} \in R^n$, the relative error in the solution to $Q\mathbf{x} = \mathbf{b}$ is equal to the relative residual, that is,
$$\frac{\|\mathbf{e}\|_2}{\|\mathbf{x}\|_2} = \frac{\|\mathbf{r}\|_2}{\|\mathbf{b}\|_2}$$

31. Let A be an $n \times n$ matrix and let Q and V be $n \times n$ orthogonal matrices. Show that:

(a) $\|QA\|_2 = \|A\|_2$ (b) $\|AV\|_2 = \|A\|_2$ (c) $\|QAV\|_2 = \|A\|_2$

32. Let A be an $m \times n$ matrix and let σ_1 be the largest singular value of A. If $\mathbf{x}$ and $\mathbf{y}$ are nonzero vectors in R^n, show that each of the following holds:

(a) $\dfrac{|\mathbf{x}^T A \mathbf{y}|}{\|\mathbf{x}\|_2 \|\mathbf{y}\|_2} \leq \sigma_1$

[*Hint:* Make use of the Cauchy-Schwarz inequality.]

(b) $\max\limits_{\mathbf{x} \neq 0,\, \mathbf{y} \neq 0} \dfrac{|\mathbf{x}^T A \mathbf{y}|}{\|\mathbf{x}\| \|\mathbf{y}\|} = \sigma_1$

33. Let A be an $m \times n$ matrix with singular value decomposition $U \Sigma V^T$. Show that
$$\min_{\mathbf{x} \neq 0} \frac{\|A\mathbf{x}\|_2}{\|\mathbf{x}\|_2} = \sigma_n$$

34. Let A be an $m \times n$ matrix with singular value decomposition $U\Sigma V^T$. Show that for any vector $\mathbf{x} \in R^n$

$$\sigma_n \|\mathbf{x}\|_2 \le \|A\mathbf{x}\|_2 \le \sigma_1 \|\mathbf{x}\|_2$$

35. Let A be a nonsingular $n \times n$ matrix and let Q be an $n \times n$ orthogonal matrix. Show that:

(a) $\text{cond}_2(QA) = \text{cond}_2(AQ) = \text{cond}_2(A)$

(b) If $B = Q^T A Q$, then $\text{cond}_2(B) = \text{cond}_2(A)$.

36. Let A be a symmetric nonsingular $n \times n$ matrix with eigenvalues $\lambda_1, \dots, \lambda_n$. Show that

$$\text{cond}_2(A) = \frac{\max\limits_{1 \le i \le n} |\lambda_i|}{\min\limits_{1 \le i \le n} |\lambda_i|}$$

5 ORTHOGONAL TRANSFORMATIONS

Orthogonal transformations are one of the most important tools in numerical linear algebra. The types of orthogonal transformations that will be introduced in this section are easy to work with and do not require much storage. Most important, processes that involve orthogonal transformations are inherently stable. For example, let $\mathbf{x} \in R^n$ and $\mathbf{x}' = \mathbf{x} + \mathbf{e}$ be an approximation to $\mathbf{x}$: if Q is an orthogonal matrix, then

$$Q\mathbf{x}' = Q\mathbf{x} + Q\mathbf{e}$$

The error in $Q\mathbf{x}'$ is $Q\mathbf{e}$. With respect to the 2-norm, the vector $Q\mathbf{e}$ is the same size as $\mathbf{e}$.

$$\|Q\mathbf{e}\|_2 = \|\mathbf{e}\|_2$$

Similarly, if $A' = A + E$, then

$$QA' = QA + QE$$

and

$$\|QE\|_2 = \|E\|_2$$

When an orthogonal transformation is applied to a vector or matrix, the error will not grow with respect to the 2-norm.

Elementary Orthogonal Transformations

By an *elementary orthogonal matrix*, we mean a matrix of the form

$$Q = I - 2\mathbf{u}\mathbf{u}^T$$

where $\mathbf{u} \in R^n$ and $\|\mathbf{u}\|_2 = 1$. To see that Q is orthogonal, note that

$$Q^T = (I - 2\mathbf{u}\mathbf{u}^T)^T = I - 2\mathbf{u}\mathbf{u}^T = Q$$

and

$$Q^T Q = Q^2 = (I - 2\mathbf{u}\mathbf{u}^T)(I - 2\mathbf{u}\mathbf{u}^T)$$
$$= I - 4\mathbf{u}\mathbf{u}^T + 4\mathbf{u}(\mathbf{u}^T \mathbf{u})\mathbf{u}^T$$
$$= I$$

Thus, if Q is an elementary orthogonal matrix, then

$$Q^T = Q^{-1} = Q$$

The matrix $Q = I - 2\mathbf{u}\mathbf{u}^T$ is completely determined by the unit vector $\mathbf{u}$. Rather than store all n^2 entries of Q, we need store only the vector $\mathbf{u}$. To compute $Q\mathbf{x}$, note that

$$Q\mathbf{x} = (I - 2\mathbf{u}\mathbf{u}^T)\mathbf{x}$$
$$= \mathbf{x} - 2\alpha\mathbf{u} \qquad \text{where} \qquad \alpha = \mathbf{u}^T \mathbf{x}$$

The matrix product QA is computed as follows:

$$QA = (Q\mathbf{a}_1, Q\mathbf{a}_2, \dots, Q\mathbf{a}_n)$$

where

$$Q\mathbf{a}_i = \mathbf{a}_i - 2\alpha_i\mathbf{u} \qquad \alpha_i = \mathbf{u}^T \mathbf{a}_i$$

Elementary orthogonal transformations can be used to obtain a QR factorization of A, and this in turn can be used to solve a linear system $A\mathbf{x} = \mathbf{b}$. As with Gaussian elimination, the elementary matrices are chosen so as to produce zeros in the coefficient matrix. To see how this is done, let us consider the problem of finding a unit vector $\mathbf{u}$ such that

$$(I - 2\mathbf{u}\mathbf{u}^T)\mathbf{x} = (\alpha, 0, \dots, 0)^T = \alpha\mathbf{e}_1$$

for a given vector $\mathbf{x} \in R^n$.

Householder Transformations

Let $H = I - 2\mathbf{u}\mathbf{u}^T$. If $H\mathbf{x} = \alpha\mathbf{e}_1$, then, since H is orthogonal, it follows that

$$|\alpha| = \|\alpha\mathbf{e}_1\|_2 = \|H\mathbf{x}\|_2 = \|\mathbf{x}\|_2$$

If we take $\alpha = \|\mathbf{x}\|_2$ and $H\mathbf{x} = \alpha\mathbf{e}_1$, then, since H is its own inverse, it follows that

$$(1) \qquad \mathbf{x} = H(\alpha\mathbf{e}_1) = \alpha(\mathbf{e}_1 - (2u_1)\mathbf{u})$$

Thus

$$x_1 = \alpha(1 - 2u_1^2)$$
$$x_2 = -2\alpha u_1 u_2$$
$$\vdots$$
$$x_n = -2\alpha u_1 u_n$$

Solving for the u_i's, we get

$$u_1 = \pm \left(\frac{\alpha - x_1}{2\alpha} \right)^{1/2}$$

$$u_i = \frac{-x_i}{2\alpha u_1} \qquad \text{for} \quad i = 2, \ldots, n$$

If we let

$$u_1 = -\left(\frac{\alpha - x_1}{2\alpha} \right)^{1/2} \qquad \text{and set} \qquad \beta = \alpha(\alpha - x_1)$$

then

$$-2\alpha u_1 = [2\alpha(\alpha - x_1)]^{1/2} = (2\beta)^{1/2}$$

It follows that

$$\mathbf{u} = \left(-\frac{1}{2\alpha u_1} \right) (-2\alpha u_1^2, x_2, \ldots, x_n)^T$$

$$= \frac{1}{\sqrt{2\beta}} (x_1 - \alpha, x_2, \ldots, x_n)^T$$

If we set $\mathbf{v} = (x_1 - \alpha, x_2, \ldots, x_n)^T$, then

$$\|\mathbf{v}\|_2^2 = (x_1 - \alpha)^2 + \sum_{i=2}^{n} x_i^2 = 2\alpha(\alpha - x_1)$$

and hence

$$\|\mathbf{v}\|_2 = \sqrt{2\beta}$$

Thus,

$$\mathbf{u} = \frac{1}{\sqrt{2\beta}} \mathbf{v} = \frac{1}{\|\mathbf{v}\|_2} \mathbf{v}$$

In summation, given a vector $\mathbf{x} \in R^n$, if we set

$$\alpha = \|\mathbf{x}\|_2, \qquad \beta = \alpha(\alpha - x_1)$$
$$\mathbf{v} = (x_1 - \alpha, x_2, \ldots, x_n)^T$$
$$\mathbf{u} = \frac{1}{\|\mathbf{v}\|_2} \mathbf{v} = \frac{1}{\sqrt{2\beta}} \mathbf{v}$$

and

$$H = I - 2\mathbf{u}\mathbf{u}^T = I - \frac{1}{\beta} \mathbf{v}\mathbf{v}^T$$

then

$$Hx = \alpha e_1$$

The matrix H formed in this way is called a *Householder transformation*. The matrix H is determined by the vector $\mathbf{v}$ and the scalar β. For any vector $\mathbf{y} \in R^n$,

$$H\mathbf{y} = \left(I - \frac{1}{\beta} \mathbf{v}\mathbf{v}^T \right) \mathbf{y} = \mathbf{y} - \left(\frac{1}{\beta} \mathbf{v}^T \mathbf{y} \right) \mathbf{v}$$

Rather than store all n^2 entries of H, we need store only $\mathbf{v}$ and β.

EXAMPLE I.   Given the vector $\mathbf{x} = (4, 4, 2)^T$, find a Householder matrix that will zero out the last two entries of $\mathbf{x}$.

SOLUTION.   Set

$$\alpha = \|\mathbf{x}\| = 6$$

$$\beta = \alpha(\alpha - x_1) = 12$$

$$\mathbf{v} = (x_1 - \alpha, x_2, x_3)^T = (-2, 4, 2)^T$$

The Householder matrix is given by

$$H = I - \frac{1}{12} \mathbf{v}\mathbf{v}^T$$

$$= \frac{1}{3} \begin{bmatrix} 2 & 2 & 1 \\ 2 & -1 & -2 \\ 1 & -2 & 2 \end{bmatrix}$$

The reader may verify that

$$H\mathbf{x} = 6e_1 \qquad \blacktriangleleft$$

Suppose now that we wish to zero out only the last $n - k$ components of a vector $\mathbf{x} = (x_1, \ldots, x_k, x_{k+1}, \ldots, x_n)^T$. To do this, we let $\mathbf{x}^{(1)} = (x_1, \ldots, x_{k-1})^T$ and $\mathbf{x}^{(2)} = (x_k, x_{k+1}, \ldots, x_n)^T$. Let $I^{(1)}$ and $I^{(2)}$ denote the $(k - 1) \times (k - 1)$ and $(n-k+1) \times (n-k+1)$ identity matrices, respectively. By the methods just described, we can construct a Householder matrix $H_k^{(2)} = I^{(2)} - (1/\beta_k)\mathbf{v}_k \mathbf{v}_k^T$ such that

$$H_k^{(2)} \mathbf{x}^{(2)} = \|\mathbf{x}^{(2)}\|_2 e_1$$

Let

$$H_k = \begin{bmatrix} I^{(1)} & O \\ O & H_k^{(2)} \end{bmatrix}$$

It follows that

$$
\begin{aligned}
H_k \mathbf{x} &= \begin{bmatrix} I^{(1)} & O \\ O & H_k^{(2)} \end{bmatrix} \begin{bmatrix} \mathbf{x}^{(1)} \\ \mathbf{x}^{(2)} \end{bmatrix} \\
&= \begin{bmatrix} I^{(1)}\mathbf{x}^{(1)} \\ H_k^{(2)}\mathbf{x}^{(2)} \end{bmatrix} \\
&= \left(x_1, \ldots, x_{k-1}, \left(\sum_{i=k}^{n} x_i^2 \right)^{1/2}, 0, \ldots, 0 \right)^T
\end{aligned}
$$

Remarks

1. The Householder matrix H_k defined above is an elementary orthogonal matrix. If we let

$$
\mathbf{v} = \begin{bmatrix} \mathbf{0} \\ \mathbf{v}_k \end{bmatrix} \qquad \text{and} \qquad \mathbf{u} = (1/\|\mathbf{v}\|)\mathbf{v}
$$

then

$$
H_k = I - \frac{1}{\beta_k}\mathbf{v}\mathbf{v}^T = I - 2\mathbf{u}\mathbf{u}^T
$$

2. H_k acts like the identity on the first $k - 1$ coordinates of any vector $\mathbf{y} \in R^n$. If $\mathbf{y} = (y_1, \ldots, y_{k-1}, y_k, \ldots, y_n)^T$, $\mathbf{y}^{(1)} = (y_1, \ldots, y_{k-1})^T$, and $\mathbf{y}^{(2)} = (y_k, \ldots, y_n)^T$, then

$$
H_k\mathbf{y} = \begin{bmatrix} I^{(1)} & O \\ O & H_k^{(2)} \end{bmatrix} \begin{bmatrix} \mathbf{y}^{(1)} \\ \mathbf{y}^{(2)} \end{bmatrix} = \begin{bmatrix} \mathbf{y}^{(1)} \\ H_k^{(2)}\mathbf{y}^{(2)} \end{bmatrix}
$$

In particular, if $\mathbf{y}^{(2)} = \mathbf{0}$, then $H_k\mathbf{y} = \mathbf{y}$.

3. It is generally not necessary to store the entire matrix H_k. It suffices to store the $n - k + 1$ vector $\mathbf{v}_k$ and the scalar β_k. ◀

EXAMPLE 2. Find a Householder matrix that zeros out the last two entries of $\mathbf{y} = (3, 4, 4, 2)^T$ while leaving the first entry unchanged.

SOLUTION. The Householder matrix will only change the last three entries of $\mathbf{y}$. These entries correspond to the vector $\mathbf{x} = (4, 4, 2)^T$ in R^3. But this is the vector whose last two entries were zeroed out in Example 1. The 3×3 Householder matrix from Example 1 can be used to form a 4×4 matrix

$$
H = \begin{bmatrix} 1 & 0 & 0 & 0 \\ 0 & \frac{2}{3} & \frac{2}{3} & \frac{1}{3} \\ 0 & \frac{2}{3} & -\frac{1}{3} & -\frac{2}{3} \\ 0 & \frac{1}{3} & -\frac{2}{3} & \frac{2}{3} \end{bmatrix}
$$

which will have the desired effect on $\mathbf{y}$. We leave it to the reader to verify that $H\mathbf{y} = (3, 6, 0, 0)^T$. ◀

We are now ready to apply Householder transformations to solve linear systems. If A is a nonsingular $n \times n$ matrix, we can use Householder transformations to reduce A to triangular form. To begin with, we can find a Householder transformation $H_1 = I - (1/\beta_1)\mathbf{v}_1\mathbf{v}_1^T$, which when applied to the first column of A will give a multiple of $\mathbf{e}_1$. Thus $H_1 A$ will be of the form

$$
\begin{pmatrix}
\times & \times & \cdots & \times \\
0 & \times & \cdots & \times \\
0 & \times & \cdots & \times \\
\vdots & & & \\
0 & \times & \cdots & \times
\end{pmatrix}
$$

We can then find a Householder transformation H_2 that will zero out the last $n - 2$ elements in the second column of $H_1 A$ while leaving the first element in that column unchanged. It follows from remark 2 that H_2 will have no effect on the first column of $H_1 A$.

$$
H_2 H_1 A =
\begin{pmatrix}
\times & \times & \times & \cdots & \times \\
0 & \times & \times & \cdots & \times \\
0 & 0 & \times & \cdots & \times \\
\vdots & & & & \\
0 & 0 & \times & \cdots & \times
\end{pmatrix}
$$

We can continue to apply Householder transformations in this fashion until we end up with an upper triangular matrix, which we will denote by R. Thus

$$ H_{n-1} \cdots H_2 H_1 A = R $$

It follows that

$$ A = H_1^{-1} H_2^{-1} \cdots H_{n-1}^{-1} R $$
$$ = H_1 H_2 \cdots H_{n-1} R $$

Let $Q = H_1 H_2 \cdots H_{n-1}$. The matrix Q is orthogonal and A can be factored into an orthogonal matrix times an upper triangular matrix:

$$ A = QR $$

Once A has been factored into a product QR, the system $A\mathbf{x} = \mathbf{b}$ is easily solved.

$$ A\mathbf{x} = \mathbf{b} $$
$$ QR\mathbf{x} = \mathbf{b} $$
$$ R\mathbf{x} = Q^T\mathbf{b} $$

(2) $$ R\mathbf{x} = H_{n-1} \cdots H_2 H_1 \mathbf{b} $$

Once $H_{n-1} \cdots H_2 H_1 \mathbf{b}$ has been calculated, the system (2) can be solved using back substitution.

Storage. The vector $\mathbf{v}_k$ can be stored in the kth column of A. Since $\mathbf{v}_k$ has $n - k + 1$ nonzero entries and there are only $n - k$ zeros in the kth column of the reduced matrix, it is necessary to store r_{kk} elsewhere. The diagonal elements of R can either be stored in an n vector or in an additional row added to A. The β_k's can also be stored in an additional row of A.

$$\begin{bmatrix} v_{11} & r_{12} & r_{13} & r_{14} \\ v_{12} & v_{22} & r_{23} & r_{24} \\ v_{13} & v_{23} & v_{33} & r_{34} \\ v_{14} & v_{24} & v_{34} & 0 \\ r_{11} & r_{22} & r_{33} & r_{44} \\ \beta_1 & \beta_2 & \beta_3 & 0 \end{bmatrix}$$

Operation Count. In solving an $n \times n$ system using Householder transformations, most of the work is done in reducing A to triangular form. The number of operations required is approximately $\frac{2}{3}n^3$ multiplications, $\frac{2}{3}n^3$ additions, and $n - 1$ square roots.

Rotations and Reflections

Often it will be desirable to have a transformation that annihilates only a single entry of a vector. In this case, it is convenient to use either a rotation or a reflection. Let us consider first the two-dimensional case.

Let

$$R = \begin{bmatrix} \cos\theta & -\sin\theta \\ \sin\theta & \cos\theta \end{bmatrix} \quad \text{and} \quad G = \begin{bmatrix} \cos\theta & \sin\theta \\ \sin\theta & -\cos\theta \end{bmatrix}$$

and let

$$\mathbf{x} = \begin{bmatrix} x_1 \\ x_2 \end{bmatrix} = \begin{bmatrix} r\cos\alpha \\ r\sin\alpha \end{bmatrix}$$

be a vector in R^2.

$$R\mathbf{x} = \begin{bmatrix} r\cos(\theta + \alpha) \\ r\sin(\theta + \alpha) \end{bmatrix} \quad \text{and} \quad G\mathbf{x} = \begin{bmatrix} r\cos(\theta - \alpha) \\ r\sin(\theta - \alpha) \end{bmatrix}$$

R represents a rotation in the plane by an angle θ. The matrix G has the effect of reflecting $\mathbf{x}$ about the line $x_2 = [\tan(\theta/2)]x_1$ (see Figure 7.5.1). If we set $\cos\theta = x_1/r$ and $\sin\theta = -x_2/r$, then

$$R\mathbf{x} = \begin{bmatrix} x_1\cos\theta - x_2\sin\theta \\ x_1\sin\theta + x_2\cos\theta \end{bmatrix} = \begin{bmatrix} r \\ 0 \end{bmatrix}$$

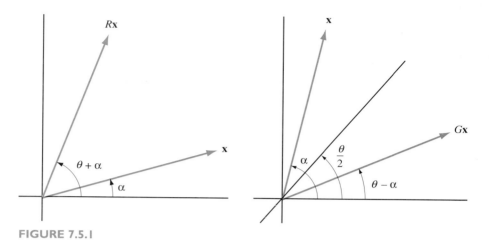

FIGURE 7.5.1

If we set $\cos\theta = x_1/r$ and $\sin\theta = x_2/r$, then

$$Gx = \begin{bmatrix} x_1\cos\theta + x_2\sin\theta \\ x_1\sin\theta - x_2\cos\theta \end{bmatrix} = \begin{bmatrix} r \\ 0 \end{bmatrix}$$

Both R and G are orthogonal. The matrix G is also symmetric. Indeed, G is an elementary orthogonal matrix. If we let $\mathbf{u} = (\sin\theta/2, -\cos\theta/2)^T$, then $G = I - 2\mathbf{u}\mathbf{u}^T$.

EXAMPLE 3. Let $\mathbf{x} = (-3, 4)^T$. To find a rotation matrix R to zero out the second coordinate of $\mathbf{x}$, set

$$r = \sqrt{(-3)^2 + 4^2} = 5$$

$$\cos\theta = \frac{x_1}{r} = -\frac{3}{5}$$

$$\sin\theta = -\frac{x_2}{r} = -\frac{4}{5}$$

and set

$$R = \begin{bmatrix} \cos\theta & -\sin\theta \\ \sin\theta & \cos\theta \end{bmatrix} = \begin{bmatrix} -\frac{3}{5} & \frac{4}{5} \\ -\frac{4}{5} & -\frac{3}{5} \end{bmatrix}$$

The reader may verify that $R\mathbf{x} = 5\mathbf{e}_1$.

To find a reflection matrix G that zeros out the second coordinate of $\mathbf{x}$, compute r and $\cos\theta$ in the same way as for the rotation matrix, but set

$$\sin\theta = \frac{x_2}{r} = \frac{4}{5}$$

and

$$G = \begin{bmatrix} \cos\theta & \sin\theta \\ \sin\theta & -\cos\theta \end{bmatrix} = \begin{bmatrix} -\frac{3}{5} & \frac{4}{5} \\ \frac{4}{5} & \frac{3}{5} \end{bmatrix}$$

The reader may verify that $G\mathbf{x} = 5\mathbf{e}_1$.   ◀

Let us now consider the n-dimensional case. Let R and G be $n \times n$ matrices with

$$r_{ii} = r_{jj} = \cos\theta \qquad\qquad g_{ii} = \cos\theta,\ g_{jj} = -\cos\theta$$
$$r_{ji} = \sin\theta,\ r_{ij} = -\sin\theta \qquad g_{ji} = g_{ij} = \sin\theta$$

and $r_{st} = g_{st} = \delta_{st}$ for all other entries of R and G. Thus R and G resemble the identity matrix except for the $ii, ij, jj,$ and ji positions. Let $c = \cos\theta$ and $s = \sin\theta$. If $\mathbf{x} \in R^n$, then

$$R\mathbf{x} = (x_1, \dots, x_{i-1}, x_i c - x_j s, x_{i+1}, \dots, x_{j-1}, x_i s + x_j c, x_{j+1}, \dots, x_n)^T$$

and

$$G\mathbf{x} = (x_1, \dots, x_{i-1}, x_i c + x_j s, x_{i+1}, \dots, x_{j-1}, x_i s - x_j c, x_{j+1}, \dots, x_n)^T$$

The transformations R and G only alter the ith and jth components of a vector. They have no effect on the other coordinates. We will refer to R as a *plane rotation* and to G as a *Givens transformation* or a *Givens reflection*. If we set

$$c = \frac{x_i}{r} \qquad \text{and} \qquad s = -\frac{x_j}{r} \qquad \left(r = \sqrt{x_i^2 + x_j^2}\right)$$

then the jth component of $R\mathbf{x}$ will be 0. If we set

$$c = \frac{x_i}{r} \qquad \text{and} \qquad s = \frac{x_j}{r}$$

then the jth component of $G\mathbf{x}$ will be 0.

EXAMPLE 4.   Let $\mathbf{x} = (5, 8, 12)^T$. Find a rotation matrix R that zeros out the third entry of $\mathbf{x}$ but leaves the second entry of $\mathbf{x}$ unchanged.

SOLUTION.   Since R will only act on x_1 and x_3, set

$$r = \sqrt{x_1^2 + x_3^2} = 13$$
$$c = \frac{x_1}{r} = \frac{5}{13}$$
$$s = -\frac{x_3}{r} = -\frac{12}{13}$$

and set

$$
R = \begin{bmatrix} c & 0 & -s \\ 0 & 1 & 0 \\ s & 0 & c \end{bmatrix} = \begin{bmatrix} \frac{5}{13} & 0 & \frac{12}{13} \\ 0 & 1 & 0 \\ -\frac{12}{13} & 0 & \frac{5}{13} \end{bmatrix}
$$

The reader may verify that $R\mathbf{x} = (13, 8, 0)^T$.   ◀

Given a nonsingular $n \times n$ matrix A, we can use either plane rotations or Givens transformations to obtain a QR factorization of A. Let G_{21} be the Givens transformation acting on the first and second coordinates, which when applied to A results in a zero in the $(2, 1)$ position. We can apply another Givens transformation, G_{31}, to $G_{21}A$ to obtain a zero in the $(3, 1)$ position. This process can be continued until the last $n - 1$ entries in the first column have been eliminated.

$$
G_{n1} \cdots G_{31} G_{21} A = \begin{bmatrix} \times & \times & \cdots & \times \\ 0 & \times & \cdots & \times \\ 0 & \times & \cdots & \times \\ \vdots & & & \\ 0 & \times & \cdots & \times \end{bmatrix}
$$

At the next step, Givens transformations $G_{32}, G_{42}, \ldots, G_{n2}$ are used to eliminate the last $n - 2$ entries in the second column. This process is continued until all elements below the diagonal have been eliminated.

$$
(G_{n,n-1}) \cdots (G_{n2} \cdots G_{32})(G_{n1} \cdots G_{21})A = R \qquad (R \text{ upper triangular})
$$

If we let $Q^T = (G_{n,n-1}) \cdots (G_{n2} \cdots G_{32})(G_{n1} \cdots G_{21})$, then $A = QR$ and the system $A\mathbf{x} = \mathbf{b}$ is equivalent to the system

$$
R\mathbf{x} = Q^T\mathbf{b}
$$

This system can be solved by back substitution.

Operation Count. The QR factorization of A using Givens transformations or plane rotations requires roughly $\frac{4}{3}n^3$ multiplications, $\frac{2}{3}n^3$ additions, and $\frac{1}{2}n^2$ square roots.

EXERCISES

1. For each of the following vectors $\mathbf{x}$, find a rotation matrix R such that $R\mathbf{x} = \|\mathbf{x}\|_2 \mathbf{e}_1$.

   (a) $\mathbf{x} = (1, 1)^T$ (b) $\mathbf{x} = (\sqrt{3}, -1)^T$ (c) $\mathbf{x} = (-4, 3)^T$

2. Given $\mathbf{x} \in R^3$, define

$$
r_{ij} = \left(x_i^2 + x_j^2 \right)^{1/2} \qquad i, j = 1, 2, 3
$$

For each of the following, determine a Givens transformation G_{ij} such that the ith and jth coordinates of $G_{ij}\mathbf{x}$ are r_{ij} and 0, respectively.

(a) $\mathbf{x} = (3, 1, 4)^T$, $i = 1$, $j = 3$
(b) $\mathbf{x} = (1, -1, 2)^T$, $i = 1$, $j = 2$
(c) $\mathbf{x} = (4, 1, \sqrt{3})^T$, $i = 2$, $j = 3$
(d) $\mathbf{x} = (4, 1, \sqrt{3})^T$, $i = 3$, $j = 2$

3. For each of the given vectors $\mathbf{x}$, find a Householder transformation such that $H\mathbf{x} = \alpha\mathbf{e}_1$, where $\alpha = \|\mathbf{x}\|_2$.

(a) $\mathbf{x} = (8, -1, -4)^T$ (b) $\mathbf{x} = (6, 2, 3)^T$ (c) $\mathbf{x} = (7, 4, -4)^T$

4. For each of the following, find a Householder transformation that zeros out the last two coordinates of the vector.

(a) $\mathbf{x} = (5, 8, 4, 1)^T$ (b) $\mathbf{x} = (4, -3, -2, -1, 2)^T$

5. Given

$$A = \begin{bmatrix} 3 & 3 & -2 \\ 1 & 1 & 1 \\ 1 & -5 & 1 \\ 5 & -1 & 2 \end{bmatrix}$$

(a) Determine the scalar β and vector $\mathbf{v}$ for the Householder matrix $H = I - (1/\beta)\mathbf{v}\mathbf{v}^T$ that zeros out the last three entries of $\mathbf{a}_1$.
(b) Without explicitly forming the matrix H, compute the product HA.

6. Given

$$A = \begin{bmatrix} 1 & 2 & -4 \\ 2 & 6 & 7 \\ -2 & 1 & 8 \end{bmatrix} \quad \text{and} \quad \mathbf{b} = \begin{bmatrix} 9 \\ 9 \\ -3 \end{bmatrix}$$

(a) Use Householder transformations to transform A into an upper triangular matrix R. Also transform the vector $\mathbf{b}$; that is, compute $\mathbf{b}^{(1)} = H_2 H_1 \mathbf{b}$.
(b) Solve $R\mathbf{x} = \mathbf{b}^{(1)}$ for $\mathbf{x}$ and check your answer by computing the residual $\mathbf{b} - A\mathbf{x}$.

7. For each of the following systems, use a Givens reflection to transform the system to upper triangular form and then solve the upper triangular system.

(a) $3x_1 + 8x_2 = 5$
    $4x_1 - x_2 = -5$

(b) $x_1 + 4x_2 = 5$
    $x_1 + 2x_2 = 1$

(c) $4x_1 - 4x_2 + x_3 = 2$
    $x_2 + 3x_3 = 2$
    $-3x_1 + 3x_2 - 2x_3 = 1$

8. Suppose that you wish to eliminate the last coordinate of a vector $\mathbf{x}$ and leave the first $n - 2$ coordinates unchanged. How many operations are necessary if

this is to be done by a Givens transformation G? A Householder transformation H? If A is an $n \times n$ matrix, how many operations are required to compute GA and HA?

9. Let $H_k = I - 2\mathbf{u}\mathbf{u}^T$ be a Householder transformation with

$$\mathbf{u} = (0, \ldots, 0, u_k, u_{k+1}, \ldots, u_n)^T$$

Let $\mathbf{b} \in R^n$ and let A be an $n \times n$ matrix. How many additions and multiplications are necessary to compute (a) $H_k\mathbf{b}$; (b) $H_k A$?

10. Let $Q^T = G_{n-k} \cdots G_2 G_1$, where each G_i is a Givens transformation. Let $\mathbf{b} \in R^n$ and let A be an $n \times n$ matrix. How many additions and multiplications are necessary to compute (a) $Q^T\mathbf{b}$; (b) $Q^T A$?

11. Let R_1 and R_2 be two 2×2 rotation matrices and let G_1 and G_2 be two 2×2 Givens transformations. What type of transformations are each of the following?

   (a) $R_1 R_2$ (b) $G_1 G_2$ (c) $R_1 G_1$ (d) $G_1 R_1$

12. Let $\mathbf{x}$ and $\mathbf{y}$ be distinct vectors in R^n with $\|\mathbf{x}\|_2 = \|\mathbf{y}\|_2$. Define

$$\mathbf{u} = \frac{1}{\|\mathbf{x} - \mathbf{y}\|_2}(\mathbf{x} - \mathbf{y}) \quad \text{and} \quad Q = I - 2\mathbf{u}\mathbf{u}^T$$

Show that:

   (a) $\|\mathbf{x} - \mathbf{y}\|_2^2 = 2(\mathbf{x} - \mathbf{y})^T\mathbf{x}$ (b) $Q\mathbf{x} = \mathbf{y}$

13. Let $\mathbf{u}$ be a unit vector in C^n and let

$$U = I - 2\mathbf{u}\mathbf{u}^H$$

   (a) Show that $\mathbf{u}$ is an eigenvector of U. What is the corresponding eigenvalue?
   (b) Let $\mathbf{z}$ be a nonzero vector in C^n that is orthogonal to $\mathbf{u}$. Show that $\mathbf{z}$ is also an eigenvector of U. What is the corresponding eigenvalue?

14. Let $A = Q_1 R_1 = Q_2 R_2$, where Q_1 and Q_2 are orthogonal and R_1 and R_2 are both upper triangular and nonsingular.

   (a) Show that $Q_1^T Q_2$ is diagonal.
   (b) How do R_1 and R_2 compare? Explain.

15. Let $A = \mathbf{x}\mathbf{y}^T$, where $\mathbf{x} \in R^m$, $\mathbf{y} \in R^n$, and both $\mathbf{x}$ and $\mathbf{y}$ are nonzero vectors. Show that A has a singular value decomposition of the form $H_1 \Sigma H_2$, where H_1 and H_2 are Householder transformations and

$$\sigma_1 = \|\mathbf{x}\| \, \|\mathbf{y}\|, \qquad \sigma_2 = \sigma_3 = \cdots = \sigma_n = 0$$

16. In constructing a Householder matrix to zero out the last $n - 1$ of a vector $\mathbf{x}$ in R^n, the scalar α must satisfy $|\alpha| = \|\mathbf{x}\|_2$. Instead of always taking $\alpha = \|\mathbf{x}\|_2$, it is customary to choose $\alpha = -\|\mathbf{x}\|_2$ whenever $x_1 > 0$. Explain the advantages of choosing α in this manner.

17. Let

$$R = \begin{bmatrix} \cos\theta & -\sin\theta \\ \sin\theta & \cos\theta \end{bmatrix}$$

Show that if θ is not an integer multiple of π, then R can be factored into a product $R = ULU$ where

$$U = \begin{bmatrix} 1 & \frac{\cos\theta-1}{\sin\theta} \\ 0 & 1 \end{bmatrix} \qquad \text{and} \qquad L = \begin{bmatrix} 1 & 0 \\ \sin\theta & 1 \end{bmatrix}$$

This type of factorization of a rotation matrix arises in applications involving wavelets and filter bases.

6 THE EIGENVALUE PROBLEM

In this section we are concerned with numerical methods for computing the eigenvalues and eigenvectors of an $n \times n$ matrix A. The first method we study is called the power method. The power method is an iterative method for finding the dominant eigenvalue of a matrix and a corresponding eigenvector. By the dominant eigenvalue we mean an eigenvalue λ_1 satisfying $|\lambda_1| > |\lambda_i|$ for $i = 2, \dots, n$. If the eigenvalues of A satisfy

$$|\lambda_1| > |\lambda_2| > \cdots > |\lambda_n|$$

then the power method can be used to compute the eigenvalues one at a time. The second method is called the QR algorithm. The QR algorithm is an iterative method involving orthogonal similarity transformations. It has many advantages over the power method. It will converge whether or not A has a dominant eigenvalue, and it calculates all the eigenvalues at the same time.

In the examples in Chapter 6 the eigenvalues were determined by forming the characteristic polynomial and finding its roots. However, this procedure is generally not recommended for numerical computations. The difficulty with this procedure is that often a small change in one or more of the coefficients of the characteristic polynomial can result in a relatively large change in the computed zeros of the polynomial. For example, consider the polynomial $p(x) = x^{10}$. The lead coefficient is 1 and the remaining coefficients are all 0. If the constant term is altered by adding -10^{-10}, we obtain the polynomial $q(x) = x^{10} - 10^{-10}$. Although the coefficients of $p(x)$ and $q(x)$ only differ by 10^{-10}, the roots of $q(x)$ all have absolute value $\frac{1}{10}$, whereas the roots of $p(x)$ are all 0. Thus, even when the coefficients of the characteristic polynomial have been determined quite accurately, the computed eigenvalues may involve significant error. For this reason, the methods presented in this section do not involve the characteristic polynomial. To see that there is some advantage to working directly with the matrix A, we must determine the effect that small changes in the entries of A have on the eigenvalues. This is done in the following theorem.

▶ **THEOREM 7.6.1** *Let A be an n × n matrix with n linearly independent eigenvectors, and let X be a matrix that diagonalizes A.*

$$X^{-1}AX = D = \begin{pmatrix} \lambda_1 & & & \\ & \lambda_2 & & \\ & & \ddots & \\ & & & \lambda_n \end{pmatrix}$$

If $A' = A + E$ and λ' is an eigenvalue of A', then

(1)
$$\min_{1 \le i \le n} |\lambda' - \lambda_i| \le \mathrm{cond}_2(X)\|E\|_2$$

▶ *Proof.* We may assume that λ' is unequal to any of the λ_i's (otherwise there is nothing to prove). Thus, if we set $D_1 = D - \lambda'I$, then D_1 is a nonsingular diagonal matrix. Since λ' is an eigenvalue of A', it is also an eigenvalue of $X^{-1}A'X$. Therefore, $X^{-1}A'X - \lambda'I$ is singular and hence $D_1^{-1}(X^{-1}A'X - \lambda'I)$ is also singular. But

$$D_1^{-1}(X^{-1}A'X - \lambda'I) = D_1^{-1}X^{-1}(A + E - \lambda'I)X$$
$$= D_1^{-1}X^{-1}EX + I$$

Therefore, -1 is an eigenvalue of $D_1^{-1}X^{-1}EX$. It follows that

$$|-1| \le \|D_1^{-1}X^{-1}EX\|_2 \le \|D_1^{-1}\|_2 \, \mathrm{cond}_2(X)\|E\|_2$$

The 2-norm of D_1^{-1} is given by

$$\|D_1^{-1}\|_2 = \max_{1 \le i \le n} |\lambda' - \lambda_i|^{-1}$$

The index i that maximizes $|\lambda' - \lambda_i|^{-1}$ is the same index that minimizes $|\lambda' - \lambda_i|$. Thus,

$$\min_{1 \le i \le n} |\lambda' - \lambda_i| \le \mathrm{cond}_2(X)\|E\|_2 \qquad ◀$$

If the matrix A is symmetric, we can choose an orthogonal diagonalizing matrix. In general, if Q is any orthogonal matrix, then

$$\mathrm{cond}_2(Q) = \|Q\|_2\|Q^{-1}\|_2 = 1$$

Hence (1) simplifies to

$$\min_{1 \le i \le n} |\lambda' - \lambda_i| \le \|E\|_2$$

Thus, if A is symmetric and $\|E\|_2$ is small, the eigenvalues of A' will be close to the eigenvalues of A.

We are now ready to talk about some of the methods for calculating the eigenvalues and eigenvectors of an $n \times n$ matrix A. The first method we will present computes an eigenvector **x** of A by successively applying A to a given vector in R^n. To see the idea behind the method, let us assume that A has n linearly independent eigenvectors $\mathbf{x}_1, \ldots, \mathbf{x}_n$ and that the corresponding eigenvalues satisfy

(2) $$|\lambda_1| > |\lambda_2| \geq \cdots \geq |\lambda_n|$$

Given an arbitrary vector $\mathbf{v}_0$ in R^n, we can write

$$\mathbf{v}_0 = \alpha_1 \mathbf{x}_1 + \cdots + \alpha_n \mathbf{x}_n$$
$$A\mathbf{v}_0 = \alpha_1 \lambda_1 \mathbf{x}_1 + \alpha_2 \lambda_2 \mathbf{x}_2 + \cdots + \alpha_n \lambda_n \mathbf{x}_n$$
$$A^2\mathbf{v}_0 = \alpha_1 \lambda_1^2 \mathbf{x}_1 + \alpha_2 \lambda_2^2 \mathbf{x}_2 + \cdots + \alpha_n \lambda_n^2 \mathbf{x}_n$$

and, in general,

$$A^k\mathbf{v}_0 = \alpha_1 \lambda_1^k \mathbf{x}_1 + \alpha_2 \lambda_2^k \mathbf{x}_2 + \cdots + \alpha_n \lambda_n^k \mathbf{x}_n$$

If we define

$$\mathbf{v}_k = A^k\mathbf{v}_0 \qquad k = 1, 2, \ldots$$

then

(3) $$\frac{1}{\lambda_1^k}\mathbf{v}_k = \alpha_1 \mathbf{x}_1 + \alpha_2 \left(\frac{\lambda_2}{\lambda_1}\right)^k \mathbf{x}_2 + \cdots + \alpha_n \left(\frac{\lambda_n}{\lambda_1}\right)^k \mathbf{x}_n$$

Since

$$\left|\frac{\lambda_i}{\lambda_1}\right| < 1 \qquad \text{for} \qquad i = 2, 3, \ldots, n$$

it follows that

$$\frac{1}{\lambda_1^k}\mathbf{v}_k \to \alpha_1 \mathbf{x}_1 \qquad \text{as} \qquad k \to \infty$$

Thus, if $\alpha_1 \neq 0$, the sequence $\{(1/\lambda_1^k)\mathbf{v}_k\}$ converges to an eigenvector $\alpha_1\mathbf{x}_1$ of A. There are some obvious difficulties with the method as it has been presented so far. We cannot compute $(1/\lambda_1^k)\mathbf{v}_k$, since λ_1 is unknown. Even if λ_1 were known, there would be difficulties because of λ_1^k approaching 0 or $\pm\infty$. Fortunately, however, we do not have to scale the sequence $\{\mathbf{v}_k\}$ using $1/\lambda_1^k$. If the $\mathbf{v}_k$'s are scaled so that we obtain unit vectors at each step, the sequence will converge to a unit vector in the direction of $\mathbf{x}_1$. The eigenvalue λ_1 can be computed at the same time. This method of computing the eigenvalue of largest magnitude and the corresponding eigenvector is called the *power method*. It may be summarized as follows.

The Power Method

Two sequences $\{\mathbf{v}_k\}$ and $\{\mathbf{u}_k\}$ are defined recursively. To start, $\mathbf{u}_0$ can be any nonzero vector in R^n. Once $\mathbf{u}_k$ has been determined, the vectors $\mathbf{v}_{k+1}$ and $\mathbf{u}_{k+1}$ are calculated as follows:

1. Set $\mathbf{v}_{k+1} = A\mathbf{u}_k$.
2. Find the coordinate j_{k+1} of $\mathbf{v}_{k+1}$ of maximum modulus.
3. Set $\mathbf{u}_{k+1} = (1/v_{j_{k+1}})\mathbf{v}_{k+1}$.

The sequence $\{\mathbf{u}_k\}$ has the property that, for $k \geq 1$, $\|\mathbf{u}_k\|_\infty = u_{j_k} = 1$. If the eigenvalues of A satisfy (2) and $\mathbf{u}_0$ can be written as a linear combination of eigenvectors $\alpha_1 \mathbf{x}_1 + \cdots + \alpha_n \mathbf{x}_n$ with $\alpha_1 \neq 0$, the sequence $\{\mathbf{u}_k\}$ will converge to an eigenvector $\mathbf{y}$ of λ_1. If k is large, then $\mathbf{u}_k$ will be a good approximation to $\mathbf{y}$ and $\mathbf{v}_{k+1} = A\mathbf{u}_k$ will be a good approximation to $\lambda_1 \mathbf{y}$. Since the j_kth coordinate of $\mathbf{u}_k$ is 1, it follows that the j_kth coordinate of $\mathbf{v}_{k+1}$ will be a good approximation to λ_1.

In view of (3), we can expect that the $\mathbf{u}_k$'s will converge to $\mathbf{y}$ at the same rate at which $(\lambda_2/\lambda_1)^k$ is converging to 0. Thus, if $|\lambda_2|$ is nearly as large as $|\lambda_1|$, the convergence will be slow.

EXAMPLE I. Let

$$A = \begin{bmatrix} 2 & 1 \\ 1 & 2 \end{bmatrix}$$

It is an easy matter to determine the exact eigenvalues of A. These turn out to be $\lambda_1 = 3$ and $\lambda_2 = 1$, with corresponding eigenvectors $\mathbf{x}_1 = (1, 1)^T$ and $\mathbf{x}_2 = (1, -1)^T$. To illustrate how the vectors generated by the power method converge, we will apply the method with $\mathbf{u}_0 = (2, 1)^T$.

$$\mathbf{v}_1 = A\mathbf{u}_0 = \begin{bmatrix} 5 \\ 4 \end{bmatrix}, \qquad \mathbf{u}_1 = \frac{1}{5}\mathbf{v}_1 = \begin{bmatrix} 1.0 \\ 0.8 \end{bmatrix}$$

$$\mathbf{v}_2 = A\mathbf{u}_1 = \begin{bmatrix} 2.8 \\ 2.6 \end{bmatrix}, \qquad \mathbf{u}_2 = \frac{1}{2.8}\mathbf{v}_2 = \begin{bmatrix} 1 \\ \dfrac{13}{14} \end{bmatrix} \approx \begin{bmatrix} 1.00 \\ 0.93 \end{bmatrix}$$

$$\mathbf{v}_3 = A\mathbf{u}_2 = \frac{1}{14}\begin{bmatrix} 41 \\ 40 \end{bmatrix}, \qquad \mathbf{u}_3 = \frac{14}{41}\mathbf{v}_3 = \begin{bmatrix} 1 \\ \dfrac{40}{41} \end{bmatrix} \approx \begin{bmatrix} 1.00 \\ 0.98 \end{bmatrix}$$

$$\mathbf{v}_4 = A\mathbf{u}_3 \approx \begin{bmatrix} 2.98 \\ 2.95 \end{bmatrix}$$

If $\mathbf{u}_3 = (1.00, 0.98)^T$ is taken as an approximate eigenvector, then 2.98 is the approximate value of λ_1. Thus, with only a few iterations, the approximation for λ_1 involves an error of only 0.02. ◀

The power method can be used to compute the eigenvalue λ_1 of largest magnitude and a corresponding eigenvector $\mathbf{y}_1$. What about the remaining eigenvalues and eigenvectors? If we could reduce the problem of finding the remaining eigenvalues of A to that of finding the eigenvalues of some $(n-1) \times (n-1)$ matrix A_1, then the power method could be applied to A_1. This can actually be done by a process called *deflation*.

Deflation

The idea behind deflation is to find a nonsingular matrix H such that HAH^{-1} is a matrix of the form

(4)

$$
\left[
\begin{array}{c|ccc}
\lambda_1 & \times & \cdots & \times \\
\hline
0 & & & \\
\vdots & & A_1 & \\
0 & & &
\end{array}
\right]
$$

Since A and HAH^{-1} are similar, they have the same characteristic polynomials. Thus, if HAH^{-1} is of the form (4), then

$$\det(A - \lambda I) = \det(HAH^{-1} - \lambda I) = (\lambda_1 - \lambda)\det(A_1 - \lambda I)$$

and it follows that the remaining $n - 1$ eigenvalues of A are the eigenvalues of A_1. The question remains: How do we find such a matrix H? Note that the form (4) requires that the first column of HAH^{-1} be $\lambda_1 \mathbf{e}_1$. The first column of HAH^{-1} is $HAH^{-1}\mathbf{e}_1$. Thus

$$HAH^{-1}\mathbf{e}_1 = \lambda_1 \mathbf{e}_1$$

or, equivalently,

$$A(H^{-1}\mathbf{e}_1) = \lambda_1(H^{-1}\mathbf{e}_1)$$

So $H^{-1}\mathbf{e}_1$ is in the eigenspace corresponding to λ_1. Thus, for some eigenvector $\mathbf{x}_1$ belonging to λ_1,

$$H^{-1}\mathbf{e}_1 = \mathbf{x}_1 \qquad \text{or} \qquad H\mathbf{x}_1 = \mathbf{e}_1$$

We must find a matrix H such that $H\mathbf{x}_1 = \mathbf{e}_1$ for some eigenvector $\mathbf{x}_1$ belonging to λ_1. This can be done by means of a Householder transformation. If $\mathbf{y}_1$ is the computed eigenvector belonging to λ_1, set

$$\mathbf{x}_1 = \frac{1}{\|\mathbf{y}_1\|_2}\mathbf{y}_1$$

Since $\|\mathbf{x}_1\|_2 = 1$, we can find a Householder transformation H such that

$$H\mathbf{x}_1 = \mathbf{e}_1$$

Because H is a Householder transformation, it follows that $H^{-1} = H$ and hence HAH is the desired similarity transformation.

Reduction to Hessenberg Form

The standard methods for finding eigenvalues are all iterative. The amount of work required in each iteration is often prohibitively high unless, initially, A is in some special form that is easier to work with. If this is not the case, the standard procedure is to reduce A to a simpler form by means of similarity transformations. Generally, Householder matrices are used to transform A into a matrix of the form

$$
\begin{pmatrix}
\times & \times & \cdots & \times & \times & \times \\
\times & \times & \cdots & \times & \times & \times \\
0 & \times & \cdots & \times & \times & \times \\
0 & 0 & \cdots & \times & \times & \times \\
\vdots & & & & & \\
0 & 0 & \cdots & \times & \times & \times \\
0 & 0 & \cdots & 0 & \times & \times
\end{pmatrix}
$$

A matrix in this form is said to be in *upper Hessenberg form*. Thus B is in upper Hessenberg form if and only if $b_{ij} = 0$ whenever $i \geq j + 2$.

A matrix A can be transformed into upper Hessenberg form in the following manner. First, choose a Householder matrix H_1 so that $H_1 A$ is of the form

$$
\begin{pmatrix}
a_{11} & a_{12} & \cdots & a_{1n} \\
\times & \times & \cdots & \times \\
0 & \times & \cdots & \times \\
\vdots & & & \\
0 & \times & \cdots & \times
\end{pmatrix}
$$

The matrix H_1 will be of the form

$$
\begin{pmatrix}
1 & 0 & \cdots & 0 \\
0 & \times & \cdots & \times \\
\vdots & & & \\
0 & \times & \cdots & \times
\end{pmatrix}
$$

and hence postmultiplication of $H_1 A$ by H_1 will leave the first column unchanged. If $A^{(1)} = H_1 A H_1$, then $A^{(1)}$ is a matrix of the form

$$
\begin{pmatrix}
a_{11}^{(1)} & a_{12}^{(1)} & \cdots & a_{1n}^{(1)} \\
a_{21}^{(1)} & a_{22}^{(1)} & \cdots & a_{2n}^{(1)} \\
0 & a_{32}^{(1)} & \cdots & a_{3n}^{(1)} \\
\vdots & & & \\
0 & a_{n2}^{(1)} & \cdots & a_{nn}^{(1)}
\end{pmatrix}
$$

Since H_1 is a Householder matrix, it follows that $H_1^{-1} = H_1$, and hence $A^{(1)}$ is similar to A. Next, a Householder matrix H_2 is chosen so that

$$H_2(a_{12}^{(1)}, a_{22}^{(1)}, \ldots, a_{n2}^{(1)})^T = (a_{12}^{(1)}, a_{22}^{(1)}, \times, 0, \ldots, 0)^T$$

The matrix H_2 will be of the form

$$\begin{bmatrix} 1 & 0 & 0 & \cdots & 0 \\ 0 & 1 & 0 & \cdots & 0 \\ 0 & 0 & \times & \cdots & \times \\ \vdots & & & & \\ 0 & 0 & \times & \cdots & \times \end{bmatrix} = \left[\begin{array}{c|c} I_2 & O \\ \hline O & X \end{array} \right]$$

Multiplication of $A^{(1)}$ on the left by H_2 will leave the first two rows and the first column unchanged.

$$H_2 A^{(1)} = \begin{bmatrix} a_{11}^{(1)} & a_{12}^{(1)} & a_{13}^{(1)} & \cdots & a_{1n}^{(1)} \\ a_{21}^{(1)} & a_{22}^{(1)} & a_{23}^{(1)} & \cdots & a_{2n}^{(1)} \\ 0 & \times & \times & \cdots & \times \\ 0 & 0 & \times & \cdots & \times \\ \vdots & & & & \\ 0 & 0 & \times & \cdots & \times \end{bmatrix}$$

Postmultiplication of $H_2 A^{(1)}$ by H_2 will leave the first two columns unchanged. Thus $A^{(2)} = H_2 A^{(1)} H_2$ is of the form

$$\begin{bmatrix} \times & \times & \times & \cdots & \times \\ \times & \times & \times & \cdots & \times \\ 0 & \times & \times & \cdots & \times \\ 0 & 0 & \times & \cdots & \times \\ \vdots & & & & \\ 0 & 0 & \times & \cdots & \times \end{bmatrix}$$

This process may be continued until we end up with an upper Hessenberg matrix

$$H = A^{(n-2)} = H_{n-2} \cdots H_2 H_1 A H_1 H_2 \cdots H_{n-2}$$

which is similar to A.

If, in particular, A is symmetric, then, since

$$
\begin{aligned}
H^T &= H_{n-2}^T \cdots H_2^T H_1^T A^T H_1^T H_2^T \cdots H_{n-2}^T \\
&= H_{n-2} \cdots H_2 H_1 A H_1 H_2 \cdots H_{n-2} \\
&= H
\end{aligned}
$$

it follows that H is tridiagonal. Thus any $n \times n$ matrix A can be reduced by similarity transformations to upper Hessenberg form. If A is symmetric, the reduction will yield a symmetric tridiagonal matrix.

We close this section by outlining one of the best methods available for computing the eigenvalues of a matrix. The method is called the QR algorithm and was presented by K. G. F. Francis in 1961.

QR Algorithm

Given an $n \times n$ matrix A, factor it into a product $Q_1 R_1$, where Q_1 is orthogonal and R_1 is upper triangular. Define

$$
A_1 = A = Q_1 R_1
$$

and

$$
A_2 = Q_1^T A Q_1 = R_1 Q_1
$$

Factor A_2 into a product $Q_2 R_2$, where Q_2 is orthogonal and R_2 is upper triangular. Define

$$
A_3 = Q_2^T A_2 Q_2 = R_2 Q_2
$$

Note that $A_2 = Q_1^T A Q_1$ and $A_3 = (Q_1 Q_2)^T A (Q_1 Q_2)$ are both similar to A. We can continue in this manner to obtain a sequence of similar matrices. In general, if

$$
A_k = Q_k R_k
$$

then A_{k+1} is defined to be $R_k Q_k$. It can be shown that under very general conditions the sequence of matrices defined in this way converges to a matrix of the form

$$
\begin{bmatrix}
B_1 & \times & \cdots & \times \\
 & B_2 & & \times \\
O & & \ddots & \\
 & & & B_s
\end{bmatrix}
$$

where the B_i's are either 1×1 or 2×2 diagonal blocks. Each 2×2 block will correspond to a pair of complex conjugate eigenvalues of A. The eigenvalues of A will be eigenvalues of the B_i's. In the case where A is symmetric, each of the A_k's will also be symmetric and the sequence will converge to a diagonal matrix.

EXAMPLE 2. Let A_1 be the matrix from Example 1. The QR factorization of A_1 requires only a single Givens transformation,

$$G_1 = \frac{1}{\sqrt{5}} \begin{bmatrix} 2 & 1 \\ 1 & -2 \end{bmatrix}$$

Thus,

$$A_2 = G_1 A G_1 = \frac{1}{5} \begin{bmatrix} 2 & 1 \\ 1 & -2 \end{bmatrix} \begin{bmatrix} 2 & 1 \\ 1 & 2 \end{bmatrix} \begin{bmatrix} 2 & 1 \\ 1 & -2 \end{bmatrix} = \begin{bmatrix} 2.8 & -0.6 \\ -0.6 & 1.2 \end{bmatrix}$$

The QR factorization of A_2 can be accomplished using the Givens transformation

$$G_2 = \frac{1}{\sqrt{8.2}} \begin{bmatrix} 2.8 & -0.6 \\ -0.6 & -2.8 \end{bmatrix}$$

It follows that

$$A_3 = G_2 A_2 G_2 \approx \begin{bmatrix} 2.98 & 0.22 \\ 0.22 & 1.02 \end{bmatrix}$$

The off-diagonal elements are getting closer to 0 after each iteration, and the diagonal elements are approaching the eigenvalues $\lambda_1 = 3$ and $\lambda_2 = 1$.   ◀

Remarks

1. Because of the amount of work required at each iteration of the QR algorithm, it is important that the starting matrix A be in either Hessenberg or symmetric tridiagonal form. If this is not the case, we should perform similarity transformations on A to obtain a matrix A_1 that is in one of these forms.

2. If A_k is in upper Hessenberg form, the QR factorization can be carried out using $n - 1$ Givens transformations.

$$G_{n,n-1} \cdots G_{32} G_{21} A_k = R_k$$

Setting

$$Q_k^T = G_{n,n-1} \cdots G_{32} G_{21}$$

we have

$$A_k = Q_k R_k$$

and

$$A_{k+1} = Q_k^T A_k Q_k$$

To compute A_{k+1}, it is not necessary to determine Q_k explicitly. We need only keep track of the $n - 1$ Givens transformations. When R_k is postmultiplied by G_{21}, the resulting matrix will have the $(2, 1)$ entry filled in. The other entries below the diagonals will all still be zero. Postmultiplying $R_k G_{21}$ by G_{32} will have the effect of filling in the $(3, 2)$ position. Postmultiplication of $R_k G_{21} G_{32}$ by G_{43} will fill in the $(4, 3)$ position, and so on. Thus the resulting matrix $A_{k+1} = R_k G_{21} G_{32} \cdots G_{n,n-1}$ will be in upper Hessenberg form. If A_1 is a symmetric tridiagonal matrix, each of the succeeding A_i's will be upper Hessenberg and symmetric. Thus $A_2, A_3, \ldots$ will all be tridiagonal.

3. As in the power method, convergence may be slow when some of the eigenvalues are close together. To speed up convergence, it is customary to introduce *origin shifts*. At the kth step a scalar α_k is chosen, and $A_k - \alpha_k I$ (rather than A_k) is decomposed into a product $Q_k R_k$. The matrix A_{k+1} is defined by

$$A_{k+1} = R_k Q_k + \alpha_k I$$

Note that

$$Q_k^T A_k Q_k = Q_k^T (Q_k R_k + \alpha_k I) Q_k = R_k Q_k + \alpha_k I = A_{k+1}$$

so A_k and A_{k+1} are similar. With the proper choice of shifts α_k, the convergence can be greatly accelerated.

4. In our brief discussion we have presented only an outline of the method. Many of the details, such as how to choose the origin shifts, have been omitted. For a more thorough discussion and a proof of convergence, see Wilkinson [27]. ◀

EXERCISES

1. Let

$$A = \begin{bmatrix} 1 & 1 \\ 1 & 1 \end{bmatrix}$$

(a) Apply one iteration of the power method with any nonzero starting vector.
(b) Apply one iteration of the QR algorithm to A.
(c) Determine the exact eigenvalues of A by solving the characteristic equation and determine the eigenspace corresponding to the largest eigenvalue. Compare your answers with those to parts (a) and (b).

2. Let

$$A = \begin{bmatrix} 2 & 1 & 0 \\ 1 & 3 & 1 \\ 0 & 1 & 2 \end{bmatrix} \quad \text{and} \quad \mathbf{u}_0 = \begin{bmatrix} 1 \\ 1 \\ 1 \end{bmatrix}$$

(a) Apply the power method to compute $\mathbf{v}_1, \mathbf{u}_1, \mathbf{v}_2, \mathbf{u}_2$, and $\mathbf{v}_3$. (Round off to two decimal places.)
(b) Determine an approximation λ_1' to the largest eigenvalue of A from the coordinates of $\mathbf{v}_3$. Determine the exact value of λ_1 and compare it to λ_1'. What is the relative error?

3. Let

$$A = \begin{bmatrix} 1 & 2 \\ -1 & -1 \end{bmatrix} \qquad \text{and} \qquad \mathbf{u}_0 = \begin{bmatrix} 1 \\ 1 \end{bmatrix}$$

(a) Compute $\mathbf{u}_1$, $\mathbf{u}_2$, $\mathbf{u}_3$, and $\mathbf{u}_4$ using the power method.

(b) Explain why the power method will fail to converge in this case.

4. Let

$$A = A_1 = \begin{bmatrix} 1 & 1 \\ 1 & 3 \end{bmatrix}$$

Compute A_2 and A_3 using the QR algorithm. Compute the exact eigenvalues of A and compare them to the diagonal elements of A_3. To how many decimal places do they agree?

5. Given

$$A = \begin{bmatrix} 5 & 2 & 2 \\ -2 & 1 & -2 \\ -3 & -4 & 2 \end{bmatrix}$$

(a) Verify that $\lambda_1 = 4$ is an eigenvalue of A and $\mathbf{y}_1 = (2, -2, 1)^T$ is an eigenvector belonging to λ_1.

(b) Find a Householder transformation H such that HAH is of the form

$$\begin{bmatrix} 4 & \times & \times \\ 0 & \times & \times \\ 0 & \times & \times \end{bmatrix}$$

(c) Compute HAH and find the remaining eigenvalues of A.

6. Let A be an $n \times n$ matrix with distinct real eigenvalues $\lambda_1, \lambda_2, \ldots, \lambda_n$. Let λ be a scalar that is not an eigenvalue of A and let $B = (A - \lambda I)^{-1}$. Show that:

(a) The scalars $\mu_j = 1/(\lambda_j - \lambda)$, $j = 1, \ldots, n$ are the eigenvalues of B.

(b) If $\mathbf{x}_j$ is an eigenvector of B belonging to μ_j, then $\mathbf{x}_j$ is an eigenvector of A belonging to λ_j.

(c) If the power method is applied to B, then the sequence of vectors will converge to an eigenvector of A belonging to the eigenvalue that is closest to λ. [The convergence will be quite rapid if λ is much closer to one λ_i than to any of the others. This method of computing eigenvectors using powers of $(A - \lambda I)^{-1}$ is called the *inverse power method*.]

7. Let $\mathbf{x} = (x_1, \ldots, x_n)^T$ be an eigenvector of A belonging to λ. If $|x_i| = \|\mathbf{x}\|_\infty$, show that:

(a) $\displaystyle\sum_{j=1}^{n} a_{ij} x_j = \lambda x_i$ (b) $\displaystyle |\lambda - a_{ii}| \le \sum_{\substack{j=1 \\ j \ne i}}^{n} |a_{ij}|$ (Gerschgorin's theorem)

8. Let A be a matrix with eigenvalues $\lambda_1, \ldots, \lambda_n$ and let λ be an eigenvalue of $A + E$. Let X be a matrix that diagonalizes A and let $C = X^{-1}EX$. Prove:

(a) For some i

$$|\lambda - \lambda_i| \leq \sum_{j=1}^{n} |c_{ij}|$$

[*Hint:* λ is an eigenvalue of $X^{-1}(A + E)X$. Apply Gerschgorin's theorem from Exercise 7.]

(b) $\displaystyle\min_{1 \leq j \leq n} |\lambda - \lambda_j| \leq \text{cond}_\infty(X)\|E\|_\infty$

9. Let $A_k = Q_k R_k$, $k = 1, 2, \ldots$ be the sequence of matrices derived from $A = A_1$ by applying the QR algorithm. For each positive integer k, define

$$P_k = Q_1 Q_2 \cdots Q_k \quad \text{and} \quad U_k = R_k \cdots R_2 R_1$$

Show that

$$P_k A_{k+1} = A P_k$$

for all $k \geq 1$.

10. Let P_k and U_k be defined as in Exercise 9. Show that:

(a) $P_{k+1} U_{k+1} = P_k A_{k+1} U_k = A P_k U_k$
(b) $P_k U_k = A^k$ and hence

$$(Q_1 Q_2 \cdots Q_k)(R_k \cdots R_2 R_1)$$

is the QR factorization of A^k.

11. Let R_k be a $k \times k$ upper triangular matrix and suppose that

$$R_k U_k = U_k D_k$$

where U_k is an upper triangular matrix with 1's on the diagonal and D_k is a diagonal matrix. Let R_{k+1} be an upper triangular matrix of the form

$$\begin{bmatrix} R_k & \mathbf{b}_k \\ \mathbf{0}^T & \beta_k \end{bmatrix}$$

where β_k is not an eigenvalue of R_k. Determine $(k+1) \times (k+1)$ matrices U_{k+1} and D_{k+1} of the form

$$U_{k+1} = \begin{bmatrix} U_k & \mathbf{x}_k \\ \mathbf{0}^T & 1 \end{bmatrix}, \qquad D_{k+1} = \begin{bmatrix} D_k & \mathbf{0} \\ \mathbf{0}^T & \beta \end{bmatrix}$$

such that

$$R_{k+1} U_{k+1} = U_{k+1} D_{k+1}$$

12. Let R be an $n \times n$ upper triangular matrix whose diagonal entries are all distinct. Let R_k denote the leading principal submatrix of R of order k and set $U_1 = (1)$.

(a) Use the result from Exercise 11 to derive an algorithm for finding the eigenvectors of R. The matrix U of eigenvectors should be upper triangular with 1's on the diagonal.

(b) Show that the algorithm requires approximately $n^3/6$ floating-point multiplications/divisions.

7 LEAST SQUARES PROBLEMS

In this section we study computational methods for finding least squares solutions to overdetermined systems. Let A be an $m \times n$ matrix with $m \geq n$ and let $\mathbf{b} \in R^m$. We consider some methods for computing a vector $\hat{\mathbf{x}}$ that minimizes $\|\mathbf{b} - A\mathbf{x}\|_2^2$.

Normal Equations

We saw in Chapter 5 that if $\hat{\mathbf{x}}$ satisfies the normal equations

$$A^T A\mathbf{x} = A^T \mathbf{b}$$

then $\hat{\mathbf{x}}$ is a solution to the least squares problem. If A is of full rank (rank n), then $A^T A$ is nonsingular and hence the system will have a unique solution. Thus, if $A^T A$ is invertible, one possible method for solving the least squares problem is to form the normal equations and then solve them using Gaussian elimination. An algorithm for doing this would have two main parts.

1. Compute $B = A^T A$ and $\mathbf{c} = A^T \mathbf{b}$.
2. Solve $B\mathbf{x} = \mathbf{c}$.

Note that forming the normal equations requires roughly $mn^2/2$ multiplications. Since $A^T A$ is nonsingular, the matrix B is positive definite. For positive definite matrices, there are reduction algorithms that require only half the usual number of multiplications. Thus the solution of $B\mathbf{x} = \mathbf{c}$ requires roughly $n^3/6$ multiplications. Most of the work then occurs in forming the normal equations rather than solving them. However, the main difficulty with this method is that in forming the normal equations we may well end up transforming the problem into an ill-conditioned one. Recall from Section 4 that, if $\mathbf{x}'$ is the computed solution to $B\mathbf{x} = \mathbf{c}$ and $\mathbf{x}$ is the exact solution, the inequality

$$\frac{1}{\text{cond}(B)} \frac{\|\mathbf{r}\|}{\|\mathbf{c}\|} \leq \frac{\|\mathbf{x} - \mathbf{x}'\|}{\|\mathbf{x}\|} \leq \text{cond}(B) \frac{\|\mathbf{r}\|}{\|\mathbf{c}\|}$$

shows how the relative error compares to the relative residual. If A has singular values $\sigma_1 \geq \sigma_2 \geq \cdots \geq \sigma_n > 0$, then $\text{cond}_2(A) = \sigma_1/\sigma_n$. The singular values of B are $\sigma_1^2, \sigma_2^2, \ldots, \sigma_n^2$. Thus

$$\text{cond}_2(B) = \frac{\sigma_1^2}{\sigma_n^2} = [\text{cond}_2(A)]^2$$

If, for example, $\text{cond}_2(A) = 100$, the relative error in the computed solution to the normal equations could be 10^4 times as large as the relative residual. For this reason we should be very careful about using the normal equations to solve least squares problems.

In Chapter 5 we saw how to use the Gram–Schmidt process to obtain a QR factorization of a matrix A with full rank. In that case the matrix Q was an $m \times n$ matrix with orthonormal columns and R was an $n \times n$ upper triangular matrix. In the following numerical method, we use Householder transformations to obtain a QR factorization of A. In this case, Q will be an $m \times m$ orthogonal matrix and R will be an $m \times n$ matrix whose subdiagonal entries are all 0.

The QR Factorization

Given an $m \times n$ matrix A of full rank, we can apply n Householder transformations to zero out all the elements below the diagonal. Thus,

$$H_n H_{n-1} \cdots H_1 A = R$$

where R is of the form

$$\begin{bmatrix} R_1 \\ O \end{bmatrix} = \begin{bmatrix} \times & \times & \times & \cdots & \times \\ & \times & \times & \cdots & \times \\ & & \times & \cdots & \times \\ & & & \ddots & \vdots \\ & & & & \times \\ & & & & \\ & & & & \end{bmatrix}$$

with nonzero diagonal. Let

$$Q^T = H_n \cdots H_1 = \begin{bmatrix} Q_1^T \\ Q_2^T \end{bmatrix}$$

where Q_1^T is an $n \times m$ matrix consisting of the first n rows of Q^T. Since

$$Q^T A = \begin{bmatrix} Q_1^T A \\ Q_2^T A \end{bmatrix} = \begin{bmatrix} R_1 \\ O \end{bmatrix}$$

it follows that $A = Q_1 R_1$. Let

$$\mathbf{c} = Q^T \mathbf{b} = \begin{bmatrix} Q_1^T \mathbf{b} \\ Q_2^T \mathbf{b} \end{bmatrix} = \begin{bmatrix} \mathbf{c}_1 \\ \mathbf{c}_2 \end{bmatrix}$$

The normal equations can be written in the form

$$R_1^T Q_1^T Q_1 R_1 \mathbf{x} = R_1^T Q_1^T \mathbf{b}$$

Since $Q_1^T Q_1 = I$ and R_1^T is nonsingular, this simplifies to

$$R_1 \mathbf{x} = \mathbf{c}_1$$

This system can be solved by back substitution. The solution $\mathbf{x} = R_1^{-1}\mathbf{c}_1$ will be the unique solution to the least squares problem. To compute the residual, note that

$$Q^T\mathbf{r} = \begin{pmatrix} \mathbf{c}_1 \\ \mathbf{c}_2 \end{pmatrix} - \begin{pmatrix} R_1 \\ O \end{pmatrix}\mathbf{x} = \begin{pmatrix} \mathbf{0} \\ \mathbf{c}_2 \end{pmatrix}$$

so that

$$\mathbf{r} = Q\begin{pmatrix} \mathbf{0} \\ \mathbf{c}_2 \end{pmatrix} \qquad \text{and} \qquad \|\mathbf{r}\|_2 = \|\mathbf{c}_2\|_2$$

In summation, if A is an $m \times n$ matrix with full rank, the least squares problem can be solved as follows:

1. Use Householder transformations to compute

$$R = H_n \cdots H_2 H_1 A \qquad \text{and} \qquad \mathbf{c} = H_n \cdots H_2 H_1 \mathbf{b}$$

2. Partition R and $\mathbf{c}$ into block form:

$$R = \begin{pmatrix} R_1 \\ O \end{pmatrix} \qquad \mathbf{c} = \begin{pmatrix} \mathbf{c}_1 \\ \mathbf{c}_2 \end{pmatrix}$$

where R_1 and $\mathbf{c}_1$ each have n rows.

3. Use back substitution to solve $R_1\mathbf{x} = \mathbf{c}_1$.

The Pseudoinverse

Now consider the case where the matrix A has rank $r < n$. The singular value decomposition provides the key to solving the least squares problem in this case. It can be used to construct a generalized inverse of A. In the case where A is a nonsingular $n \times n$ matrix with singular value decomposition $U\Sigma V^T$, the inverse is given by

$$A^{-1} = V\Sigma^{-1}U^T$$

More generally, if $A = U\Sigma V^T$ is an $m \times n$ matrix of rank r, the matrix Σ will be an $m \times n$ matrix of the form

$$\Sigma = \left(\begin{array}{c|c} \Sigma_1 & O \\ \hline O & O \end{array} \right) = \left(\begin{array}{cccc|c} \sigma_1 & & & & \\ & \sigma_2 & & & O \\ & & \ddots & & \\ & & & \sigma_r & \\ \hline & & O & & O \end{array} \right)$$

and we can define

(1)
$$A^+ = V\Sigma^+ U^T$$

where Σ^+ is the $n \times m$ matrix

$$\Sigma^+ = \left[\begin{array}{c|c} \Sigma_1^{-1} & O \\ \hline O & O \end{array} \right] = \left[\begin{array}{ccc|c} \dfrac{1}{\sigma_1} & & & \\ & \ddots & & O \\ & & \dfrac{1}{\sigma_r} & \\ \hline & O & & O \end{array} \right]$$

Equation (1) gives a natural generalization of the inverse of a matrix. The matrix A^+ defined by (1) is called the *pseudoinverse* of A.

It is also possible to define A^+ by its algebraic properties. These properties are given in the following four conditions.

The Penrose Conditions

1. $AXA = A$
2. $XAX = X$
3. $(AX)^T = AX$
4. $(XA)^T = XA$

If A is an $m \times n$ matrix, we claim that there is a unique $n \times m$ matrix X that satisfies these conditions. Indeed, if we choose $X = A^+ = V\Sigma^+ U^T$, then it is easily verified that X satisfies all four conditions. We leave this as an exercise for the reader. To show uniqueness, suppose that Y also satisfies the Penrose conditions. By successively applying these conditions, we can argue as follows:

$X = XAX$	(2)	$Y = YAY$	(2)
$= A^T X^T X$	(4)	$= YY^T A^T$	(3)
$= (AYA)^T X^T X$	(1)	$= YY^T (AXA)^T$	(1)
$= (A^T Y^T)(A^T X^T)X$		$= Y(Y^T A^T)(X^T A^T)$	
$= YAXAX$	(4)	$= YAYAX$	(3)
$= YAX$	(1)	$= YAX$	(1)

Therefore, $X = Y$. Thus A^+ is the unique matrix satisfying the four Penrose conditions. These conditions are often used to define the pseudoinverse, and A^+ is often referred to as the *Moore–Penrose pseudoinverse*.

To see how the pseudoinverse can be used in solving least squares problems, let us first consider the case where A is an $m \times n$ matrix of rank n. Thus Σ is of the form

$$\Sigma = \left[\begin{array}{c} \Sigma_1 \\ O \end{array} \right]$$

where Σ_1 is a nonsingular $n \times n$ diagonal matrix. The matrix $A^T A$ is nonsingular and

$$(A^T A)^{-1} = V(\Sigma^T \Sigma)^{-1} V^T$$

The solution to the normal equations is given by

$$\begin{aligned}
\mathbf{x} &= (A^T A)^{-1} A^T \mathbf{b} \\
&= V(\Sigma^T \Sigma)^{-1} V^T V \Sigma^T U^T \mathbf{b} \\
&= V(\Sigma^T \Sigma)^{-1} \Sigma^T U^T \mathbf{b} \\
&= V \Sigma^+ U^T \mathbf{b} \\
&= A^+ \mathbf{b}
\end{aligned}$$

Thus, if A has full rank, $A^+ \mathbf{b}$ is the solution to the least squares problem. What about the case where A has rank $r < n$? In this case there are infinitely many solutions to the least squares problem. The following theorem shows that not only is $A^+ \mathbf{b}$ a solution, but it is also the minimal solution with respect to the 2-norm.

▶ **THEOREM 7.7.1** *If A is an $m \times n$ matrix of rank $r < n$ with singular value decomposition $U \Sigma V^T$, then the vector*

$$\mathbf{x} = A^+ \mathbf{b} = V \Sigma^+ U^T \mathbf{b}$$

minimizes $\|\mathbf{b} - A\mathbf{x}\|_2^2$. Moreover, if $\mathbf{z}$ is any other vector that minimizes $\|\mathbf{b} - A\mathbf{x}\|_2^2$, then $\|\mathbf{z}\|_2 > \|\mathbf{x}\|_2$.

▶*Proof.* Let $\mathbf{x}$ be a vector in R^n and define

$$\mathbf{c} = U^T \mathbf{b} = \begin{bmatrix} \mathbf{c}_1 \\ \mathbf{c}_2 \end{bmatrix} \qquad \text{and} \qquad \mathbf{y} = V^T \mathbf{x} = \begin{bmatrix} \mathbf{y}_1 \\ \mathbf{y}_2 \end{bmatrix}$$

where $\mathbf{c}_1$ and $\mathbf{y}_1$ are vectors in R^r. Since U^T is orthogonal, it follows that

$$\begin{aligned}
\|\mathbf{b} - A\mathbf{x}\|_2^2 &= \|U^T \mathbf{b} - \Sigma(V^T \mathbf{x})\|_2^2 \\
&= \|\mathbf{c} - \Sigma \mathbf{y}\|_2^2 \\
&= \left\| \begin{bmatrix} \mathbf{c}_1 \\ \mathbf{c}_2 \end{bmatrix} - \begin{bmatrix} \Sigma_1 & O \\ O & O \end{bmatrix} \begin{bmatrix} \mathbf{y}_1 \\ \mathbf{y}_2 \end{bmatrix} \right\|_2^2 \\
&= \left\| \begin{bmatrix} \mathbf{c}_1 - \Sigma_1 \mathbf{y}_1 \\ \mathbf{c}_2 \end{bmatrix} \right\|_2^2 \\
&= \|\mathbf{c}_1 - \Sigma_1 \mathbf{y}_1\|_2^2 + \|\mathbf{c}_2\|_2^2
\end{aligned}$$

Since $\mathbf{c}_2$ is independent of $\mathbf{x}$, it follows that $\|\mathbf{b} - A\mathbf{x}\|^2$ will be minimal if and only if

$$\|\mathbf{c}_1 - \Sigma_1 \mathbf{y}_1\| = 0$$

Thus $\mathbf{x}$ is a solution to the least squares problem if and only if $\mathbf{x} = V\mathbf{y}$, where $\mathbf{y}$ is a vector of the form

$$\begin{bmatrix} \Sigma_1^{-1}\mathbf{c}_1 \\ \mathbf{y}_2 \end{bmatrix}$$

In particular,

$$\mathbf{x} = V \begin{bmatrix} \Sigma_1^{-1}\mathbf{c}_1 \\ \mathbf{0} \end{bmatrix}$$

$$= V \begin{bmatrix} \Sigma_1^{-1} & O \\ O & O \end{bmatrix} \begin{bmatrix} \mathbf{c}_1 \\ \mathbf{c}_2 \end{bmatrix}$$

$$= V\Sigma^{+}U^{T}\mathbf{b}$$

$$= A^{+}\mathbf{b}$$

is a solution. If $\mathbf{z}$ is any other solution, $\mathbf{z}$ must be of the form

$$V\mathbf{y} = V \begin{bmatrix} \Sigma_1^{-1}\mathbf{c}_1 \\ \mathbf{y}_2 \end{bmatrix}$$

where $\mathbf{y}_2 \neq \mathbf{0}$. It follows then that

$$\|\mathbf{z}\|^2 = \|\mathbf{y}\|^2 = \|\Sigma_1^{-1}\mathbf{c}_1\|^2 + \|\mathbf{y}_2\|^2 > \|\Sigma_1^{-1}\mathbf{c}_1\|^2 = \|\mathbf{x}\|^2 \qquad \blacktriangleleft$$

If the singular value decomposition $U\Sigma V^{T}$ of A is known, it is a simple matter to compute the solution to the least squares problem. If $U = (\mathbf{u}_1, \ldots, \mathbf{u}_m)$ and $V = (\mathbf{v}_1, \ldots, \mathbf{v}_n)$, then, defining $\mathbf{y} = \Sigma^{+}U^{T}\mathbf{b}$, we have

$$y_i = \frac{1}{\sigma_i}\mathbf{u}_i^{T}\mathbf{b} \qquad i = 1, \ldots, r \qquad (r = \text{rank of } A)$$

$$y_i = 0 \qquad i = r+1, \ldots, n$$

and hence

$$A^{+}\mathbf{b} = V\mathbf{y} = \begin{bmatrix} v_{11}y_1 + v_{12}y_2 + \cdots + v_{1r}y_r \\ v_{21}y_1 + v_{22}y_2 + \cdots + v_{2r}y_r \\ \vdots \\ v_{n1}y_1 + v_{n2}y_2 + \cdots + v_{nr}y_r \end{bmatrix}$$

$$= y_1\mathbf{v}_1 + y_2\mathbf{v}_2 + \cdots + y_r\mathbf{v}_r$$

Thus, the solution $\mathbf{x} = A^{+}\mathbf{b}$ can be computed in two steps:

1. Set $y_i = (1/\sigma_i)\mathbf{u}_i^{T}\mathbf{b}$ for $i = 1, \ldots, r$.
2. Let $\mathbf{x} = y_1\mathbf{v}_1 + \cdots + y_r\mathbf{v}_r$.

We conclude this section by outlining a method for computing the singular values of a matrix. We saw in the last section that the eigenvalues of a symmetric matrix are relatively insensitive to perturbations in the matrix. The same is true for the singular values of an $m \times n$ matrix. If two matrices A and B are close, their singular values must also be close. More precisely, if A has the singular values $\sigma_1 \geq \sigma_2 \geq \cdots \geq \sigma_n$ and B has the singular values $\omega_1 \geq \omega_2 \geq \cdots \geq \omega_n$, then

$$|\sigma_i - \omega_i| \leq \|A - B\|_2 \qquad i = 1, \ldots, n$$

(see Datta [17], p. 560). Thus, in computing the singular values of a matrix A, we need not worry that small changes in the entries of A will cause drastic changes in the computed singular values.

The problem of computing singular values can be simplified using orthogonal transformations. If A has singular value decomposition $U \Sigma V^T$ and $B = HAP^T$, where H is an $m \times m$ orthogonal matrix and P is an $n \times n$ orthogonal matrix, then B has singular value decomposition $(HU)\Sigma(PV)^T$. The matrices A and B will have the same singular values, and if B has a much simpler structure than A, it should be easier to compute its singular values. Indeed, G.H. Golub and W. Kahan have shown that A can be reduced to upper bidiagonal form and the reduction can be carried out using Householder transformations.

Bidiagonalization

Let H_1 be a Householder transformation that annihilates all the elements below the diagonal in the first column of A. Let P_1 be a Householder transformation such that postmultiplication of $H_1 A$ by P_1 annihilates the last $n - 2$ entries of the first row of $H_1 A$ while leaving the first column unchanged.

$$H_1 A P_1 = \begin{pmatrix} \times & \times & 0 & \cdots & 0 \\ 0 & \times & \times & \cdots & \times \\ & & \vdots & & \\ 0 & \times & \times & \cdots & \times \end{pmatrix}$$

The next step is to apply a Householder transformation H_2 that annihilates the elements below the diagonal in the second column of $H_1 A P_1$ while leaving the first row and column unchanged.

$$H_2 H_1 A P_1 = \begin{pmatrix} \times & \times & 0 & \cdots & 0 \\ 0 & \times & \times & \cdots & \times \\ 0 & 0 & \times & \cdots & \times \\ & & \vdots & & \\ 0 & 0 & \times & \cdots & \times \end{pmatrix}$$

$H_2 H_1 A P_1$ is then postmultiplied by a Householder transformation P_2 that annihilates the last $n - 3$ elements in the second row while leaving the first two columns and the first row unchanged.

$$H_2 H_1 A P_1 P_2 = \begin{bmatrix} \times & \times & 0 & 0 & \cdots & 0 \\ 0 & \times & \times & 0 & \cdots & 0 \\ 0 & 0 & \times & \times & \cdots & \times \\ & & & & \vdots & \\ 0 & 0 & \times & \times & \cdots & \times \end{bmatrix}$$

We continue in this manner until we obtain a matrix

$$B = H_n \cdots H_1 A P_1 \cdots P_{n-2}$$

of the form

$$\begin{bmatrix} \times & \times & & & \\ & \times & \times & & \\ & & \ddots & \ddots & \\ & & & \times & \times \\ & & & & \times \end{bmatrix}$$

Since $H = H_n \cdots H_1$ and $P^T = P_1 \cdots P_{n-2}$ are orthogonal, it follows that B has the same singular values as A.

The problem has now been simplified to that of finding the singular values of an upper bidiagonal matrix B. We could at this point form the symmetric tridiagonal matrix $B^T B$ and then compute its eigenvalues using the QR algorithm. The problem with this approach is that in forming $B^T B$ we would still be squaring the condition number, and consequently our computed solution would be much less reliable. The method we outline produces a sequence of bidiagonal matrices $B_1, B_2, \ldots$ that converges to a diagonal matrix Σ. The method involves applying a sequence of Givens transformations to B alternately on the right- and left-hand sides.

The Golub--Reinsch Algorithm

Let

$$R_k = \begin{bmatrix} I_{k-1} & O & O \\ O & G(\theta_k) & O \\ O & O & I_{n-k-1} \end{bmatrix}$$

and

$$
L_k = \begin{bmatrix} I_{k-1} & O & O \\ O & G(\varphi_k) & O \\ O & O & I_{n-k-1} \end{bmatrix}
$$

The 2×2 matrices $G(\theta_k)$ and $G(\varphi_k)$ are given by

$$
G(\theta_k) = \begin{bmatrix} \cos\theta_k & \sin\theta_k \\ \sin\theta_k & -\cos\theta_k \end{bmatrix} \qquad \text{and} \qquad G(\varphi_k) = \begin{bmatrix} \cos\varphi_k & \sin\varphi_k \\ \sin\varphi_k & -\cos\varphi_k \end{bmatrix}
$$

for some angles θ_k and φ_k. The matrix $B = B_1$ is first multiplied on the right by R_1. This will have the effect of filling in the $(2, 1)$ position.

$$
B_1 R_1 = \begin{bmatrix} \times & \times & & & & \\ \times & \times & \times & & & \\ & & \times & & & \\ & & & \ddots & \times & \\ & & & & \times \end{bmatrix}
$$

Next L_1 is chosen so as to annihilate the element filled in by R_1. It will also have the effect of filling in the $(1, 3)$ position. Thus

$$
L_1 B_1 R_1 = \begin{bmatrix} \times & \times & \times & & \\ & \times & \times & & \\ & & & \ddots & \\ & & & & \times \\ & & & & \times \end{bmatrix}
$$

R_2 is chosen so as to annihilate the $(1, 3)$ entry. It will fill in the $(3, 2)$ entry of $L_1 B_1 R_1$. Next L_2 annihilates the $(3, 2)$ entry and fills in the $(2, 4)$ entry, and so on.

$$
\begin{bmatrix} \times & \times & & & & \\ & \times & \times & & & \\ & \times & \times & \times & & \\ & & & \ddots & & \\ & & & & \times & \\ & & & & \times \end{bmatrix} \qquad \begin{bmatrix} \times & \times & & & & \\ & \times & \times & \times & & \\ & & \times & \times & & \\ & & & \ddots & & \\ & & & & \times & \\ & & & & \times \end{bmatrix}
$$

$$
L_1 B_1 R_1 R_2 \qquad\qquad\qquad L_2 L_1 B_1 R_1 R_2
$$

We continue this process until we end up with a new bidiagonal matrix,

$$B_2 = L_{n-1} \cdots L_1 B_1 R_1 \cdots R_{n-1}$$

Why should we be any better off with B_2 than B_1? It can be shown that, if the first transformation R_1 is chosen correctly, $B_2^T B_2$ will be the matrix obtained from $B_1^T B_1$ by applying one iteration of the QR algorithm with shift. The same process can now be applied to B_2 to obtain a new bidiagonal matrix B_3 such that $B_3^T B_3$ would be the matrix obtained by applying two iterations of the QR algorithm with shifts to $B_1^T B_1$. Even though the $B_i^T B_i$'s are never computed, we know that with the proper choice of shifts these matrices will converge rapidly to a diagonal matrix. The B_i's then must also converge to a diagonal matrix Σ. Since each of the B_i's has the same singular values as B, the diagonal elements of Σ will be the singular values of B. The matrices U and V^T can be determined by keeping track of all the orthogonal transformations.

Only a brief sketch of the algorithm has been given. To include more would have been beyond the scope of this book. For complete details of the algorithm and an ALGOL program, we refer the reader to the paper by Golub and Reinsch in [28], p. 135.

EXERCISES

1. Find the solution **x** to the least squares problem given that $A = QR$ in each of the following.

(a) $Q = \begin{bmatrix} \dfrac{1}{\sqrt{2}} & \dfrac{1}{\sqrt{2}} \\ \dfrac{1}{\sqrt{2}} & -\dfrac{1}{\sqrt{2}} \\ 0 & 0 \end{bmatrix}$, $\qquad R = \begin{bmatrix} 1 & 1 \\ 0 & 1 \end{bmatrix}$, $\qquad \mathbf{b} = \begin{bmatrix} 1 \\ 1 \\ 1 \end{bmatrix}$

(b) $Q = \begin{bmatrix} 1 & 0 & 0 \\ 0 & \dfrac{1}{\sqrt{2}} & -\dfrac{1}{\sqrt{2}} \\ 0 & \dfrac{1}{\sqrt{2}} & \dfrac{1}{\sqrt{2}} \\ 0 & 0 & 0 \end{bmatrix}$, $\qquad R = \begin{bmatrix} 1 & 1 & 0 \\ 0 & 1 & 1 \\ 0 & 0 & 1 \end{bmatrix}$, $\qquad \mathbf{b} = \begin{bmatrix} 1 \\ 3 \\ 1 \\ 2 \end{bmatrix}$

(c) $Q = \begin{bmatrix} 1 & 0 & 0 \\ 0 & \dfrac{1}{\sqrt{2}} & -\dfrac{1}{\sqrt{2}} \\ 0 & \dfrac{1}{\sqrt{2}} & \dfrac{1}{\sqrt{2}} \end{bmatrix}$, $\qquad R = \begin{bmatrix} 1 & 1 \\ 0 & 1 \\ 0 & 0 \end{bmatrix}$, $\qquad \mathbf{b} = \begin{bmatrix} 1 \\ \sqrt{2} \\ -\sqrt{2} \end{bmatrix}$

(d) $Q = \begin{pmatrix} \frac{1}{2} & \frac{1}{\sqrt{2}} & 0 & \frac{1}{2} \\ \frac{1}{2} & 0 & \frac{1}{\sqrt{2}} & -\frac{1}{2} \\ \frac{1}{2} & 0 & -\frac{1}{\sqrt{2}} & -\frac{1}{2} \\ \frac{1}{2} & -\frac{1}{\sqrt{2}} & 0 & \frac{1}{2} \end{pmatrix}$, $R = \begin{pmatrix} 1 & 1 & 0 \\ 0 & 1 & 1 \\ 0 & 0 & 1 \\ 0 & 0 & 0 \end{pmatrix}$, $b = \begin{pmatrix} 2 \\ -2 \\ 0 \\ 2 \end{pmatrix}$

2. Let

$$A = \begin{bmatrix} D \\ E \end{bmatrix} = \begin{pmatrix} d_1 \\ & d_2 \\ & & \ddots \\ & & & d_n \\ \hline e_1 \\ & e_2 \\ & & \ddots \\ & & & e_n \end{pmatrix} \quad \text{and} \quad b = \begin{pmatrix} b_1 \\ b_2 \\ \vdots \\ b_{2n} \end{pmatrix}$$

Use the normal equations to find the solution **x** to the least squares problem.

3. Given

$$A = \begin{pmatrix} 1 & 2 \\ 1 & 3 \\ 1 & 2 \\ 1 & -1 \end{pmatrix}, \quad b = \begin{pmatrix} -3 \\ 10 \\ 3 \\ 6 \end{pmatrix}$$

(a) Use Householder transformations to reduce A to the form

$$\begin{bmatrix} R_1 \\ O \end{bmatrix} = \begin{pmatrix} \times & \times \\ 0 & \times \\ 0 & 0 \\ 0 & 0 \end{pmatrix}$$

and apply the same transformations to **b**.

(b) Use the results from part (a) to find the least squares solution to $A\mathbf{x} = \mathbf{b}$.

4. Let

$$A = \begin{pmatrix} 1 & 1 \\ \epsilon & 0 \\ 0 & \epsilon \end{pmatrix}$$

where ϵ is a small scalar.

(a) Determine the singular values of A exactly.

(b) Suppose that ϵ is sufficiently small that $1+\epsilon^2$ gets rounded off to 1 on your calculator. Determine the eigenvalues of the calculated $A^T A$ and compare the square roots of these eigenvalues to your answers in part (a).

5. Show that the pseudoinverse A^+ satisfies the four Penrose conditions.

6. Let B be any matrix that satisfies Penrose conditions (1) and (3) and let $\mathbf{x} = B\mathbf{b}$. Show that $\mathbf{x}$ is a solution to the normal equations $A^T A\mathbf{x} = A^T \mathbf{b}$.

7. If $\mathbf{x} \in R^m$, we can think of $\mathbf{x}$ as an $m \times 1$ matrix. Define

$$X = \frac{1}{\|\mathbf{x}\|_2^2} \mathbf{x}^T$$

where X is a $1 \times m$ matrix. Show that X and $\mathbf{x}$ satisfy the four Penrose conditions and consequently that

$$\mathbf{x}^+ = X = \frac{1}{\|\mathbf{x}\|_2^2} \mathbf{x}^T$$

8. Show that if A is a $m \times n$ matrix of rank n, then $A^+ = (A^T A)^{-1} A^T$.

9. Let A be an $m \times n$ matrix and let $\mathbf{b} \in R^m$. Show that $\mathbf{b} \in R(A)$ if and only if

$$\mathbf{b} = AA^+\mathbf{b}$$

10. Let A be an $m \times n$ matrix with singular value decomposition $U\Sigma V^T$ and suppose that A has rank r, where $r < n$. Let $\mathbf{b} \in R^m$. Show that a vector $\mathbf{x} \in R^n$ minimizes $\|\mathbf{b} - A\mathbf{x}\|_2$ if and only if

$$\mathbf{x} = A^+\mathbf{b} + c_{r+1}\mathbf{v}_{r+1} + \cdots + c_n\mathbf{v}_n$$

where $c_{r+1}, \ldots, c_n$ are scalars.

11. Let

$$A = \begin{bmatrix} 1 & 1 \\ 1 & 1 \\ 0 & 0 \end{bmatrix}$$

Determine A^+ and verify that A and A^+ satisfy the four Penrose conditions (see Example 1 of Section 5).

12. Given

$$A = \begin{bmatrix} 1 & 2 \\ -1 & -2 \end{bmatrix} \qquad \text{and} \qquad \mathbf{b} = \begin{bmatrix} 6 \\ -4 \end{bmatrix}$$

(a) Compute the singular value decomposition of A and use it to determine A^+.

(b) Use A^+ to find a least squares solution to the system $A\mathbf{x} = \mathbf{b}$.

(c) Find all solutions to the least squares problem $A\mathbf{x} = \mathbf{b}$.

13. Show each of the following:

(a) $(A^+)^+ = A$

(b) $(AA^+)^2 = AA^+$

(c) $(A^+A)^2 = A^+A$

14. Let $A_1 = U\Sigma_1 V^T$ and $A_2 = U\Sigma_2 V^T$, where

$$\Sigma_1 = \begin{pmatrix} \sigma_1 & & & & & & \\ & \ddots & & & & & \\ & & \sigma_{r-1} & & & & \\ & & & 0 & & & \\ & & & & \ddots & & \\ & & & & & 0 & \end{pmatrix}$$

and

$$\Sigma_2 = \begin{pmatrix} \sigma_1 & & & & & & \\ & \ddots & & & & & \\ & & \sigma_{r-1} & & & & \\ & & & \sigma_r & & & \\ & & & & 0 & & \\ & & & & & \ddots & \\ & & & & & & 0 \end{pmatrix}$$

and $\sigma_r = \epsilon > 0$. What are the values of $\|A_1 - A_2\|_F$ and $\|A_1^+ - A_2^+\|_F$? What happens to these values as we let $\epsilon \to 0$?

15. Let $A = XY^T$, where X is an $m \times r$ matrix, Y^T is an $r \times n$ matrix, and $X^T X$ and $Y^T Y$ are both nonsingular. Show that the matrix

$$B = Y(Y^T Y)^{-1}(X^T X)^{-1} X^T$$

satisfies the Penrose conditions and hence must equal A^+. Thus A^+ can be determined from any factorization of this form.

MATLAB EXERCISES

Sensitivity of Linear Systems

In these exercises we are concerned with the numerical solution of linear systems of equations. The entries of the coefficient matrix A and the right-hand side **b** may often contain small errors due to limitations in the accuracy of the data. Even if there are no errors in either A or $\mathbf{b}$, roundoff errors will occur when their entries are translated into the finite-precision number system of the computer. Thus we generally expect that the coefficient matrix and the right-hand side will involve small errors. The system that the computer solves then is a slightly perturbed version of the original system. If the original system is very sensitive, its solution could differ greatly from the solution to the perturbed system.

Generally, a problem is well-conditioned if the perturbations in the solutions are on the same order as the perturbations in the data. A problem is ill-conditioned if the changes in the solutions are much greater than the changes in the data. How well- or ill-conditioned a problem is depends on how the size of the perturbations in the solution compares to the size of the perturbations in the data. For linear systems this depends on how close the coefficient matrix is to a matrix of lower rank. The conditioning of a system can be measured using the condition number of the matrix. This can be computed using the MATLAB function **cond**. MATLAB computations are carried out with 16 significant digits of accuracy. You will lose digits of accuracy depending on how sensitive the system is. The greater the condition number, the more digits of accuracy that will be lost.

1. Set

$$A = \textbf{round}(10 * \textbf{rand}(6))$$

$$\mathbf{s} = \textbf{ones}(6, 1)$$

$$\mathbf{b} = A * \mathbf{s}$$

The solution to the linear system $A\mathbf{x} = \mathbf{b}$ is clearly $\mathbf{s}$. Solve the system using the MATLAB \ operation. Compute the error $\mathbf{x} - \mathbf{s}$ (since $\mathbf{s}$ consists entirely of 1's, this is the same as $\mathbf{x} - \mathbf{1}$). Now perturb the system slightly. Set

$$t = 1.0\text{e}{-}12, \qquad E = \textbf{rand}(6) - 0.5, \qquad \mathbf{r} = \textbf{rand}(6, 1) - 0.5$$

and set

$$M = A + t * E, \qquad \mathbf{c} = \mathbf{b} + t * \mathbf{r}$$

Solve the perturbed system $M\mathbf{z} = \mathbf{c}$ for $\mathbf{z}$. Compare the solution $\mathbf{z}$ to the solution of the original system by computing $\mathbf{z} - \mathbf{1}$. How does the size of the perturbation in the solution compare to the size of the perturbations in A and $\mathbf{b}$? Repeat the perturbation analysis with $t = 1.0\text{e}{-}04$ and $t = 1.0\text{e}{-}02$. Is the system $A\mathbf{x} = \mathbf{b}$ well-conditioned? Explain. Use MATLAB to compute the condition number of A.

2. If a vector $\mathbf{y} \in R^n$ is used to construct an $n \times n$ Vandermonde matrix V, then V will be nonsingular provided that $y_1, y_2, \ldots, y_n$ are all distinct.

(a) Construct a Vandermonde system by setting

$$\mathbf{y} = \textbf{rand}(6, 1) \qquad \text{and} \qquad V = \textbf{vander}(\mathbf{y})$$

Generate vectors $\mathbf{b}$ and $\mathbf{s}$ in R^6 by setting

$$\mathbf{b} = \textbf{sum}(V')' \qquad \text{and} \qquad \mathbf{s} = \textbf{ones}(6, 1)$$

If V and $\mathbf{b}$ had been computed in exact arithmetic, then the exact solution to $V\mathbf{x} = \mathbf{b}$ would be $\mathbf{s}$. Why? Explain. Solve $V\mathbf{x} = \mathbf{b}$ using the \ operation. Compare the computed solution $\mathbf{x}$ to the exact solution $\mathbf{s}$ using the MATLAB **format long**. How many significant digits were lost? Determine the condition number of V.

(b) The Vandermonde matrices become increasingly ill-conditioned as the dimension n increases. Even for small values of n we can make the matrix ill-conditioned by taking two of the points close together. Set

$$x(2) = x(1) + 1.0\mathrm{e}{-}12$$

and recompute V using the new value of $x(2)$. For the new matrix V, set $\mathbf{b} = \mathbf{sum}(V')'$ and solve the system $V\mathbf{z} = \mathbf{b}$. How many digits of accuracy were lost? Compute the condition number of V.

3. Construct a matrix C as follows. Set

$$A = \mathbf{round}(100 * \mathbf{rand}(4))$$

$$L = \mathbf{tril}(A, -1) + \mathbf{eye}(4)$$

$$C = L * L'$$

(a) The matrix C is a nice matrix in that it is a symmetric matrix with integer entries and its determinant is equal to 1. Use MATLAB to verify these claims. Why do we know ahead of time that the determinant will equal 1? In theory, the entries of the exact inverse should all be integers. Why? Explain. Does this happen computationally? Compute $D = \mathbf{inv}(C)$ and check its entries using **format long**. Compute $C * D$ and compare it to **eye**(4).

(b) Set

$$\mathbf{r} = \mathbf{ones}(4, 1) \qquad \text{and} \qquad \mathbf{b} = \mathbf{sum}(C')'$$

In exact arithmetic the solution to the system $C\mathbf{x} = \mathbf{b}$ should be $\mathbf{r}$. Compute the solution using \ and display the answer using **format long**. How many digits of accuracy were lost? We can perturb the system slightly by taking e to be a small scalar such as $1.0\mathrm{e}{-}12$ and then replacing the right-hand side of the system by

$$\mathbf{b1} = \mathbf{b} + e * [1, -1, 1, -1]'$$

Solve the perturbed system first for the case $e = 1.0\mathrm{e}{-}12$ and then for the case $e = 10\mathrm{e}{-}06$. In each case compare your solution $\mathbf{x}$ to the original solution by displaying $\mathbf{x} - \mathbf{1}$. Compute $\mathbf{cond}(C)$. Is C ill-conditioned? Explain.

4. The $n \times n$ Hilbert matrix H is defined by

$$h(i, j) = 1/(i + j - 1) \qquad i, j = 1, 2, \ldots, n$$

It can be generated using the MATLAB function **hilb**. The Hilbert matrix is notoriously ill-conditioned. It is often used in examples to illustrate the dangers of matrix computations. The MATLAB function **invhilb** gives the exact inverse of the Hilbert matrix. For the cases $n = 6, 8, 10, 12$ construct H and $\mathbf{b}$ so that $H\mathbf{x} = \mathbf{b}$ is a Hilbert system whose solution in exact arithmetic should be **ones**$(n, 1)$. For each of the cases, determine the solution $\mathbf{x}$ to the system using **invhilb** and examine $\mathbf{x}$ using **format long**. How many digits of accuracy were lost in each case? Compute the condition number of each of the Hilbert matrices. How does the condition number change as n increases?

Sensitivity of Eigenvalues

If A is an $n \times n$ matrix and X is a matrix that diagonalizes A, then the sensitivity of the eigenvalues of A depends on the condition number of X. If A is defective, the condition number for the eigenvalue problem will be infinite. For more on the sensitivity of eigenvalues, see Wilkinson [27], Chapter 2.

5. Use MATLAB to compute the eigenvalues and eigenvectors of a random 6×6 matrix B. Compute the condition number of the matrix of eigenvectors. Is the eigenvalue problem well-conditioned? Perturb B slightly by setting

$$B1 = B + 1.0e - 04 * \mathbf{rand}(6)$$

Compute the eigenvalues and compare them to the exact eigenvalues of B.

6. Set

$$A = \mathbf{round}(10 * \mathbf{rand}(5)); A = A + A'$$

$$[X, D] = \mathbf{eig}(A)$$

Compute **cond**(X) and $X^T X$. What type of matrix is X? Is the eigenvalue problem well-conditioned? Explain. Perturb A by setting

$$A1 = A + 1.0e{-}06 * \mathbf{rand}(5)$$

Calculate the eigenvalues of $A1$ and compare them to the eigenvalues of A.

7. Set $A = \mathbf{magic}(4)$ and $t = \mathbf{trace}(A)$. The scalar t should be an eigenvalue of A and the remaining eigenvalues should add up to zero. Why? Explain. Use MATLAB to verify that $A - tI$ is singular. Compute the eigenvalues of A and a matrix X of eigenvectors. Determine the condition numbers of A and X. Is the eigenvalue problem well-conditioned? Explain. Perturb A by setting

$$A1 = A + 1.0e{-}04 * \mathbf{rand}(4)$$

How do the eigenvalues of $A1$ compare to those of A?

8. Set

$$A = \mathbf{diag}(10 : -1 : 1) + 10 * \mathbf{diag}(\mathbf{ones}(1, 9), 1)$$

$$[X, D] = \mathbf{eig}(A)$$

Compute the condition number of X. Is the eigenvalue problem well-conditioned? Ill-conditioned? Explain. Perturb A by setting

$$A1 = A; \qquad A1(10, 1) = 0.1$$

Compute the eigenvalues of $A1$ and compare them to the eigenvalues of A.

9. Construct a matrix A as follows:

$$A = \mathbf{diag}(11 : -1 : 1, -1);$$
$$\text{for } j = 0 : 11$$
$$\quad A = A + \mathbf{diag}(12 - j : -1 : 1, j);$$
$$\text{end}$$

(a) Compute the eigenvalues of A and the value of the determinant of A. Use the MATLAB function **prod** to compute the product of the eigenvalues. How does the value of the product compare to the determinant?

(b) Compute the eigenvectors of A and the condition number for the eigenvalue problem. Is the problem well-conditioned? Ill-conditioned? Explain.

(c) Set

$$A1 = A + 1.0e{-}04 * \mathbf{rand}(\mathbf{size}(A))$$

Compute the eigenvalues of $A1$. Compare them to the eigenvalues of A by computing

$$\mathbf{sort}(\mathbf{eig}(A1)) - \mathbf{sort}(\mathbf{eig}(A))$$

and displaying the result using **format long**.

Householder Transformations

A Householder matrix is an $n \times n$ orthogonal matrix of the form $I - \frac{1}{b}\mathbf{vv}^T$. For any given nonzero vector $\mathbf{x} \in R^n$, it is possible to choose b and $\mathbf{v}$ so that $H\mathbf{x}$ will be a multiple of $\mathbf{e}_1$.

10. (a) In MATLAB the simplest way to compute a Householder matrix to zero out entries of a given vector $\mathbf{x}$ is to compute the QR factorization of $\mathbf{x}$. Thus, if we are given a vector $\mathbf{x} \in R^n$, then the MATLAB command

$$[H, R] = \mathbf{qr}(\mathbf{x})$$

will compute the desired Householder matrix H. Compute a Householder matrix H to zero out the last three entries of $\mathbf{e} = \mathbf{ones}(4, 1)$. Set

$$C = [\mathbf{e}, \mathbf{rand}(4, 3)]$$

Compute $H * \mathbf{e}$ and $H * C$.

(b) We can also compute the vector $\mathbf{v}$ and the scalar b that determine the Householder transformation that zeros out entries of a given vector. To do this for a given vector $\mathbf{x}$, we would set

$$a = \mathbf{norm}(\mathbf{x});$$

$$\mathbf{v} = \mathbf{x}; \qquad v(1) = v(1) - a$$

$$b = a * (a - x(1))$$

Construct $\mathbf{v}$ and b in this way for the vector $\mathbf{e}$ from part (a). If $K = I - \frac{1}{b}\mathbf{vv}^T$, then

$$K\mathbf{e} = \mathbf{e} - \left(\frac{\mathbf{v}^T \mathbf{e}}{b}\right)\mathbf{v}$$

Compute both of these quantities using MATLAB and verify that they are equal. How does $K\mathbf{e}$ compare to $H\mathbf{e}$ from part (a)? Compute also $K*C$ and $C - \mathbf{v}*((\mathbf{v}'*C)/b)$ and verify that the two are equal.

11. Set

$$\mathbf{x1} = (1:5)'; \qquad \mathbf{x2} = [1, 3, 4, 5, 9]'; \qquad \mathbf{x} = [\mathbf{x1}; \mathbf{x2}]$$

Construct a Householder matrix of the form

$$H = \begin{bmatrix} I & O \\ O & K \end{bmatrix}$$

where K is a 5×5 Householder matrix that zeros out the last four entries of $\mathbf{x2}$. Compute the product $H\mathbf{x}$.

Rotations and Reflections

12. To plot $y = \sin(x)$, we must define vectors of x and y values and then use the **plot** command. This can be done as follows:

$$\mathbf{x} = 0:0.1:6.3; \qquad \mathbf{y} = \sin(\mathbf{x});$$

$$\mathbf{plot(x, y)}$$

(a) Let us define a rotation matrix and use it to rotate the graph of $y = \sin(x)$. Set

$$t = \mathbf{pi}/4; \qquad c = \cos(t); \qquad s = \sin(t); \qquad R = [c, -s; s, c]$$

To find the rotated coordinates, set

$$Z = R*[\mathbf{x}; \mathbf{y}]; \qquad \mathbf{x1} = Z(1, :); \qquad \mathbf{y1} = Z(2, :);$$

The vectors $\mathbf{x1}$ and $\mathbf{y1}$ contain the coordinates for the rotated curve. Set

$$\mathbf{w} = [0, 5]; \qquad \mathbf{axis('square')}$$

and plot $\mathbf{x1}$ and $\mathbf{y1}$ using the MATLAB command

$$\mathbf{plot(x1, y1, w, w)}$$

By what angles has the graph been rotated and in what direction?

(b) Keep all your variables from part (a) and set

$$G = [c, s; s, -c]$$

The matrix G represents a Givens reflection. To determine the reflected coordinates, set

$$Z = G*[\mathbf{x}; \mathbf{y}]; \qquad \mathbf{x2} = Z(1, :); \qquad \mathbf{y2} = Z(2, :);$$

Plot the reflected curve using the MATLAB command

$$\mathbf{plot(x2, y2, w, w)}$$

The curve $y = \sin(x)$ has been reflected about a line through the origin making an angle of $\pi/8$ with the x axis. To see this, set

$$\mathbf{w1} = \left[0, 6.3 * \cos(t/2)\right]; \qquad \mathbf{z1} = \left[0, 6.3 * \sin(t/2)\right];$$

and plot the new axis and both curves using the MATLAB command

$$\mathbf{plot(x, y, x2, y2, w1, z1)}$$

(c) Use the rotation matrix R from part (a) to rotate the curve $y = -\sin(x)$. Plot the rotated curve. How does the graph compare to that of the curve from part (b)? Explain.

Singular Value Decomposition

13. Given

$$A = \begin{bmatrix} 4 & 5 & 2 \\ 4 & 5 & 2 \\ 0 & 3 & 6 \\ 0 & 3 & 6 \end{bmatrix}$$

Enter the matrix A in MATLAB and compute its singular values by setting $\mathbf{s} = \mathbf{svd}(A)$.

(a) How can the entries of $\mathbf{s}$ be used to determine the values $\|A\|_2$ and $\|A\|_F$? Compute these norms by setting $p = \mathbf{norm}(A)$ and $q = \mathbf{norm}(A, \text{'fro'})$ and compare your results to $s(1)$ and $\mathbf{norm(s)}$.

(b) To obtain the full singular value decomposition of A, set

$$[U, D, V] = \mathbf{svd}(A)$$

Compute the closest matrix of rank 1 to A by setting

$$B = s(1) * U(:, 1) * V(:, 1)'$$

How are the row vectors of B related to the two distinct row vectors of A?

(c) The matrices A and B should have the same 2-norm. Why? Explain. Use MATLAB to compute $\|B\|_2$ and $\|B\|_F$. In general, for a rank 1 matrix the 2-norm and the Frobenius norm should be equal. Why? Explain.

14. Set

$$A = \mathbf{round}(10 * \mathbf{rand}(10, 5)) \qquad \text{and} \qquad \mathbf{s} = \mathbf{svd}(A)$$

(a) Use MATLAB to compute $\|A\|_2$, $\|A\|_F$, $\mathrm{cond}_2(A)$ and compare your results to $s(1)$, $\mathbf{norm(s)}$, $s(1)/s(5)$, respectively.

(b) Set

$$[U, D, V] = \mathbf{svd}(A); \qquad D(5, 5) = 0; \qquad B = U * D * V'$$

The matrix B should be the closest matrix of rank 4 to A (where distance is measured in terms of the Frobenius norm). Compute $\|A\|_2$ and $\|B\|_2$. How do these values compare? Compute and compare the Frobenius norms of the two matrices. Compute also $\|A - B\|_F$ and compare the result to $s(5)$. Set $r = \mathbf{norm}(s(1 : 4))$ and compare the result to $\|B\|_F$.

(c) Use MATLAB to construct a matrix C that is the closest matrix of rank 3 to A with respect to the Frobenius norm. Compute $\|C\|_2$ and $\|C\|_F$. How do these values compare with the computed values for $\|A\|_2$ and $\|A\|_F$? Set

$$p = \mathbf{norm}(s(1 : 3)) \qquad \text{and} \qquad q = \mathbf{norm}(s(4 : 5))$$

Compute $\|C\|_F$ and $\|A - C\|_F$ and compare your results to p and q.

15. Set

$$A = \mathbf{rand}(8, 4) * \mathbf{rand}(4, 6), \qquad [\, U, \ D, \ V \,] = \mathbf{svd}(A)$$

(a) What is the rank of A? Use the column vectors of V to generate two matrices $V1$ and $V2$ whose columns form orthonormal bases for $R(A^T)$ and $N(A)$, respectively. Set

$$P = V2 * V2', \quad \mathbf{r} = P * \mathbf{rand}(6, 1), \quad \mathbf{w} = A' * \mathbf{rand}(8, 1)$$

If $\mathbf{r}$ and $\mathbf{w}$ had been computed in exact arithmetic, they would be orthogonal. Why? Explain. Use MATLAB to compute $\mathbf{r}^T \mathbf{w}$.

(b) Use the column vectors of U to generate two matrices $U1$ and $U2$ whose column vectors form orthonormal bases for $R(A)$ and $N(A^T)$, respectively. Set

$$Q = U2 * U2', \quad \mathbf{y} = Q * \mathbf{rand}(8, 1), \quad \mathbf{z} = A * \mathbf{rand}(6, 1)$$

Explain why $\mathbf{y}$ and $\mathbf{z}$ would be orthogonal if all computations were done in exact arithmetic. Use MATLAB to compute $\mathbf{y}^T \mathbf{z}$.

(c) Set $X = \mathbf{pinv}(A)$. Use MATLAB to verify the four Penrose conditions:

   (i) $AXA = A$
   (ii) $XAX = X$
   (iii) $(AX)^T = AX$
   (iv) $(XA)^T = XA$

(d) Compute and compare AX and $U1(U1)^T$. Had all computations been done in exact arithmetic the two matrices would be equal. Why? Explain.

Gerschgorin Circles

16. With each $A \in R^{n \times n}$ we can associate n closed circular disks in the complex plane. The ith disk is centered at a_{ii} and has radius

$$r_i = \sum_{\substack{j=1 \\ j \neq i}}^{n} |a_{ij}|$$

Each eigenvalue of A is contained in at least one of the disks (see Exercise 7, Chapter 7, Section 6).

(a) Set

$$A = \mathbf{round}(10 * \mathbf{rand}(5))$$

Compute the radii of the Gerschgorin disks of A and store them in a vector **r**. To plot the disks, we must parameterize the circles. This can be done by setting

$$t = [0 : 0.1 : 6.3]';$$

We can then generate two matrices X and Y whose columns contain the x and y coordinates of the circles. First we initialize X and Y to zero by setting

$$X = \mathbf{zeros}(\mathbf{length}(t), 5); \qquad Y = X;$$

The matrices can then be generated using the following commands:

```
for i = 1 : 5
    X(:, i) = r(i) * cos(t) + real(A(i, i));
    Y(:, i) = r(i) * sin(t) + imag(A(i, i));
end
```

Set $\mathbf{e} = \mathbf{eig}(A)$ and plot the eigenvalues and the disks with the command

$$\mathbf{plot}(X, Y, \mathbf{real}(\mathbf{e}), \mathbf{imag}(\mathbf{e}), \text{'}x\text{'})$$

If everything is done correctly, all the eigenvalues of A should lie within the union of the circular disks.

(b) If k of the Gerschgorin disks form a connected domain in the complex plane that is isolated from the other disks, then exactly k of the eigenvalues of the matrix will lie in that domain. Set

$$B = [3 \quad 0.1 \quad 2; \quad 0.1 \quad 7 \quad 2; \quad 2 \quad 2 \quad 50]$$

  (i) Use the method described in part (a) to compute and plot the Gerschgorin disks of B.

 (ii) Since B is symmetric, its eigenvalues are all real and so must all lie on the real axis. Without computing the eigenvalues, explain why B must have exactly one eigenvalue in the interval $[46, 54]$. Multiply the first two rows of B by 0.1 and then multiply the first two columns by 10. We can do this in MATLAB by setting

$$D = \mathbf{diag}([0.1, 0.1, 1]) \qquad \text{and} \qquad C = D * B / D$$

The new matrix C should have the same eigenvalues as B. Why? Explain. Use C to find intervals containing the other two eigenvalues. Compute and plot the Gerschgorin disks for C.

(iii) How are the eigenvalues of C^T related to the eigenvalues of B and C? Compute and plot the Gerschgorin disks for C^T. Use one of the rows of C^T to find an interval containing its largest eigenvalue.

Distribution of Condition Numbers and Eigenvalues of Random Matrices

17. We can generate a random symmetric 10×10 matrix by setting

$$A = \mathbf{rand}(10); \; A = (A + A')/2$$

Since A is symmetric, its eigenvalues are all real. The number of positive eigenvalues can be calculated by setting

$$y = \mathbf{sum}(\mathbf{eig}(A) > 0)$$

(a) For $j = 1, 2, \ldots, 100$, generate a random symmetric 10×10 matrix and determine the number of positive eigenvalues. Denote the number of positive eigenvalues of the jth matrix by $y(j)$. Set $\mathbf{x} = 0 : 10$ and determine the distribution of the $\mathbf{y}$ data by setting $\mathbf{n} = \mathbf{hist}(\mathbf{y}, \mathbf{x})$. Determine the mean of the $y(j)$ values using the MATLAB command $\mathbf{mean}(\mathbf{y})$. Use the MATLAB command $\mathbf{hist}(\mathbf{y}, \mathbf{x})$ to generate a plot of the histogram.

(b) We can generate a random symmetric 10×10 matrix whose entries are in the interval $[-1, 1]$ by setting

$$A = 2 * \mathbf{rand}(10) - 1; \qquad A = (A + A')/2$$

Repeat part (a) using random matrices generated in this manner. How does the distribution of the $\mathbf{y}$ data compare to the one obtained in part (a)?

18. A nonsymmetric matrix A may have complex eigenvalues. We can determine the number of eigenvalues of A that are both real and positive using the MATLAB commands

$$\mathbf{e} = \mathbf{eig}(A)$$

$$y = \mathbf{sum}(\mathbf{e} > 0 \ \& \ \mathbf{imag}\,(\mathbf{e}) == 0)$$

Generate 100 random nonsymmetric 10×10 matrices. For each matrix, determine the number of positive real eigenvalues and store that number as an entry of a vector $\mathbf{z}$. Determine the mean of the $z(j)$ values and compare it to the mean computed in part (a) of Exercise 17. Determine the distribution and plot the histogram.

19. (a) Generate 100 random 5×5 matrices and compute the condition number of each matrix. Determine the mean of the condition numbers and plot the histogram of the distribution.

(b) Repeat part (a) using 10×10 matrices. Compare your results to those obtained in part (a).

CHAPTER TEST

In each of the following, answer *true* if the statement is always true and *false* otherwise. In the case of a true statement, explain or prove your answer. In the case of a false statement, give an example to show that the statement is not always true.

1. If a, b, and c are floating-point numbers , then

$$fl(fl(a + b) + c) = fl(a + fl(b + c))$$

2. The computation of $A(BC)$ requires the same number of floating-point operations as the computation of $(AB)C$.

3. If A is a nonsingular matrix and a numerically stable algorithm is used to compute the solution to a system $A\mathbf{x} = \mathbf{b}$, then the relative error in the computed solution will always be small.

4. If A is a symmetric matrix and a numerically stable algorithm is used to compute the eigenvalues of $A\mathbf{x} = \mathbf{b}$, then the relative error in the computed eigenvalues should always be small.

5. If A is a nonsymmetric matrix and a numerically stable algorithm is used to compute the eigenvalues of $A\mathbf{x} = \mathbf{b}$, then the relative error in the computed eigenvalues should always be small.

6. If both A^{-1} and the LU factorization of an $n \times n$ matrix A have already been computed, then it is more efficient to solve a system $A\mathbf{x} = \mathbf{b}$ by multiplying $A^{-1}\mathbf{b}$, rather than solving $LU\mathbf{x} = \mathbf{b}$ using forward and back substitution.

7. If A is a symmetric matrix, $\|A\|_1 = \|A\|_\infty$.

8. If A is an $m \times n$ matrix, $\|A\|_2 = \|A\|_F$.

9. If the coefficient matrix A in a least squares problem has dimensions $m \times n$ and rank n, then the three methods of solution discussed in Section 7, the normal equations, the QR factorization, and the singular value decomposition, will all compute highly accurate solutions.

10. If two $m \times n$ matrices A and B are close in the sense that $\|A - B\|_2 < \epsilon$ for some small positive number ϵ, then their pseudoinverses will also be close; that is, $\|A^+ - B^+\|_2 < \delta$, for some small positive number δ.

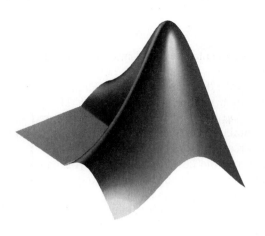

APPENDIX

MATLAB

MATLAB is an interactive program for matrix computations. The original version of MATLAB, short for *matrix laboratory*, was developed by Cleve Moler from the Linpack and Eispack software libraries. Over the years MATLAB has undergone a series of expansions and revisions. Today it is the leading software for scientific computations. The professional version of MATLAB is distributed by the Math Works, Inc. of Natick, Massachusetts.

In addition to widespread use in industrial and engineering settings, MATLAB has also become a standard instructional tool for undergraduate linear algebra courses. A Student Edition of MATLAB is available at a price affordable to undergraduates.

Another highly recommended resource for teaching linear algebra with MATLAB is *ATLAST Computer Exercises for Linear Algebra*, (see [9]). This manual contains MATLAB-based exercises and projects for linear algebra and a collection of MATLAB utilities (M-files) that help students to visualize linear algebra concepts. The M-files are available for download from the ATLAST Web page:

http://www.umassd.edu/SpecialPrograms/ATLAST/

THE MATLAB DESKTOP DISPLAY

At start-up, MATLAB will display a desk top with three windows. The window on the right is the command window where MATLAB commands are entered and executed. The window on the top left displays either the MATLAB Launch Pad or the Workspace Browser, depending on which button has been toggled. The Launch Pad can be used to search for tools and demos and then start them up by a double-click of the mouse. The Workspace Browser allows you to view and make changes to the contents of the workspace.

The window on the lower left displays either the Command History or the Current Directory Browser, depending on which button has been toggled. The Command History allows you view a log of all the functions that have been entered in the command window. To reexecute a command, just click on it to highlight and then double-click to execute. You can also reexecute and edit a command from the command window using the arrow keys. From the command window you can use the up arrow to recall previous commands. The commands can then be edited using the left and right arrow keys. Press the Enter key of your computer to execute the edited command. The Current Directory Browser allows you to view MATLAB and other files and to perform file operations such as opening and editing or searching for files.

Any of the MATLAB windows can be closed by clicking on the $\times$ in the upper-right corner of the window. To detach a window from the MATLAB desktop, click on the arrow that is next to the $\times$ in the upper-right corner of the window.

BASIC DATA ELEMENTS

The basic elements that MATLAB uses are matrices. Once the matrices have been entered or generated, the user can quickly perform sophisticated computations with a minimal amount of programming.

Entering matrices in MATLAB is easy. To enter the matrix

$$\begin{bmatrix} 1 & 2 & 3 & 4 \\ 5 & 6 & 7 & 8 \\ 9 & 10 & 11 & 12 \\ 13 & 14 & 15 & 16 \end{bmatrix}$$

type

$$A = [1 \quad 2 \quad 3 \quad 4; \quad 5 \quad 6 \quad 7 \quad 8; \quad 9 \quad 10 \quad 11 \quad 12; \quad 13 \quad 14 \quad 15 \quad 16]$$

or the matrix could be entered one row at a time:

$$A = [\, 1 \quad 2 \quad 3 \quad 4$$
$$5 \quad 6 \quad 7 \quad 8$$
$$9 \quad 10 \quad 11 \quad 12$$
$$13 \quad 14 \quad 15 \quad 16\,]$$

Once a matrix has been entered you can edit it in two ways. From the command window you can redefine any entry with a MATLAB command. For example, the command $A(1, 3) = 5$ will change the third entry in the first row of A to 5. You can also edit the entries of a matrix from the Workspace Browser. To change the $(1, 3)$ entry of A using the Workspace Browser, we first locate A in the Name column of the browser and then click on the array icon to the left of A to open an array display of the matrix. To change the $(1, 3)$ entry to a 5, click on the corresponding cell of the array and enter 5.

Row vectors of equally spaced points can be generated using MATLAB's : operation. The command $\mathbf{x} = 2:6$ generates a row vector with integer entries going from 2 to 6.

$$\mathbf{x} =$$

$$2 \quad 3 \quad 4 \quad 5 \quad 6$$

It is not necessary to use integers or to have a step size of 1. For example, the command $\mathbf{x} = 1.2:0.2:2$ will produce

$$\mathbf{x} =$$

$$1.2000 \quad 1.4000 \quad 1.6000 \quad 1.8000 \quad 2.0000$$

SUBMATRICES

To refer to a submatrix of the matrix A entered earlier, use the : to specify the rows and columns. For example, the submatrix consisting of the entries in the second two rows of columns 2 through 4 is given by $A(2:3, 2:4)$. Thus the statement

$$C = A(2:3, 2:4)$$

generates

$$C =$$
$$\begin{array}{ccc} 6 & 7 & 8 \\ 10 & 11 & 12 \end{array}$$

If the colon is used by itself for one of the arguments, either all the rows or all the columns of the matrix will be included. For example, $A(:, 2:3)$ represents the submatrix of A consisting of all the elements in the second and third columns, and $A(4, :)$ denotes the fourth row vector of A. We can generate a submatrix using nonadjacent rows or columns by using vector arguments to specify which rows and columns are to be included. For example, to generate a matrix whose entries are those that appear only in the first and third rows and second and fourth columns of A, set

$$E = A([1, 3], [2, 4])$$

The result will be

$$E =$$
$$\begin{array}{cc} 2 & 4 \\ 10 & 12 \end{array}$$

GENERATING MATRICES

We can also generate matrices using built-in MATLAB functions. For example, the command

$$B = \mathbf{rand}(4)$$

will generate a 4×4 matrix whose entries are random numbers between 0 and 1. Other functions that can be used to generate matrices are **eye, zeros, ones,**

magic, **hilb**, **pascal**, **toeplitz**, **compan**, and **vander**. To build triangular or diagonal matrices, we can use the MATLAB functions **triu**, **tril**, and **diag**.

The matrix building commands can be used to generate blocks of partitioned matrices. For example, the MATLAB command

$$E = [\ \mathbf{eye}(2),\ \mathbf{ones}(2,3);\ \mathbf{zeros}(2),\ [1\!:\!3;\quad 3\!:\!-1\!:\!1]\ ]$$

will generate the matrix

$$E =$$
$$
\begin{matrix}
1 & 0 & 1 & 1 & 1 \\
0 & 1 & 1 & 1 & 1 \\
0 & 0 & 1 & 2 & 3 \\
0 & 0 & 3 & 2 & 1
\end{matrix}
$$

MATRIX ARITHMETIC

Matrix arithmetic in MATLAB is straightforward. We can multiply our original matrix A times B simply by typing $A * B$. The sum and difference of A and B are given by $A + B$ and $A - B$, respectively. The transpose of A is given by A'. If $\mathbf{c}$ represents a vector in R^4, the solution to the linear system $A\mathbf{x} = \mathbf{c}$ can be computed by setting

$$\mathbf{x} = A\backslash\mathbf{c}$$

Powers of matrices are easily generated. The matrix A^5 is computed in MATLAB by typing $A\hat{}5$. We can also perform operations elementwise by preceding the operand by a period. For example, if $W = [1\quad 2;\quad 3\quad 4]$, then $W\hat{}2$ results in

$$\text{ans} =$$
$$
\begin{matrix}
7 & 10 \\
15 & 22
\end{matrix}
$$

while $W.\hat{}2$ will give

$$\text{ans} =$$
$$
\begin{matrix}
1 & 4 \\
9 & 16
\end{matrix}
$$

MATLAB FUNCTIONS

To compute the eigenvalues of a square matrix A, we need only type **eig**(A). The eigenvectors and eigenvalues can be computed by setting

$$[X \quad D] = \mathbf{eig}(A)$$

Similarly, we can compute the determinant, inverse, condition number, norm, and rank of a matrix with simple one-word commands. Matrix factorizations such as the LU, QR, Cholesky, Schur decomposition, and singular value decomposition can be computed with a single command. For example, the command

$$[Q \quad R] = \mathbf{qr}(A)$$

will produce an orthogonal (or unitary) matrix Q and an upper triangular matrix R, with the same dimensions as A, such that $A = QR$.

PROGRAMMING FEATURES

MATLAB has all the flow control structures that you would expect in a high-level language, including **for** loops, **while** loops, and **if** statements. This allows the user to write his or her own MATLAB programs and to create additional MATLAB functions. It should be noted that MATLAB prints out automatically the result of each command unless the command line ends in a semicolon. *When using loops, we recommend ending each command with a semicolon to avoid printing all the results of the intermediate computations.*

M-FILES

It is possible to extend MATLAB by adding your own programs. MATLAB programs are all given the extension **.m** and are referred to as *M-files*. There are two basic types of M-files.

Script files are files that contain a series of MATLAB commands. All the variables used in these commands are global and, consequently, the values of these variables in your MATLAB session will change every time you run the script file. For example, if you wanted to determine the nullity of a matrix, you could create a script file **nullity.m** containing the following commands:

$$\mathbf{[m,n]} = \mathbf{size}(A);$$

$$\mathbf{nuldim} = \mathbf{n} - \mathbf{rank}(A)$$

Entering the command **nullity** would cause the two lines of code in the script file to be executed. The disadvantage of determining the nullity this way is that the matrix must be named A and the values of the variable m and n may get changed in your MATLAB session. An alternative would be to create a *function file*.

Function files begin with a function declaration statement of the form

$$\mathbf{function}\ [\mathbf{oarg1},\ \ldots,\mathbf{oargj}] = \mathbf{fname}(\mathbf{inarg1},\ \ldots,\mathbf{inargk})$$

All the variables used in the function M-file are local. When you call a function file, only the values of the output variables will change in your MATLAB session. For example, we could create a function file **nullity.m** to compute the nullity of a matrix as follows:

```
function k = nullity(A)
% The command nullity(A) computes the dimension
% of the nullspace of A.
[m,n] = size(A);
k = n - rank(A);
```

The lines beginning with a **%** are comments that are not executed. These lines will be displayed whenever you type **help nullity** in a MATLAB session. Once the function is saved it can be used in a MATLAB session in the same way that we use built-in MATLAB functions. For example, if we set

$$B = [1\ 2\ 3;\ 4\ 5\ 6;\ 7\ 8\ 9];$$

and then enter the command

$$n = \mathbf{nullity}(B)$$

MATLAB will return the answer: $\mathbf{n} = 1$.

The M-files that you develop should be kept in a directory that can be added to the *MATLAB path*, the list of directories that MATLAB searches for M-files. To have your directories automatically appended to the MATLAB path at the start of a MATLAB session, create an M-file **startup.m** that includes commands to be executed at start-up. To append a directory to the MATLAB path, include a line in the **startup** file of the form

$$\mathbf{addpath}\ \text{dirlocation}$$

For example, if you are working on a PC and the linear algebra files that you created are in drive **c** in a subdirectory **linalg** of the MATLAB directory, then, if you add the line

$$\mathbf{addpath\ c:\backslash MATLAB\backslash linalg}$$

to the MATLAB start-up file, MATLAB will automatically preappend the **linalg** directory to its search path at start-up. On Windows platforms, the **startup.m** file should be placed in the **tools\local** subdirectory of your root MATLAB directory.

RELATIONAL AND LOGICAL OPERATORS

MATLAB has six relational operators that are used for comparisons of scalars or elementwise comparisons of arrays. These operators are:

Relational Operators

$<$	less than
$<=$	less than or equal
$>$	greater than
$>=$	greater than or equal
$==$	equal
$\sim=$	not equal

Given two $m \times n$ matrices A and B, the command

$$C = A < B$$

will generate an $m \times n$ matrix consisting of zeros and ones. The (i, j) entry will be equal to 1 if and only if $a_{ij} < b_{ij}$. For example, suppose that

$$A = \begin{pmatrix} -2 & 0 & 3 \\ 4 & 2 & -5 \\ -1 & -3 & 2 \end{pmatrix}$$

The command $A >= 0$ will generate

ans =
$$\begin{matrix} 0 & 1 & 1 \\ 1 & 1 & 0 \\ 0 & 0 & 1 \end{matrix}$$

There are three logical operators as shown in the following table:

Logical Operators

&	AND
|	OR
~	NOT

These logical operators regard any nonzero scalar as corresponding to TRUE and 0 as corresponding to FALSE. The operator & corresponds to the logical AND. If a and b are scalars, the expression $a \& b$ will equal 1 if a and b are both nonzero (TRUE) and 0 otherwise. The operator $|$ corresponds to the logical OR. The expression $a|b$ will have the value 0 if both a and b are 0 and otherwise it will be equal to 1. The operator $\sim$ corresponds to the logical NOT. For a scalar a it takes on the value 1 (TRUE) if $a = 0$ (FALSE) and the value 0 (FALSE) if $a \neq 0$ (TRUE).

For matrices these operators are applied elementwise. Thus, if A and B are both $m \times n$ matrices, then $A \& B$ is a matrix of zeros and ones whose ij th entry is $a(i, j) \& b(i, j)$. For example, if

$$A = \begin{pmatrix} 1 & 0 & 1 \\ 0 & 1 & 1 \\ 0 & 0 & 1 \end{pmatrix} \quad \text{and} \quad B = \begin{pmatrix} -1 & 2 & 0 \\ 1 & 0 & 3 \\ 0 & 1 & 2 \end{pmatrix}$$

then

$$A \& B = \begin{pmatrix} 1 & 0 & 0 \\ 0 & 0 & 1 \\ 0 & 0 & 1 \end{pmatrix}, \quad A|B = \begin{pmatrix} 1 & 1 & 1 \\ 1 & 1 & 1 \\ 0 & 1 & 1 \end{pmatrix}, \quad \sim A = \begin{pmatrix} 0 & 1 & 0 \\ 1 & 0 & 0 \\ 1 & 1 & 0 \end{pmatrix}$$

The relational and logical operators are often used in **if** statements.

COLUMNWISE ARRAY OPERATORS

MATLAB has a number of functions that, when applied to either a row or column vector **x**, return a single number. For example, the command **max(x)** will compute the maximum entry of **x**, and the command **sum(x)** will return the value of the sum of the entries of **x**. Other functions of this form are **min**, **prod**, **mean**, **all**, and **any**. When used with a matrix argument, these functions are applied to each column vector and the results are returned as a row vector. For example, if

$$A = \begin{bmatrix} -3 & 2 & 5 & 4 \\ 1 & 3 & 8 & 0 \\ -6 & 3 & 1 & 3 \end{bmatrix}$$

then

$$\mathbf{min}(A) = (-6, 2, 1, 0)$$
$$\mathbf{max}(A) = (1, 3, 8, 4)$$
$$\mathbf{sum}(A) = (-8, 8, 14, 7)$$
$$\mathbf{prod}(A) = (18, 18, 40, 0)$$

GRAPHICS

If **x** and **y** are vectors of the same length, the command **plot(x, y)** will produce a plot of all the (x_i, y_i) pairs, and each point will be connected to the next by a line segment. If the x coordinates are taken close enough together, the graph should resemble a smooth curve. The command **plot(x, y, 'x')** will plot the ordered pairs with x's but will not connect the points.

For example, to plot the function $f(x) = \dfrac{\sin x}{x + 1}$ on the interval [0, 10], set

$$\mathbf{x} = 0 : 0.2 : 10 \qquad \text{and} \qquad \mathbf{y} = \sin(\mathbf{x}). / (\mathbf{x} + 1)$$

The command **plot(x, y)** will generate the graph of the function. To compare the graph to that of $\sin x$, we could set $\mathbf{z} = \sin(\mathbf{x})$ and use the command

$$\mathbf{plot(x, y, x, z)}$$

to plot both curves at the same time, as in Figure A.1.

It is also possible to do more sophisticated types of plots in MATLAB, including polar coordinates, three-dimensional surfaces, and contour plots.

SYMBOLIC TOOLBOX

In addition to doing numeric computations, it is also possible to do symbolic calculations using MATLAB's symbolic toolbox. The symbolic toolbox allows us to manipulate symbolic expressions. It can be used to solve equations, differentiate and integrate functions, and for symbolic matrix operations.

MATLAB's **sym** command can be used to turn any MATLAB data structure into a symbolic object. For example, the command **sym('t')** will turn the string '**t**'

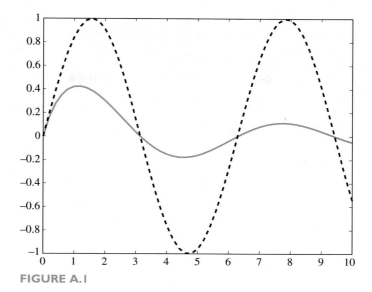

into a symbolic variable **t**, and the command **sym(hilb(3))** will produce the symbolic version of the 3×3 Hilbert matrix written in the form

$$\left[\, 1, \ \tfrac{1}{2}, \ \tfrac{1}{3} \,\right]$$
$$\left[\, \tfrac{1}{2}, \ \tfrac{1}{3}, \ \tfrac{1}{4} \,\right]$$
$$\left[\, \tfrac{1}{3}, \ \tfrac{1}{4}, \ \tfrac{1}{5} \,\right]$$

We can create a number of symbolic variables at once using the **syms** command. For example, the command

$$\text{\textbf{syms a b c}}$$

creates three symbolic variables **a**, **b**, and **c**. If we then set

$$A = \text{\textbf{[a, b, c; b, c, a; c, a, b]}}$$

the result will be the symbolic matrix

$$A =$$

$$\left[\, \text{\textbf{a, b, c}} \,\right]$$
$$\left[\, \text{\textbf{b, c, a}} \,\right]$$
$$\left[\, \text{\textbf{c, a, b}} \,\right]$$

The MATLAB command **subs** can be used to substitute an expression or value for a symbolic variable. For example, the command **subs(A,c,3)** will substitute 3 for each occurrence of **c** in the symbolic matrix A. Multiple substitutions are also possible. The command

$$\text{\textbf{subs(A, [a,b,c], [a-1,b+1,3])}}$$

will substitute **a** -1, **b** $+1$, and 3 for **a, b**, and **c** in the matrix A.

The standard matrix operations $*, \char94, +, -, '$ all work for symbolic matrices and also for combinations of symbolic and numeric matrices. If an operation involves two matrices and one of them is symbolic, the result will be a symbolic matrix. For example, the command

$$\texttt{sym(hilb(3)) + eye(3)}$$

will produce the symbolic matrix

$$\left[2, \tfrac{1}{2}, \tfrac{1}{3} \right]$$
$$\left[\tfrac{1}{2}, \tfrac{4}{3}, \tfrac{1}{4} \right]$$
$$\left[\tfrac{1}{3}, \tfrac{1}{4}, \tfrac{6}{5} \right]$$

Standard MATLAB matrix commands such as

$$\texttt{det, eig, inv, null, trace, sum, prod, poly}$$

all work for symbolic matrices; however, others such as

$$\texttt{rref, orth, rank, norm}$$

do not. Likewise, none of the standard matrix factorizations are possible for symbolic matrices.

HELP FACILITY

MATLAB includes a HELP facility that provides help on all MATLAB features. To access MATLAB's help browser, click on the help button in the toolbar (this is the button with the symbol: ?) or type **helpbrowser** in the command window. You can also access HELP by selecting it from the View menu. The help facility gives information on getting started with MATLAB and on using and customizing the desktop. It lists and describes all the MATLAB functions, operations, and commands.

You can also obtain help information on any of the MATLAB commands directly from the command window. Simply enter **help** followed by the name of the command. For example, the MATLAB command **eig** is used to compute eigenvalues. For information on how to use this command, you could either find the command using the help browser or simply type **help eig** in the command window.

From the command window, you also can obtain help on any MATLAB operator. Simply type **help** followed by the name of the operator. To do this, you need to know the name that MATLAB gives to the operator. You can obtain a complete list of all operator names by entering **help** followed by any operator symbol. For example, to obtain help on the backslash operation, first type **help **. MATLAB will respond by displaying the list of all operator names. The backslash operator is listed as **mldivide** (short for matrix left divide). To find out how the operator works, simply type **help mldivide**.

CONCLUSIONS

MATLAB is a powerful tool for matrix computations that is also user friendly. The fundamentals can be mastered easily and, consequently, students are able to begin numerical experiments with only a minimal amount of preparation. Indeed, the material in this appendix together with the online help facility should be enough to get you started.

The MATLAB exercises at the end of each chapter are designed to enhance understanding of linear algebra. The exercises do not assume familiarity with MATLAB. Often specific commands are given to guide the reader through the more complicated MATLAB constructions. Consequently, you should be able to work through all the exercises without resorting to additional MATLAB books or manuals.

While this appendix summarizes the features of MATLAB that are relevant to an undergraduate course in linear algebra, many other advanced capabilities have not been discussed. References [14] and [22] describe MATLAB in greater detail.

Bibliography

A LINEAR ALGEBRA AND MATRIX THEORY

[1] Brualdi, Richard A., and Herbert J. Ryser, *Combinatorial Matrix Theory*. New York: Cambridge University Press, 1991.

[2] Carlson, David, Charles R. Johnson, David C. Lay, A. Duane Porter, Ann Watkins, and William Watkins, eds., *Resources for Teaching Linear Algebra*. Washington, D.C.: MAA, 1997.

[3] Gantmacher, F. R., *The Theory of Matrices*, 2 vols. New York: Chelsea Publishing Co., 1960.

[4] Hill, David R., and David E. Zitarelli, *Linear Algebra Labs with MATLAB*. Upper Saddle River, N.J.: Prentice Hall, 1994.

[5] Horn, Roger A., and Charles R. Johnson, *Matrix Analysis*. New York: Cambridge University Press, 1985.

[6] Horn, Roger A., and Charles R. Johnson, *Topics in Matrix Analysis*. New York: Cambridge University Press, 1991.

[7] Keith, Sandra, *Visualizing Linear Algebra Using Maple*. Upper Saddle River, N.J.: Prentice Hall, 2001.

[8] Lancaster, Peter, and M. Tismenetsky, *The Theory of Matrices with Applications*, 2nd ed. New York: Academic Press, 1985.

[9] Leon, Steven J., Eugene Herman, and Richard Faulkenberry, *ATLAST Computer Exercises for Linear Algebra*. Upper Saddle River, N.J.: Prentice Hall, 1997.

[10] Ortega, James M., *Matrix Theory: A Second Course*. New York: Plenum Press, 1987.

B APPLIED AND NUMERICAL LINEAR ALGEBRA

[11] Anderson, E., Z. Bai, C. Bischof, J. Demmel, J. Dongarra, J. Du Croz, A. Greenbaum, S. Hammarling, A. McKenney, S. Ostrouchov, and D. Sorenson, *LAPACK Users' Guide*, 2nd ed. Philadelphia: SIAM, 1995.

[12] Bellman, Richard, *Introduction to Matrix Analysis*, 2nd ed. New York: McGraw-Hill Book Co., 1970.

[13] Björck, Åke, *Numerical Methods for Least Squares Problems*. Philadelphia: SIAM, 1996.

[14] Coleman, Thomas F., and Charles Van Loan, *Handbook for Matrix Computations*. Philadelphia: SIAM, 1988.

[15] Conte, S. D., and C. deBoor, *Elementary Numerical Analysis: An Algorithmic Approach*, 3rd ed. New York: McGraw-Hill Book Co., 1980.

[16] Dahlquist, G., and A. Bjorck, *Numerical Methods*. Englewood Cliffs, N.J.: Prentice Hall, 1974.

[17] Datta, Biswa Nath, *Numerical Linear Algebra and Applications*. Pacific Grove, Calif.: Brooks/Cole Publishing Co., 1995.

[18] Demmel, James W., *Applied Numerical Linear Algebra*, Philadelphia: SIAM, 1997.

[19] Fletcher, T. J., *Linear Algebra Through Its Applications*. New York: Van Nostrand Reinhold, 1972.

[20] Golub, Gene H., and Charles F. Van Loan, *Matrix Computations*, 3rd ed. Baltimore: Johns Hopkins University Press, 1996.

[21] Greenbaum, Anne, *Iterative Methods for Solving Linear Systems*. Philadelphia: SIAM, 1997.

[22] Higham, Desmond J., and Nicholas J. Higham, *MATLAB Guide*. Philadelphia: SIAM, 2000.

[23] Parlett, B. N., *The Symmetric Eigenvalue Problem*. Englewood Cliffs, N.J.: Prentice Hall, 1980.

[24] Saad, Yousef, *Iterative Methods for Sparse Linear Systems*. Boston: PWS Publishing Co., 1996.

[25] Trefethen, Loyd N., *Numerical Linear Algebra*. Philadelphia: SIAM, 1997.

[26] Watkins, David S., *Fundamentals of Matrix Computation*. New York: John Wiley & Sons, 1991.

[27] Wilkinson, J. H., *The Algebraic Eigenvalue Problem*. New York: Oxford University Press, 1965.

[28] Wilkinson, J. H., and C. Reinsch, *Handbook for Automatic Computation*, Vol. II: *Linear Algebra*. New York: Springer-Verlag, 1971.

C BOOKS OF RELATED INTEREST

[29] Chiang, Alpha C., *Fundamental Methods of Mathematical Economics*. New York: McGraw-Hill Book Co., 1967.

[30] Courant, R., and D. Hilbert, *Methods of Mathematical Physics*, Vol. I. New York: Wiley-Interscience, 1953.

[31] Edwards, Allen L., *Multiple Regression and the Analysis of Variance and Covariance*. New York: W. H. Freeman and Co., 1985.

[32] Gander, Walter, and Jiří Hřebíček, *Solving Problems in Scientific Computing Using Maple and MATLAB*, 2nd ed., Berlin: Springer-Verlag, 1995.

[33] Higham, N., *Accuracy and Stability of Numerical Algorithms*. Philadelphia: SIAM, 1996.

[34] Rivlin, T. J., *The Chebyshev Polynomials*. New York: Wiley-Interscience, 1974.

[35] Van Loan, Charles, *Computational Frameworks for the Fast Fourier Transform*. Philadelphia: SIAM, 1992.

References [4], [9], [14], and [22] contain information on MATLAB. Reference [9] may be used as a companion volume to this book (see the Preface for more details and for information on how to obtain the ATLAST collection of M-files for linear algebra). Extended bibliographies are included in the following references: [3], [5], [17], [20], [23], [24], and [27].

ANSWERS TO SELECTED EXERCISES

Chapter 1

SECTION I

1. (a) $(11, 3)$; (b) $(4, 1, 3)$; (c) $(-2, 0, 3, 1)$; (d) $(-2, 3, 0, 3, 1)$

2. (a) $\begin{bmatrix} 1 & -3 \\ 0 & 2 \end{bmatrix}$; (b) $\begin{bmatrix} 1 & 1 & 1 \\ 0 & 2 & 1 \\ 0 & 0 & 3 \end{bmatrix}$; (c) $\begin{bmatrix} 1 & 2 & 2 & 1 \\ 0 & 3 & 1 & -2 \\ 0 & 0 & -1 & 2 \\ 0 & 0 & 0 & 4 \end{bmatrix}$

3. (a) One solution. The two lines intersect at the point $(3, 1)$.
   (b) No solution. The lines are parallel.
   (c) Infinitely many solutions. Both equations represent the same line.
   (d) No solution. Each pair of lines intersect in a point; however, there is no point that is on all three lines.

4. (a) $\left[\begin{array}{cc|c} 1 & 1 & 4 \\ 1 & -1 & 2 \end{array}\right]$; (b) $\left[\begin{array}{cc|c} 1 & 2 & 4 \\ -2 & -4 & 4 \end{array}\right]$; (c) $\left[\begin{array}{cc|c} 2 & -1 & 3 \\ -4 & 2 & -6 \end{array}\right]$; (d) $\left[\begin{array}{cc|c} 1 & 1 & 1 \\ 1 & -1 & 1 \\ -1 & 3 & 3 \end{array}\right]$

6. (a) $(1, -2)$; (b) $(3, 2)$; (c) $(\frac{1}{2}, \frac{2}{3})$; (d) $(1, 1, 2)$; (e) $(-3, 1, 2)$; (f) $(-1, 1, 1)$;
   (g) $(1, 1, -1)$; (h) $(4, -3, 1, 2)$

7. (a) $(2, -1)$; (b) $(-2, 3)$ 8. (a) $(-1, 2, 1)$; (b) $(3, 1, -2)$

SECTION 2

1. Row echelon form: (a), (c), (d), (g), and (h); reduced row echelon form: (c), (d), and (g)

2. (a) Inconsistent;   (b) consistent $(4, -1)$;   (c) consistent, infinitely many solutions;
   (d) consistent $(4, 5, 2)$;   (e) inconsistent;   (f) consistent, $(5, 3, 2)$

3. (a) $(-2, 5, 3)$;   (b) $\emptyset$;   (c) $\{(2+3\alpha, \alpha, -2) \mid \alpha \text{ real}\}$; (d) $\{(5-2\alpha-\beta, \alpha, 4-3\beta, \beta) \mid \alpha, \beta \text{ real}\}$;
   (e) $\{(3 - 5\alpha + 2\beta, \alpha, \beta, 6) \mid \alpha, \beta \text{ real}\}$;   (f) $\{(\alpha, 2, -1) \mid \alpha \text{ real}\}$

4. (a) x_1, x_2, x_3 are lead variables.   (c) x_1, x_3 are lead variables and x_2 is a free variable.
   (e) x_1, x_4 are lead variables and x_2, x_3 are free variables.

5. (a) $(5, 1)$;   (b) inconsistent;   (c) $(0, 0)$;   (d) $\left\{ \left(\dfrac{5 - \alpha}{4}, \dfrac{1 + 7\alpha}{8}, \alpha \right) \middle| \alpha \text{ real} \right\}$;
   (e) $\{(8 - 2\alpha, \alpha - 5, \alpha)\}$; (f) inconsistent; (g) inconsistent; (h) inconsistent; (i) $(0, \frac{3}{2}, 1)$;
   (j) $\{(2 - 6\alpha, 4 + \alpha, 3 - \alpha, \alpha)\}$;   (k) $\{(\frac{15}{4} - \frac{5}{8}\alpha - \beta, -\frac{1}{4} - \frac{1}{8}\alpha, \alpha, \beta)\}$;   (l) $\{(1 + \frac{2}{7}\alpha, \frac{3}{7}\alpha, \alpha)\}$

6. (a) $(0, -1)$;   (b) $\{(\frac{3}{4} - \frac{5}{8}\alpha, -\frac{1}{4} - \frac{1}{8}\alpha, \alpha, 3) \mid \alpha \text{ is real}\}$; (c) $\{(0, \alpha, -\alpha)\}$;   (d) $\{\alpha(-\frac{4}{3}, 0, \frac{1}{3}, 1)\}$

8. $a \neq -2$ 9. $\beta = 2$ 10. (a) $a = 5, b = 4$;   (b) $a = 5, b \neq 4$

11. (a) $(-2, 2)$;   (b) $(-7, 4)$ 12. (a) $(-3, 2, 1)$;   (b) $(2, -2, 1)$

13. $x_1 = 280, x_2 = 230, x_3 = 350, x_4 = 590$ 17. $x_1 = 1, \ x_2 = 3, \ x_3 = 6, \ x_4 = 6$

18. 6 moles N_2, 18 moles H_2, 21 moles O_2

19. All three should be equal, i.e., $x_1 = x_2 = x_3$.

20. (a) $(5, 3, -2)$;   (b) $(2, 4, 2)$;   (c) $(2, 0, -2, -2, 0, 2)$

SECTION 3

1. (a) $\begin{bmatrix} 6 & 2 & 8 \\ -4 & 0 & 2 \\ 2 & 4 & 4 \end{bmatrix}$;   (b) $\begin{bmatrix} 4 & 1 & 6 \\ -5 & 1 & 2 \\ 3 & -2 & 3 \end{bmatrix}$;   (c) $\begin{bmatrix} 3 & 2 & 2 \\ 5 & -3 & -1 \\ -4 & 16 & 1 \end{bmatrix}$;

   (d) $\begin{bmatrix} 3 & 5 & -4 \\ 2 & -3 & 16 \\ 2 & -1 & 1 \end{bmatrix}$;   (f) $\begin{bmatrix} 5 & 5 & 8 \\ -10 & -1 & -9 \\ 15 & 4 & 6 \end{bmatrix}$;   (h) $\begin{bmatrix} 5 & -10 & 15 \\ 5 & -1 & 4 \\ 8 & -9 & 6 \end{bmatrix}$

2. (a) $\begin{bmatrix} 15 & 19 \\ 4 & 0 \end{bmatrix}$;   (c) $\begin{bmatrix} 19 & 21 \\ 17 & 21 \\ 8 & 10 \end{bmatrix}$;   (d) $\begin{bmatrix} 36 & 10 & 56 \\ 10 & 3 & 16 \end{bmatrix}$;   (f) $\begin{bmatrix} 6 & 4 & 8 & 10 \\ -3 & -2 & -4 & -5 \\ 9 & 6 & 12 & 15 \end{bmatrix}$

   (b) and (e) are not possible.

3. (a) 3×3;   (b) 1×2

4. (a) $\begin{bmatrix} 3 & 2 \\ 2 & -3 \end{bmatrix} \begin{bmatrix} x_1 \\ x_2 \end{bmatrix} = \begin{bmatrix} 1 \\ 5 \end{bmatrix}$; (b) $\begin{bmatrix} 1 & 1 & 0 \\ 2 & 1 & -1 \\ 3 & -2 & 2 \end{bmatrix} \begin{bmatrix} x_1 \\ x_2 \\ x_3 \end{bmatrix} = \begin{bmatrix} 5 \\ 6 \\ 7 \end{bmatrix}$;

   (c) $\begin{bmatrix} 2 & 1 & 1 \\ 1 & -1 & 2 \\ 3 & -2 & -1 \end{bmatrix} \begin{bmatrix} x_1 \\ x_2 \\ x_3 \end{bmatrix} = \begin{bmatrix} 4 \\ 2 \\ 0 \end{bmatrix}$.

10. $A = A^2 = A^3 = A^n$ 11. $A^{2n} = I$, $A^{2n+1} = A$ 13. (a) $\mathbf{b} = 2\mathbf{a}_1 + \mathbf{a}_2$

14. (a) inconsistent; (b) consistent; (c) inconsistent

28. Monday, 575; Tuesday, 936; Wednesday, 457.8; Thursday, 1105; Friday, 457.8

29. 4500 married, 5500 single

31. (b) 0 walks of length 2 from V_2 to V_3 and 3 walks of length 2 from V_2 to V_5;
    (c) 6 walks of length 3 from V_2 to V_3 and 2 walks of length 3 from V_2 to V_5

32. (a) $A = \begin{bmatrix} 0 & 1 & 0 & 1 & 0 \\ 1 & 0 & 1 & 1 & 0 \\ 0 & 1 & 0 & 0 & 0 \\ 1 & 1 & 0 & 0 & 1 \\ 0 & 0 & 0 & 1 & 0 \end{bmatrix}$;

(c) 5 walks of length 3 from V_2 to V_4 and 7 walks of length 3 or less

33. $b = a_{22} - \dfrac{a_{12}a_{21}}{a_{11}}$

SECTION 4

1. (a) Type I;   (b) not an elementary matrix;   (c) type III;   (d) type II

3. (a) $\begin{bmatrix} -2 & 0 \\ 0 & 1 \end{bmatrix}$;   (b) $\begin{bmatrix} 1 & 0 & 0 \\ 0 & 0 & 1 \\ 0 & 1 & 0 \end{bmatrix}$;   (c) $\begin{bmatrix} 1 & 0 & 0 \\ 0 & 1 & 0 \\ 0 & 2 & 1 \end{bmatrix}$

4. (a) $\begin{bmatrix} 0 & 0 & 1 \\ 0 & 1 & 0 \\ 1 & 0 & 0 \end{bmatrix}$;   (b) $\begin{bmatrix} 1 & -3 \\ 0 & 1 \end{bmatrix}$;   (c) $\begin{bmatrix} \frac{1}{2} & 0 & 0 \\ 0 & 1 & 0 \\ 0 & 0 & 1 \end{bmatrix}$

5. (a) $E = \begin{bmatrix} 1 & 0 & 0 \\ 0 & 1 & 0 \\ 1 & 0 & 1 \end{bmatrix}$;   (b) $F = \begin{bmatrix} 1 & 0 & 0 \\ 0 & 1 & -1 \\ 0 & 0 & 1 \end{bmatrix}$

6. (a) $E_1 = \begin{bmatrix} 1 & 0 & 0 \\ -3 & 1 & 0 \\ 0 & 0 & 1 \end{bmatrix}$;   (b) $E_2 = \begin{bmatrix} 1 & 0 & 0 \\ 0 & 1 & 0 \\ -2 & 0 & 1 \end{bmatrix}$;   (c) $E_3 = \begin{bmatrix} 1 & 0 & 0 \\ 0 & 1 & 0 \\ 0 & 1 & 1 \end{bmatrix}$

7. (a) $\begin{bmatrix} 1 & 0 \\ 3 & 1 \end{bmatrix} \begin{bmatrix} 3 & 1 \\ 0 & 2 \end{bmatrix}$,   (c) $\begin{bmatrix} 1 & 0 & 0 \\ 3 & 1 & 0 \\ -2 & 2 & 1 \end{bmatrix} \begin{bmatrix} 1 & 1 & 1 \\ 0 & 2 & 3 \\ 0 & 0 & 3 \end{bmatrix}$

8. (b) (i) $(0, -1, 1)^T$,   (ii) $(-4, -2, 5)^T$,   (iii) $(0, 3, -2)^T$

9. (a) $\begin{bmatrix} 0 & 1 \\ 1 & 1 \end{bmatrix}$; (b) $\begin{bmatrix} 3 & -5 \\ -1 & 2 \end{bmatrix}$; (c) $\begin{bmatrix} -4 & 3 \\ \frac{3}{2} & -1 \end{bmatrix}$; (d) $\begin{bmatrix} \frac{1}{3} & 0 \\ -1 & \frac{1}{3} \end{bmatrix}$;

(e) $\begin{bmatrix} 1 & -1 & 0 \\ 0 & 1 & -1 \\ 0 & 0 & 1 \end{bmatrix}$; (f) $\begin{bmatrix} 3 & 0 & -5 \\ 0 & \frac{1}{3} & 0 \\ -1 & 0 & 2 \end{bmatrix}$; (g) $\begin{bmatrix} 2 & -3 & 3 \\ -\frac{3}{5} & \frac{6}{5} & -1 \\ -\frac{2}{5} & -\frac{1}{5} & 0 \end{bmatrix}$; (h) $\begin{bmatrix} -\frac{1}{2} & -1 & -\frac{1}{2} \\ -2 & -1 & -1 \\ \frac{3}{2} & 1 & \frac{1}{2} \end{bmatrix}$

10. (a) $\begin{bmatrix} -1 & 0 \\ 4 & 2 \end{bmatrix}$; (b) $\begin{bmatrix} -8 & 5 \\ -14 & 9 \end{bmatrix}$ 11. (a) $\begin{bmatrix} 20 & -5 \\ -34 & 7 \end{bmatrix}$; (c) $\begin{bmatrix} 0 & -2 \\ -2 & 2 \end{bmatrix}$

SECTION 5

1. (a) $\begin{bmatrix} I & A^{-1} \end{bmatrix}$; (b) $\begin{bmatrix} I \\ A^{-1} \end{bmatrix}$; (c) $\begin{bmatrix} A^TA & A^T \\ A & I \end{bmatrix}$; (d) $AA^T + I$; (e) $\begin{bmatrix} I & A^{-1} \\ A & I \end{bmatrix}$

3. (a) $A\mathbf{b}_1 = \begin{bmatrix} 3 \\ 3 \end{bmatrix}$, $A\mathbf{b}_2 = \begin{bmatrix} 4 \\ -1 \end{bmatrix}$;

(b) $\begin{bmatrix} 1 & 1 \end{bmatrix} B = \begin{bmatrix} 3 & 4 \end{bmatrix}$, $\begin{bmatrix} 2 & -1 \end{bmatrix} B = \begin{bmatrix} 3 & -1 \end{bmatrix}$; (c) $AB = \begin{bmatrix} 3 & 4 \\ 3 & -1 \end{bmatrix}$

4. (a) $\left[\begin{array}{cc|cc} 3 & 1 & 1 & 1 \\ 3 & 2 & 1 & 2 \\ \hline 1 & 1 & 1 & 1 \\ 1 & 2 & 1 & 1 \end{array}\right]$; (b) $\left[\begin{array}{cc|cc} 1 & 1 & 1 & 1 \\ 0 & 1 & 0 & 0 \\ \hline 3 & 1 & 1 & 1 \\ 0 & 1 & 0 & 1 \end{array}\right]$; (c) $\left[\begin{array}{cc|cc} 2 & 2 & 2 & 2 \\ 2 & 4 & 2 & 2 \\ \hline 3 & 1 & 1 & 1 \\ 3 & 2 & 1 & 2 \end{array}\right]$; (d) $\left[\begin{array}{cc|cc} 1 & 2 & 1 & 1 \\ 1 & 1 & 1 & 1 \\ \hline 3 & 2 & 1 & 2 \\ 3 & 1 & 1 & 1 \end{array}\right]$

5. (b) $\left[\begin{array}{ccc|c} 0 & 2 & 0 & -2 \\ 8 & 5 & 8 & -5 \\ 3 & 2 & 3 & -2 \\ \hline 5 & 3 & 5 & -3 \end{array}\right]$; (d) $\left[\begin{array}{cc} 3 & -3 \\ 2 & -2 \\ 1 & -1 \\ \hline 5 & -5 \\ 4 & -4 \end{array}\right]$

12. $A^2 = \begin{bmatrix} B & O \\ O & B \end{bmatrix}$, $A^4 = \begin{bmatrix} B^2 & O \\ O & B^2 \end{bmatrix}$

13. (a) $\begin{bmatrix} O & I \\ I & O \end{bmatrix}$; (b) $\begin{bmatrix} I & O \\ -B & I \end{bmatrix}$

CHAPTER TEST

1. False 2. True 3. True 4. False 5. False 6. False 7. False 8. False 9. True 10. True

Chapter 2

SECTION 1

1. (a) $\det(M_{21}) = -8$, $\det(M_{22}) = -2$, $\det(M_{23}) = 5$; (b) $A_{21} = 8$, $A_{22} = -2$, $A_{23} = -5$

2. (a) and (c) are nonsingular.

3. (a) 1; (b) 4; (c) 0; (d) 58; (e) -39; (f) 0; (g) 8; (h) 20

4. (a) 2; (b) -4; (c) 0; (d) 0

5. $-x^3 + ax^2 + bx + c$ 6. $\lambda = 6$ or -1

SECTION 2

1. (a) -24; (b) 30; (c) -1 2. (a) 10; (b) 20

3. (a), (e), and (f) are singular while (b), (c), and (d) are nonsingular.

4. $c = 5$ or -3 7. (a) 20; (b) 108; (c) 160; (d) $\frac{5}{4}$

8. (a) -6; (c) 6; (e) 1 11. $\det(A) = u_{11}u_{22}u_{33}$

SECTION 3

1. (a) $\det(A) = -7$, adj $A = \begin{bmatrix} -1 & -2 \\ -3 & 1 \end{bmatrix}$, $A^{-1} = \begin{bmatrix} \frac{1}{7} & \frac{2}{7} \\ \frac{3}{7} & -\frac{1}{7} \end{bmatrix}$;

(c) $\det(A) = 3$, adj $A = \begin{bmatrix} -3 & 5 & 2 \\ 0 & 1 & 1 \\ 6 & -8 & -5 \end{bmatrix}$, $A^{-1} = \frac{1}{3}$ adj A

2. (a) $(\frac{5}{7}, \frac{8}{7})$; (b) $(\frac{11}{5}, -\frac{4}{5})$; (c) $(4, -2, 2)$; (d) $(2, -1, 2)$; (e) $(-\frac{2}{3}, \frac{2}{3}, \frac{1}{3}, 0)$

3. $-\frac{3}{4}$

4. $(\frac{1}{2}, -\frac{3}{4}, 1)^T$

5. (a) $\det(A) = 0$, so A is singular.

(b) adj $A = \begin{bmatrix} -1 & 2 & -1 \\ 2 & -4 & 2 \\ -1 & 2 & -1 \end{bmatrix}$ and A adj $A = \begin{bmatrix} 0 & 0 & 0 \\ 0 & 0 & 0 \\ 0 & 0 & 0 \end{bmatrix}$

9. (a) $\det(\mathrm{adj}(A)) = 8$ and $\det(A) = 2$; (b) $A = \begin{bmatrix} 1 & 0 & 0 & 0 \\ 0 & 4 & -1 & 1 \\ 0 & -6 & 2 & -2 \\ 0 & 1 & 0 & 1 \end{bmatrix}$

14. DO YOUR HOMEWORK.

CHAPTER TEST

1. True 2. False 3. False 4. True 5. False 6. True 7. True 8. True 9. False 10. True

Chapter 3

SECTION 1

1. (a) $\|\mathbf{x}_1\| = 10$, $\|\mathbf{x}_2\| = \sqrt{17}$; (b) $\|\mathbf{x}_3\| = 13 < \|\mathbf{x}_1\| + \|\mathbf{x}_2\|$

2. (a) $\|\mathbf{x}_1\| = \sqrt{5}$, $\|\mathbf{x}_2\| = 3\sqrt{5}$; (b) $\|\mathbf{x}_3\| = 4\sqrt{5} = \|\mathbf{x}_1\| + \|\mathbf{x}_2\|$

7. If $\mathbf{x} + \mathbf{y} = \mathbf{x}$ for all $\mathbf{x}$ in the vector space, then $\mathbf{0} = \mathbf{0} + \mathbf{y} = \mathbf{y}$.

8. If $\mathbf{x} + \mathbf{y} = \mathbf{x} + \mathbf{z}$, then $-\mathbf{x} + (\mathbf{x} + \mathbf{y}) = -\mathbf{x} + (\mathbf{x} + \mathbf{z})$ and the conclusion follows using axioms 1, 2, 3, and 4.

11. V is not a vector space. Axiom 6 does not hold.

SECTION 2

1. (a) and (c) are subspaces; (b), (d) and (e) are not.

2. (b) and (c) are subspaces; (a) and (d) are not.

3. (a), (d), and (e) are subspaces; (b) (c) and (f) are not.

4. (a) $\{(0, 0)^T\}$;   (b) Span$((-2, 1, 0, 0)^T, (3, 0, 1, 0)^T)$;   (c) Span$((1, 1, 1)^T)$;
   (d) Span$((-5, 0, -3, 1)^T, (-1, 1, 0, 0)^T)$

5. Only the set in part (c) is a subspace of P_4.

6. (a), (b), and (d) are subspaces.

9. (a), (c), and (e) are spanning sets.

10. (a) and (b) are spanning sets.

14. (b) and (c)

SECTION 3

1. (a) and (e) are linearly independent; (b), (c), and (d) are linearly dependent.

2. (a) and (e) are linearly independent; (b), (c), and (d) are not.

3. (a) and (b) are all of 3-space;   (c) a plane through $(0, 0, 0)$;
   (d) a line through $(0, 0, 0)$;   (e) a plane through $(0, 0, 0)$

4. (a) linearly independent;   (b) linearly independent;   (c) linearly dependent

6. (a) and (b) are linearly dependent while (c) and (d) are linearly independent.

9. When α is an odd multiple of $\pi/2$. If the graph of $y = \cos x$ is shifted to the left or right by an odd multiple of $\pi/2$, we obtain the graph of either $\sin x$ or $-\sin x$.

SECTION 4

1. Only in parts (a) and (e) do they form a basis.

2. Only in part (a) do they form a basis.

3. (c) 2   4. 1

5. (c) 2;   (d) a plane through $(0, 0, 0)$ in 3-space

6. (b) $\{(1, 1, 1)^T\}$, dimension 1;   (c) $\{(1, 0, 1)^T, (0, 1, 1)^T\}$, dimension 2

7. basis $\{(1, 1, 0, 0)^T, (\mathrm{i}, -\mathrm{i}, \mathrm{i}, 0)^T, (0, 2, 0, 1)^T\}$ 11 $\{x^2 + 2, x + 3\}$

12. (a) $\{E_{11}, E_{22}\}$;   (b) $\{E_{11}, E_{21}, E_{22}\}$;   (d) $\{E_{12}, E_{21}, E_{22}\}$;
   (e) $\{E_{11}, E_{22}, E_{21} + E_{12}\}$

13. 2   14. (a) 3;   (b) 3;   (c) 2;   (d) 2

15. (a) $\{x, x^2\}$;   (b) $\{x - 1, (x - 1)^2\}$;   (c) $\{x(x - 1)\}$

SECTION 5

1. (a) $\begin{bmatrix} 1 & -1 \\ 1 & 1 \end{bmatrix}$;   (b) $\begin{bmatrix} 1 & 2 \\ 2 & 5 \end{bmatrix}$;   (c) $\begin{bmatrix} 0 & 1 \\ 1 & 0 \end{bmatrix}$

2. (a) $\begin{bmatrix} \frac{1}{2} & \frac{1}{2} \\ -\frac{1}{2} & \frac{1}{2} \end{bmatrix}$; (b) $\begin{bmatrix} 5 & -2 \\ -2 & 1 \end{bmatrix}$; (c) $\begin{bmatrix} 0 & 1 \\ 1 & 0 \end{bmatrix}$

3. (a) $\begin{bmatrix} \frac{5}{2} & \frac{7}{2} \\ -\frac{1}{2} & -\frac{1}{2} \end{bmatrix}$; (b) $\begin{bmatrix} 11 & 14 \\ -4 & -5 \end{bmatrix}$; (c) $\begin{bmatrix} 2 & 3 \\ 3 & 4 \end{bmatrix}$

4. $[\mathbf{x}]_E = (-1, 2)^T$, $[\mathbf{y}]_E = (5, -8)^T$, $[\mathbf{z}]_E = (-1, 5)^T$

5. (a) $\begin{bmatrix} 2 & 0 & -1 \\ -1 & 2 & -1 \\ 0 & -1 & 1 \end{bmatrix}$; (b) $(1, -4, 3)^T$; (c) $(0, -1, 1)^T$; (d) $(2, 2, -1)^T$

6. (a) $\begin{bmatrix} 1 & -1 & -2 \\ 1 & 1 & 0 \\ 1 & 0 & 1 \end{bmatrix}$; (b) $\begin{bmatrix} 7 \\ 5 \\ -2 \end{bmatrix}$

7. $\mathbf{w}_1 = (5, 9)^T$ and $\mathbf{w}_2 = (1, 4)^T$ 8. $\mathbf{u}_1 = (0, -1)^T$ and $\mathbf{u}_2 = (1, 5)^T$

9. (a) $\begin{bmatrix} 2 & 2 \\ -1 & 1 \end{bmatrix}$; (b) $\begin{bmatrix} \frac{1}{4} & -\frac{1}{2} \\ \frac{1}{4} & \frac{1}{2} \end{bmatrix}$ 10. $\begin{bmatrix} 1 & -1 & 0 \\ 0 & 1 & -1 \\ 0 & 0 & 1 \end{bmatrix}$

SECTION 6

2. (a) 3;   (b) 3;   (c) 2

3. (a) $\mathbf{u}_2$, $\mathbf{u}_4$, $\mathbf{u}_5$ are the column vectors of U corresponding to the free variables. $\mathbf{u}_2 = 2\mathbf{u}_1$, $\mathbf{u}_4 = 5\mathbf{u}_1 - \mathbf{u}_3$, $\mathbf{u}_5 = -3\mathbf{u}_1 + 2\mathbf{u}_3$

4. (a) consistent;   (b) inconsistent;   (e) consistent

5. (a) infinitely many solutions;   (c) unique solution

8. rank of $A = 3$; $\dim N(B) = 1$ 18. (b) $n - 1$

23. If $\mathbf{x}_j$ is a solution to $A\mathbf{x} = \mathbf{e}_j$ for $j = 1, \ldots, m$ and $X = (\mathbf{x}_1, \mathbf{x}_2, \ldots, \mathbf{x}_m)$, then $AX = I_m$.

CHAPTER TEST

1. True 2. False 3. False 4. False 5. True 6. True 7. False 8. True 9. True 10. False

Chapter 4

SECTION 1

1. (a) reflection about x_2 axis;   (b) reflection about the origin;   (c) reflection about the line $x_2 = x_1$;
(d) the length of the vector is halved;   (e) projection onto x_2 axis

4. $(7, 18)^T$

5. All except (c) are linear transformations from R^3 into R^2.

6. (b) and (c) are linear transformations from R^2 into R^3.

7. (a), (b), and (d) are linear transformations.

8. (a) and (c) are linear transformations from P_2 into P_3.

9. $L(e^x) = e^x - 1$ and $L(x^2) = x^3/3$.

10. (a) and (c) are linear transformations from $C[0, 1]$ into R^1.

16. (a) $\ker(L) = \{0\}$, $L(R^3) = R^3$;
    (c) $\ker(L) = \text{Span}(e_2, e_3)$, $L(R^3) = \text{Span}((1, 1, 1)^T)$

17. (a) $L(S) = \text{Span}(e_2, e_3)$;   (b) $L(S) = \text{Span}(e_1, e_2)$

18. (a) $\ker(L) = P_1$, $L(P_3) = \text{Span}(x^2, x)$;   (c) $\ker(L) = \text{Span}(x^2 - x)$, $L(P_3) = P_2$

22. The operator in part (a) is one-to-one and onto.

SECTION 2

1. (a) $\begin{bmatrix} -1 & 0 \\ 0 & 1 \end{bmatrix}$;   (b) $\begin{bmatrix} -1 & 0 \\ 0 & -1 \end{bmatrix}$;   (c) $\begin{bmatrix} 0 & 1 \\ 1 & 0 \end{bmatrix}$;   (d) $\begin{bmatrix} \frac{1}{2} & 0 \\ 0 & \frac{1}{2} \end{bmatrix}$;   (e) $\begin{bmatrix} 0 & 0 \\ 0 & 1 \end{bmatrix}$

2. (a) $\begin{bmatrix} 1 & 1 & 0 \\ 0 & 0 & 0 \end{bmatrix}$;   (b) $\begin{bmatrix} 1 & 0 & 0 \\ 0 & 1 & 0 \end{bmatrix}$;   (c) $\begin{bmatrix} -1 & 1 & 0 \\ 0 & -1 & 1 \end{bmatrix}$

3. (a) $\begin{bmatrix} 0 & 0 & 1 \\ 0 & 1 & 0 \\ 1 & 0 & 0 \end{bmatrix}$;   (b) $\begin{bmatrix} 1 & 0 & 0 \\ 1 & 1 & 0 \\ 1 & 1 & 1 \end{bmatrix}$;   (c) $\begin{bmatrix} 0 & 0 & 2 \\ 3 & 1 & 0 \\ 2 & 0 & -1 \end{bmatrix}$

4. (a) $(0, 0, 0)^T$;   (b) $(2, -1, -1)^T$;   (c) $(-15, 9, 6)^T$

5. (a) $\begin{bmatrix} \frac{1}{\sqrt{2}} & \frac{1}{\sqrt{2}} \\ -\frac{1}{\sqrt{2}} & \frac{1}{\sqrt{2}} \end{bmatrix}$;   (b) $\begin{bmatrix} 0 & 1 \\ 1 & 0 \end{bmatrix}$;   (c) $\begin{bmatrix} \sqrt{3} & -1 \\ 1 & \sqrt{3} \end{bmatrix}$;   (d) $\begin{bmatrix} 0 & 1 \\ 0 & 0 \end{bmatrix}$

6. $\begin{bmatrix} 1 & 0 \\ 0 & 1 \\ 1 & 1 \end{bmatrix}$ 7. (b) $\begin{bmatrix} 0 & 0 & 1 \\ 0 & 1 & -1 \\ 1 & -1 & 0 \end{bmatrix}$

8. (a) $\begin{bmatrix} 1 & 1 & 1 \\ 2 & 0 & 1 \\ 0 & -2 & -1 \end{bmatrix}$;   (b) (i) $7y_1 + 6y_2 - 8y_3$,   (ii) $3y_1 + 3y_2 - 3y_3$,   (iii) $y_1 + 5y_2 + 3y_3$

9. (a) square; (b) (i) contraction by a factor $\frac{1}{2}$, (ii) clockwise rotation by $45°$, (iii) translation 2 units to the right and 3 units down

10. (a) $\begin{bmatrix} \frac{1}{2} & -\frac{\sqrt{3}}{2} \\ \frac{\sqrt{3}}{2} & \frac{1}{2} \end{bmatrix}$;   (b) $\begin{bmatrix} 1 & 0 & -3 \\ 1 & 1 & 5 \\ 0 & 0 & 1 \end{bmatrix}$;   (d) $\begin{bmatrix} -1 & 0 & 0 \\ 0 & 1 & 2 \\ 0 & 0 & 1 \end{bmatrix}$

11. $\begin{bmatrix} 1 & \frac{1}{2} \\ 1 & 0 \end{bmatrix}$ 12. $\begin{bmatrix} 1 & \frac{1}{2} & \frac{1}{2} \\ -2 & 0 & 0 \end{bmatrix}$;   (a) $\begin{bmatrix} \frac{1}{2} \\ -2 \end{bmatrix}$;   (d) $\begin{bmatrix} 5 \\ -8 \end{bmatrix}$

13. $\begin{bmatrix} 1 & 1 & 0 \\ 0 & 1 & 2 \\ 0 & 0 & 1 \end{bmatrix}$ 16. (a) $\begin{bmatrix} -1 & -3 & 1 \\ 0 & 2 & 0 \end{bmatrix}$;   (c) $\begin{bmatrix} 2 & -2 & -4 \\ -1 & 3 & 3 \end{bmatrix}$

SECTION 3

1. For the matrix A, see the answers to Exercise 1 of Section 4.2.

(a) $B = \begin{bmatrix} 0 & 1 \\ 1 & 0 \end{bmatrix}$;   (b) $B = \begin{bmatrix} -1 & 0 \\ 0 & -1 \end{bmatrix}$;   (c) $B = \begin{bmatrix} 1 & 0 \\ 0 & -1 \end{bmatrix}$;   (d) $B = \begin{bmatrix} \frac{1}{2} & 0 \\ 0 & \frac{1}{2} \end{bmatrix}$;

(e) $B = \begin{bmatrix} \frac{1}{2} & \frac{1}{2} \\ \frac{1}{2} & \frac{1}{2} \end{bmatrix}$

2. (a) $\begin{bmatrix} 1 & 1 \\ -1 & -3 \end{bmatrix}$;   (b) $\begin{bmatrix} 1 & 0 \\ -4 & -1 \end{bmatrix}$ 3. (a) $\begin{bmatrix} 1 & 1 & 0 \\ 1 & 0 & 1 \\ 0 & 1 & 1 \end{bmatrix}$;   (b) $\begin{bmatrix} 2 & -1 & -1 \\ -1 & 2 & -1 \\ -1 & -1 & 2 \end{bmatrix}$

4. $V = \begin{bmatrix} 1 & 1 & 0 \\ 1 & 2 & -2 \\ 1 & 0 & 1 \end{bmatrix}$, $B = \begin{bmatrix} 0 & 0 & 0 \\ 0 & 1 & 0 \\ 0 & 0 & 1 \end{bmatrix}$

5. (a) $\begin{bmatrix} 0 & 0 & 2 \\ 0 & 1 & 0 \\ 0 & 0 & 2 \end{bmatrix}$;   (b) $\begin{bmatrix} 0 & 0 & 0 \\ 0 & 1 & 0 \\ 0 & 0 & 2 \end{bmatrix}$;   (c) $\begin{bmatrix} 1 & 0 & 1 \\ 0 & 1 & 0 \\ 0 & 0 & 1 \end{bmatrix}$;   (d) $a_1 x + a_2 2^n (1 + x^2)$

6. (a) $\begin{bmatrix} 1 & 0 & 0 \\ 0 & 1 & 1 \\ 0 & 1 & -1 \end{bmatrix}$;   (b) $\begin{bmatrix} 0 & 0 & 0 \\ 0 & 0 & 1 \\ 0 & 1 & 0 \end{bmatrix}$;   (c) $\begin{bmatrix} 0 & 0 & 0 \\ 0 & 1 & 0 \\ 0 & 0 & -1 \end{bmatrix}$

CHAPTER TEST

1. False 2. True 3. True 4. False 5. False 6. True 7. True 8. True 9. True 10. False

Chapter 5

SECTION I

1. (a) $0°$;   (b) $90°$

2. (a) $\sqrt{14}$ (scalar projection), $(2, 1, 3)^T$ (vector projection);   (b) $0, \mathbf{0}$;

(c) $\dfrac{14\sqrt{13}}{13}$, $(\frac{42}{13}, \frac{28}{13})^T$;   (d) $\dfrac{8\sqrt{21}}{21}$, $(\frac{8}{21}, \frac{16}{21}, \frac{32}{21})^T$

3. (a) $\mathbf{p} = (3, 0)^T$, $\mathbf{x} - \mathbf{p} = (0, 4)^T$, $\mathbf{p}^T(\mathbf{x} - \mathbf{p}) = 3 \cdot 0 + 0 \cdot 4 = 0$;

(c) $\mathbf{p} = (3, 3, 3)^T$, $\mathbf{x} - \mathbf{p} = (-1, 1, 0)^T$, $\mathbf{p}^T(\mathbf{x} - \mathbf{p}) = -1 \cdot 3 + 1 \cdot 3 + 0 \cdot 3 = 0$

5. $(1.8, 3.6)$ 6. $(1.4, 3.8)$ 7. 0.4

8. (a) $2x + 4y + 3z = 0$;   (c) $z - 4 = 0$ 9. $\frac{5}{3}$ 10. $\frac{8}{7}$

18. The correlation matrix with entries rounded to two decimal places is $\begin{bmatrix} 1.00 & -0.04 & 0.41 \\ -0.04 & 1.00 & 0.87 \\ 0.41 & 0.87 & 1.00 \end{bmatrix}$.

SECTION 2

1. (a) $\{(3, 4)^T\}$ basis for $R(A^T)$, $\{(-4, 3)^T\}$ basis for $N(A)$, $\{(1, 2)^T\}$ basis for $R(A)$, $\{(-2, 1)^T\}$ basis for $N(A^T)$;

   (d) basis for $R(A^T)$: $\{(1, 0, 0, 0)^T, (0, 1, 0, 0)^T (0, 0, 1, 1)^T\}$,
   basis for $N(A)$: $\{(0, 0, -1, 1)^T\}$,
   basis for $R(A)$: $\{(1, 0, 0, 1)^T, (0, 1, 0, 1)^T (0, 0, 1, 1)^T\}$,
   basis for $N(A^T)$: $\{(1, 1, 1, -1)^T\}$

2. (a) $\{(1, 1, 0)^T, (-1, 0, 1)^T\}$

3. (b) The orthogonal complement is spanned by $(-5, 1, 3)^T$.

4. $\{(-1, 2, 0, 1)^T, (2, -3, 1, 0)^T\}$ is one basis for $S^{\perp}$.

5. (a) $\mathbf{N} = (8, -2, 1)^T$;   (b) $8x - 2y + z = 7$

9. $\dim N(A) = n - r$, $\dim N(A^T) = m - r$

SECTION 3

1. (a) $(2, 1)^T$;   (c) $(1.6, 0.6, 1.2)^T$

2. (1a) $\mathbf{p} = (3, 1, 0)^T$, $\mathbf{r} = (0, 0, 2)^T$ (1c) $\mathbf{p} = (3.4, 0.2, 0.6, 2.8)^T$,
   $\mathbf{r} = (0.6, -0.2, 0.4, -0.8)^T$

3. (a) $\{(1 - 2\alpha, \alpha)^T \mid \alpha$ real$\}$;   (b) $\{(2 - 2\alpha, 1 - \alpha, \alpha)^T \mid \alpha$ real$\}$

4. (a) $\mathbf{p} = (1, 2, -1)^T$, $\mathbf{b} - \mathbf{p} = (2, 0, 2)^T$;   (b) $\mathbf{p} = (3, 1, 4)^T$, $\mathbf{p} - \mathbf{b} = (-5, -1, 4)^T$

5. (a) $y = 1.8 + 2.9x$ 6. $0.55 + 1.65x + 1.25x^2$

13. The least squares circle will have center $(0.58, -0.64)$ and radius 2.73 (answers rounded to two decimal places).

SECTION 4

1. $\|\mathbf{x}\|_2 = 2$, $\|\mathbf{y}\|_2 = 6$, $\|\mathbf{x} + \mathbf{y}\|_2 = 2\sqrt{10}$ 2. (a) $\theta = \dfrac{\pi}{4}$; $\mathbf{p} = (\tfrac{4}{3}, \tfrac{1}{3}, \tfrac{1}{3}, 0)^T$

3. (b) $\|\mathbf{x}\| = 1$, $\|\mathbf{y}\| = 3$ 4. (a) 0;   (b) 5;   (c) 7;   (d) $\sqrt{74}$

7. (a) 1;   (b) $\dfrac{1}{\pi}$;   (c) $\tfrac{1}{6}$ 8. (a) $\dfrac{\pi}{6}$;   (b) $\mathbf{p} = \tfrac{3}{2}x$ 11. (a) $\dfrac{\sqrt{10}}{2}$;   (b) $\dfrac{\sqrt{34}}{4}$

15. (a) $\|\mathbf{x}\|_1 = 7$, $\|\mathbf{x}\|_2 = 5$, $\|\mathbf{x}\|_\infty = 4$;
    (b) $\|\mathbf{x}\|_2 = \sqrt{6}$, $\|\mathbf{x}\|_\infty = 2$;
                   $\|\mathbf{x}\|_2$ $\|\mathbf{x}\|_\infty = 1$

16. $\|\mathbf{x} - \mathbf{y}\|_1 = 5$, $\|\mathbf{x} - \mathbf{y}\|_2 = 3$, $\|\mathbf{x} - \mathbf{y}\|_\infty = 2$

26. (a) not a norm;   (b) norm;   (c) norm

SECTION 5

1. (a) and (d)

2. (b) $\mathbf{x} = -\dfrac{\sqrt{2}}{3}\mathbf{x}_1 + \dfrac{5}{3}\mathbf{x}_2$, $\|\mathbf{x}\| = \left[\left(-\dfrac{\sqrt{2}}{3}\right)^2 + \left(\dfrac{5}{3}\right)^2\right]^{1/2} = \sqrt{3}$

3. $\mathbf{p} = (\frac{23}{18}, \frac{41}{18}, \frac{8}{9})^T$, $\mathbf{p} - \mathbf{x} = (\frac{5}{18}, \frac{5}{18}, -\frac{10}{9})^T$

4. (b) $c_1 = y_1 \cos\theta + y_2 \sin\theta$, $c_2 = -y_1 \sin\theta + y_2 \cos\theta$

6. (a) 15;   (b) $\|\mathbf{u}\| = 3$, $\|\mathbf{v}\| = 5\sqrt{2}$;   (c) $\dfrac{\pi}{4}$

9. (b)   (i) 0,   (ii) $-\dfrac{\pi}{2}$,   (iii) 0,   (iv) $\dfrac{\pi}{8}$

19. (b)   (i) $(2, -2)^T$,   (ii) $(5, 2)^T$,   (iii) $(3, 1)^T$

20. (a) $P = \begin{pmatrix} \frac{1}{2} & \frac{1}{2} & 0 & 0 \\ \frac{1}{2} & \frac{1}{2} & 0 & 0 \\ 0 & 0 & \frac{1}{2} & \frac{1}{2} \\ 0 & 0 & \frac{1}{2} & \frac{1}{2} \end{pmatrix}$ 21. (b) $Q = \begin{pmatrix} \frac{1}{2} & -\frac{1}{2} & 0 & 0 \\ -\frac{1}{2} & \frac{1}{2} & 0 & 0 \\ 0 & 0 & \frac{1}{2} & -\frac{1}{2} \\ 0 & 0 & -\frac{1}{2} & \frac{1}{2} \end{pmatrix}$

25. (b) $\|1\| = \sqrt{2}$, $\|x\| = \dfrac{\sqrt{6}}{3}$;   (c) $l(x) = \dfrac{9}{7}x$

SECTION 6

1. (a) $\left\{ \left(-\dfrac{1}{\sqrt{2}}, \dfrac{1}{\sqrt{2}}\right)^T, \left(\dfrac{1}{\sqrt{2}}, \dfrac{1}{\sqrt{2}}\right)^T \right\}$;

   (b) $\left\{ \left(\dfrac{2}{\sqrt{5}}, \dfrac{1}{\sqrt{5}}\right)^T, \left(-\dfrac{1}{\sqrt{5}}, \dfrac{2}{\sqrt{5}}\right)^T \right\}$

2. (a) $\begin{bmatrix} -\dfrac{1}{\sqrt{2}} & \dfrac{1}{\sqrt{2}} \\ \dfrac{1}{\sqrt{2}} & \dfrac{1}{\sqrt{2}} \end{bmatrix} \begin{bmatrix} \sqrt{2} & \sqrt{2} \\ 0 & 4\sqrt{2} \end{bmatrix}$;   (b) $\begin{bmatrix} \dfrac{2}{\sqrt{5}} & -\dfrac{1}{\sqrt{5}} \\ \dfrac{1}{\sqrt{5}} & \dfrac{2}{\sqrt{5}} \end{bmatrix} \begin{bmatrix} \sqrt{5} & 4\sqrt{5} \\ 0 & 3\sqrt{5} \end{bmatrix}$

3. $\left\{ (\frac{1}{3}, \frac{2}{3}, -\frac{2}{3})^T, (\frac{2}{3}, \frac{1}{3}, \frac{2}{3})^T, (-\frac{2}{3}, \frac{2}{3}, \frac{1}{3})^T \right\}$

4. $u_1(x) = \dfrac{1}{\sqrt{2}}$, $u_2(x) = \dfrac{\sqrt{6}}{2}x$, $u_3(x) = \dfrac{3\sqrt{10}}{4}\left(x^2 - \dfrac{1}{3}\right)$

5. (a) $\left\{ \dfrac{1}{3}(2, 1, 2)^T, \dfrac{\sqrt{2}}{6}(-1, 4, -1)^T \right\}$;

   (b) $Q = \begin{bmatrix} \dfrac{2}{3} & \dfrac{-\sqrt{2}}{6} \\ \dfrac{1}{3} & \dfrac{2\sqrt{2}}{3} \\ \dfrac{2}{3} & \dfrac{-\sqrt{2}}{6} \end{bmatrix}$; $R = \begin{bmatrix} 3 & \dfrac{5}{3} \\ 0 & \dfrac{\sqrt{2}}{3} \end{bmatrix}$;   (c) $\mathbf{x} = \begin{bmatrix} 9 \\ -3 \end{bmatrix}$

6. (b) $\begin{bmatrix} \dfrac{3}{5} & -\dfrac{4}{5\sqrt{2}} \\ \dfrac{4}{5} & \dfrac{3}{5\sqrt{2}} \\ 0 & \dfrac{1}{\sqrt{2}} \end{bmatrix} \begin{bmatrix} 5 & 1 \\ 0 & 2\sqrt{2} \end{bmatrix}$;   (c) $(2.1, 5.5)^T$

7. $\left\{\left(-\dfrac{1}{\sqrt{2}}, \dfrac{1}{\sqrt{2}}, 0, 0\right)^T, \left(\dfrac{\sqrt{2}}{3}, \dfrac{\sqrt{2}}{3}, -\dfrac{\sqrt{2}}{2}, \dfrac{\sqrt{2}}{6}\right)^T\right\}$

9. $\left\{\left(\dfrac{4}{5}, \dfrac{2}{5}, \dfrac{2}{5}, \dfrac{1}{5}\right)^T, \left(\dfrac{1}{5}, -\dfrac{2}{5}, -\dfrac{2}{5}, \dfrac{4}{5}\right)^T, \left(0, \dfrac{1}{\sqrt{2}}, -\dfrac{1}{\sqrt{2}}, 0\right)^T\right\}$

SECTION 7

1. (a) $T_4 = 8x^4 - 8x^2 + 1$, $T_5 = 16x^5 - 20x^3 + 5x$;
   (b) $H_4 = 16x^4 - 48x^2 + 12$, $H_5 = 32x^5 - 160x^3 + 120x$

2. $p_1(x) = x$, $p_2(x) = x^2 - \dfrac{4}{\pi} + 1$

4. $p(x) = (\sinh 1) P_0(x) + \dfrac{3}{e} P_1(x), +5\left(\sinh 1 - \dfrac{3}{e}\right) P_2(x)$, $p(x) \approx 0.9963 + 1.1036x + 0.5367x^2$

6. (a) $U_0 = 1$, $U_1 = 2x$, $U_2 = 4x^2 - 1$

11. $p(x) = (x - 2)(x - 3) + (x - 1)(x - 3) + 2(x - 1)(x - 2)$

13. $1 \cdot f\left(-\dfrac{1}{\sqrt{3}}\right) + 1 \cdot f\left(\dfrac{1}{\sqrt{3}}\right)$

14. (a) degree 3 or less; (b) the formula gives the exact answer for the first integral. The approximate value for the second integral is 1.5, while the exact answer is $\dfrac{\pi}{2}$.

CHAPTER TEST

1. False 2. False 3. False 4. False 5. True 6. False 7. True 8. True 9. True 10. False

Chapter 6

SECTION I

1. (a) $\lambda_1 = 5$, the eigenspace is spanned by $(1, 1)^T$,
   $\lambda_2 = -1$, the eigenspace is spanned by $(1, -2)^T$;

   (b) $\lambda_1 = 3$, the eigenspace is spanned by $(4, 3)^T$,
   $\lambda_2 = 2$, the eigenspace is spanned by $(1, 1)^T$;

   (c) $\lambda_1 = \lambda_2 = 2$, the eigenspace is spanned by $(1, 1)^T$,

   (d) $\lambda_1 = 3 + 4i$, the eigenspace is spanned by $(2i, 1)^T$,
   $\lambda_2 = 3 - 4i$, the eigenspace is spanned by $(-2i, 1)^T$;

   (e) $\lambda_1 = 2 + i$, the eigenspace is spanned by $(1, 1 + i)^T$,
   $\lambda_2 = 2 - i$, the eigenspace is spanned by $(1, 1 - i)^T$;

   (f) $\lambda_1 = \lambda_2 = \lambda_3 = 0$, the eigenspace is spanned by $(1, 0, 0)^T$;

   (g) $\lambda_1 = 2$, the eigenspace is spanned by $(1, 1, 0)^T$,
   $\lambda_2 = 1$, the eigenspace is spanned by $(1, 0, 0)^T$, $(0, 1, -1)^T$;

   (h) $\lambda_1 = 1$, the eigenspace is spanned by $(1, 0, 0)^T$,
   $\lambda_2 = 4$, the eigenspace is spanned by $(1, 1, 1)^T$,
   $\lambda_3 = -2$, the eigenspace is spanned by $(-1, -1, 5)^T$;

   (i) $\lambda_1 = 2$, the eigenspace is spanned by $(7, 3, 1)^T$,
   $\lambda_2 = 1$, the eigenspace is spanned by $(3, 2, 1)^T$,
   $\lambda_3 = 0$, the eigenspace is spanned by $(1, 1, 1)^T$;

(j) $\lambda_1 = \lambda_2 = \lambda_3 = -1$, the eigenspace is spanned by $(1, 0, 1)^T$;

(k) $\lambda_1 = \lambda_2 = 2$, the eigenspace is spanned by $\mathbf{e}_1$ and $\mathbf{e}_2$,
$\lambda_3 = 3$, the eigenspace is spanned by $\mathbf{e}_3$,
$\lambda_4 = 4$, the eigenspace is spanned by $\mathbf{e}_4$;

(l) $\lambda_1 = 3$, the eigenspace is spanned by $(1, 2, 0, 0)^T$,
$\lambda_2 = 1$, the eigenspace is spanned by $(0, 1, 0, 0)^T$,
$\lambda_3 = \lambda_4 = 2$, the eigenspace is spanned by $(0, 0, 1, 0)^T$

8. β is an eigenvalue of B if and only if $\beta = \lambda - \alpha$ for some eigenvalue λ of A.

11. $\lambda_1 = 6$, $\lambda_2 = 2$

24. $\lambda_1 \mathbf{x}^T \mathbf{y} = (A\mathbf{x})^T \mathbf{y} = \mathbf{x}^T A^T \mathbf{y} = \lambda_2 \mathbf{x}^T \mathbf{y}$

SECTION 2

1. (a) $\begin{bmatrix} c_1 e^{2t} + c_2 e^{3t} \\ c_1 e^{2t} + 2c_2 e^{3t} \end{bmatrix}$; (b) $\begin{bmatrix} -c_1 e^{-2t} - 4c_2 e^t \\ c_1 e^{-2t} + c_2 e^t \end{bmatrix}$; (c) $\begin{bmatrix} 2c_1 + c_2 e^{5t} \\ c_1 - 2c_2 e^{5t} \end{bmatrix}$;

(d) $\begin{bmatrix} -c_1 e^t \sin t + c_2 e^t \cos t \\ c_1 e^t \cos t + c_2 e^t \sin t \end{bmatrix}$; (e) $\begin{bmatrix} -c_1 e^{3t} \sin 2t + c_2 e^{3t} \cos 2t \\ c_1 e^{3t} \cos 2t + c_2 e^{3t} \sin 2t \end{bmatrix}$;

(f) $\begin{bmatrix} -c_1 + c_2 e^{5t} + c_3 e^t \\ -3c_1 + 8c_2 e^{5t} \\ c_1 + 4c_2 e^{5t} \end{bmatrix}$

2. (a) $\begin{bmatrix} e^{-3t} + 2e^t \\ -e^{-3t} + 2e^t \end{bmatrix}$; (b) $\begin{bmatrix} e^t \cos 2t + 2e^t \sin 2t \\ e^t \sin 2t - 2e^t \cos 2t \end{bmatrix}$; (c) $\begin{bmatrix} -6e^t + 2e^{-t} + 6 \\ -3e^t + e^{-t} + 4 \\ -e^t + e^{-t} + 2 \end{bmatrix}$;

(d) $\begin{bmatrix} -2 - 3e^t + 6e^{2t} \\ 1 + 3e^t - 3e^{2t} \\ 1 + 3e^{2t} \end{bmatrix}$

4. $y_1(t) = 15e^{-0.24t} + 25e^{-0.08t}$, $y_2(t) = -30e^{-0.24t} + 50e^{-0.08t}$

5. (a) $\begin{bmatrix} -2c_1 e^t - 2c_2 e^{-t} + c_3 e^{\sqrt{2}t} + c_4 e^{-\sqrt{2}t} \\ c_1 e^t + c_2 e^{-t} - c_3 e^{\sqrt{2}t} - c_4 e^{-\sqrt{2}t} \end{bmatrix}$; (b) $\begin{bmatrix} c_1 e^{2t} + c_2 e^{-2t} - c_3 e^t - c_4 e^{-t} \\ c_1 e^{2t} - c_2 e^{-2t} + c_3 e^t - c_4 e^{-t} \end{bmatrix}$

6. $y_1(t) = -e^{2t} + e^{-2t} + e^t$; $y_2(t) = e^{2t} - e^{-2t} + 2e^t$

8. $x_1(t) = \cos t + 3 \sin t + \dfrac{1}{\sqrt{3}} \sin \sqrt{3}t$, $x_2(t) = \cos t + 3 \sin t - \dfrac{1}{\sqrt{3}} \sin \sqrt{3}t$

10. (a) $m_1 x_1''(t) = -kx_1 + k(x_2 - x_1)$, $m_2 x_2''(t) = -k(x_2 - x_1) + k(x_3 - x_2)$, $m_3 x_3''(t) = -k(x_3 - x_2) - kx_3$;

(b) $\begin{bmatrix} 0.1 \cos 2\sqrt{3}t + 0.9 \cos \sqrt{2}t \\ -0.2 \cos 2\sqrt{3}t + 1.2 \cos \sqrt{2}t \\ 0.1 \cos 2\sqrt{3}t + 0.9 \cos \sqrt{2}t \end{bmatrix}$

11. $p(\lambda) = (-1)^n(\lambda^n - a_{n-1}\lambda^{n-1} - \cdots - a_1\lambda - a_0)$

SECTION 3

8. (b) $\alpha = 2$;   (c) $\alpha = 3$ or $\alpha = -1$;   (d) $\alpha = 1$

19. The transition matrix and steady-state vector for the Markov chain are

$$\begin{bmatrix} 0.80 & 0.30 \\ 0.20 & 0.70 \end{bmatrix} \qquad \mathbf{x} = \begin{bmatrix} 0.60 \\ 0.40 \end{bmatrix}$$

In the long run we would expect 60% of the employees to be enrolled.

20. (a) $A = \begin{bmatrix} 0.70 & 0.20 & 0.10 \\ 0.20 & 0.70 & 0.10 \\ 0.10 & 0.10 & 0.80 \end{bmatrix}$

(c) The membership of all three groups will approach 100,000 as n gets large.

23. (b) $\begin{bmatrix} e & e \\ 0 & e \end{bmatrix}$

24. (a) $\begin{bmatrix} 3 - 2e & 1 - e \\ -6 + 6e & -2 + 3e \end{bmatrix}$;   (c) $\begin{bmatrix} e & -1 + e & -1 + e \\ 1 - e & 2 - e & 1 - e \\ -1 + e & -1 + e & e \end{bmatrix}$

25. (a) $\begin{bmatrix} e^{-t} \\ e^{-t} \end{bmatrix}$;   (b) $\begin{bmatrix} -3e^t - e^{-t} \\ e^t + e^{-t} \end{bmatrix}$;   (c) $\begin{bmatrix} 3e^t - 2 \\ 2 - e^{-t} \\ e^{-t} \end{bmatrix}$

SECTION 4

1. (a) $\|\mathbf{z}\| = 6$, $\|\mathbf{w}\| = 3$, $\langle \mathbf{z}, \mathbf{w} \rangle = -4 + 4i$, $\langle \mathbf{w}, \mathbf{z} \rangle = -4 - 4i$;
   (b) $\|\mathbf{z}\| = 4$, $\|\mathbf{w}\| = 7$, $\langle \mathbf{z}, \mathbf{w} \rangle = -4 + 10i$, $\langle \mathbf{w}, \mathbf{z} \rangle = -4 - 10i$

2. (b) $\mathbf{z} = 4\mathbf{z}_1 + 2\sqrt{2}\mathbf{z}_2$

3. (a) $\mathbf{u}_1^H \mathbf{z} = 4 + 2i$, $\mathbf{z}^H \mathbf{u}_1 = 4 - 2i$, $\mathbf{u}_2^H \mathbf{z} = 6 - 5i$, $\mathbf{z}^H \mathbf{u}_2 = 6 + 5i$;
   (b) $\|\mathbf{z}\| = 9$

4. (b) and (f) are Hermitian while (b), (c), (e), and (f) are normal.

11. (b) $\|U\mathbf{x}\|^2 = (U\mathbf{x})^H U\mathbf{x} = \mathbf{x}^H U^H U\mathbf{x} = \mathbf{x}^H \mathbf{x} = \|\mathbf{x}\|^2$

12. U is unitary, since $U^H U = (I - 2\mathbf{u}\mathbf{u}^H)^2 = I - 4\mathbf{u}\mathbf{u}^H + 4\mathbf{u}(\mathbf{u}^H\mathbf{u})\mathbf{u}^H = I$.

20. $\lambda_1 = 1$, $\lambda_2 = -1$, $\mathbf{u}_1 = \left(\frac{1}{\sqrt{2}}, \frac{1}{\sqrt{2}} \right)^T$, $\mathbf{u}_2 = \left(-\frac{1}{\sqrt{2}}, \frac{1}{\sqrt{2}} \right)^T$,

$$A = 1 \begin{bmatrix} \frac{1}{2} & \frac{1}{2} \\ \frac{1}{2} & \frac{1}{2} \end{bmatrix} + (-1) \begin{bmatrix} \frac{1}{2} & -\frac{1}{2} \\ -\frac{1}{2} & \frac{1}{2} \end{bmatrix}$$

SECTION 5

2. (a) $\sigma_1 = \sqrt{10}$, $\sigma_2 = 0$;   (b) $\sigma_1 = 3$, $\sigma_2 = 2$;   (c) $\sigma_1 = 4$, $\sigma_2 = 2$;
   (d) $\sigma_1 = 3$, $\sigma_2 = 2$, $\sigma_3 = 1$. The matrices U and V are not unique. The reader may check his or her answers by multiplying out $U\Sigma V^T$.

3. (b) rank of $A = 2$, $\|A\|_2 = 3$, $A' = \begin{bmatrix} 1.2 & -2.4 \\ -0.6 & 1.2 \end{bmatrix}$

4. (a) $\text{cond}_2(A) = 10$;

   (b) closest matrix of rank 2 is $\begin{bmatrix} -2 & 8 & 20 \\ 14 & 19 & 10 \\ 0 & 0 & 0 \end{bmatrix}$,

   closest matrix of rank 1 is $\begin{bmatrix} 6 & 12 & 12 \\ 8 & 16 & 16 \\ 0 & 0 & 0 \end{bmatrix}$

5. (a) basis for $R(A^T)$: $\{\mathbf{v}_1 = (\frac{2}{3}, \frac{2}{3}, \frac{1}{3})^T, \mathbf{v}_2 = (-\frac{2}{3}, \frac{1}{3}, \frac{2}{3})^T\}$;

   basis for $N(A)$: $\{\mathbf{v}_3 = \frac{1}{3}, -\frac{2}{3}, \frac{2}{3}^T\}$

SECTION 6

1. (a) $\begin{bmatrix} 3 & -\frac{5}{2} \\ -\frac{5}{2} & 1 \end{bmatrix}$;   (b) $\begin{bmatrix} 2 & \frac{1}{2} & -1 \\ \frac{1}{2} & 3 & \frac{3}{2} \\ -1 & \frac{3}{2} & 1 \end{bmatrix}$

3. (a) $Q = \dfrac{1}{\sqrt{2}} \begin{bmatrix} 1 & 1 \\ 1 & -1 \end{bmatrix}$, $\dfrac{(x')^2}{4} + \dfrac{(y')^2}{12} = 1$, ellipse;

   (d) $Q = \dfrac{1}{\sqrt{2}} \begin{bmatrix} 1 & 1 \\ -1 & 1 \end{bmatrix}$, $\left(y' + \dfrac{\sqrt{2}}{2}\right)^2 = -\dfrac{\sqrt{2}}{2}(x' - \sqrt{2})$ or

   $(y'')^2 = -\dfrac{\sqrt{2}}{2}x''$, parabola

6. (a) positive definite;   (b) indefinite;   (d) negative definite;   (e) indefinite

7. (a) minimum;   (b) saddle point;   (c) saddle point;   (f) local maximum

SECTION 7

1. (a) $\det(A_1) = 2$, $\det(A_2) = 3$, positive definite;
   (b) $\det(A_1) = 3$, $\det(A_2) = -10$, not positive definite;
   (c) $\det(A_1) = 6$, $\det(A_2) = 14$, $\det(A_3) = -38$, not positive definite;
   (d) $\det(A_1) = 4$, $\det(A_2) = 8$, $\det(A_3) = 13$, positive definite

2. $a_{11} = 3$, $a_{22}^{(1)} = 2$, $a_{33}^{(2)} = \frac{4}{3}$

3. (a) $\begin{bmatrix} 1 & 0 \\ \frac{1}{2} & 1 \end{bmatrix} \begin{bmatrix} 4 & 0 \\ 0 & 9 \end{bmatrix} \begin{bmatrix} 1 & \frac{1}{2} \\ 0 & 1 \end{bmatrix}$;

   (b) $\begin{bmatrix} 1 & 0 \\ -\frac{1}{3} & 1 \end{bmatrix} \begin{bmatrix} 9 & 0 \\ 0 & 1 \end{bmatrix} \begin{bmatrix} 1 & -\frac{1}{3} \\ 0 & 1 \end{bmatrix}$;

   (c) $\begin{bmatrix} 1 & 0 & 0 \\ \frac{1}{2} & 1 & 0 \\ \frac{1}{4} & -1 & 1 \end{bmatrix} \begin{bmatrix} 16 & 0 & 0 \\ 0 & 2 & 0 \\ 0 & 0 & 4 \end{bmatrix} \begin{bmatrix} 1 & \frac{1}{2} & \frac{1}{4} \\ 0 & 1 & -1 \\ 0 & 0 & 1 \end{bmatrix}$;

(d) $\begin{bmatrix} 1 & 0 & 0 \\ \frac{1}{3} & 1 & 0 \\ -\frac{2}{3} & 1 & 1 \end{bmatrix} \begin{bmatrix} 9 & 0 & 0 \\ 0 & 3 & 0 \\ 0 & 0 & 2 \end{bmatrix} \begin{bmatrix} 1 & \frac{1}{3} & -\frac{2}{3} \\ 0 & 1 & 1 \\ 0 & 0 & 1 \end{bmatrix}$

4. (a) $\begin{bmatrix} 2 & 0 \\ 1 & 3 \end{bmatrix} \begin{bmatrix} 2 & 1 \\ 0 & 3 \end{bmatrix}$;   (b) $\begin{bmatrix} 3 & 0 \\ -1 & 1 \end{bmatrix} \begin{bmatrix} 3 & -1 \\ 0 & 1 \end{bmatrix}$;

(c) $\begin{bmatrix} 4 & 0 & 0 \\ 2 & \sqrt{2} & 0 \\ 1 & -\sqrt{2} & 2 \end{bmatrix} \begin{bmatrix} 4 & 2 & 1 \\ 0 & \sqrt{2} & -\sqrt{2} \\ 0 & 0 & 2 \end{bmatrix}$;

(d) $\begin{bmatrix} 3 & 0 & 0 \\ 1 & \sqrt{3} & 0 \\ -2 & \sqrt{3} & \sqrt{2} \end{bmatrix} \begin{bmatrix} 3 & 1 & -2 \\ 0 & \sqrt{3} & \sqrt{3} \\ 0 & 0 & \sqrt{2} \end{bmatrix}$

SECTION 8

1. (a) $\lambda_1 = 4$, $\lambda_2 = -1$, $\mathbf{x}_1 = (3, 2)^T$;
   (b) $\lambda_1 = 8$, $\lambda_2 = 3$, $\mathbf{x}_1 = (1, 2)^T$;
   (c) $\lambda_1 = 7$, $\lambda_2 = 2$, $\lambda_3 = 0$, $\mathbf{x}_1 = (1, 1, 1)^T$

2. (a) $\lambda_1 = 3$, $\lambda_2 = -1$, $\mathbf{x}_1 = (3, 1)^T$;
   (b) $\lambda_1 = 2 = 2\exp(0)$, $\lambda_2 = -2 = 2\exp(\pi i)$, $\mathbf{x}_1 = (1, 1)^T$;
   (c) $\lambda_1 = 2 = 2\exp(0)$, $\lambda_2 = -1 + \sqrt{3}i = 2\exp\left(\dfrac{2\pi i}{3}\right)$, $\lambda_3 = -1 - \sqrt{3}i = 2\exp\left(\dfrac{4\pi i}{3}\right)$,
   $\mathbf{x}_1 = (4, 2, 1)^T$

3. $x_1 = 70,000$, $x_2 = 56,000$, $x_3 = 44,000$ 4. $x_1 = x_2 = x_3$

5. $(I - A)^{-1} = I + A + \cdots + A^{m-1}$

6. (a) $(I - A)^{-1} = \begin{bmatrix} 1 & -1 & 3 \\ 0 & 0 & 1 \\ 0 & -1 & 2 \end{bmatrix}$;

(b) $A^2 = \begin{bmatrix} 0 & -2 & 2 \\ 0 & 0 & 0 \\ 0 & 0 & 0 \end{bmatrix}$, $A^3 = \begin{bmatrix} 0 & 0 & 0 \\ 0 & 0 & 0 \\ 0 & 0 & 0 \end{bmatrix}$

7. (b) and (c) are reducible.

CHAPTER TEST

1. False 2. False 3. False 4. False 5. False 6. False 7. True 8.True 9. Ture 10. False

Chapter 7

SECTION I

1. (a) 0.231×10^4;   (b) 0.326×10^2;   (c) 0.128×10^{-1};   (d) 0.824×10^5

2. (a) $\epsilon = -2$; $\delta \approx -8.7 \times 10^{-4}$;   (b) $\epsilon = 0.04$; $\delta \approx 1.2 \times 10^{-3}$;
   (c) $\epsilon = 3.0 \times 10^{-5}$; $\delta \approx 2.3 \times 10^{-3}$;   (d) $\epsilon = -31$; $\delta \approx -3.8 \times 10^{-4}$

3. (a) 0.10101×2^5;   (b) 0.10100×2^{-1};   (c) 0.10111×2^4;   (d) -0.11010×2^{-3}

4. (a) $10,420$, $\epsilon = -0.0018$, $\delta \approx -1.7 \times 10^{-7}$;   (b) 0, $\epsilon = -8$, $\delta = -1$;
   (c) 1×10^{-4}, $\epsilon = 5 \times 10^{-5}$, $\delta = 1$;   (d) $82,190$, $\epsilon = 25.7504$, $\delta \approx 3.1 \times 10^{-4}$

5. (a) 0.1043×10^6;   (b) 0.1045×10^6;   (c) 0.1045×10^6

6. $\epsilon = (0.00001)_2 = \frac{1}{32}$

7. 23

SECTION 2

1. $A = \begin{bmatrix} 1 & 0 & 0 \\ 2 & 1 & 0 \\ -3 & 2 & 1 \end{bmatrix} \begin{bmatrix} 1 & 1 & 1 \\ 0 & 2 & -1 \\ 0 & 0 & 3 \end{bmatrix}$ 2. (a) $(2, -1, 3)^T$;   (b) $(1, -1, 3)^T$;   (c) $(1, 5, 1)^T$

3. (a) n^2 multiplications and $n(n-1)$ additions;   (b) n^3 multiplications and $n^2(n-1)$ additions;
   (c) $(AB)\mathbf{x}$ requires $n^3 + n^2$ multiplications and $n^3 - n$ additions; $A(B\mathbf{x})$ requires $2n^2$ multiplications and $2n(n-1)$ additions.

4. (b) (i) 156 multiplications and 105 additions,   (ii) 47 multiplications and 24 additions,
   (iii) 100 multiplications and 60 additions

8. $5n - 4$ multiplications/divisions, $3n - 3$ additions/subtractions

9. (a) $[(n-j)(n-j+1)]/2$ multiplications; $[(n-j-1)(n-j)]/2$ additions;
   (c) It requires on the order of $\frac{2}{3}n^3$ additional multiplications/divisions to compute A^{-1} given the LU factorization.

SECTION 3

1. (a) $(1, 1, -2)$;   (b) $\begin{bmatrix} 0 & 0 & 1 \\ 1 & 0 & 0 \\ 0 & 1 & 0 \end{bmatrix} \begin{bmatrix} 1 & 0 & 0 \\ 2 & 1 & 0 \\ 0 & 3 & 1 \end{bmatrix} \begin{bmatrix} 1 & 2 & -2 \\ 0 & 1 & 8 \\ 0 & 0 & -23 \end{bmatrix}$

2. (a) $(1, 2, 2)$;   (b) $(4, -3, 0)$;   (c) $(1, 1, 1)$

3. $P = \begin{bmatrix} 0 & 0 & 1 \\ 1 & 0 & 0 \\ 0 & 1 & 0 \end{bmatrix}$, $L = \begin{bmatrix} 1 & 0 & 0 \\ \frac{1}{2} & 1 & 0 \\ -\frac{1}{2} & -\frac{1}{3} & 1 \end{bmatrix}$,

   $U = \begin{bmatrix} 2 & 4 & -6 \\ 0 & 6 & 9 \\ 0 & 0 & 5 \end{bmatrix}$, $\mathbf{x} = \begin{bmatrix} 6 \\ -\frac{1}{2} \\ 1 \end{bmatrix}$

4. $P = Q = \begin{bmatrix} 0 & 1 \\ 1 & 0 \end{bmatrix}$, $PAQ = LU = \begin{bmatrix} 1 & 0 \\ \frac{1}{2} & 1 \end{bmatrix} \begin{bmatrix} 4 & 2 \\ 0 & 2 \end{bmatrix}$

5. (a) $\hat{\mathbf{c}} = P\mathbf{c} = (-4, 6)^T$, $\mathbf{y} = L^{-1}\hat{\mathbf{c}} = (-4, 8)^T$, $\mathbf{z} = U^{-1}\mathbf{y} = (-3, 4)^T$;   (b) $\mathbf{x} = Q\mathbf{z} = (4, -3)^T$

6. (b) $P = \begin{bmatrix} 0 & 0 & 1 \\ 0 & 1 & 0 \\ 1 & 0 & 0 \end{bmatrix}$, $Q = \begin{bmatrix} 0 & 0 & 1 \\ 1 & 0 & 0 \\ 0 & 1 & 0 \end{bmatrix}$,

$L = \begin{bmatrix} 1 & 0 & 0 \\ -\frac{1}{2} & 1 & 0 \\ \frac{1}{2} & \frac{2}{3} & 1 \end{bmatrix}$, $U = \begin{bmatrix} 8 & 6 & 2 \\ 0 & 6 & 3 \\ 0 & 0 & 2 \end{bmatrix}$

7. Error $\dfrac{-2000\epsilon}{0.6} \approx -3333\epsilon$. If $\epsilon = 0.001$, then $\delta = -\frac{2}{3}$.

8. $(1.667, 1.001)$ 9. $(5.002, 1.000)$ 10. $(5.001, 1.001)$

SECTION 4

1. (a) $\|A\|_F = \sqrt{2}$, $\|A\|_\infty = 1$, $\|A\|_1 = 1$;   (b) $\|A\|_F = 5$, $\|A\|_\infty = 5$, $\|A\|_1 = 6$;
   (c) $\|A\|_F = \|A\|_\infty = \|A\|_1 = 1$;   (d) $\|A\|_F = 7$, $\|A\|_\infty = 6$, $\|A\|_1 = 10$;
   (e) $\|A\|_F = 9$, $\|A\|_\infty = 10$, $\|A\|_1 = 12$

2. 2   4. $\|I\|_1 = \|I\|_\infty = 1$, $\|I\|_F = \sqrt{n}$

6. (a) 10;   (b) $(-1, 1, -1)^T$

15. (a) $\|Ax\|_\infty \le \|Ax\|_2 \le \|A\|_2 \|x\|_2 \le \sqrt{n} \|A\|_2 \|x\|_\infty$

17. $\text{cond}_\infty A = 400$ 18. (a) $(-0.48, 0.8)$;   (b) $(-2.902, 2.0)$

19. $\text{cond}_\infty(A) = 28$ 21. (a) $A_n^{-1} = \begin{pmatrix} 1-n & n \\ n & -n \end{pmatrix}$;   (b) $\text{cond}_\infty A_n = 4n$;

   (c) $\lim_{n\to\infty} \text{cond}_\infty A_n = \infty$

22. $\sigma_1 = 8$, $\sigma_2 = 8$, $\sigma_3 = 4$

23. (a) $\mathbf{r} = (-0.06, 0.02)^T$ and the relative residual is 0.012;
   (b) 20;   (d) $\mathbf{x} = (1, 1)^T$, $\|\mathbf{x} - \mathbf{x}'\|_\infty = 0.12$

24. $\text{cond}_1(A) = 6$ 25. 0.3

26. (a) $\|\mathbf{r}\|_\infty = 0.10$, $\text{cond}_\infty(A) = 32$;   (b) 0.64;
   (c) $\mathbf{x} = (12.50, 4.26, 2.14, 1.10)^T$, $\delta = 0.04$

SECTION 5

1. (a) $\begin{bmatrix} \dfrac{1}{\sqrt{2}} & \dfrac{1}{\sqrt{2}} \\ -\dfrac{1}{\sqrt{2}} & \dfrac{1}{\sqrt{2}} \end{bmatrix}$;   (b) $\begin{bmatrix} \dfrac{\sqrt{3}}{2} & -\dfrac{1}{2} \\ \dfrac{1}{2} & \dfrac{\sqrt{3}}{2} \end{bmatrix}$;   (c) $\begin{bmatrix} -\dfrac{4}{5} & \dfrac{3}{5} \\ -\dfrac{3}{5} & -\dfrac{4}{5} \end{bmatrix}$

2. (a) $\begin{bmatrix} \dfrac{3}{5} & 0 & \dfrac{4}{5} \\ 0 & 1 & 0 \\ \dfrac{4}{5} & 0 & -\dfrac{3}{5} \end{bmatrix}$;   (b) $\begin{bmatrix} \dfrac{1}{\sqrt{2}} & -\dfrac{1}{\sqrt{2}} & 0 \\ -\dfrac{1}{\sqrt{2}} & -\dfrac{1}{\sqrt{2}} & 0 \\ 0 & 0 & 1 \end{bmatrix}$;

$$
\text{(c)} \begin{bmatrix} 1 & 0 & 0 \\ 0 & \dfrac{1}{2} & \dfrac{\sqrt{3}}{2} \\ 0 & \dfrac{\sqrt{3}}{2} & -\dfrac{1}{2} \end{bmatrix}; \quad \text{(d)} \begin{bmatrix} 1 & 0 & 0 \\ 0 & -\dfrac{\sqrt{3}}{2} & \dfrac{1}{2} \\ 0 & \dfrac{1}{2} & \dfrac{\sqrt{3}}{2} \end{bmatrix}
$$

3. $H = I - \dfrac{1}{\beta}\mathbf{v}\mathbf{v}^T$ for the given β and $\mathbf{v}$.

    (a) $\beta = 9$, $\mathbf{v} = (-1, -1, -4)^T$;   (b) $\beta = 7$, $\mathbf{v} = (-1, 2, 3)^T$;

    (c) $\beta = 18$, $\mathbf{v} = (-2, 4, -4)^T$

4. (a) $\beta = 9$, $\mathbf{v} = (0, -1, 4, 1)^T$;   (b) $\beta = 15$, $\mathbf{v} = (0, 0, -5, -1, 2)^T$

5. (a) $\beta = 18$, $\mathbf{v} = (-3, 1, 1, 5)^T$;   (b) $HA = \begin{bmatrix} 6 & 0 & 6 \\ 0 & 2 & 0 \\ 0 & -4 & -6 \\ 0 & 4 & -6 \end{bmatrix}$

6. (a) $H_2 H_1 A = R$, where $H_i = I - \dfrac{1}{\beta_i}\mathbf{v}_i \mathbf{v}_i^T$, $i = 1, 2$, and $\beta_1 = 6$, $\beta_2 = 5$.

$$
\mathbf{v}_1 = \begin{bmatrix} -2 \\ 2 \\ -2 \end{bmatrix}, \quad \mathbf{v}_2 = \begin{bmatrix} 0 \\ -1 \\ 3 \end{bmatrix}, \quad R = \begin{bmatrix} 3 & 4 & -2 \\ 0 & 5 & 10 \\ 0 & 0 & -5 \end{bmatrix},
$$

    $\mathbf{b}^{(1)} = H_2 H_1 \mathbf{b} = (11, 5, 5)^T$;

    (b) $\mathbf{x} = (-1, 3, -1)^T$

7. (a) $G = \begin{bmatrix} \dfrac{3}{5} & \dfrac{4}{5} \\ \dfrac{4}{5} & -\dfrac{3}{5} \end{bmatrix}$, $\mathbf{x} = \begin{bmatrix} -1 \\ 1 \end{bmatrix}$

8. It takes three multiplications, two additions, and one square root to determine H. It takes four multiplications/divisions, one addition, and one square root to determine G. The calculation of GA requires $4n$ multiplications and $2n$ additions, while the calculation of HA requires $3n$ multiplications/divisions and $3n$ additions.

9. (a) $n - k + 1$ multiplications/divisions, $2n - 2k + 1$ additions;

    (b) $n(n - k + 1)$ multiplications/divisions, $n(2n - 2k + 1)$ additions

10. (a) $4(n - k)$ multiplications/divisions, $2(n - k)$ additions;

    (b) $4n(n - k)$ multiplications, $2n(n - k)$ additions

11. (a) Rotation;   (b) rotation;   (c) Givens transformation;

    (d) Givens transformation

SECTION 6

1. (a) $\mathbf{u}_1 = \begin{bmatrix} 1 \\ 1 \end{bmatrix}$;   (b) $A_2 = \begin{bmatrix} 2 & 0 \\ 0 & 0 \end{bmatrix}$;

    (c) $\lambda_1 = 2$, $\lambda_2 = 0$; the eigenspace corresponding to λ_1 is spanned by $\mathbf{u}_1$.

2. (a) $\mathbf{v}_1 = \begin{bmatrix} 3 \\ 5 \\ 3 \end{bmatrix}$, $\mathbf{u}_1 = \begin{bmatrix} 0.6 \\ 1.0 \\ 0.6 \end{bmatrix}$, $\mathbf{v}_2 = \begin{bmatrix} 2.2 \\ 4.2 \\ 2.2 \end{bmatrix}$, $\mathbf{u}_2 = \begin{bmatrix} 0.52 \\ 1.00 \\ 0.52 \end{bmatrix}$, $\mathbf{v}_3 = \begin{bmatrix} 2.05 \\ 4.05 \\ 2.05 \end{bmatrix}$;

(b) $\lambda_1' = 4.05$;   (c) $\lambda_1 = 4$, $\delta = 0.0125$

3. (b) A has no dominant eigenvalue.

4. $A_2 = \begin{bmatrix} 3 & -1 \\ -1 & 1 \end{bmatrix}$, $A_3 = \begin{bmatrix} 3.4 & 0.2 \\ 0.2 & 0.6 \end{bmatrix}$, $\lambda_1 = 2 + \sqrt{2} \approx 3.414$, $\lambda_2 = 2 - \sqrt{2} \approx 0.586$

5. (b) $H = I - \dfrac{1}{\beta}\mathbf{v}\mathbf{v}^T$, where $\beta = \frac{1}{3}$ and $\mathbf{v} = (-\frac{1}{3}, -\frac{2}{3}, \frac{1}{3})^T$;

(c) $\lambda_2 = 3$, $\lambda_3 = 1$, $HAH = \begin{bmatrix} 4 & 0 & 3 \\ 0 & 5 & -4 \\ 0 & 2 & -1 \end{bmatrix}$

SECTION 7

1. (a) $(\sqrt{2}, 0)^T$;   (b) $(1 - 3\sqrt{2}, 3\sqrt{2}, -\sqrt{2})^T$;   (c) $(1, 0)^T$;   (d) $(1 - \sqrt{2}, \sqrt{2}, -\sqrt{2})^T$

2. $x_i = \dfrac{d_i b_i + e_i b_{n+i}}{d_i^2 + e_i^2}$, $i = 1, \ldots, n$

3. (a) $\sigma_1 = \sqrt{2 + \epsilon^2}$, $\sigma_2 = \epsilon$;   (b) $\lambda_1' = 2$, $\lambda_2' = 0$, $\sigma_1' = \sqrt{2}$, $\sigma_2' = 0$

10. $A^+ = \begin{bmatrix} \frac{1}{4} & \frac{1}{4} & 0 \\ \frac{1}{4} & \frac{1}{4} & 0 \end{bmatrix}$

11. (a) $A^+ = \begin{bmatrix} \frac{1}{10} & -\frac{1}{10} \\ \frac{2}{10} & -\frac{2}{10} \end{bmatrix}$;   (b) $A^+\mathbf{b} = \begin{bmatrix} 1 \\ 2 \end{bmatrix}$;   (c) $\left\{ \mathbf{y} \mid \mathbf{y} = \begin{bmatrix} 1 \\ 2 \end{bmatrix} + \alpha \begin{bmatrix} -2 \\ 1 \end{bmatrix} \right\}$

13. $\|A_1 - A_2\|_F = \epsilon$, $\|A_1^+ - A_2^+\|_F = 1/\epsilon$. As $\epsilon \to 0$, $\|A_1 - A_2\|_F \to 0$ and $\|A_1^+ - A_2^+\|_F \to \infty$.

CHAPTER TEST

1. False 2. False 3. False 4. True 5. False 6. False 7. True 8. False 9. False 10. False

INDEX